T0238633

Communications
in Computer and Information Science   1244

Commenced Publication in 2007
Founding and Former Series Editors:
Simone Diniz Junqueira Barbosa, Phoebe Chen, Alfredo Cuzzocrea,
Xiaoyong Du, Orhun Kara, Ting Liu, Krishna M. Sivalingam,
Dominik Ślęzak, Takashi Washio, Xiaokang Yang, and Junsong Yuan

More information about this series at http://www.springer.com/series/7899

Mayank Singh · P. K. Gupta ·
Vipin Tyagi · Jan Flusser ·
Tuncer Ören · Gianluca Valentino (Eds.)

Advances in Computing and Data Sciences

4th International Conference, ICACDS 2020
Valletta, Malta, April 24–25, 2020
Revised Selected Papers

 Springer

Editors
Mayank Singh
University of KwaZulu-Natal
Durban, South Africa

Vipin Tyagi
Jaypee University of Engineering
and Technology
Guna, Madhya Pradesh, India

Tuncer Ören
University of Ottawa
Ottawa, ON, Canada

P. K. Gupta
Jaypee University of Information
Technology
Waknaghat, Himachal Pradesh, India

Jan Flusser
Institute of Information Theory
and Automation
Prague, Czech Republic

Gianluca Valentino
University of Malta
Valletta, Malta

ISSN 1865-0929   ISSN 1865-0937 (electronic)
Communications in Computer and Information Science
ISBN 978-981-15-6633-2   ISBN 978-981-15-6634-9 (eBook)
https://doi.org/10.1007/978-981-15-6634-9

This Springer imprint is published by the registered company Springer Nature Singapore Pte Ltd.
The registered company address is: 152 Beach Road, #21-01/04 Gateway East, Singapore 189721, Singapore

Preface

Computing techniques like Big Data, Cloud Computing, Machine learning, Internet of Things (IoT), etc. are playing a key role in the processing of data and retrieving of advanced information. Several state-of-art techniques and computing paradigms have been proposed based on these techniques. This volume contains papers presented at 4th International Conference on Advances in Computing and Data Sciences (ICACDS 2020) organized during April 24–25, 2020, by the Faculty of Information & Communication Technology, University of Malta, Malta. Due to the COVID-19 pandemic, ICACDS 2020 was organized virtually. The conference was organized specifically to help bring together researchers, academicians, scientists, and industry experts and to derive benefits from the advances of next generation computing technologies in the areas of Advanced Computing and Data Sciences.

The Program Committee of ICACDS 2020 is extremely grateful to the authors who showed an overwhelming response to the call for papers, with over 354 papers submitted in the two tracks of Advanced Computing and Data Sciences. All submitted papers went through double-blind peer-review process, and finally 46 papers were accepted for publication in Springer CCIS series. We are thankful to the reviewers for their efforts in finalizing the high-quality papers.

The conference featured many distinguished personalities like Prof. J. S. P. Rai, Vice Chancellor at Jaypee University of Engineering and Technology, India; Prof. Alfred J. Vella, Rector at the University of Malta, Malta; Prof. Ing. Saviour Zammit, Pro-Rector at the University of Malta, Malta; Dr. Neeraj Saxena, Advisor at the Policy & Academic Planning Bureau, AICTE, India; Prof. Shailendra Mishra, Majmaah University, Saudi Arabia; Prof. Viranjay M. Srivastava, University of KwaZulu-Natal, South Africa; Prof. Adrian Muscat, University of Malta, Malta; Prof. Ram Bilas Pachori, IIT Indore, India; Prof. Arun Sharma, Indira Gandhi Delhi Technical University for Women, India; Prof. Shashi Kant Dargar, University of KwaZulu-Natal, South Africa; Prof. Prathmesh Churi, NMIMS University, India; among many others. We are very grateful for the participation of all speakers in making this conference a memorable event.

The Organizing Committee of ICACDS 2020 is indebted to Prof. Ing. Carl James Debono, Dean Faculty of ICT, University of Malta, Malta, for the confidence that he gave to us during organization of this international conference, and all faculty members and staff of Faculty of Information & Communication Technology, University of Malta, for their support in organizing the conference and for making it a grand success.

We would also like to thank Mr. Sameer Kumar Jasra, University of Malta, Malta; Mr. Hemant Gupta, Carleton University, Canada; Mr. Nishant Gupta, MGM CoET, India; Mr. Arun Agarwal, Delhi University, India; Mr. Kunj Bihari Meena, JUET Guna, India; Dr. Neelesh Jain, JUET Guna, India; Dr. Nilesh Patel, JUET Guna, India; Dr. Vibhash Yadav, REC Banda, India; Dr. Sandhya Tarar, GBU Noida, India; Mr. Abhishek Dixit; Mr. Vipin Deval from Tallinn University of Technology, Estonia;

Ms. Kriti Tyagi, JUET Guna, India; Mr. Rohit Kapoor, SK Info Techies, India; Mr. Akshay Chaudhary; Ms. Akansha Singh; Ms. Neha Agarwal; and Mr. Tarun Pathak, Consilio Intelligence Research Lab, India; for their support.

Our sincere thanks to Consilio Intelligence Research Lab, India; the GISR Foundation, India; SK Info Techies, India; Print Canvas, India; and VGeekers, India; for sponsoring the event.

April 2020

Mayank Singh
P. K. Gupta
Vipin Tyagi
Jan Flusser
Tuncer Ören
Gianluca Valentino

Organization

Steering Committee

Alexandre Carlos Brandão Ramos	UNIFEI, Brazil
Mohit Singh	Georgia Institute of Technology, USA
H. M. Pandey	Edge Hill University, UK
M. N. Hooda	BVICAM, India
S. K. Singh	IIT BHU, India
Jyotsna Kumar Mandal	University of Kalyani, India
Ram Bilas Pachori	Indian Institute of Technology Indore, India

Chief Patron

Alfred Vella	University of Malta, Malta

Patron

Saviour Zammit	University of Malta, Malta

Honorary Chair

Carl J. Debono	University of Malta, Malta

General Chairs

Jan Flusser	Institute of Information Theory and Automation, Czech Republic
Gianluca Valentino	University of Malta, Malta
Mayank Singh	University of KwaZulu-Natal, South Africa

Advisory Board Chairs

Shailendra Mishra	Majmaah University, Saudi Arabia
P. K. Gupta	JUIT, India
Vipin Tyagi	JUET, India

Technical Program Committee Chair

Tuncer Ören	University of Ottawa, Canada

Program Chairs

Viranjay M. Srivastava	University of KwaZulu-Natal, South Africa
Ling Tok Wang	National University of Singapore, Singapore
Ulrich Klauck	Aalen University, Germany
Anup Girdhar	Sedulity Group, India
Arun Sharma	IGDTUW, India

Conference Chair

Lalit Garg	University of Malta, Malta

Conference Co-chairs

Alexiei Dingli	University of Malta, Malta
John Abela	University of Malta, Malta

Convener

Sameer Kumar Jasra	University of Malta, Malta

Co-conveners

Sandhya Tarar	Gautam Buddha University, India
Prathamesh Churi	NMIMS, India
Shikha Badhani	Delhi University, India
Lavanya Sharma	Amity University, India
Arun Agarwal	Delhi University, India
Hemant Gupta	Carleton University, Canada
Gaurav Agarwal	Inderprastha Engineering College, India
Sahil Verma	Lovely Professional University, India
Kavita	Lovely Professional University, India
Rakesh Saini	DIT University, India

Organizing Chairs

Peter Xuereb	University of Malta, Malta
Michel Camilleri	University of Malta, Malta
Conrad Attard	University of Malta, Malta
Lucienne May Bugeja	University of Malta, Malta

Organizing Co-chairs

Abhishek Dixit	Tallinn University of Technology, Estonia
Vibhash Yadav	REC Banda, India
Nishant Gupta	MGMCoET, India

Organizing Secretary

Akshay Kumar   Consilio Intelligence Research Lab, India

Creative Head

Tarun Pathak   Consilio Intelligence Research Lab, India

Organizing Committee

Lucienne May Bugeja   University of Malta, Malta
Conrad Attard   University of Malta, Malta
Michel Camilleri   University of Malta, Malta
Lalit Garg   University of Malta, Malta
Gianluca Valentino   University of Malta, Malta
Sameer Kumar Jasra   University of Malta, Malta
Ila Tewari Jarsa   University of Malta, Malta
Peter Xuereb   University of Malta, Malta
Reuben Farrugia   University of Malta, Malta
Akansha Singh   Consilio Intelligence Research Lab, India
Neha Agarwal   Consilio Intelligence Research Lab, India
Kriti Tyagi   JUET, India
Rohit Kapoor   SK Info Techies, India

Sponsored by

Consilio Intelligence Research Lab, India

Co-sponsored by

GISR Foundation, India
Print Canvas, India
SK Info Techies, India
VGeekers, India

Contents

Data Sciences

Advanced Computing

A Computer Vision Based Approach for the Analysis of Acuteness of Garbage

Chitransh Bose[(⊠)], Siddheshwar Pathak, Ritik Agarwal, Vikas Tripathi, and Ketan Joshi

Graphic Era Deemed to be University, Dehradun, India
chitransh0211@gmail.com, pathak.siddheshwar@gmail.com, agarwalritik91@gmail.com, vikastripathi.be@gmail.com, ketanjoshi4477@gmail.com

Abstract. As the population is increasing rapidly day by day the pollution level is also increasing significantly. Several campaigns like Swachh Bharat Abhiyaan (SBA) are aiming to reduce the pollution level. Our approach is to use computer vision technique to classify the garbage based on its severity. For this we have rated garbage on a scale of 1 to 5 with 5 as cleanest and 1 as the dirtiest. To achieve our aim, we have used Faster-RCNN Inception v2 model, and have procured an accuracy of 89.14% using SVM and 89.68% using CNN in detecting different classes of garbage.

Keywords: Garbage detection · CNN · SBA · SVM · kNN · Random Forest

1 Introduction

Computer vision plays a key role in the object detection and classification. It is a field that aims to detect and analyze the objects in the same way as the human vision does to extract the information from the images and videos. As the tracking of garbage and its classification as per the quantity is an important task for its timely disposal and maintaining hygiene in the society computer vision based approach can be very helpful in the detection of severity of garbage and classify them according to the need of urgency of their disposal. The work of municipal corporations also reduces to much extent, if the acuteness of the garbage is known by some automated systems.

According to the study of 60 major cities of India it was estimated that around 4,059 tons of plastic waste are generated per day and more than 1.50 lakh metric tons of waste is being produced in the county everyday [1]. According to [2] India was the third largest producer of garbage in the world in 2017 and by 2050 the generation of waste in India is expected to reach the mark of 436 million tons. According to a blog by Samar Lahiry [3], in 7,935 towns and cities of India over 377 million urban people live there and a solid waste of 62 million tonnes is generated per annum. As the generation of garbage is in such a huge amount therefore the areas having not considerable amount of littering garbage got neglected as there is no proper solid mechanism for its tracking and its detection as garbage. To deal with this problem, a computer vision based automated system is required.

M. Singh et al. (Eds.): ICACDS 2020, CCIS 1244, pp. 3–11, 2020.
https://doi.org/10.1007/978-981-15-6634-9_1

In this paper, we have proposed a framework that uses computer vision approach for detecting and classifying the garbage on the basis of their quantity. For this we have used machine learning algorithms like CNN and SVM. Although some work on the field of garbage detection has been already done [4] but none of them took the acuteness of garbage into the consideration. Our approach detects and classifies the garbage and rates it on the scale of 5 (1 being the least and 5 as the highest rating).

Rest of the paper is structured as follows: Sect. 2 consists of the previous works in the field of garbage detection. In Sect. 3, our approach to analyze the garbage has been discussed. Section 4 gives the result and discussion of our proposed work. And in Sect. 5 we have concluded our work with its future aspects.

2   Literature Review

With increase in amount of garbage in recent years, a lot of work in the field of garbage detection has also been carried out. Some works include waste management system for smart cities while some are in the area of garbage classification. Also there were several approaches based on computer visions, using IOT, etc. [5, 6]. In [4], Rad et al. has proposed a computer vision system that localizes and classify littering wastes on the streets. In this they have classified garbage as cigarettes, leaves, bottles, cans, etc. Manikandan et al. presented a smart waste management system for managing the city wastes. In this they have proposed a system that sense the gases emitted from the dustbins and send the information of the garbage to the municipality on regular basis [3]. Arebey et al. in [6] has used matrix feature extraction approach for solid waste bin level detection. The features extracted were used as an input for the multi-layer perceptron and KNN classifiers. Bobulski et al. [7] proposed a system for the classification of plastic waste using Image Processing and Convolution Neural Network. Their proposed system classifies plastic waste as polythene terephthalate, high-density polyethylene, polypropylene and polystyrene. In [8] Bhor et al. also presented a sensor based garbage management system using IR sensor so that they could detect the level of garbage present in the dustbin so that this information could be used to notify the authorities when the garbage level reaches its maximum. In [9], Peng et al. have proposed an intelligent garbage bin. The system is based on Narrow Band Internet of Things (NB-IOT) technology. In their work they have performed the garbage detection and classification using infrared sensors and odor sensors and NB-IOT module for the transmission of information obtained. Shah et al. [10] has given a CNN based system for the detection of potholes and garbage. They used CNN for auto-identification of the image that the user captures. Anagnostopoulos et al. [11] in their work have presented a survey of ICT-enabled models for waste management. In this, they have focused on the adoption of smart devices that work as a key enabling technology in the field of waste management. Agarwal et al. [12] in their paper have discussed various initiatives taken in India in the area of waste management. This paper also discusses various scopes for improving waste management techniques in India. Patil et al. [13] in their paper discussed the various hazards associated with the health-care waste management. They also identified the shortcomings in the current system. Along with this, they summarized the management and handling rules of biomedical wastes.

Machine learning algorithms like Convolution Neural Network (CNN) and Support Vector Machine (SVM) plays a vital role in the field of object detection and classification as they provide the result with higher accuracy and precision. In several works of classification these algorithms have provided very good results. Pathak, et al. has highlighted the state of art approaches based on deep CNN [14]. In [15], authors have combined the SVM and CNN for the classification of the animals. Danadas et al. [16] compared the kNN and SVM algorithms by classifying the water quality status. Based on the result, they obtained the higher accuracy with SVM which is 92.40% using the linear kernel. The accuracy they obtained through kNN by keeping the value of k as 7.

In this paper, we have used CNN because the patterns that are complex and difficult to perceive can easily be extracted using CNN. In comparison to CNN we have worked on various algorithms like Random Forests, KNN but we found that SVM gave the best results among all.

3 Methodology

Image classification involves the extraction of features and on the basis of the extracted features the algorithm classifies the image to the suitable class. The architecture of our proposed framework is shown in Fig. 1. In our approach, we have taken images as an input. Firstly, the images are taken and are separated into the testing and training data in the ratio of 7:3 or 70% and 30%. After that the training data is converted into gray scale and fed to the image embedder for the extraction of features and the embedding of images. For image embedding, we have used Inception V3 model which is a Google's pre-trained image recognition model. It uses Convolution Neural Network in the backend for the feature extraction. After this, the model is trained using different machine learning algorithms. In our framework, we have used Support Vector Machine (SVM) and Convolution Neural Network (CNN), k-nearest neighbors (kNN) and Random Forest. For the CNN, the images are taken as input and it is passed in the first convolution layer. The image is dot product with the feature vector and the result is stored in feature map and a convolution is formed. After that, pooling is done. In our work, we have used max-pooling which chooses the maximum value from the feature map and thus reduces the size of image. We are using 3 convolution layers in our work out of which two are having 32 batch sizes and the last one has 64 batch size. The activation function used is Rectified Linear Unit (ReLU). ReLU is an activation function that is used for converting all negative values to positive values. It converts values to either 0 or the other positive value. The equation of the ReLU is given by the Eq. 1.

$$y = \max(0, x). \tag{1}$$

After the pooling in first layer, the output of first convolution is passed to second convolution layer and the process is repeated and the output of second is passed to third convolution layer. After the third layer the output is flattened and converted to a 1-D vector which is thus passed to the fully connected layer. This layer is to detect a feature and preserve its value which is thus matched with the features of all the class and the prediction is made and the output is passed to the output layer. The working of convolution layer is shown in Fig. 2. In case of SVM, firstly the data is fed as an input. After that the points which are closest to the classes (support vectors) are located. After that the distance between support vectors and the hyperplane is computed. Thus the distance computed is maximized and the decision boundary is computed. Figure 3

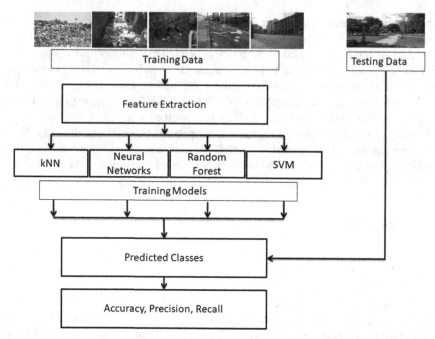

Fig. 1. A Proposed framework for the classification of garbage.

shows the SVM classification using two classes. After the model is trained using CNN or SVM, the test image is input for predicting the class of the test image. The kernel used is radial basis function. The Eq. 2 shows the radial basis function (RBF) kernel function.

$$k\left(x, x'\right) = e^{\|x - x'\|^2} \tag{2}$$

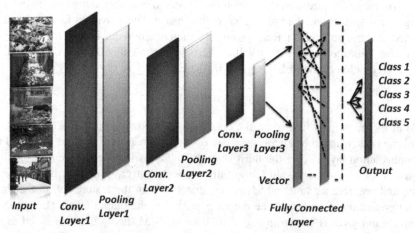

Fig. 2. Structure of Convolution Neural Network.

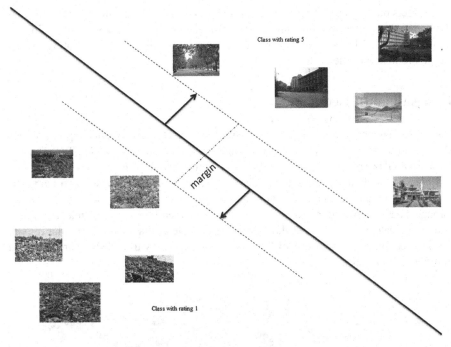

Fig. 3. SVM classification using 2 classes.

Algorithm:

1. Divide data into training and testing.
2. Convert the images into gray scale.
3. Input the images into image embedder.
4. Dot product the image and feature descriptor.
5. Pass the feature map into pooling layer.
6. Select the values from the feature map according to the specified pooling method.
7. The new feature map is passed to another convolution layer and the process is repeated for 3 convolution layers.
8. Pass the result into flattening layer.
9. Train the model using the desired ML algorithm.
10. Predict the class.
11. Output the prediction.

The other algorithms used are kNN, Decision Tree and Random Forest. In kNN classification is done by calculating the 'k' nearest neighbors. For this the distance between the features is calculated and the mode of 'k' nearest neighbors is taken. It uses Manhattan distance for finding the neighbors which is given by Eq. 3.

$$\sum_{i=1}^{k} |x_i - y_i| \tag{3}$$

In Decision Tree algorithm uses the probability which is calculated for the information gains or entropies of parent and child for the prediction of classes. Random Forests are the extension of Decision Tree as it contains multiple trees and their voting is used for the prediction of the class.

4 Result and Discussion

The model have been trained using Convolution Neural Network and Support Vector Machine classifiers, on a system having configuration as Intel Xeon (R) CPU E3-1231 v3, 3.40x8 GHz processor with 16 GB RAM on the dataset of garbage which consists of 5 classes which are rated on the scale of 1 to 5, 1 being the dirtiest and 5 being the cleanest. Some sample images from our dataset are shown in Fig. 4. Since, there was no garbage dataset as per our requirement was available therefore we have created our own dataset and categorized it as per the amount of garbage available in a particular region. For the creation of dataset some of the images are clicked by us and some are collected from internet [17–20].

CLASS 1   CLASS 2   CLASS 3

CLASS 4   CLASS 5

Fig. 4. Sample images of used dataset (Source: pixaby.com [17], unsplash.com [18], shutterstock.com [19], istockphoto.com [20])

Each class is having an average of 100 images. We have trained our model in 110 epochs and 25000 iterations. 30% data is used for testing purpose and 70% is used for training purpose and 20% of this training data has been used for validation. We have used Faster R-CNN as it gives comparatively good results in comparison to other embedders like VGG-16, VGG-19. The comparative analysis shows that VGG-16 gives

64% accuracy with Neural Networks and 66% with SVM, similarly VGG-19 gives 63.8% accuracy for Neural Networks and 66% for SVM whereas Faster R-CNN provides accuracy of 89.68% for Neural Networks and 89.14% for SVM.

Table 1. Statistics representing the accuracy, precision and recall obtained through SVM.

Class	Accuracy (%)	Precision (%)	Recall (%)
1	94.8	82.1	97.0
2	86.5	65.5	68.3
3	81.9	56.5	41.9
4	89.0	75.0	67.9
5	93.5	77.8	93.3
Average	**89.14**	**71.38**	**73.68**

Table 1 depicts the comparative accuracy, precision and recall of each individual class and the average of each of them obtained by using SVM classifier. The average accuracy obtained by SVM is 89.14%. Table 2 shows the accuracy, precision and recall of these classes with an average of each of them obtained by using CNN classifier. Accuracy is the ratio of the true results to the total number of results. The average accuracy procured by the CNN is 89.68%. Table 3 shows the comparative result of accuracy, precision and recall obtained by using kNN algorithm and Table 4 similarly shows the results of classes using Random Forest.

Table 2. Statistics representing the accuracy, precision and recall obtained through CNN.

Class	Accuracy (%)	Precision (%)	Recall (%)
1	94.2	80.0	97.0
2	89.0	69.7	76.7
3	82.6	61.1	35.5
4	88.4	71.0	71.0
5	94.2	81.8	90.0
Average	**89.68**	**72.72**	**74.04**

The accuracy, precision and recall obtained for the classes 2, 3 and 4 are comparatively less because the acuteness of garbage in these 3 classes has very less distinction. The classes 2, 3, 4 all are a part of moderate garbage, thus there is some level of ambiguity in determining these classes, thus reducing the overall precision and recall in both the cases i.e. SVM and CNN. But our framework is capable enough to predict these 3 classes with very high accuracy ranging from 81% to 89%. The classes 1 and 5 can clearly be classified.

Table 3. Statistics representing the accuracy, precision and recall obtained through kNN.

Class	Accuracy (%)	Precision (%)	Recall (%)
1	90.1	69.0	87.9
2	82.6	50.0	66.7
3	76.2	29.6	26.7
4	84.3	65.2	44.1
5	90.1	85.0	75.6
Average	**84.66**	**59.76**	**60.2**

Table 4. Statistics representing the accuracy, precision and recall obtained through Random Forest.

Class	Accuracy (%)	Precision (%)	Recall (%)
1	91.9	75.0	81.8
2	81.4	47.4	60.0
3	78.5	33.3	23.3
4	81.4	52.9	52.9
5	94.2	88.6	90.7
Average	**85.48**	**59.44**	**61.74**

Also, the images are converted into gray scale and then are trained. And we have such a high accuracy thus this system is capable enough to detect the garbage in the absence of light also. Moreover the images used are covering the distant areas also which proves that our model detects garbage at a distance also.

5   Conclusion

In this paper, we have proposed a computer vision based approach for the detection and classification of acuteness of garbage. Our approach is capable enough of detecting the garbage and classifying them according to its severity with an accuracy of 89.14% using the SVM algorithm and an accuracy of 89.68% using CNN. There is a scope of increasing these accuracies if the features of classes 2, 3 and 4 are enhanced or some other algorithm is used that can recognize them more distinctively and proficiently. This work has an active application in the campaigns like SBA, and can be very helpful for the municipal corporations to detect the garbage and collect them. This work can also be incorporated with previous works of garbage classifications in which garbage were classified on the basis of type of garbage like bottle, leaf, etc. Also, there is scope of incorporating an alert system that will send the notification to the concerned authority with the image and rating of the garbage area.

References

1. https://www.indiatoday.in/india/story/india-s-trash-bomb-80-of-1-5-lakh-metric-tonne-daily-garbage-remains-exposed-untreated-1571769-2019-07-21
2. https://swachhindia.ndtv.com/waste-management-india-drowning-garbage-2147
3. https://www.downtoearth.org.in/blog/waste/india-s-challenges-in-waste-management-56753
4. Rad, M.S., et al.: A computer vision system to localize and classify wastes on the streets. In: Liu, M., Chen, H., Vincze, M. (eds.) ICVS 2017. LNCS, vol. 10528, pp. 195–204. Springer, Cham (2017). https://doi.org/10.1007/978-3-319-68345-4_18
5. Manikandan, R., Jamunadevi, S., Ajeyanthi, A., Divya, M., Keerthana, D.: An analysis of garbage mechanism for smart cities (2019)
6. Arebey, M., Hannan, M.A., Begum, R.A., Basri, H.: Solid waste bin level detection using gray level co-occurrence matrix feature extraction approach (2012)
7. Bobulski, J., Kubanek, M.: Waste classification system using image processing and convolutional neural networks. In: Rojas, I., Joya, G., Catala, A. (eds.) IWANN 2019. LNCS, vol. 11507, pp. 350–361. Springer, Cham (2019). https://doi.org/10.1007/978-3-030-20518-8_30
8. Morajkar, P., Bhor, V., Pandya, D., Deshpande, A., Gurav, M.: Smart garbage management system. Int. J. Eng. Res. Technol. (IJERT) 4(03) (2015). https://doi.org/10.17577/IJERTV4IS031175
9. Pan, P., et al.: An intelligent garbage bin based on NB-IOT research mode. In: IEEE International Conference of Safety Produce Informatization (IICSPI), Chongqing, China (2018)
10. Shah, B., Singh, T., Patil, A., Ambadekar, S.: Pothole and garbage detection using convolution neural networks. In: 2nd International Conference on Advances in Science & Technology (ICAST) 2019 on 8th, 9th April 2019 by K J Somaiya Institute of Engineering & Information Technology, Mumbai, India (2019)
11. Anagnostopoulos, T., Zaslavsky, A., Kolomvatsos, K.: Challenges and opportunities of waste management in IoT-enabled smart cities: a survey. IEEE Trans. Sustain. Comput. 2(3), 275–289 (2017)
12. Agarwal, R., Chaudhary, M., Singh, J.: Waste management initiatives in india for human well being. Eur. Sci. J. ESJ 11(10) (2015). http://eujournal.org/index.php/esj/article/view/5715
13. Patil, A., Shekdarf, A.: Health-care waste management in India. J. Environ. Manag. 63(2), 211–220 (2002). https://doi.org/10.1006/jema.2001.0453
14. Pathak, A.R., Pandey, M., Rautaray, S., Pawar, K.: Assessment of object detection using deep convolutional neural networks. In: Bhalla, S., Bhateja, V., Chandavale, A.A., Hiwale, A.S., Satapathy, S.C. (eds.) Intelligent Computing and Information and Communication. AISC, vol. 673, pp. 457–466. Springer, Singapore (2018). https://doi.org/10.1007/978-981-10-7245-1_45
15. Manohar, N., Kumar, Y.H.S., Rani, R., Kumar, G.H.: Convolutional neural network with SVM for classification of animal images. In: Sridhar, V., Padma, M.C., Rao, K.A.R. (eds.) Emerging Research in Electronics, Computer Science and Technology. LNEE, vol. 545, pp. 527–537. Springer, Singapore (2019). https://doi.org/10.1007/978-981-13-5802-9_48
16. Danades, A., Pratama, D., Anggraini, D., Anggriani, D.: Comparison of accuracy level K-Nearest Neighbor algorithm and Support Vector Machine algorithm in classification water quality status. In: 2016 6th International Conference on System Engineering and Technology (ICSET) (2016)
17. https://pixabay.com/images/search/garbage/
18. https://unsplash.com/s/photos/waste
19. https://www.shutterstock.com/search/garbage
20. https://www.istockphoto.com/in/photos/garbage?mediatype=photography&phrase=garbage&sort=mostpopular#close

The Moderating Effect of Demographic Factors Acceptance Virtual Reality Learning in Developing Countries in the Middle East

Malik Mustafa[1]([✉]), Sharf Alzubi[2], and Marwan Alshare[1]

[1] Gulf College, Seeb, Sultanate of Oman
malikjawarneh@gmail.com, marwan@gulfcollege.edu.om
[2] Jordan University of Science and Technology, Ar-Ramtha, Jordan
sharaf_alzoubi@yahoo.com

Abstract. Innovations of technology keep expanding particularly within the sector of virtual reality and this have sparked competition, transforming the manner of businesses operation. This has stimulated the acceptance towards virtual reality learning in developing nations in the Middle East. Accordingly, the factors impacting consumer acceptance of virtual reality learning are examined in this study, which will further expand the current knowledge particularly on what motivates individuals to utilize virtual reality. A quantitative strategy supports this study and the Unified Theory of Acceptance and Use of Technology (UTAUT) was utilized in deciding the components influencing the reception of people towards virtual generated reality learning. An online survey was performed in the Middle Eastern developing countries to gather data from sample obtained through the technique of snowball sampling. The 432 valid obtained responses were analyzed using SPSS. Scale reliability, normality, correlation and multiple linear regressions were tested for conceptual model establishment. The model was tested for fit by comparing the observed results from the survey tool. The results show that the intent of a person to accept virtual reality learning was significantly impacted by (according to their succession of influencing strength), Execution Expectancy, Effort Expectancy, Social Influence, Facilitating Conditions, Personal Innovativeness (PInn). This examination clarifies how segment factors and factors sway the reception of virtual reality learning administrations in developing countries. This consequently will greatly contribute to increased acceptance level of virtual reality learning in these regions. Furthermore, behavioral intention was significantly impacted by Personal Innovativeness (PInn) on actual acceptance Use behavior. Hence, educational bodies in the Middle East should consider investing massively in virtual reality learning and in other innovations of information technology to increase their support towards efficient service delivery while also increasing the services of virtual reality learning.

Keywords: Virtual Reality · UTAUT · Performance Expectancy · Effort Expectancy · Personal Innovativeness · Learning

© Springer Nature Singapore Pte Ltd. 2020
M. Singh et al. (Eds.): ICACDS 2020, CCIS 1244, pp. 12–23, 2020.
https://doi.org/10.1007/978-981-15-6634-9_2

1   Introduction

The conventional classroom method has been changed by the technological evolutions particularly with respect to combination, appropriation and correspondence related to educating and learning. The use of computer mediated communication (CMC) has significantly enhanced availability and quality in education sectors, taking education to the succeeding level [1]. As reported by several studies [2–5], the study syllabi or programs all over the world are advocators of learning via CMC. Relevantly, Virtual Reality (VR) is part of computer mediated communication (CMC) due to its technology involvement which includes computer in addition to other devices of electronic communication in the construction of virtual environment (VE). VE comprises a situation that allows the participation of user in real time application in circumstances fabricated by computer technology. According to [6], the fabricated illusions may have high identicalness with Virtual Reality (VR).

The literature is demonstrating a serious scarcity of studies that explore virtual reality learning in the Middle Eastern developing countries. As such, this study will attempt to explore various middle eastern countries, which fits the criteria, in terms of this issue. This study will hopefully bridge the identified research gap within the context of the Middle East nations. Accordingly, the available literature on the acceptance of virtual reality learning will be reviewed. Then, a conceptual model will be constructed with the use of UTAUT, and the most significant factors impacting use intention and behavior of consumer to accept virtual reality learning will consequently be determined.

2   Research Background

Countless researches have been carried out to examine the factors that influence Virtual Reality learning acceptance. Such researches involve the application of diverse theories and models as described below:

Blended learning (BL) are still progressing in its early stages in the United Arab Emirates [7]. Somehow, interest and acceptance towards this learning type within this region is expanding. In addition, the university under scrutiny provides its largely authorize graduate projects and programs utilizing the arrangement of BL which allows synchronous virtual classrooms, face-to-face sessions, and asynchronous self-study. Accordingly, this study qualitatively examines the perceptions of students regarding their involvement related with BL model in their fairly traditional cultural background. This research likewise evaluated these students' suggestions for course structure design which would satisfy their necessities as adult learners while additionally improving their experience of learning. In such a manner, an aggregate of 21 alumni students participated in their study and the results show a large positive perspective in regards to involvement in a strong positive effect on female empowerment. The most striking topics for viable instructional methodologies inferred the significance of student-cantered practices, particularly as for joined ventures and understudy directed tasks.

3   Theoretical Background and Conceptual Model

This research recommends a reasonable model-based based on UTAUT with one added factor and the purpose is to inspect at the variables that influence acceptance of user of virtual reality learning in countries located in the Middle East region. Accordingly, the constructs included in the proposed model will be detailed in this section. Aside from the factors that might impact the acceptance of user of virtual reality learning, one more construct is added into UTUAT with regards to development and establishing Middle Eastern nations. In this specific circumstance, users of virtual reality are explained in the model (Fig. 1).

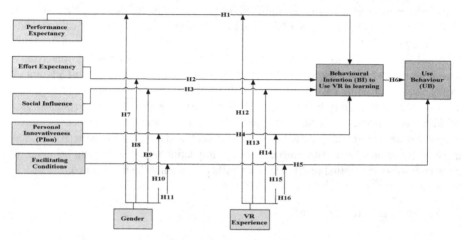

Fig. 1. Conceptual model

4   Methodology

This study applies quantitative research strategy which means that it will provide a numerical measurement as well as analysis of the factors that impact acceptance. Data are obtained through the use of survey questionnaires. This study will ascertain if the independent factors significantly affect the acknowledgment of VR administrations.

4.1   Sampling and Location

The legitimacy of the questionnaire poll utilized in this examination was resolved through a pilot study carried out on 45 respondents from developing countries. In selecting the sample, this study used the convenience sampling method. In order to be eligible for this study, the respondents must be at least 20 years old. The survey was carried out online in Middle East countries. Accordingly, the technique of snowball sampling was used to obtain the data. Via the self-administered questionnaire, this study obtained 432 completed cases.

4.2 Measurement of Variables

The survey applied in this study involves two segments, where the first covers the items on the respondent's personal data, while the other covers the items that represent the variables selected. The second segment of the questionnaire consists of 25 statements where each is provided with five-point ordinal scale for estimating conduct goal of acknowledgment of computer-generated virtual reality (3 statements) and Use Behavior (3 statements) and the factors chose including performance expectancy (4 statements), effort expectancy (3 statements), social impact (4 statements), facilitating conditions (4 statements) and Personal Innovativeness (PInn) (4 statements). The items were adapted from [8, 9].

4.3 Data Analysis

This research has utilized Statistical Package for Social Science for Windows (SPSS for Windows Version 13.0) to obtain the final result from information analysis. A descriptive analysis investigation was carried out to examine the respondents' profile. Frequency and percentage rates were utilized in this test. Different tests utilized in this investigation incorporate a trial of Pearson moment correlation, Independent sample T-test, Chi-square test of independence and multiple linear regressions. As for the essential level, this test has chosen to implement the probability level of 5%.

5 Research and Findings

This section presents respondents profile, Goodness of Measures for the construct validity, and Exploratory Factory Analysis (EFA), as shown below:

5.1 Profile of the Respondents

There were 432 respondents from Middle Eastern countries partaking in this study, where they responded to the questionnaire online. The details of these respondents' personal data are shown in the Table 1.

As Table 1 highlighted, virtual reality learning is dominated by males at 55.1%. In addition, the biggest percentage of respondents comprises those aged between 20 and 30 (38.0%) followed by those age between 31 and 41 (25.0%), and between 42 and 52 (22.2%). Meanwhile, the smallest portion of respondents (14.8%) was of the age of 53 above. In the context of virtual reality learning, 12.0% of respondents indicated "No experience", 19.3% had between 1 and 2 years of experience, 19.0% had between 3 and 4 years of experience 3–4, 17.5% had between 5 and 6 years of experience, 14.8% had between 6 and 7 years of experience, while 17.4% had more than 8 years of experience. The table also provides the data on the background of education of the respondents as follows: 14.3% had up to Elementary School education, whereby 19.4% had up to High School education, 31.7% held College degree, and 35.3% held Graduate degree. The construct validity of all factors is evidenced by the factor analysis outcomes.

Table 1. Profile of respondents (N = 432)

Measure	Item	Frequency	Percentage (%)	Cumulative %
Gender	Male	238	55.1	55.1
	Female	194	44.9	100
Age	20–30	164	38.0	83.0
	31–41	108	25.0	63.0
	42–52	96	22.2	85.2
	53 and Above	64	14.8	100
Experience	No experience	58	12.0	12.0
	1–2 Years	86	19.3	31.3
	3–4 Years	82	19.0	50.3
	5–6 Years	80	17.5	67.8
	6–7 Years	64	14.8	82.6
	8–above	80	17.4	100
Educational background	Elementary school	62	14.3	14.3
	High school	84	19.4	33.7
	College degree	134	31.0	64.7
	Graduate degree	152	35.3	100
Total		432	100%	

In this regard, the dependability of the score size of the sample is determined. For the reason, a dependability test was completed to analyze and re-examine the sample. Next, the elements factors are for the most part additionally tested. This will affirm their inner consistency, through the utilization of Cronbach's alpha.

Meanwhile, an iterative procedure was done to decide the reliability quality of the scale. Affirmation was established on whether the elimination of certain item would cause the reliability of the scale to increase. In the situation that it is so, the item will be eliminated and the analysis is re-executed. Somehow, following the advice of [10], no elimination would be done of it results only in small increases, and since the alpha attribution qualities for all factors in this study are more prominent than 0.7 (refer Table 2), no elimination was done.

As shown in Table 2. The 24 items used in this study underwent several constant treatments of reliability testing. Furthermore, as shown by the statistical figures of these items, for Compatibility, the Cronbach's alpha scores are at 0.726 at least, implying the adequateness of reliability of the whole construct.

Table 2. Cronbach's Alpha Test Results of the Framework Study

Constructs	No of item	Item deleted	Alpha
Performance Expectancy (PF)	4	Null	0771
Effort Expectancy (EE)	3	Null	0.775
Social Influence (SI)	4	Null	0.771
Facilitating Conditions (FC)	4	Null	0.726
Personal Innovativeness (PInn)	3	Null	0.762
Behavioral Intention (BI)	3	Null	0.751
Use Behavior (UB)	3	Null	0.771

5.2 Descriptive Analysis

A specific descriptive test analysis, presented in Table 3, depicts the overall situation of the universities in Middle East. This analysis produces the mean, the standard deviation, maximum and minimum of the constructs. Meanwhile, to make the five-point Likert scale easily construable, three categories are applied in this study as follows: scores less 2.33 [4/3 + lowest rate (1)] were considered as low; scores of 3.67 [highest value (5) − 4/3] were considered as high and those in the middle table were referred as moderate.

As can be seen from Table 3, the base estimation value of majority of the hypotheses was 1.00 while the maximum was 5.00; these surely imply the base minimum and maximum levels degrees of the Likert scale that this study is utilizing.

Furthermore, as shown by the obtained data, social influences have the most supreme mean worth of 3.7876 while its standard deviation is 3.7346. Meanwhile, Facilitating

Table 3. Descriptive statistics analysis, reliability factors (α) and correlations (N = 432)

Constructs	M	SD	1	2	3	4	5	6	α
Performance Expectancy	3.7823	3.5432	0.76	0.57**	0.45**	0.55**	0.34**	0.75**	0.75
Effort Expectancy	3.7653	3.7857		0.46**	0.42**	0.54**	0.39**	0.66**	0.81
Social Influence	3.7876	3.7346			0.37**	0.62**	0.58**	0.56**	0.72
Facilitating Condition	3.7649	3.8346				0.58**	0.62**	0.69**	0.75
Personal Innovativeness	3.7894	3.7349					0.77**	0.52**	0.76
Behavioral Intention	3.7124	3.6576						0.77**	0.91
Use Behavior	3.7834	3.8218							0.73

Conditions obtained the minimum mean value of 3.7649 while its standard deviation is 3.8346. It can thus be said that the three categories of mean values were greater than 2.33 and below (low), but less than 3.67 (moderate), and 3.67 and above (high) which means that the respondents had the inclination to show an elevated level of perceptions as far as performance expectancy, effort expectancy, social influence, facilitating conditions, personal innovativeness (PInn), behavioral Intention, and use behavioral. Table 3 has further details.

5.3   Evaluation of the Model Quality for PLS-SEM

The information was analyzed using the product package Smart-PLS, Version 2.0 M3 [11]. In fact, as indicated [12] the use of Smart-PLS is common particularly in the domain of marketing and management science. Usually, a model PLS is evaluated then construed in two phases [13, 14]. During the first stage, test is carried out on the measurement model (outer model). This will assure the model's validity and reliability. Furthermore, an assessment was performed on the estimation properties of multi-item develops. These incorporate focalized legitimacy, discriminant legitimacy, and unwavering quality. For the purpose, corroborative factor examination (CFA) is completed. During the second stage, analysis is performed on the structural model. Here, R square, effect size, the prescient importance of the model, the GoF and path coefficient are evaluated using.

5.4   Original Study Model

In the initial study model, there were 25 insightful estimation measurement items (manifest variable or display indicator) for eight factors (latent variables factors). These include five independent reliable or antecedent variables factors, two moderating factors and one dependent variable. Here, fourteen relationships between them were proposed in accordance with the proposed hypotheses. Figure 2 provides the details.

5.5   R Square (R^2)

PLS-SEM has been chosen in this study in the appraisal of the structural assessment model. As instructed in the work of [15], the preliminary criteria in surveying the PLS-SEM structural model include the R^2 esteems just as the level and noteworthiness of the way coefficients. Approach for PLS-SEM is prediction-oriented and it is primarily used for depicting the endogenous latent variable using the exogenous latent variables.

The degree of R^2 is directed by the exploration discipline given. For example, shopper conduct, R^2 of (.20) is regarded as high, while R^2 of (.75) would be considered as high in the domain of achievement driver module. In general, for studies in marketing, the value of R^2 of (.75), (.50), or (.25) for endogenous inert factors in the basic model is individually translated as generous, moderate, or frail. R^2 worth can be utilized in the assessment of the structural model's quality. The value depicts the variation in the endogenous variable by method for the exogenous factors.

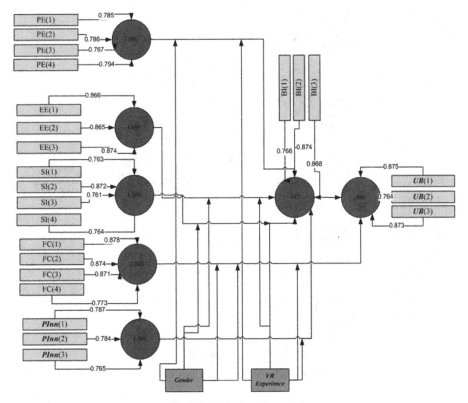

Fig. 2. Original study model

As can be observed in Fig. 3, R^2 of Behavior Intention (BI) is (.934) which implies that Performance Expectancy (PE), Effort Expectancy (EE), Social Influence (SI), Facilitating Conditions (FC), Personal Innovativeness (PInn), collectively make up (93.4%) of the adjustment behavior aim, demonstrating significance. Furthermore, the R^2 variables of Use Behavioral (UB) is (.865) which indicates that Behavior Intention (BI) makes up (86.5%) of the adjustment in Use Behavioral (UB), demonstrating significance as well.

5.6 Hypotheses Testing – Regression

Briefly stated, the prescient model encompasses 15.6% of the difference in Behavioral Intention (BI), and this is clarified through Performance Expectancy (PE), Effort Expectancy (EE), Social Influence (SI), and Personal Innovativeness (PInn) (PR). Furthermore, the model makes up 65.7% of the variance in Use Behavior (UB), and UB is directly depicted by Facilitating Conditions (FC). Overall variance (65.8%) depicted in Use Behavior (UB) is directly described by behavioral intention (BI).

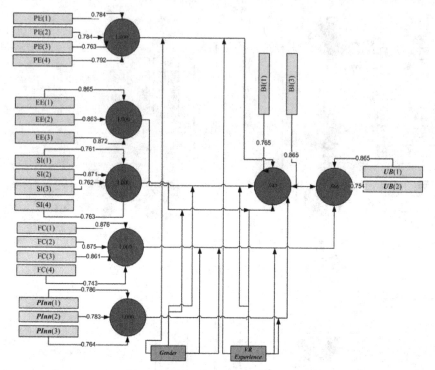

Fig. 3. Path coefficient and R^2 values

Accordingly, Fig. 4 provides the details of the prescient models with R^2 and way of coefficients in the exploration model. The variables of each factor that are considered as predicting m-banking acceptance amongst citizens the most are discussed in the ensuing section.

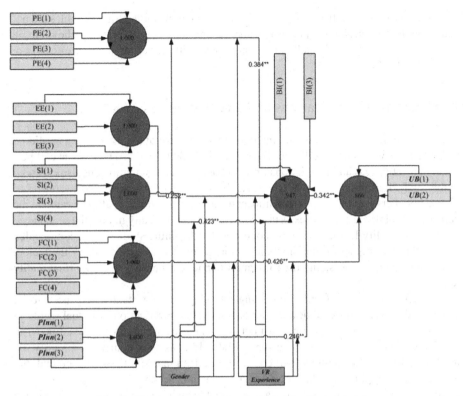

Fig. 4. PLS bootstrapping (t-values) for the Study Model

Table 4. Summary of various regressions factors for variables forecasting behavioral intentions (with moderating effects)

Constructs	B	T	Sig. (p)
PE × Gender	−.0768	−2.210	.047*
EE × Gender	.0540**	1.325	.187
SI × Gender	.031	.873	.382
PInn × Gender	.056*	1.514	.114
FC × Gender	.045**	.368	.713
PE × VR Experience	−.129	−1.162	.248
EE × VR Experience	.113	.992	.322
SI × VR Experience	.051**	.472	.636
PInn × VR Experience	−.135	−1.142	.286
FC × VR Experience	.112**	.865	.637

As can be construed from the results, the addition of the moderating variables did not cause the model's predictive power to improve significantly. Accordingly, the multiple relapses for factors that foretell social intentions (with direct moderating effects) are summarized in Table 4.

6 Conclusion, Implications and Recommendations

The impact of factors on behavioral intention to accept virtual reality learning in Middle East countries was successfully scrutinized in this study research. The influence of personal innovativeness and its relationship with the behavioral intention are also explored in this study. It is anticipated that this research study will contribute to the awareness of how factors Personal Innovativeness impact the acceptance virtual reality learning in the context of Middle East countries as this consequently will greatly impact the acceptance level virtual reality learning in these countries. Forthcoming studies are recommended to look into the impact of demographic factors as moderators, particularly the factors of Income and awareness, and their potential impact on the reception of virtual reality learning.

Furthermore, all the factors impacting the adoption of virtual reality need to be scrutinized. In the context of developing countries, the introduction of virtual reality in learning services and applications must not neglect these contributing factors. Likewise, providers of virtual reality learning services should consistently strive to simplify the needs of users, in this context, learners. As for education institutions, it is crucial that they establish applications that could satisfy clients' needs. These institutions must consistently prove their capacity in making available high-quality applications equipped with secure value-added services.

Moreover, both education institutions and clients will reap benefits from the increased acceptance of virtual reality in learning applications. For these institutions, expenses related to the establishment of more platforms for virtual reality learning will be decreased, whereas for clients, they will need to make less effort while also saving money and time. The acceptance of virtual reality is increasing and platforms for virtual reality learning are now being offered by many institutes. Such positive progression generates a great opening to education institutions to reach those with low-income and low education level.

References

1. Hiltz, S.R., Turoff, M.: Video plus virtual classroom for distance education: experience with graduate courses. In: Invited Paper for Conference on Distance Education in DoD, National Defense University, February 1993
2. Paulsen, M.F., Rekkedal, T.: The electronic college: selected articles from the EKKO project. NKI Forlaget (1990)
3. Hsu, E.Y., Hiltz, S.R.: Management gaming on a computer mediated conferencing system: a case of collaborative learning through computer conferencing. In: Proceedings of the Twenty-Fourth Annual Hawaii International Conference on System Sciences, vol. 4, pp. 367–371. IEEE, January 1991

4. Weedman, J.: Task and non-task functions of a computer conference used in professional eduction: a measure of flexibility. Int. J. Man Mach. Stud. **34**(2), 303–318 (1991)
5. Harasim, L.M.: Learning Networks: A Field Guide to Teaching and Learning Online. MIT Press, Cambridge (1995)
6. Radvansky, B.A., Dombeck, D.A.: An olfactory virtual reality system for mice. Nat. Commun. **9**(1), 839 (2018)
7. Tamim, R.M.: Blended learning for learner empowerment: voices from the middle east. J. Res. Technol. Educ. **50**(1), 70–83 (2018)
8. Agarwal, R., Prasad, J.: A conceptual and operational definition of personal innovativeness in the domain of information technology. Inf. Syst. Res. **9**(2), 204–215 (1998). https://doi.org/10.1287/isre.9.2.204
9. Venkatesh, V., Morris, M.G., Davis, G.B., Davis, F.D.: User acceptance of information technology: toward a unified view. MIS Q. 425–478 (2003)
10. Nunnally, J.C., Bernstein, I.H.: Psychometric theory (1978)
11. Ringle, C.M., Wende, S., Will, A.: Smart PLS 2.0 M3, University of Hamburg: Book Smart Pls, 2, M3 (2005)
12. Henseler, J., Ringle, C.M., Sinkovics, R.R.: The use of partial least squares path modeling in international marketing. Adv. Int. Mark. (2009). https://doi.org/10.1108/S1474-7979(2009)0000020014
13. Hair, J.F., Sarstedt, M., Ringle, C.M., Mena, J.A.: An assessment of the use of partial least squares structural equation modeling in marketing research. J. Acad. Mark. **40**, 414–433 (2012)
14. Fernandes, V.: (Re)discovering the PLS approach in management science. Management (France) (2012)
15. Hair, J., Black, W., Babin, B., Anderson, R., Tatham, R.: Multivariate Data Analysis, 6th edn. Pearson Prentice Hall, Upper Saddle River (2006)

Table Tennis Forehand and Backhand Stroke Recognition Based on Neural Network

Kristian Dokic[1]($\boxtimes$) (iD), Tomislav Mesic[2] (iD), and Marko Martinovic[3] (iD)

[1] Polytechnic in Pozega, Pozega, Croatia
kdjokic@vup.hr
[2] Algebra University College, Zagreb, Croatia
tomislav.mesic@racunarstvo.hr
[3] College of Slavonski Brod, Slavonski Brod, Croatia
Marko.Martinovic@vusb.hr

Abstract. In the last few years, microcontroller producers started to produce SoC boards that are not only used to collect data from implemented sensors but also can be used for small neural networks implementation. The goal of this paper is to analyses the possibility of simple neural network implementation for sports monitoring but we will try to use the state of art technologies on that field. Sport monitoring devices can be used in most sports, but in this paper, the device that can recognize forehand and backhand strokes in table tennis will be developed. This task is not so complicated for development but the focus will be on the flexibility and possibility of using this system for other sports. According to the final test results in laboratory conditions, the system that has been developed is 96% accurate in table tennis forehand and backhand stroke recognition. Finally, in our implementation trained neural network was transferred to microcontroller and this approach opens some new possibilities that can be developed in future versions.

Keywords: Table tennis · Neural network · Stroke recognition

1 Introduction

Advances in sensor development and data analysis opened possibilities for wearable sensors usage in various fields. One of that field is sport and in the last two decades many devices were developed that sportsmen can wear without activity disrupting. Sensors and microcontrollers/SoC become so small and powerful that they can be integrated into body-worn accessories. Accelerometers and gyroscopes have special place within wearable sensor because they are cheap and do not need software 'training' or 'patterns' programming before usage [1]. Machine learning and artificial intelligence development also supported wearable sensors and devices usage because there are many algorithms that can be easy used for sensor data analysis today.

In this paper, wearable sensors that can be used for sports monitoring will be analyzed. After that, the focus will be moved on the methods and systems for tennis and table tennis

© Springer Nature Singapore Pte Ltd. 2020
M. Singh et al. (Eds.): ICACDS 2020, CCIS 1244, pp. 24–35, 2020.
https://doi.org/10.1007/978-981-15-6634-9_3

monitoring. Finally, the device based on the microcontroller board will be developed with the main goal - table tennis forehand and backhand stroke recognition. The state of art technologies will be used (Tensor Flow 2, Tensor Flow Lite, Arduino Nano 33 BLE Sense board, p5.js JavaScript library) and stroke recognition will be done with an artificial neural network. It is not so complicated for development but in the available literature, only one paper deals with neural networks usage for the described task. On the other hand, our implementation includes neural network transfer to the microcontroller and this approach opens some new possibilities.

2 Wearable Sensors

There are dozens of wearable sensors available today that can be used for human body monitoring. Park et al. analyzed wearable sensors from different angles and proposed taxonomy for wearables. There are different keys that can be used to classify wearable sensors so Park et al. proposed next:

a) functionality (single function and multi-functional)
b) Type (active and passive)
c) Deployment mode (invasive and non-invasive, where non-invasive can be classified as in body contact and no body contact)
d) Communication mode (wired and wireless)
e) Field of use (health, public safety, entertainment, military, information processing, acoustic sensing, pressure sensing and position tracking) [2].

In this section, some of them will be shortly described.

Tao et al. have analyzed sensors that can be used for gait analysis and they mentioned accelerometer as a sensor that can measure acceleration along its sensitive axis, gyroscope as an angular velocity sensor and magneto resistive sensor that can estimate changes in the orientation of a body concerning the magnetic North. They also mentioned flexible goniometer that can be used to measure a relative rotation between human body segments and electromagnetic tracking system that can be used to determine the positions and orientations of the object concerning the electromagnetic transmitter. Same authors also mentioned sensing fabric that is based on piezo resistive, piezoelectric or piezo capacitive materials based on polymers and finally force and electromyography sensors. Force sensors can be used to measure GRF, the force exerted by the ground on a body in contact with it, and electromyography sensors can detect voltage potentials to provide information on the timing and intensity of muscle contraction [3].

Morris et al. have described textile-based devices for the real-time analysis of sweat pH and sodium levels. They suggest that the system has applications in sports performance and training analysis but it can be expanded to other health monitoring applications [4].

Ryan et al. have suggested that wearable devices and sensors are becoming more available to the population and they described heart rate monitor, accelerometer/gyroscope, temperature monitor, GPS (global positioning satellite) and pedometer [5].

James and Petrone analyzed sensors and wearable technologies in sport and they suggest next categorization:

a) Load and Pressure Measurement,
b) Inertial Sensors,
c) Optical and Other Sensors,
d) Angle and Displacement Sensors,
e) Garment and Apparel [6].

In this paper Arduino Nano 33 BLE Sense board has been used because it has a LSM6DS3 module that includes 3D accelerometer and 3D gyroscope. This sensor belongs to Inertial Sensors group. Diaz et al. [7]. LSM6DS3 module is implemented in a plastic land grid array package and it can be seen in the Fig. 1.

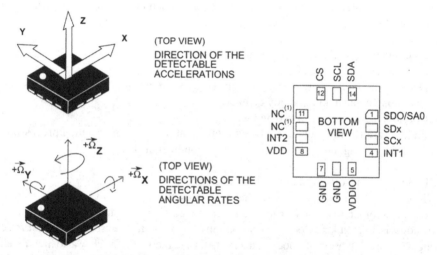

Fig. 1. LSM6DS3 module (figure from datasheets)

3 Table Tennis and Tennis Monitoring Systems

Table tennis is "invented in England in the early days of the 20th century and was originally called Ping-Pong. The name table tennis was adopted in 1921–22 when the old Ping-Pong Association formed in 1902 was revived" [8]. A player starts to play by serving the ball to other player and the rest of the game includes two kinds of strokes: forehand and backhand. Difference between these two strokes is the same in tennis and table tennis, and "is the side of the body where you make contact with the ball. With forehand ground strokes, the ball is struck off to the player's dominant side. For example, if you are right-hand dominant, you contact the ball off to the right side of your body when hitting a forehand" [9].

Table tennis is a fast sport, and many papers deal with table tennis monitoring. Wong et al. developed a system that can be used to identify table tennis ball from match scenes. Their system uses artificial neural network for classification and it can be used for fast and precise decision-making over the validity of a table tennis service. They claim that developed algorithms can be used in other sports for verification purposes in key umpiring decisions [10–12].

Gastinger et al. designed, developed and constructed a monitoring table tennis net to minimize mistakes in net adjusting as well as to detect ball-net contacts during service. Their device makes it possible to standardize net tension and can provide similarly playing conditions worldwide [13].

Hyunju and Sangchul have developed game controller based on a 3-axis accelerometer sensor that is part of every modern smartphone. Their project and paper are interesting because they used data from the accelerometer to decide if the user move racket as a forehand, backend or a service shot. They did not use neural networks but they achieve the successful hitting rate between 97% and 98% without errors. They used a smartphone as a handheld controller for tennis but the main strokes are the same for both games [14].

Ahmadi et al. have developed the wearable device for skill assessment of a tennis player during the first serve. Their device contains three gyroscope sensors and they are used to determine the upper arm rotation, shoulder rotation, and wrist flexion of a tennis player. Authors have had problems with gyroscopes speed but they introduced simulated gyroscopes to rich their goal. Finally, they proved that measured performance during the first tennis serve could be used to classify the athletes [15].

Pei et al. have developed a device that can be inserted in tennis racket and can recognize tennis strokes. It acquires more than 98% accuracy for shot detection and 96% accuracy for stroke recognition (forehand and backhand). They do not use neural networks but they use a product of one angular velocity component and one gravity component to decide is shot forehand or backhand. They used a sensor that has a 3-axis accelerometer with a range of ±16g, and a 3-axis gyroscope with a range of $\pm2000°$/s [16].

Connaghan et al. have described an approach to automatically index a tennis match with a device that is attached to a tennis player's forearm and it can index a tennis match based on strokes played. They used accelerometers, magnetometers and gyroscopes to detect tennis events and they use Support Vector Machine (SVM) classifiers and K-means nearest-neighbor clustering to classify tennis strokes (forehand, backhand and serves) [17].

Wang et al. presented method based on Support Vector Machine for table tennis stroke recognition. They used velocity and acceleration data to recognize five different stroke actions (forehand chop, flat push, backhand chop, forehand stroke and smash). They tried to use K-Nearest Neighbors method but it had lower recognition rate. With SVM they reached almost 97% recognition rate. Fifteen table tennis players were included in the research and 50 strokes were recorded for every player [18].

Blank et al. used 3D accelerometer and 3D gyroscope for stroke detection and classification, but they put sensor in rackets. They compared multiple classifiers but RBF SVM had best performing classifier with accuracy of almost 97%. They collected more than 3 thousand ball contacts but about two thousand ware labeled as valid strokes. They used eight stroke categories [19].

Chunyu used only data from 3D accelerometer that was fixed on the player wrist. He reached the accuracy of more than 98% in left/right movement and two stroke types identification. Author used KNN and decision tree algorithm for classification [20].

Ebner and Findling found a SVM classification method as the most effective for tennis stroke classification. They used sensors on a wrist and on the racket and they reached around 98% accuracy but when they made model with one player data. With user independent model they reached around 90% accuracy. They used accelerometer and gyroscope data for eight different strokes recognition [21].

Liu et al. collected acceleration and angular velocity data from sensors placed on upper arm, lower arm and back. They reduced feature dimensions with PCA and recognized strokes with SVM. They reached more than 97% accuracy and they used five stroke categories (forehand drive, forehand chop, block shot, smash and backhand chop). They collected 270 ball contacts [22].

Finally, Fu et al. reached 95% recognition accuracy but they used acceleration, angular velocity and magnetic field intensity data from Huawei smart watch. They collected data with smart phone that was connected with smart watch over Bluetooth connection. More than 2 thousand strokes were collected and eight different strokes were recognized (Backhand Drive, Backhand Dial, Backhand Twist, Backhand Chop, Forehand Drive, Forehand Attack, Forehand Pick, Forehand Chop). Authors used convolutional neural network for classification [23].

4 Hardware Part

In this paper, Arduino Nano 33 BLE Sense board is used. This microcontroller development board was announced in July 2019 [24]. It is powered by Nordic nRF52840 processor that contains a Cortex M4F with 1 MB flash memory and 256 Kb SRAM. Clock speed is 64 MHz and it has many interfaces (USB, I2S, I2C, SPI, UART). Bluetooth 5 wireless connectivity is supported by NINA B306 module and the board is packed with lots of sensors so it can be used without additional sensors for most applications. These sensors are an accelerometer, gyroscope, magnetometer, pressure sensor, temperature and humidity sensor, proximity sensor, light sensor, color sensor and microphone. It can be seen on the Fig. 2 [25].

The key component on Nano 33 BLE Sense board for our application is LSM9DS1, 3D accelerometer, 3D gyroscope and 3D magnetometer. The LSM9DS1 has a linear acceleration full scale of $\pm2g/\pm4g/\pm8/\pm16g$ and an angular rate of $\pm245/\pm500/\pm2000$ dps. It also includes an I^2C bus interface as well as SPI standard interface. This sensor is in a plastic land grid array package (LGA) on the board [26].

Salazar et al. suggested that wearable monitoring sport systems use different wireless communication standards for transmission between sensor device and storage device (IrDA, MICS, Zigbee, Bluetooth and 802.11g) [27]. In this paper Bluetooth is used for transmission. The Nano 33 BLE Sense board includes Bluetooth 5 wireless module, so it needs only a power source to work. On Fig. 3, a player hand with the wristband can be seen and under the wristband is a breadboard with Arduino Nano 33 BLE Sense. The

Fig. 2. Arduino Nano 33 BLE Sense

white cable that can be seen is a micro USB cable that is connected to the 2000 mAh power bank.

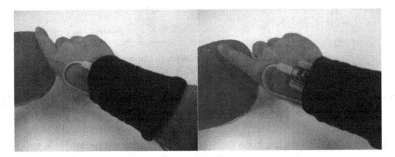

Fig. 3. Arduino Nano 33 BLE Sense under the wristband

5   Software Part

Software development is divided into four steps. They are:

a) Program for Arduino Nano 33 board to collect data
b) Program for a personal computer to convert collected data to CSV files
c) Neural network training
d) Program for Arduino Nano 33 board that use trained neural network

The first part was developing the program for Arduino Nano 33 board to collect data. Offline Arduino integrated development environment has been used for that and source code can be found on GitHub (https://github.com/kristian1971/tabletennis/tree/master/ArduinoIDE). After component initialization program waits in the loop for a strong movement and then it starts to measure acceleration and angular velocity and send it to a personal computer over Bluetooth connection.

On the personal computer that was connected with Arduino Nano 33 board over Bluetooth connection, JavaScript library called p5.js has been used. It is an open-source library that enables easy connecting to Arduino Nano 33 board and easy converting collected data to CSV files. On the Fig. 4, a part of the screen before Bluetooth connecting can be seen. source code can be found on GitHub (https://github.com/kristian1971/tab letennis/tree/master/JavaScript).

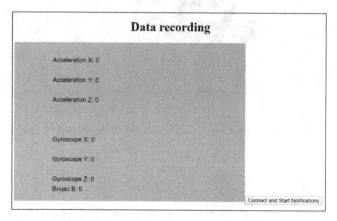

Fig. 4. Screen before Bluetooth connection

Neural network training has been done on Google Colaboratory service, and Jupyther notebook can be found on the GitHub (https://github.com/kristian1971/tabletennis/tree/master/Python). TensorFlow 2.0 has been used and high-level Keras API. A neural network with two hidden layers has been built, with 50 and 15 neurons, and *relu* and *softmax* functions have been used as activation functions. The trained model was converted to Tensor Flow Lite format as well as to constant byte array with a model that can be imported and used in Arduino IDE.

Finally, the last part was developing the program for Arduino Nano 33 board that use trained neural network. Offline Arduino integrated development environment has been used for that and source code can be found on GitHub (https://github.com/kristian1971/tabletennis/tree/master/ArduinoIDE).

6 Procedure and Results

In the first part, only ten forehand and ten backhand strokes have been recorded. Every stroke record consists of three hundred values because there were three accelerometers, and three gyroscope values in every sample and there were fifty samples per stroke. These twenty samples have been combined in two files. The first file named *forehand.csv* includes ten forehand records and the second file named *backhand.csv* includes ten backhand records. These files have been uploaded to Google Colaboratory for neural network build and train. Google Colaboratory can import many libraries that can be used for data visualization and Matplotlib has been used to display data from these CSV files graphically. On the Figs. 5 and 6, acceleration values for forehand and backhand strokes can be seen and on the Figs. 7 and 8, gyroscope values for both strokes can be seen. Values from all ten forehand strokes are displayed.

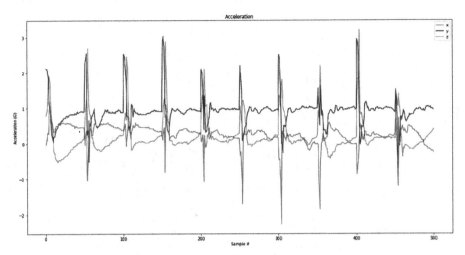

Fig. 5. Forehand acceleration values for all strokes

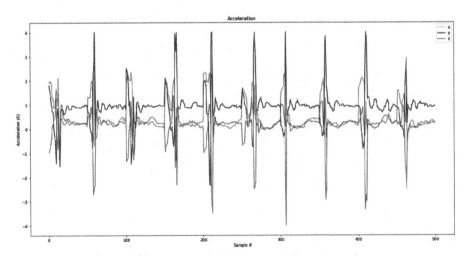

Fig. 6. Backhand acceleration values for all strokes

Finally, Arduino Nano 33 BLE Sense board with transferred neural network has been tested in laboratory conditions that include two series of fifty real time forehand and backhand strokes but with another player. Results from the implemented neural network have been saved and analyzed and can be seen in Table 1 and Fig. 9.

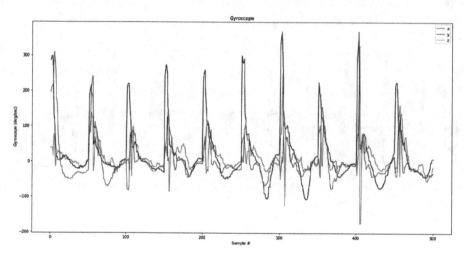

Fig. 7. Forehand gyroscope values for all strokes

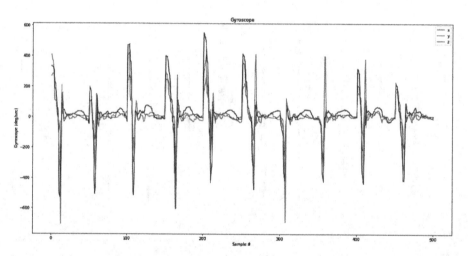

Fig. 8. Backhand gyroscope values for all strokes

Table 1. Accuracy of shot detection

	Sample	Accuracy %
Forehand	50	100
Backhand	50	96

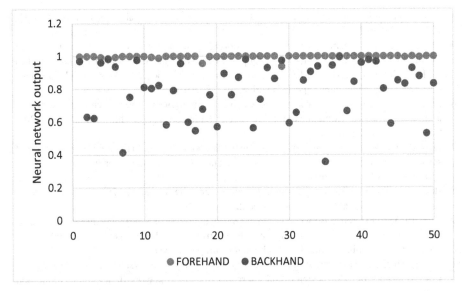

Fig. 9. Neural network output histogram

7 Discussion and Conclusion

It is obvious that the neural network is not precise with backhand stroke recognition (red points). Backhand stroke is more complicated for right-hand players because it consists of more arm moves. Pei et al. also had a little bit better accuracy with forehand stroke type in tennis [16].

This research was more focused on the possibility of using the state of art technologies, and we used too few data to train a reliable neural network. On the other hand, results are promising and in the future, more players and more samples should be used. Connaghan et al. used magnetometer data in their research, and they suggest that it can rise accuracy so in the future research magnetometer data should be used, too [17]. Interestingly, the usability level of used libraries and technologies is so high that the most code used in this research is forked examples adjusted for our specific purpose.

Finally, in our implementation trained neural network was transferred to microcontroller and this approach opens some new possibilities that can be developed in future versions. Neural network transfer was possible because Arduino Nano 33 BLE Sense board includes 3D accelerometer, 3D gyroscope, Bluetooth and it is supported by TensorFlow Lite Micro.

References

1. Martin, A.J.: Sensors and computing systems in smart clothing. In: Smart Clothes and Wearable Technology, pp. 183–204. Elsevier (2009)
2. Park, S., Chung, K., Jayaraman, S.: Wearables: fundamentals, advancements, and a roadmap for the future. In: Wearable Sensors, pp. 1–23. Elsevier (2014)

3. Tao, W., Liu, T., Zheng, R., Feng, H.: Gait analysis using wearable sensors. Sensors **12**, 2255–2283 (2012)
4. Morris, D., et al.: Wearable sensors for monitoring sports performance and training. In: 2008 5th International Summer School and Symposium on Medical Devices and Biosensors (2008)
5. Li, R.T., Kling, S.R., Salata, M.J., Cupp, S.A., Sheehan, J., Voos, J.E.: Wearable performance devices in sports medicine. Sports Health **8**, 74–78 (2016)
6. James, D.A., Petrone, N.: Sensors and Wearable Technologies in Sport: Technologies, Trends and Approaches for Implementation. Springer, Heidelberg (2016). https://doi.org/10.1007/978-981-10-0992-1
7. Diaz, E.M., Ahmed, D.B., Kaiser, S.: A review of indoor localization methods based on inertial sensors. In: Geographical and Fingerprinting Data to Create Systems for Indoor Positioning and Indoor/Outdoor Navigation, pp. 311–333. Elsevier (2019)
8. Barna, V.: Table tennis. Encyclopedia Britannica Inc., 18 December 2019. https://www.britannica.com/sports/table-tennis. Accessed 22 Dec 2019
9. Howard, M.: Difference Between Forehand & Backhand Stroke. https://healthyliving.azcentral.com/difference-between-forehand-backhand-stroke-1897.html. Accessed 12 Oct 2019
10. Wong, P.: Developing an intelligent table tennis umpiring system (2007)
11. Wong, P.K.C.: Developing an intelligent table tennis umpiring system: identifying the ball from the scene. In: 2008 Second Asia International Conference on Modelling & Simulation (AMS) (2008)
12. Wong, P., Dooley, L.: Tracking table tennis balls in real match scenes for umpiring applications. Br. J. Math. Comput. Sci. **1**, 228–241 (2011)
13. Gastinger, R., Litzenberger, S., Sabo, A.: Design, development and construction of a monitoring table tennis net. Procedia Eng. **13**, 297–303 (2011)
14. Cho, H., Kim, S., Baek, J., Fisher, P.S.: Motion recognition with smart phone embedded 3-axis accelerometer sensor. In: 2012 IEEE International Conference on Systems, Man, and Cybernetics (SMC) (2012)
15. Ahmadi, A., Rowlands, D., James, D.A.: Towards a wearable device for skill assessment and skill acquisition of a tennis player during the first serve. Sports Technol. **2**, 129–136 (2009)
16. Pei, W., Wang, J., Xu, X., Wu, Z., Du, X.: An embedded 6-axis sensor based recognition for tennis stroke. In: 2017 IEEE International Conference on Consumer Electronics (ICCE) (2017)
17. Connaghan, D., Kelly, P., O'Connor, N.E., Gaffney, M., Walsh, M., O'Mathuna, C.: Multi-sensor classification of tennis strokes. In: Sensors 2011. IEEE (2011)
18. Wang, H., Li, L., Chen, H., Li, Y., Qiu, S., Gravina, R.: Motion recognition for smart sports based on wearable inertial sensors. In: Mucchi, L., Hämäläinen, M., Jayousi, S., Morosi, S. (eds.) BODYNETS 2019. LNICST, vol. 297, pp. 114–124. Springer, Cham (2019). https://doi.org/10.1007/978-3-030-34833-5_10
19. Blank, P., Houndefinedbach, J., Schuldhaus, D., Eskofier, B.M.: Sensor-based stroke detection and stroke type classification in table tennis. In: Proceedings of the 2015 ACM International Symposium on Wearable Computers, New York, NY, USA (2015)
20. Chunyu, Y.: Application of accelerometer in table tennis action recognition. In: 5th International Conference on Electrical & Electronics Engineering and Computer Science, Beijing (2018)
21. Ebner, C.J., Findling, R.D.: Tennis stroke classification: comparing wrist and racket as IMU sensor position (2019)
22. Liu, R., et al.: Table tennis stroke recognition based on body sensor network. In: Montella, R., Ciaramella, A., Fortino, G., Guerrieri, A., Liotta, A. (eds.) IDCS 2019. LNCS, vol. 11874, pp. 1–10. Springer, Cham (2019). https://doi.org/10.1007/978-3-030-34914-1_1

23. Fu, Z., Shu, K.-I., Zhang, H.: Ping pong motion recognition based on smart watch. In: 3rd International Conference on Mechatronics Engineering and Information Technology (ICMEIT 2019) (2019)
24. Arduino team: The Arduino Nano 33 BLE and BLE Sense are officially available!, Arduino, 31 July 2019. https://blog.arduino.cc/2019/07/31/the-arduino-nano-33-ble-and-ble-sense-are-officially-available/. Accessed 22 Nov 2019
25. Raj, A.: Arduino Nano 33 BLE Sense Review - What's New and How to Get Started?, Circuit-Digest, 08 November 2019. https://circuitdigest.com/microcontroller-projects/arduino-nano-33-ble-sense-board-review-and-getting-started-guide. Accessed 11 Dec 2019
26. STMicroelectronics: iNEMO inertial module: 3D accelerometer, 3D gyroscope, 3D magnetometer. https://content.arduino.cc/assets/Nano_BLE_Sense_lsm9ds1.pdf. Accessed 22 Nov 2019
27. Salazar, A.J., Silva, A.S., Borges, C.M., Correia, M.V.: An initial experience in wearable monitoring sport systems. In: Proceedings of the 10th IEEE International Conference on Information Technology and Applications in Biomedicine (2010)

An Effective Vision Based Framework for the Identification of Tuberculosis in Chest X-Ray Images

Tejasvi Ghanshala[1], Vikas Tripathi[2(✉)], and Bhaskar Pant[2]

[1] University of British Columbia, Vancouver, Canada
tejasviuniversity18@gmail.com
[2] Graphic Era Deemed to be University, Dehradun, India
vikastripathi.be@gmail.com, pantbhaskar2@gmail.com

Abstract. Tuberculosis is an infection that influences numerous individuals worldwide. While treatment is conceivable, it requires an exact conclusion first. Especially in developing countries there are by and large accessible X-beam machines, yet frequently the radiological aptitude is missing for precisely surveying the pictures. An automated vision based framework that could play out this undertaking rapidly and inexpensively could radically improve the capacity to analyze and at last treat the sickness. In this paper we propose image analysis based framework using various machine learning techniques like SVM, kNN, Random Forest and Neural Network for effective identification of tuberculosis. The proposed framework using neural network was able to classify better than other classifiers to detect Tuberculosis and achieves accuracy of 80.45% .

Keywords: Image processing · Neural Network · Tuberculosis · Chest X-Ray · Machine learning

1 Introduction

Tuberculosis (TB) is a potent killer of over millions of people all around the world. *Mycobacterium tuberculosis* is an etiological agent of TB and according to WHO in 2017, the estimated number of TB incidents was 10.0 million, comprising 3.2 million incidences in women and one million cases in children. Almost 1.6 million deaths were caused from TB including 0.4 million death of people with HIV [1]. At present, TB is the second leading cause of death from an infectious agent worldwide, after the Human Immunodeficiency Virus (HIV) [2]. The diagnostic and cure of tuberculosis is an obstacle that many researchers have been trying to overcome from several decades.

Diagnosis of pulmonary infection needs a combination of radiographic study, appropriate microbiological problems and clinical awareness. Chest X-ray is an inexpensive technique that is used for rapid detection of pulmonary abnormalities [3]. The radiologist faces a challenging situation during diagnosis of the patient with the pulmonary infection because various infections shares similar sign and symptoms.

© Springer Nature Singapore Pte Ltd. 2020
M. Singh et al. (Eds.): ICACDS 2020, CCIS 1244, pp. 36–45, 2020.
https://doi.org/10.1007/978-981-15-6634-9_4

The study of image processing will be helpful to identify the minors that the radiologist misses during the X-ray investigation. Drastically improving quantitative performances of image processing tools and algorithm is used for the recognition, detection and segmentation of features. Deep learning jointly with language and image learning tasks helps in knowledge-guided transfer learning, in image captioning and visual question answering [4].

In this paper we have showcased computer aided image analysis for easily identifying and discriminating tuberculosis from normal person's chest X-ray images.

Further paper is structured as follows: Sect. 2 consists of the literature work related to our work. In Sect. 3, our framework for the identification of tuberculosis is mentioned. Section 4 gives the result and discussion of our proposed work. And in Sect. 5 we have concluded our work with its future aspects.

2   Literature Review

Investigation of Pulmonary infection has been a diagnostic challenge since decades and the misdiagnosed patients have suffered from the ill treatment course. A case report of Cupples & Blackie and a case study on 105 patients done by the Dr. Barnes at USC School of Medicine, Los Angeles have suggested that clinical and radiographic features of pulmonary infections may sometimes be indistinguishable [5, 6]. A case study on a 22-year-old man was done by Pinto et al., reported that there can be occasional situations when pulmonary tuberculosis with respiratory failure can masquerade as pneumonia, and entertained as pneumonia by the physicians [7]. Similar case report of 16-year-old female was presented by Singh et al. [8]. The girl was suffering from non-productive cough, significant weight loss, joint pains, nocturnal fever and fatigue. The primary treatment for cough and fever was done by physician, and the differential diagnosis for pneumocystis pneumonia, viral/fungal pneumonia or miliary tuberculosis was performed with HIV test. After the demise of the patient, disease was diagnosed as Tuberculosis after seeing the result of Nucleic Acid Amplification [Xpert™ MTB/RIF]. But unfortunately, the misdiagnosis had killed the little angel [8]. The proper, accurate and inexpensive diagnosis is needed to distinguish between the pulmonary infections. The radiologist and physicians primarily rely on Chest X-Ray report for diagnosis of pulmonary infection. Rohmah et al., have presented an image processing-based approach to identify pulmonary tuberculosis. They tried to reduce the waiting time of diagnosis result. They have used the minimum distance classifier as classification method to detect the tuberculosis through chest X-ray images [9]. Poornimadevi et al., implemented an automated approach by using registration-based segmentation methods for detecting tuberculosis [10]. Fatima et al., have also developed an automated tuberculosis detection technique [11]. Parveen & Sathik, have used unsupervised fuzzy c-means classification learning algorithm for detection of pneumonia infection [12]. Sharma et al., have used a novel approach for detecting pneumonia clouds in chest X-rays [13]. Stephen et al., have deployed several data augmentation algorithms and improved the accuracy of CNN model for detecting pneumonia [14]. They have showcased the usage of deep neural network for detection of pneumonia.

Since inception of machine learning algorithms, application are of these algorithms have been grown exponentially. In today's world almost everywhere we can find application of machine learning like smart city development, medical analysis, business, education etc. In medical field itself various algorithms have been proposed like SVM, kNN, Decision tree and its variations etc. for disease diagnostics. Fatima et al. have presented a comprehensive survey on usage of machine learning algorithms for identification and analysis of various diseases like heart, liver, hepatitis etc. [15]. In [16] authors have presented comparative analysis between the kNN and SVM algorithms. CNN have been efficiently used for activity detection in given video shown in [17].

The main motive of this paper is to provide an effective mechanism to identify TB in given Chest X-Ray images. As show cased on above literature survey, it is still a challenging task to identify disease in given X-Ray images. Further for identification of suitable machine learning algorithm our survey suggest that comparative analysis should be used to show case effectiveness of any specific algorithm for TB analysis.

3   Methodology

Machine learning has a wide variety of applications like cancer detection, helmet detection etc. We proposed a framework that will detect chest tuberculosis using neural network. Our framework is shown in Fig. 1. As shown in Fig. 1, images are fed in framework then features are extracted from images further training and testing is performed.

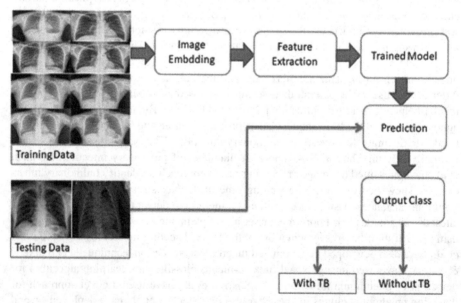

Fig. 1. A framework proposed for identification of Tuberculosis in chest X-Ray.

To get training and testing images, first we divide our database in the ratio of 7:3 and used 70% of the data for training, in training first we applied image embedding on our database which has two classes with TB and without TB, with TB class depicts the people which are suffering from tuberculosis. Image embedding is used to calculate a feature vector for each image. It calculates features by reading an image and then uploading them to a remote server or evaluates them locally by using different machine models. Further we used these extracted features to train the neural network model which will be further used for prediction.

We have used different algorithms to compare our results. Image embedding has been done using Google's pre-trained Inception V-3 model. It uses CNN for the feature extraction. It takes images as input and passes it to the first convolution layer. The images are then dot product with the feature descriptor and a feature map is obtained. This feature map is further sent to pooling layer; in our case max-pooling has been used. The feature map is thus reduced and a convolution is formed. The process is repeated for the entire convolution layer and then the resultant feature map is converted to 1-D vector. The output is thus sent to ANN and the classification is done. In case of SVM the classification is done with the help of hyperplanes. The aim is to maximize the distance or margin between the supporting hyperplanes and to obtain the optimal decision boundaries. The third algorithm used is kNN in which the classification is based on finding the 'k' nearest neighbors. These k neighbors are calculated and the mode of the class of these k neighbors is chosen as the required class. And the last algorithm used is Random forest. This algorithm is the cluster of multiple decision trees that provide output by calculating the information gain or entropies of parent and child classes. After that the required class is chosen by voting the output of each decision tree.

4   Result and Discussion

To achieve results and deployment of machine learning algorithms, the experiments were run on Machine in which configuration includes Windows machine with Intel(R) Core (TM) i5-6200U CPU @ 2.30 GHz × 4 with 6 GB RAM. The chest x-ray image dataset is taken form [18] and separated into two different classes. As per benchmark settings used for analysis, we divided the dataset into the ratio of 7:3, 70% of the images are used for training and 30% of the images are used for testing, validation size is kept 20% and we have used batch size of 10. The model was created using three convolution layers with filter size of 3 and number of filters used is 16, 32 and 64 respectively. To extract features for classification, flattening is used to convert multi-dimensional array into 1-D vector with layer size of 128, the model was trained with 1000 iterations and calculated epoch is 28. Some sample images of our dataset are shown in Fig. 2.

Figure 3 and Fig. 4 shows comparative ROC curve for both the classes respectively, both ROC curve contains comparison between different machine learning algorithms. These curve shows that performance of each algorithm is quite stable.

Table 1 includes proportional scrutiny in the form of various statistics like F1 score, recall and precision for both the classes when neural network classification have been deployed. We can see that the recall value of normal people is better than abnormal people means it is detecting normal people with more accuracy than abnormal people but the precision of detecting the abnormal people is more than precision of detecting normal people which is a good sign. Further results can be improved by incorporating

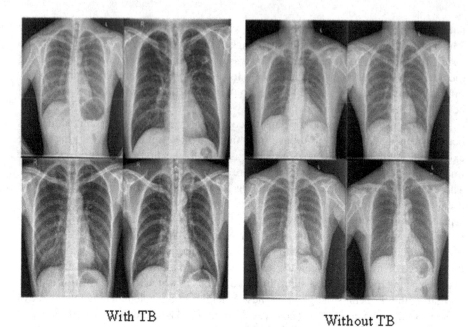

With TB   Without TB

Fig. 2. Sample Chest X- Ray Images of TB and without TB

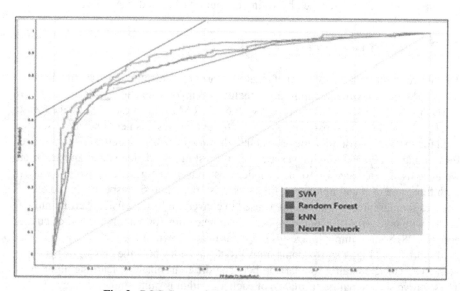

Fig. 3. ROC Curve of various algorithms for class TB

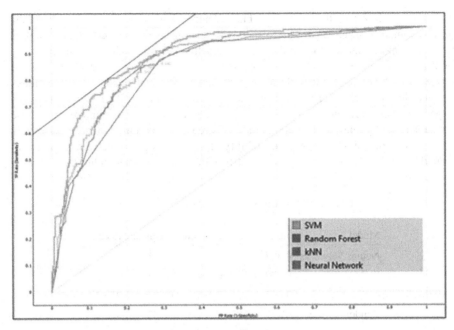

Fig. 4. ROC Curve of various algorithms for class without TB

more images on training and testing. Since images are limited and especially TB images are comparatively different in structure, framework is not able to classify all images of TB accurately however it is still giving efficient results.

Table 1. Comparison between F1-Score, Precision, Recall of different classes.

Class	F1-Score (%)	Precision (%)	Recall (%)
With TB	78.9	87.0	72.0
Without TB	82.0	76.0	89.0
Average	80.45	81.5	80.5

Further to show case the suitability of neural network for analysis of TB and Non TB images, comparisons have been performed with other classification techniques. Table 2 includes proportional scrutiny in the form of recall, precision, AUC, CA and F1-score for different machine learning algorithms. As we can see that Neural Networks provides most efficient results as compared to other techniques. Although results of all techniques are very close to each other however it is clearly visible from table that neural network outperform other techniques in each parameter. Average of all technique show case that deviation of values in different parameters are stable in nature hence framework is effective. Figure 5 and 6 demonstrate stability of framework and effectiveness of neural network. It is clearly visible that neural network outperforms other techniques.

Table 2. Proportional scrutiny of different models.

Model	AUC	CA	F1-score	Precision	Recall
Neural Networks	0.894	0.811	0.811	0.811	0.811
SVM	0.870	0.795	0.790	0.793	0.796
kNN	0.853	0.796	0.795	0.803	0.792
Random Forest	0.851	0.782	0.782	0.782	0.782
Average	0.867	0.796	0.795	0.797	0.795

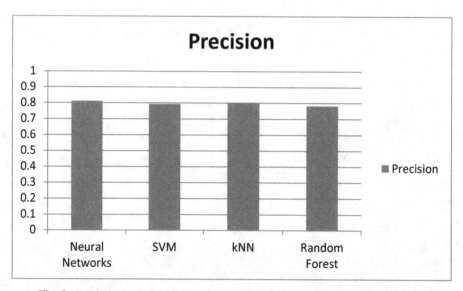

Fig. 5. Precision Statistics of various machine learning algorithms over TB dataset

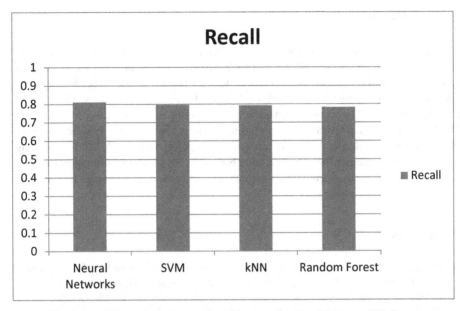

Fig. 6. Recall Statistics of various machine learning algorithms over TB dataset

For more insight of result Table 3 is presented. Table 3 includes true positive, true negative, false negative, false positive values for both the classes. The overall accuracy for detecting normal people is 88.7% and detecting abnormal people is 72.2%. As we can see in table (row 1 with TB) the false positive score is 28, which means that framework has detected x-ray of people having TB as normal people which is not correct similarly row 2 also provide valuable information about effectiveness and limitations of framework. This limitation of efficiency can be corrected by increasing the dataset.

Table 3. Statistics representing the True Positive, True Negative, False Positive, False Negative values along with accuracy and average accuracy.

Class	True Positive	True Negative	False Positive	False Negative	Accuracy
With TB	87	73	28	11	88.7%
Without TB	73	87	11	28	72.2%
Average accuracy					80.45%

5 Conclusion

In the current study we have proposed a novel framework to classify chest X-ray image data into two categories i.e. with TB and without TB. Various classification algorithms have been compared to showcase that neural network is most suited for identification of TB from given X-ray image. The framework is capable to classify all the studied classes with high accuracy. The accuracy can be increased in future by adding more images to the TB class and removing unwanted objects from the chest X-ray images. It is the first study where the investigation between the tuberculosis is performed which is differentiating the X-Rays according to the presence and absence of TB. The discrimination between the pulmonary diseases, pneumonia, tuberculosis, lung cancer is a complex task and our work can be used as a base for this.

References

1. World Health Organization. Global tuberculosis report 2017: World Health Organization (2017). https://www.who.int/tb/publications/global_report/gtbr2017_main_text.pdf
2. Zumla, A., George, A., Sharma, V., Herbert, N.: WHO's 2013 global report on tuberculosis: successes, threats, and opportunities. Lancet 382(9907), 1765–1767 (2013)
3. Franquet, T.: Imaging of pneumonia: trends and algorithms. Eur. Respir. J. 18(1), 196–208 (2001)
4. Wang, X., Peng, Y., Lu, L., Lu, Z., Bagheri, M., Summers, R.M.: Chestx-ray8: hospital-scale chest x-ray database and benchmarks on weakly-supervised classification and localization of common thorax diseases. In: Proceedings of the IEEE Conference on Computer Vision and Pattern Recognition, pp. 2097–2106 (2017)
5. Cupples, J.B., Blackie, S.P.: Granulomatous Pneumocystis carinii pneumonia mimicking tuberculosis. Arch. Pathol. Lab. Med. 113, 1281–1284 (1989)
6. Barnes, P.F., Steele, M.A., Young, S.M., Vachon, L.A.: Tuberculosis in patients with human immunodeficiency virus infection: how often does it mimic Pneumocystis carinii pneumonia. Chest 102(2), 428–432 (1992)
7. Pinto, L.M., Shah, A.C., Shah, K.D., Udwadia, Z.F.: Pulmonary tuberculosis masquerading as community acquired pneumonia. Respir. Med. CME 4(3), 138–140 (2011)
8. Singh, K., Hyatali, S., Giddings, S., Singh, K., Bhagwandass, N.: Miliary tuberculosis presenting with ards and shock: a case report and challenges in current management and diagnosis. Case reports in critical care (2017)
9. Rohmah, R.N., Susanto, A., Soesanti, I., Tjokronagoro, M.: Computer Aided Diagnosis for lung tuberculosis identification based on thoracic X-ray. In: 2013 International Conference Information Technology and Electrical Engineering (ICITEE), pp. 73–78 (2013)
10. Poornimadevi, C.S., Sulochana, H.: Automatic detection of pulmonary tuberculosis using image processing techniques. In: 2016 International Conference on Wireless Communications, Signal Processing and Networking (WiSPNET), pp. 798–802, March 2016
11. Fatima, S., Shah, S.I.A., Samad, M.Z.: Automated tuberculosis detection and analysis using CXR's images. Int. J. Comput. Electr. Eng. 10(4), 284–290 (2018)
12. Parveen, N., Sathik, M.M.: Detection of pneumonia in chest X-ray images. J. X-ray Sci. Technol. 19(4), 423–428 (2011)
13. Sharma, A., Raju, D., Ranjan, S.: Detection of pneumonia clouds in chest X-ray using image processing approach. In: Nirma University International Conference on Engineering (NUiCONE), pp. 1–4, 23 November 2017

14. Stephen, O., Sain, M., Maduh, U.J., Jeong, D.U.: An efficient deep learning approach to pneumonia classification in healthcare. J. Healthcare Eng. (2019)
15. Fatima, M., Pasha, M.: Survey of machine learning algorithms for disease diagnostic. J. Intell. Learn. Syst. Appl. **9**(01), 1 (2017)
16. Kim, J.I.N.H.O., Kim, B.S., Savarese, S.: Comparing image classification methods: K-nearest-neighbor and support-vector-machines. In: Proceedings of the 6th WSEAS International Conference on Computer Engineering and Applications, and Proceedings of the 2012 American conference on Applied Mathematics, vol. 1001, p. 48109–2122 (2002)
17. Bose, C., Sharma, D., Tripathi, V., Singh, A., Pandey, P.: A framework for analyzing the exercise and athletic activities. In: 2019 6th International Conference on Computing for Sustainable Global Development (INDIACom), pp. 121–124, March 2019
18. Jaeger, S., et al.: Automatic tuberculosis screening using chest radiographs. IEEE Trans. Med. Imaging **33**(2), 233–245 (2014). https://doi.org/10.1109/TMI.2013.2284099. PMID: 24108713

User Assisted Clustering Based Key Frame Extraction

Nisha P. Shetty[✉] and Tushar Garg

Department of Information and Communication Technology, Manipal Institute of Technology, Manipal Academy of Higher Education, Manipal 576104, India
nisha.pshetty@manipal.edu, tusag31@gmail.com

Abstract. Our study proposes a novel method of key frame extraction, useful for video data. Video summarization indicates condensing the amount of data that must be examined to retrieve any noteworthy information from the video. Video summarization [1] proves to be a challenging problem as the content of video varies significantly from each other. Further significant human labor is required to manually summarize video. To tackle this issue, this paper proposes an algorithm that summarizes video without prior knowledge. Video summarization is not only useful in saving time but might represent some features which may not be caught by a human at first sight. A significant difficulty is the lack of a pre-defined dataset as well as a metric to evaluate the performance of a given algorithm. We propose a modified version of the harvesting representative frames of a video sequence for abstraction. The concept is to quantitatively measure the difference between successive frames by computing the respective statistics including mean, variation and multiple standard deviations. Then only those frames are considered that are above a predefined threshold of standard deviation. The proposed methodology is further enhanced by making it user interactive, so a user will enter the keyword about the type of frames he desires. Based on input keyword, frames are extracted from the Google Search API and compared with video frames to get desired frames.

Keywords: Video skimming · K-means · Google images scraping · Key frames

1 Introduction

Video content analysis is an integral part of today's fast-paced world. Be it the security forces using it for detection of theft or crimes and scientific researchers using it for mobility; it's found a varied usage in every field. Any video analysis process is performed on the original frames. The consecutive frames in a video differ in a slight manner, mostly projecting redundant information. Therefore, to speed up the tedious process of video analysis, key frame extraction [2, 3] which represents the core information of the entire video in a miniscule frameset is showcased. Thus, owing to the reduced amount of video information, low bit rate channel transmission, reduction in the physical memory space, and faster and better video viewing experience can be provided. We define key frames as those frames which represent a significant change from one frame to another.

© Springer Nature Singapore Pte Ltd. 2020
M. Singh et al. (Eds.): ICACDS 2020, CCIS 1244, pp. 46–55, 2020.
https://doi.org/10.1007/978-981-15-6634-9_5

Considering an example of night CCTV footage of a bank, we may define key frames to be those that provide information of suspicious activity. Such frames must be extracted from what may be hours of video content containing little to no activity.

1.1 Need for Video Summarization in Today's World

Video summarization proves to be significantly useful in a modern day context. For example, a Football Committee wants to decide players for next season for which they need to evaluate the performance of players by going through all the matches for the previous year. The solution to such problems is key frame extraction as it helps to summarize a video effectively narrowing down hours of video into important snippets. Video analysis can also be used for generating highlights of a match which is a manual process right now. Moreover, the security forces would be an advantage as CCTV footage of a surveillance camera can be analyzed in minutes by using key frames rather than going through entire footage as time is an important factor in case of crimes. Video summarization can also be used to deal with the problem of people not wearing helmets in India by installing CCTV cameras at various junctions or traffic stops and using a dynamic system of video analysis to identify people not wearing helmets.

1.2 Drawbacks of Current Methods [4]

The comparison based method chronologically matches each frame with the previously mined key frame for feature differences. This method is very effortless and easy to instigate. However, the extracted key frames contain only local properties and may generate superfluous information and also there is a high chance of the algorithm to get stuck in the local optimum. The reference frame based method espouses the algorithms to spawn a virtual reference frame and harmonizes each frame with it. Based on this comparison key frames are decided. Nonetheless, the accuracy of this model solely depends on the selection of reference frame which is more sort of a hit and trial process. However, these algorithms are not only computationally expensive but also strongly rely on highly specific heuristic rules. Most importantly none of them considers the user input as this has become an important aspect of video analysis as everyone user has a different perspective. So some content may be useful for one user but could be worthless for another user.

2 Related Works

Janko Calic and Ebroul Izquierdo [5] developed a key frame extraction method using difference metrics curve simplification by discrete contour evolution algorithm. They suggested the extension of their work to further improve by using faster discrete curve simplification.

Lei Zhang and Paul Bao [6] developed a wavelet-based edge detection scheme by scale multiplication. They compare their algorithm with Canny edge detection and LOG algorithms and found relatively better results with their algorithm.

Yi Liu and Xu Cheng [7] experimented on the application of wavelet multi resolution technology in medical image analysis and found their technique superior to Canny and Mallat. The problem they are facing with is setting the appropriate threshold value.

B V Patel and B B Meshram [8] conducted a survey on extracting useful information from the video. They emphasize on developing a user interactive process for video analysis.

3   Methodology

The algorithm is a 5 step pipeline where the input is a video formatted in the .3GP format and the output is a set of key frames as shown in Fig. 1. If the user wants specific results then he can give an input about the type of frame he needs and further processing will be done according to it.

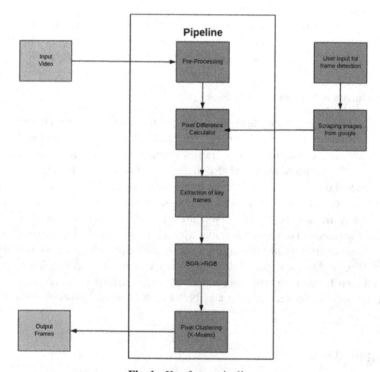

Fig. 1. Key frame pipeline

3.1   Preprocessing with Frame Skipping

The first step in the extraction of key frames is Frame Skipping [9]. It mainly refers to selecting frames at predefined intervals from an input source video. In the sample test video, we started with 1300 frames. After applying a simple algorithm to skip frames after every second of video we end up with 128 frames. Hence for each 1 s of video, we are able to generate a single frame to summarize the content. To perform extraction the OpenCV library [10] - a library aimed at computer vision and video processing were utilized. To get every second frame of video frames-per-second count is extracted.

The frames which satisfies the following equation are picked for analysis:

$$frame\_number \% frame\_per\_second \qquad (1)$$

If the video has a frame-per-second count as 10 fps then a frame for every 10 frames is extracted. This process effectively compresses a video down into much fewer data and non-reductant data.

3.2 Extracting Images from Google

This is an optional stage (Fig. 2) if a user is not satisfied with the results of video summarization. Then he can go for extracting images from google which pertain to his requirements, so then the frames are extracted from video based on similarity from that images. Basically, sometimes the user is keener on particular contents of video then the video as a whole. Scraping from Google is done using google images API, it enables the user to extract images from google image search with the particular keywords provided by the user. It has the limit of extracting 100 images at a time. The scraping code is written in python and user can give other attributes such as the color of the images according to his needs. Now instead of comparing every frame with the previous frame, the frames are compared with the scrapped images and then clustered which gives us the desired key frames.

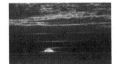   

Fig. 2. Sunset images extracted from google

3.3 Greyscale Conversion and Pixel Difference Calculation

The next stage in the extraction of key frames is to quantitatively compute the difference between two successive frames. The input to the pixel difference calculator is the set of frames chosen by the frame skipping a stage. Then iterate through all the chosen frames and convert them to a greyscale image. Every single pixel is defined by a triple RGB value. OpenCV defines a luminosity function to convert the frame to greyscale. The greyscale [11] is performed on images because color information doesn't aid in identification of important edges or other intricate features. In terms of time complexity, let make an postulation that administering a three-channel color image takes three times as long as treating a grayscale image or possibly four times as long, so for scrutinizing thousands of images from a database, it is imperative to save processing time by resizing images, exploring only portions of images, and/or abolishing color channels we don't need. Shrinking processing time by a factor of three to four makes a significant difference.

The luminosity function is as follows:

$$\text{Luminosity} = 0.21\text{R} + 0.72\text{G} + 0.07\text{B} \tag{2}$$

Simultaneously scale the frames and apply a Gaussian blur [12, 13] (Fig. 3). These stages remove redundant information and make the pixel difference calculation more robust to noise. This is now the modified frame. A Gaussian blur effect is typically generated by convolving an image with a kernel of Gaussian values. The Gaussian blur is a type of image blurring filter that uses a Gaussian function (which also expresses the normal distribution in statistics) for calculating the transformation to apply to each pixel in the image.

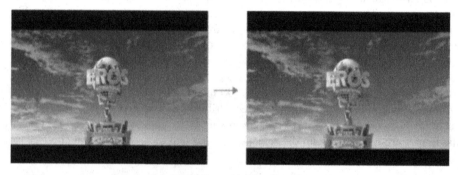

Fig. 3. Scaling with Gaussian blur

The equation of a Gaussian function is:

$$G(x) = \frac{1}{2\pi\sigma^2}e^{-\left(\frac{x^2+y^2}{2\sigma^2}\right)} \tag{3}$$

Further, perform a simple subtract function - provided by the OpenCV library between each modified frame and the preceding one. Then store each modified frame along with the computed difference. Finally, the computed statistics such as the mean, median and standard deviation for the list of computed differences. The formulae for computing the same are given as

$$\bar{x} = \frac{x1 + x2 + \ldots + x_n}{n} \tag{4}$$

$$\sigma = \sqrt{\frac{1}{N}\sum_{i=1}^{N}(x_i - \mu)^2} \tag{5}$$

3.4 K-Means Clustering [14, 15]

In this stage, frames having a computed difference above a predefined threshold is defined to be a key frame. The predefined threshold is calculated using the computed standard deviation. The formula for computing the threshold is as follows.

$$\text{Threshold} = \sigma * 1.85 + \mu \text{ (without user input)} \tag{6}$$

$$\text{Threshold} = \sigma * 0.75 + \mu \text{ (with user input)} \tag{7}$$

All the frames corresponding to the value bigger than the threshold value are clustered using the K-Means approach. Clustering is a very effective technique used in many areas like pattern recognition and information retrieval. K-Means clustering intents to partition n observations into k clusters ensuring each observation belongs to the cluster whose mean value is closest to it. The algorithm extracts a predefined number of dominant colors from an image by using K-Means on the pixels and then returning the centroids of the colors. The algorithm assigns each pixel in the frame to the closest cluster. An output of the clustering is saved into a new file in the folder defined by the user.

4 Results

To gage the operation of the projected technique the algorithm is tested on 5 source videos obtained from YouTube. The obtained output key-frame and expected key frame are discussed here.

4.1 Example 1

The first video, Surveillance.3GP (Fig. 4 and Fig. 5), is a video of women trapped in a building whose activities are captured on the security camera of the building. In this video, she is moving from one room to other. The algorithm extracts only frames which show her movement of moving from one room to another not the movements of her walking in the room. Four shots of this video are below:

Fig. 4. Surveillance.3GP without user input

Fig. 5. Surveillance.3GP with user input

4.2 Example 2

The second video, Candy.3GP (Fig. 6 and Fig. 7), is a video of slow camera movement, in which little gems combine to form shapes moving from one place to another. As camera movement is small, changes between frames are not rapid so according to the algorithm, most of the frames are treated as redundant which is not accurate. Shots of this video are below:

Fig. 6. Candy.3GP without user input

 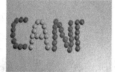  

Fig. 7. Candy.3GP with user input

4.3 Example 3

The third video, Timelapse.3GP (Fig. 8 and Fig. 9), is a video of fast camera motion. The original video is a time lapse that covers two days' worth of data in just 5 min. The key-frames extracted are appropriate in that all different transitions scenes in the video are extracted. The algorithm obtains the best results with this video. Shots of this video are below:

Fig. 8. Timelapse.3GP without user input

Fig. 9. Timelapse.3GP with user input

4.4 Example 4

The fourth video, Effects.3GP (Fig. 10 and Fig. 11), is a video of special effects(artificial editing effects which can also give rise to huge changes between frames) in which a boy produces different color lighting strokes using his hand movement and other devices. Unfortunately, key-frames extracted using the algorithm cannot reproduce the event. Shots of this video are below:

Fig. 10. Effects.3GP without user input

Fig. 11. Effects.3GP with user input

4.5 Example 5

The fifth video, Starwars.3GP (Fig. 12 and Fig. 13), is a video of scrolling captions, in which a small episode of Star wars is displayed in scrolling text. However, the algorithm does not achieve the expected result of extracting the key frames that display the plot and instead only the title sequence has been extracted. Shots of this video are below:

Fig. 12. Scrolling.3GP without user input

Fig. 13. Scrolling.3GP with user input

5   Conclusion and Future Work

Here, a method for programmed extraction of key-frames for video summarization is exhibited that computes the difference between two consecutive frames present in a video sequence or computes the difference between images extracted from google image search upon user assistance and frames present in the video, subsequently clusters key frames. Results on the test video show a good summary of a shot while maintaining a low execution time. The experimental results show that maintaining a good threshold value can give quite accurate results. As in the case of example videos, it is seen that algorithm can produce a summary of entire video content in just 5-10 key frames. For example in the case of surveillance video, key frames are the ones in which there is movement from one room to another which is what extracted by our algorithm. If a user is not satisfied with the result, he may opt for a user-assisted video summarization by giving the algorithm particular keywords which describe the desired content. Based on the keywords images are retrieved from Google and compared with video frames. Slow camera movement gives the more accurate result of user input. However, the proposed approach does not work well for long videos owing to the significant processing time. CUDA [16] can be utilized to reduce execution time for a faster summary of video content. Video summarization can be extended for filtering spam or vague contents from Facebook and other network sites. In the future, we would like to produce a more robust model using parallel computing and better analysis of user input.

References

1. Varghese, J., Nair, K.N.R.: An algorithmic approach for general video summarization. In: 2015 Fifth International Conference on Advances in Computing and Communications (ICACC), Kochi, pp. 7–11 (2015)
2. Raikwar, S.C., Bhatnagar, C., Jalal, A.S.: A framework for key frame extraction from surveillance video. In: 2014 International Conference on Computer and Communication Technology (ICCCT), Allahabad, pp. 297–300 (2014)
3. Luo, Y., Zhou, H., Tan, Q., Chen, X., Yun, M.: Key frame extraction of surveillance video based on moving object detection and image similarity. Pattern Recogn. Image Anal. **28**(2), 225–231 (2018). https://doi.org/10.1134/S1054661818020190
4. Sujatha, C., Mudenagudi, U.: A study on keyframe extraction methods for video summary. In: 2011 International Conference on Computational Intelligence and Communication Networks, Gwalior, pp. 73–77 (2011)
5. Calic, J., Izuierdo, E.: Efficient key-frame extraction and video analysis. In: Proceedings. International Conference on Information Technology: Coding and Computing, pp. 28–33, April 2002
6. Zhang, L., Bao, P.: Edge detection by scale multiplication in wavelet domain. Pattern Recog. Lett. **23**(14), 1771–1784 (2002)
7. Liu, Y., Cheng, X.: The application of wavelet multiresolution technology in medical image analysis. In: 2007 International Conference on Wavelet Analysis and Pattern Recognition, Beijing, pp. 489–493 (2007). https://doi.org/10.1109/icwapr.2007.4420719
8. Patel, B.V., Meshram, B.B.: Content based video retrieval systems. Int. J. UbiComp **3**(2), 13–30 (2012)
9. Fadlallah, F.A., Khalifa, O.O., Abdalla, A.H.: Video streaming based on frames skipping and interpolation techniques. In: 2016 International Conference on Computer and Communication Engineering (ICCCE), Kuala Lumpur, pp. 475–479 (2016)
10. OpenCV. https://opencv.org/
11. Saravanan, C.: Color image to grayscale image conversion. In: 2010 Second International Conference on Computer Engineering and Applications, Bali Island, pp. 196–199 (2010)
12. Flusser, J., Suk, T., Farokhi, S., Höschl, C.: Recognition of images degraded by gaussian blur. In: Azzopardi, G., Petkov, N. (eds.) CAIP 2015. LNCS, vol. 9256, pp. 88–99. Springer, Cham (2015). https://doi.org/10.1007/978-3-319-23192-1_8
13. Gedraite, E.S., Hadad, M.: Investigation on the effect of a Gaussian Blur in image filtering and segmentation. In: Proceedings ELMAR-2011, Zadar, pp. 393–396 (2011)
14. Kanungo, T., Mount, D.M., Netanyahu, N.S., Piatko, C.D., Silverman, R., Wu, A.Y.: An efficient k-means clustering algorithm: analysis and implementation. IEEE Trans. Pattern Anal. Mach. Intell. **24**(7), 881–892 (2002)
15. Wang, S., et al.: K-means clustering with incomplete data. IEEE Access **7**, 69162–69171 (2019)
16. Zheng, R., Yao, C., Jin, H., Zhu, L., Zhang, Q., Deng, W.: Parallel key frame extraction for surveillance video service in a smart city. PloS One **10**, e0135694 (2015)

A Threat Towards the Neonatal Mortality

Kumari Deepika$^{(\boxtimes)}$ and Santosh Chowhan$^{(\boxtimes)}$

Symbiosis Institute of Computer Studies and Research, Symbiosis
International (Deemed University), Atur Centre, Model Colony, Pune 411016, India
{kumari.deepika,santosh.chowhan}@sicsr.ac.in

Abstract. This paper focusses on review and an analysis of the observational studies or case control studies for identification of the threats to the neonatal stage as it is the critical phase for the adaption of the extrauterine life so that significant risk factors are deduced and aiming at the reduction in the mortality of neonatal. Vulnerabilities with respect to both maternal as well as neonatal threatened the survival of neonates. These threats impact economically, socially, psychologically and physical. Predictive analytics to be done through the techniques of machine learning. Supervised learning techniques are adopted for analysis of threats to the neonatal stage. This process splits in two sets: from a give data set, unknown dependencies are to be estimated for training the model and outputs of the system is to be predicted or tested by using estimated dependencies. In future, model can be designed to predict the threats and diseases.

Keywords: Threats · Neonatal mortality · Machine learning

1 Introduction

Pollution is a serious concern globally that impact the life of neonates. Exposure to it through different way like as occupation, in polluted working environments, different agents or pollutants are being consumed by the workers. It impacts adversely to their health as well as their offspring in their womb. The list of some pollutants with their side effects is shown in Table 1 [1].

Not only limited to the pollution, but also vulnerabilities in the different parameters of socio demographic factors like income, education level, nutrition so on threaten the survival of neonates. The list of nutrition and their deficiencies is shown in Table 2 [2].

Below is the list of adverse outcomes on neonates

- Preterm Birth
- Birth Defects
- Still Birth
- Mortality

As one threats may cause the occurrence of another threat that will lead to the mortality of the maternal as well as neonatal. There is a close association between the mothers and their child growing in their wombs so whatever consumed or absorbed by

© Springer Nature Singapore Pte Ltd. 2020
M. Singh et al. (Eds.): ICACDS 2020, CCIS 1244, pp. 56–65, 2020.
https://doi.org/10.1007/978-981-15-6634-9_6

Table 1. Pollutants with their side effects

Agents	Effects of Maternal exposure to foetus
Organic Solvents in General	Fetal Loss and Birth Defects
Benzene	Low Birth Weight and Fetal Loss
Tetrachloroethylene Toluene	Fetal Loss
Some Ethylene glycol ethers and their acetates	Fetal loss and Birth Defects
Inorganic Mercury (Hg)	Fetal Loss
Lead	Fetal Loss, Preterm Birth, Low Birth Weight, Birth Defects
Pesticides	Childhood leukemia, fetal loss, birth defects, preterm birth, reduced fetal growth
Nitrous Oxide	Reduced birth weight, fetal loss
Carbon Monoxide	Preterm Birth, Intrauterine death

Table 2. Micronutrients and their impact on their neonates

Micronutrients	Impact
Vitamin B2, Niacin	Risk of malformations of the urogenital tract
Vitamin B12, Vitamin B6 Zinc	Preterm Birth Higher Neural Tube Defects risk
Iron, Magnesium, Vitamin C	Lower risk of cleft formation
Vitamin A	Risk of inadequate lung maturation in preterm birth
Folate	Higher risk of Neural Tube Defects
Folic Acid	Spina Bifida, Anencephaly
B-Vitamins	Risk of malformations of the cardiovascular system
Iodine	Low Birth Weight
Iron, Magnesium, Niacin	Higher risk of Neural Tube Defect(s)
Protein	Low Birth Weight, Abortion, Prematurity

mother is also communicated to their offspring and leaves its impact on them. Hence Vulnerabilities in parameters directly or indirectly threatens towards the adaptation of the neonates to their extrauterine life.

2   Literature Review

From the neonatal perspective [3, 4] proposed the risk factors on neonatal threats. [5] considered the various risk factors that impact the life of neonates from the neonatal

Table 3. Birth weight and mortality relationship

	7 Days	7–28 Days	≥28 Days
2500+ g	Neonatal Health	Infant Health	Infant Health
1500–2499 g	Neonatal Health	Neonatal Health	Infant Health
0–1499 g	Maternal Health	Maternal Health	Maternal Health

perspective. [6] analyzed the association of birth weight and contribution of health factors from both maternal as well as neonatal towards the mortality of neonates (Table 3).

By reviewing the above-mentioned papers these are the following threats that affect the neonatal mortality from the neonate's perspective

- Prematurity & Low birth weight
  Prematurity is the tenure of the gestation before a neonate is mature enough to be born. Gestation is measured in number of weeks. This condition is generally, before 37 weeks. Low Birth weight is having a weight < 2,500 grams at the time of birth.
- Birth Asphyxia & Birth trauma
  Birth Asphyxia is the indication of an asphyxia at the time of birth due to inadequate of oxygen and respiration. Birth injury is an injury to the baby while delivering. It is also known as birth trauma (Fig. 1).

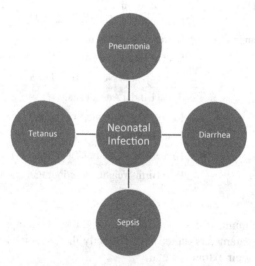

Fig. 1. Types of Neonatal Infections

- Neonatal Infections
- Other noncommunicable diseases
- Ill-defined or cause unknown
- Congenital Anomalies
- Injuries

The transmission medium from maternal to neonates is contact while during delivery and breastfeeding (Fig. 2).

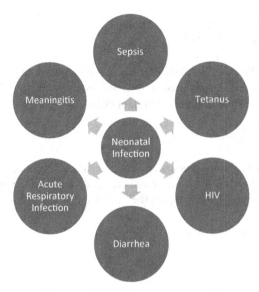

Fig. 2. Disease due to neonatal infections

From the maternal perspective [7] proposed the risk factors on neonatal threats from the maternal perspective. [8] considered the various risk factors that impact the life of neonates from the maternal perspective.

Below is the list of threats from the maternal aspects for the neonatal mortality

- Birth Asphyxia- is the indication of an asphyxia at the time of birth due to inadequate of oxygen and respiration [9, 10].
- Prematurity- is the tenure of the gestation before a neonate is mature enough to be born. Gestation is measured in number of weeks. This condition is generally before 37 weeks [11].
- Occupation- Consumption of pollutants in abundance poses the threats towards mortality [12, 13].
- Multiple gestation- it is the count of foetus carried in the maternal womb [14–16].
- Maternal Disease- it may act as an agent that can be the source for congenital disease. Below is the list of severe agents.

  - Chronic hypertension- When the Systolic level of blood > 120 mm Hg and the diastolic pressure > 80 mm Hg then it is the stage of high blood pressure. [17] analyzed the impact of maternal blood pressure on neonates. Following are the outcomes

Intrauterine Fetal Death (IUFD)
Intrauterine Growth Restriction (IUGR)
Still Birth
Preterm Baby
Perinatal/Neonatal Mortality and Morbidity
Low Birth Weight (<2500 g)
Necrotizing Entrocolitis (NEC)
Retinopathy of Prematurity (ROP)/Retrolental Fibroplasia (RLF)/Terry Syndrome
Hypoxic-ischemic Encephalopathy (HIE) caused by Birth Asphyxia result in
Cerebral Palsy (CP) disabilities and infant death.
Fetal Asphyxia
Meconium Aspiration

– Diabetes before or during pregnancy-When the sugar level of blood > 200 mg/dL
(11.1 mmol/L) then it is the stage of diabetes. [18] analyzed the impact of maternal
diabetes on neonates. Following are the outcomes

Low Birth Weight
Over Birth Weight
Macrosomia
Apgar Score
Jaundice
Intrauterine fetal death and perinatal asphyxia
Premature birth
Polycythemia, hyperviscosity, and hyperbilirubinemia
Birth Injuries
Hypoglycemia
Neonatal respiratory distress
Hypertrophic cardiomyopathy
congenital malformations

– Heart Disease- Heart is a muscular organ of having size approximate to size of an
average fist. It works like a pump, and is situated at left of the center in Chest. Heart
disease is described as the different kinds of conditions that may have adverse effect
on the heart function. These different conditions include:

Coronary artery type of heart disease-it may have adverse effect on the arteries of
the heart
Valvular type of heart disease- it may have adverse effect on the normal function
of the valves that is responsible for the inflow and outflow of the blood to the heart.
Cardiomyopathy type heart disease-it may have effect on squeezing of the heart
muscles
Heart infections-in case of heart infections, some structural problems developed in
the heart before birth i.e. congenital heart disease.

[19] analyzed the impact of maternal heart disease on neonates. Following are the outcomes

Congenital heart disease
Neonatal mortality
Term neonates having low birth weight i.e. <2500 g
Neonatal Intensive Care Unit Admission
Preterm neonates having ≤36 gestational age
stillbirth

- Asthma- An asthma can cause the conditions in which a patient airway become narrow, inflamed and swell that makes difficult to breathe. In some of the cases asthma attack may be life threatening. The common symptoms of asthma are coughing, wheezing, difficulty in normal breathing etc.
- Epilepsy- a type of disorder that make cause the disturbance in nerve cell activity of the brain. It can cause Seizures. This disease is usually diagnosed by self. In case a person starts showing abnormal behavior and sometimes the loss of consciousness during seizure are the symptoms of epilepsy.
- Thyroid (hyper or hypothyroidism)- Thyroid is a gland of small size and in the shape of butterfly found in the mid part of the lower neck. The primary function of gland is to control the metabolism of our body. In general, it produces T3 and T4 hormones that control the metabolism and also tells the body cell that the amount of energy to use. When thyroid functions in proper way the right amount of hormones will be maintained that is needed to keep the metabolism of the body to function at a good rate. In case the thyroid starts producing the excess hormones so the body has to use energy faster than normal such condition is called hyperthyroidism and whenever it produces less amount of hormones the body has to use energy slower than the normal and such condition is called hypothyroidism. There are many different reasons that are responsible for thyroid as below:

Thyroiditis-this is the condition in which the thyroid gland is inflamed. As a result, it produces relatively the lower amount of hormones.
Hashimoto's thyroiditis- it is generally hereditary.
Postpartum thyroiditis- postpartum is the phase after birth giving and this type of thyroid is generally found in women after giving birth. It is not a long term but a temporary condition.
Iodine deficiency – iodine is a food supplement that can control the thyroid. Iodine deficiency can cause the hypothyroidism.
Non-functioning thyroid gland-in some cases the thyroid gland doesn't function at all so no production of hormones. If this is not treated properly it can severely affect the child and will make them physically and mentally retarded.

- Polycystic ovarian disease- Poly Cystic Ovarian Syndrome (PCOS) is a condition of disorder in hormones that causes enlargement of ovaries with small cysts [20]. Analyzed the impact of PCOS on neonates. Following are the outcomes.

Special care Nursery (SCN) admission
Cesarean Delivery
Birth Weight < 2500 g and Birth Weight > 4000 g
Small for Gestation Age (SGA)
Preterm Birth
Stillbirth
Apgar Score at 5 min <7
Perinatal mortality

– Any others [21–23].

- Infectious diseases during pregnancy- maternal infections can be transmitted to the child either when the mother delivers the child or when she feeds her baby.

  – Malaria
  – HIV aids
  – hepatitis (A, B or C)
  – Zika virus
  – Renal infection
  – Bladder infection
  – Pneumonia
  – Any Other [24, 25].

- Socio/Demographic Factors- The maternal lifestyle can also have the impact on neonates. Following are the parameters.

  – Education Level
  – Income
  – Nutrition [7, 13, 26].

The development of an embryo can be disturbed and, it causes birth defects through different agents termed as teratogens. They are categorized into different classes as follows (Fig. 3):

The list of teratogens and their effect is shown in Table 4 [27].

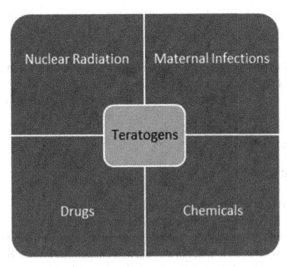

Fig. 3. Classes of Teratogens

Table 4. Teratogens with their impact

Teratogens	Impact
Infection	
HIV	HIV Neonatal Transmission
Cytomegalovirus	Lungs, Infected Kidney, Liver
Rubella	Heart Defects, Congenital Rubella Syndrome, Cataracts, Deafness
Parvovirus, Varicella, Mumps	Nerve Deafness
Herpes Simplex, Syphilis	Microcephaly, Mental Retardation
Toxoplasmosis	Central Nervous System Lesions
Drugs	
Anaesthesia	Structural Deformities, Miscarriages
Dilantin, valproic Acid	Microcephaly, Cleftpalate, Hean Malformations, Retardation
Cocaine	Addiction, Prematurity, Retardation
DES (Diethylstilbestrol)	Genital Deformities, Vaginal Cancer
Barbiturates	Retardation, Hean Defects, Microcephaly
Thalidomide	Deformed limbs (Phocomelia)
Alcohol	Facial Defects, Mental Retardation, Microcephaly
Ionizing Radiation	
Nuclear Radiation or X rays	Mental Retardation, Microcephaly, Central Nervous System Disorders
Chemicals	
Dioxin	Miscarriage, Physical Deformities
Smoke	Low Birth Weight, Miscarriage, Prematurity
Cadmium, Methyl Mercury, Lead	Neurological Disorders, Miscarriages, Mental Retardation

3 Result

By reviewing the literature of many papers, the different threats of neonatal mortality from the maternal as well as neonate's perspective are analyzed and listed out. To identify the issues that has to be occurred in future is to be done using the machine learning techniques. By the use of Machine Learning techniques, predictive analysis is to be performed for this case. Prediction of this case can be done on the basis of the two factors i.e. mother side and infant side factors.

4 Conclusion and Future Scope

The neonatal mortality depends on the several factors occurred on maternal as well neonatal side. Various threats on neonatal mortality are observed and analyzed. In future we can design a predictive modelling with these variables using different machine learning techniques. List of threats are identified based on case control studies or observational studies towards the neonatal stage as it is the vital phase of their extrauterine life for their survival. It focused on the mitigation in their mortality from both the aspects mother as well as neonates.

5 Discussion

The different threats of neonatal mortality have been found out based on the study of different literatures. The contribution of different threats can be further explored by using different machine learning techniques. There are three categories of techniques. They are supervised learning technique, unsupervised learning techniques and reinforcement technique. It will help in reduction of the mortality of neonate and its impact will also mitigate several factors economically, socially and psychologically.

References

1. Taskinen, H.: Occupational exposure to chemicals and reproductive health. In: Reproductive and Developmental Toxicology, pp. 949–959 (2011)
2. Biesalski Hans, K., Jana, T.: Micronutrients in the life cycle: requirements and sufficient supply. NFS J. **11**, 1–11 (2018)
3. Adatara, P., et al.: Risk factors associated with neonatal sepsis: a case study at a specialist hospital in Ghana. Sci. World J. 1–8 (2019)
4. Schuchat, A., et al.: Risk factors and opportunities for prevention of early-onset neonatal sepsis: a multicenter case-control study. Pediatrics **105**(1), 21–26 (2000)
5. Beck, D.F.S.P.: Care of the new born. Save the Children US (2004)
6. Underestimation of Infant Mortality Rates in One Republic of the Former Soviet Union, vol. 111, no. 5, pp. e596–e600 (2003)
7. Garcia, L.P., Fernandes, C.M., Traebert, J.: Risk factors for neonatal death in the capital city with the lowest infant mortality rate in Brazil. Jornal de Pediatria, p. S0021755717307489 (2018)

8. Irles, C., et al.: Estimation of neonatal intestinal perforation associated with necrotizing entero-colitis by machine learning reveals new key factors. Int. J. Environ. Res. Public Health **15**, 2509 (2018)

9. Lee, A.C., et al.: Risk factors for neonatal mortality due to birth asphyxia in southern nepal: a prospective, community-based cohort study. PEDIATRICS **121**(5), e1381–e1390 (2008)

10. Ensing, S., Groenendaal, F., Eskes, M., Abu-Hanna, A., Mol, B.W., Ravelli, A.: 537: maternal and neonatal risk factors for asphyxia related perinatal mortality at term. Am. J. Obstet. Gynecol. **212**(1), S268–S269 (2015)

11. Marchant, T., et al.: Neonatal mortality risk associated with preterm birth in East Africa, adjusted by weight for gestational age: individual participant level meta-analysis. PLoS Med. **9**(8), e1001292 (2012)

12. Türker, G.: Handbook of Fertility. The Effect of Heavy Metals on Preterm Mortality and Morbidity, pp. 45–59 (2015)

13. Delnord, M., Zeitlin, J.: Epidemiology of late preterm and early term births – an international perspective. Seminars in Fetal and Neonatal Medicine, p. S1744165X18301069 (2018)

14. Seikku, L., et al.: Asphyxia, neurologic morbidity, and perinatal mortality in early-term and postterm birth. PEDIATRICS **137**(06), e20153334–e20153334 (2016)

15. Kolobo, H.A., Chaka, T.E., Kassa, R.T.: Determinants of neonatal mortality among new-borns admitted to neonatal intensive care unit Adama, Ethiopia: a case–control study. J. Clin. Neonatol. **8**(4), 232–237 (2019)

16. Abdullah, A., Hort, K., Butu, Y., Simpson, L.: Risk factors associated with neonatal deaths: a matched case–control study in Indonesia. Glob. Health Action **9**(1), 30445 (2016)

17. Bridwell, M., et al.: Hypertensive disorders in pregnancy and maternal and neonatal outcomes in Haiti: the importance of surveillance and data collection. BMC Pregnancy Childbirth **19**, 1–11 (2019)

18. Gonzalez-Quintero, V.H., et al.: The impact of glycemic control on neonatal outcome in singleton pregnancies complicated by gestational diabetes. Diab. Care **30**(3), 467–470 (2007)

19. Ojiyi, E.E., et al.: Pregnancy outcomes of maternal heart disease: a cosmopolitan experience. Women's Wellness Res. Center **2**(1), 12–21 (2020)

20. McDonnell, R., Hart, R.J.: Pregnancy-related outcomes for women with polycystic ovary syndrome. Women's Health, p. 174550571773197 (2017)

21. Peng, S., et al.: A nested case-control study indicating heavy metal residues in meconium associate with maternal gestational diabetes mellitus risk. Environ. Health **14**(1), 19 (2015)

22. Lee, A.C.: Risk Factors for Birth Asphyxia Mortality in a Community-based setting in Southern Nepal. Gary L. Darmstadt, Luke C. Mullany, United States (2007)

23. Gardosi, J., Madurasinghe, V., Williams, M., Malik, A., Francis, A.: Maternal and fetal risk factors for stillbirth: population based study. BMJ **346**, f108–f108 (2013)

24. Brahmanandan, M., Murukesan, L., Nambisan, B., Salmabeevi, S.: Risk factors for perina-tal mortality: a case control study from Thiruvananthapuram, Kerala, India. Int. J. Reprod. Contracept. Obstet. Gynecol. **6**(6), 2452 (2017)

25. Gilman, S.: Pan-Lancashire Neonatal Mortality Review (2013)

26. Alemu, A., Melaku, G., Abera, G.B., Damte, A.: Prevalence and associated factors of perinatal asphyxia among newborns in Dilla University referral hospital, Southern Ethiopia – 2017. Pediatr. Health Med. Ther. **10**, 69–74 (2019)

27. Tulchinsky, T.H.: Environmental and occupational health. In: The New Public Health, pp. 471–533 (2014)

Digital Marketing Effectiveness Using Incrementality

Shubham Gupta and Sneha Chokshi[✉]

Research and Development, MiQ Digital India, Bangalore 560001, KA, India
{shubhamgupta,sneha}@miqdigital.com
http://www.wearemiq.com

Abstract. Digital marketing is one of the fastest-growing advertising channels and crossed the $330 billion mark in 2019. With exponentially increasing budgets, measuring the impact of marketing investments and driving effectiveness becomes essential for brands. The complexity of the digital ad-tech ecosystem is constantly evolving with brands running marketing activities across multiple channels, new targeting capabilities, and different formats. Due to this intricacy, traditional digital measurement metrics like cost per click, return on investment, cost per conversion, etc. just scratch the surface while measuring the actual impact of marketing strategies remains unsettled. We bridged this gap in marketing measurement by using the incremental lift as a metric to measure the impact of a marketing strategy. Incrementality testing is a mathematical approach to differentiate between correlation and causation. We formulated the Viewability Lift method by applying the concepts of A/B testing which can be implemented in the digital marketing ecosystem. In this method, we measure the effectiveness of an ad by comparing the users who are exposed to an ad versus users that are not exposed to an ad. Our methodology covers concepts of test environment setup, randomization, bias handling, hypothesis testing, primary output and understanding different ways of using this output. We used this output for digital marketing strategy planning and campaign optimizations leading to improved campaign efficiency.

Keywords: Digital marketing · Incrementality · Incremental lift · Statistics · Optimization · Hypothesis testing · Artificial intelligence · Machine learning · Marketing measurement · A/B Testing

1 Introduction

The dynamic nature of the digital marketing industry leads to constant advancements within its ad-tech ecosystem in terms of new channels, multiple ad formats, and targeting capabilities. Due to this, approaches to measuring the effectiveness or success of digital marketing need to keep evolving. In digital marketing, effectiveness is measured by various metrics called KPI or Key Performance Indicators. The traditional digital marketing KPIs are Cost per Click, Return

© Springer Nature Singapore Pte Ltd. 2020
M. Singh et al. (Eds.): ICACDS 2020, CCIS 1244, pp. 66–75, 2020.
https://doi.org/10.1007/978-981-15-6634-9_7

on Investment, Cost per Acquisition, Conversion Rate, and Ad Viewed Rate [1]. These KPIs are calculated by various modeling techniques like Conversion Attribution models [2], Last Touch Attribution [3], Revenue Attribution [4] and Multitouch Attribution [5].

In the current digital marketing scenario, one brand runs marketing on multiple channels like email, SEO, programmatic, digital out of home, etc. and the above mentioned KPI's and attribution model alone fall short in measuring the effectiveness of a single channel [6]. This is because the above mentioned KPIs do not measure causality and fails to consider impact by exposure to brand ads on multiple channels and if the user would have made a purchase even without seeing an ad. That's why the industry started focusing on the KPI of Incrementality to measure effectiveness. There are various methods in the industry for calculating Incrementality like Geo Experiments [7], Ghost Ads [8] and A/B Testing using charity ads [9]. These methods require significant resources from the experiment setup and cost point of view. We formulated the Viewability Lift method by applying concepts of A/B testing and using data signals collected while running a digital campaign, that does not require major changes in campaign setup, is cost-effective and hence can be used as an always-on solution.

In the next section, we define incrementality and demonstrate how to calculate incremental lift using data signals collected from the marketing campaign. Section 3 is on statistical inference methods used to validate and build confidence in the results generated in Sect. 2. Section 4 covers results from one of the digital marketing campaigns on which we tested this viewability method. Section 5 concludes with observations drawn from tests till date and limitations of the approach.

2   Incrementality Testing

Incrementality testing [10] is a mathematical approach to measure the causal impact of ad investments. In simple words, incrementality is that additional impact in your campaign performance which happens because of your marketing budget. The need for incrementality is because of the marketing industry's lack of a unified measurement solution that can differentiate between an organic conversion compared to conversion because of the advertiser's marketing budget. Although traditional KPIs like CVR, CTR, CPA [1], are still popular in the industry, these can not quantify if the user purchase action on a website is because of the advertisement or was he going to make the purchase anyway?

2.1   Incrementality Testing

Incrementality testing is achieved from the concept of measuring the impact of a variation from baseline by analyzing and comparing the performance between two user groups (Fig. 1):

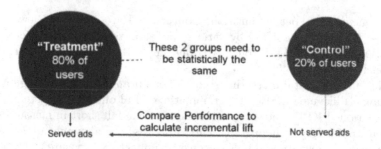

Fig. 1. Test vs Control groups

- Users exposed to advertiser Ad
- Users not exposed to advertiser Ad

The difference in the conversion performance is what we say as conversion lift [11]. There are various techniques to achieve the split in the audience like conventional A/B Testing, Ghost Ads [8], and Geo-Experiments [7]. The approach we took in this paper considers the viewability [12] as a measurement tool to identify when ads have been served to a consumer but not viewed for whatever reason.

2.2 Viewability Lift Method for Incrementality Testing

In digital marketing, the way ads are delivered in the ad-tech ecosystem, they will either be above the fold (viewable to the user) or below the fold (not viewable to the user, unless the user scrolls down to see), refer Fig. 2 below. Every digital marketing campaign always has some ads which do not end up being viewed by the user and are deemed as budget wastage. This is because, campaign targeting strategies are set to reach a specific audience group, demographic, geography or browsing behavior while managing optimal ad serving cost ranges [13]. This data point of the ad viewed or not viewed by a user is leveraged as an input in our approach to create the control and test groups. By using the ad viewability data point as a data asset, this approach does not require extra cost for experiment setup and data collection. Additionally, the campaign can run the business as usual and measurement can be done after the campaign as an always-on solution. Due to its cost-effective nature, implementation can be scaled easily. Unlike in other Controlled experiments [14], where we manually split control and test groups, we are here synthetically creating users who saw the ad (inView users) and users who didn't see the ad (out view users) from the same campaign targeting data.

Here you can treat users who have not been served to a single viewed ad as the control group/outview group and the users who were served at least one ad which was viewed as test group/inView group.

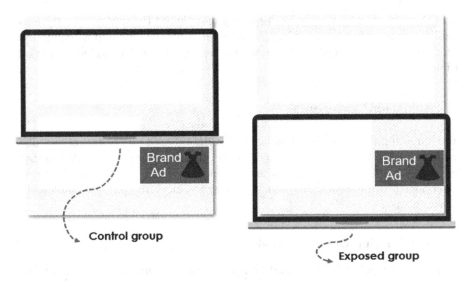

Fig. 2. Viewability as a measurement approach

After the ad serving, campaign data is collected and users are tagged to Test and Control groups and Incremental response rate is calculated as below,

$$\text{test performance} = \frac{\text{users converted in test group}}{\text{users in test group}} = \frac{\text{inView users converted}}{\text{Total inView users}}$$

$$= \text{inView Conversion Rate} \tag{1}$$

$$\text{control performance} = \frac{\text{users converted in control group}}{\text{users in control group}} = \frac{\text{outView users converted}}{\text{Total outView users}}$$

$$= \text{outView Conversion Rate} \tag{2}$$

$$\text{Incremental Lift} = \left[\frac{\text{test performance}}{\text{control performance}} - 1\right] * 100$$

$$= \left[\frac{\text{inView Conversion Rate}}{\text{outView Conversion Rate}} - 1\right] * 100 \tag{3}$$

$$\text{Incremental Response Rate (IRR)} = \text{treatment conversion rate} - \text{control conversion rate} \tag{4}$$

In the next section, we will be discussing more about the analysis stage which mainly focuses on calculation of uplift based on incrementality and its validation using statistical techniques like hypothesis testing [15] and confidence interval estimation.

3 Statistical Inference

After targeting the users and generating the data, based on the techniques discussed in Sect. 2.1, the main focus now will be mostly on how to validate whether there was an uplift (3) through the incremental approach or not. The following subsections provide a building ground on how to calculate uplift and build confidence in that.

3.1 Formulating the Hypothesis

In the experiment we have defined two population groups, inView Users (Test Group/Exposed to Ads) and outView Users (Control Group/Not exposed to Ads). Of these two groups, the output of the experiment is whether the user converted or not. The target variable, i.e., the conversion, follows a binomial distribution,

$$p(C) = \binom{n}{x} p^x q^{n-x} \tag{5}$$

Where,

$x =$ no of success out of 'n' finite trials
$p =$ the probability of success for a single trial or the (hypothesized) population proportion
$q =$ the probability of failure for a single trial or the (hypothesized) population proportion such that $p + q = 1$
$n =$ number of trials

Using the Central Limit Theorem [16], for a large number of samples we can assume the binomial distribution of the target variable to be virtually identical to standard normal distribution.

Now, when we know the probability distribution of our target variable let's define the population proportion for hypothesis testing. If C is the number of converted users in a population N, the *population proportion* P or *conversion rate* (CVR) can be defined as $P = \frac{C}{N}$. Similarly, the corresponding *sample proportion* is given by $p = \frac{c}{n}$.

Step 1: Stating the Null hypothesis and Alternative Hypothesis

- H_o (**Null Hypothesis**): There is no significant difference in the conversion rates of two groups, i.e., $p_1 \approx p_2$
- H_a (**Alternate Hypothesis**): There is a significant difference between the conversion rates of two groups, i.e., $p1 \neq p2$

Where, $p1 = $ inView CVR and $p2 = $ outView CVR

Note:- Hypothesis constitutes a two -Tailed test. The null hypothesis will be rejected if the proportion from test group is too big or if it is too small.

Step 2: Specifying the Level of Significance

In our methodology, we mostly use confidence intervals of 99%, 95% and 90% depending on the size of the data. The values of significance level at various confidence interval are shown below (Table 1):-

Table 1. Significance Level of different Confidence Intervals

Confidence interval	α (Significance level)
99%	1%
95%	5%
90%	9%

Step 3: Defining the test Statistic

Since, the target variable, C, follows standard normal distribution according to Central Limit Theorem, we are using Z-test at various confidence intervals shown above. Generally, Z-test for two population is given by:-

$$Z = \frac{(p_1 - p_2)}{\sqrt{\frac{p_1 q_1}{n_1} + \frac{p_2 q_2}{n_2}}} \tag{6}$$

where,

$$p_1 = \text{inView CVR} = \frac{c_1}{n_1} = \frac{\text{inView users converted}}{\text{Total inView users}}$$

$$q_1 = 1 - p_1$$

$$p_2 = \text{outView CVR} = \frac{c_2}{n_2} = \frac{\text{outView users converted}}{\text{Total outView users}}$$

$$q_2 = 1 - p_2$$

After defining the test statistic, we calculate the value Z_o on the data obtained through techniques from Sect. 2.1 and obtain the p-value from the standard normal table.

Step 4: Decision Rule

The p-value is the probability of getting an extreme result from the sample obtained assuming the null hypothesis is true.

Generally,

1. If p-value $\geq \alpha$, we accept the Null Hypothesis
2. If p-value $< \alpha$, we accept the Alternate Hypothesis and reject the null

3.2 Calculations for Interval Estimation

Since this is a two-tail test so, let's say, for $\mathbb{Z}_{\alpha/2} = 0.05/2$, the critical value is 1.96 and confidence interval for the difference in two population proportions can be stated by:

$$(p_1 - p_2) - \mathbb{Z}_{\alpha/2}(\sqrt{\frac{p_1 q_1}{n_1} + \frac{p_2 q_2}{n_2}}) < (p_1 - p_2) < (p_1 - p_2) + \mathbb{Z}_{\alpha/2}(\sqrt{\frac{p_1 q_1}{n_1} + \frac{p_2 q_2}{n_2}}) \tag{7}$$

From (6),

$$\text{Lower Bound} = [\text{inView CVR} - \text{outView CVR}]$$

$$-\mathbb{Z}_{\alpha/2}\left[\frac{\text{inViewCVR}(1 - \text{inViewCVR})}{n_1} + \frac{\text{outViewCVR}(1 - \text{outViewCVR})}{n_2}\right] \tag{8}$$

$$\text{Upper Bound} = [\text{inView CVR} - \text{outView CVR}]$$

$$+\mathbb{Z}_{\alpha/2}\left[\frac{\text{inViewCVR}(1 - \text{inViewCVR})}{n_1} + \frac{\text{outViewCVR}(1 - \text{outViewCVR})}{n_2}\right] \tag{9}$$

4   Results

We ran the Viewability lift framework for a direct-to-consumer digital photography products retailer to determine the value of their digital ad campaign. Five different strategies were running within its digital ad campaign to drive incremental customers over a duration of 1 month. Using the data obtained from incremental framework mentioned in Sect. 2 and applying the statistical inference explained above, we tried to validate the lifts obtained for digital campaigns for this client (Fig. 3).

	Total	Inview	Outview
Users Reached	9863336	4806034	5057302
Convertors	36722	20373	16349
Conversion Rate	0.37%	0.42%	0.32%

Fig. 3. Campaign data

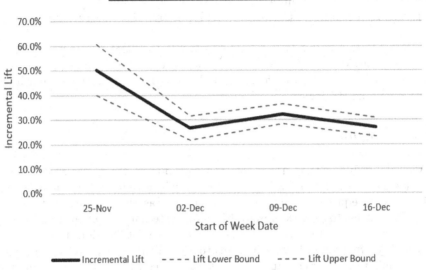

Fig. 4. Incremental CVR lift

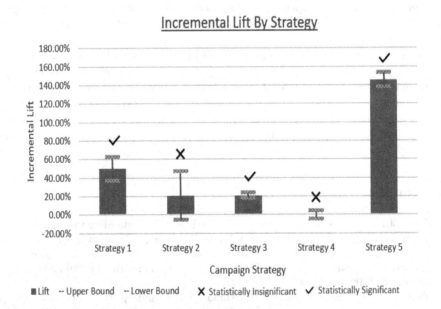

Fig. 5. Incremental CVR lift

Duration of the campaign: 1 Month.

Overall Incremental Lift Achieved: 31.1% statistically significant lift achieved with 2.36% margin of error at 95% confidence interval.

Total Incremental Customer Driven by the Campaign = 4836 (Fig. 4).

Following the campaign, we analyzed results at strategy level, see Fig. 5 below, to identify strategies driving incremental customers. These results were further used for activation by focusing on high performing strategies and build effective bidding strategies.

5   Conclusion

As Viewability lift method uses ad-viewed data points into a data asset, we are not buying ad-slots for control group after all which reduces experiment costs, and does not require splitting our audience pool as the measurement can be done after the campaign. Because of it's zero waste nature, this method is cost effective and can be used as an always on measurement solution. However, the View-ability lift method faces challenges to have Test and Control groups with exactly similar behaviors because of biases introduced by bidding algorithms and ad-serving ecosystem. These biases can be minimized depending on the depth of data and activation integration you have with your client and ad-serving platform. We can further gain confidence on the results via Hypothesis testing as explained in this paper. Using incremental lift in campaign strategy planning and optimizations has helped us understand true impact of the ad campaign and achieve higher campaign efficiency in driving incremental customers for our clients.

Lift Tests are a great way to understand incremental performance of the ad campaign but without the necessary transparency and control, you can end up seeing biased results. While a lot of marketers might be already aware of the basic nuances of Lift tests, a continuous debate on different methodologies and their pros and cons ensues. No single approach is perfect but as a marketer, it is essential that you are aware of these nuances of the method of the lift test you are using.

References

1. Saura, J.R., Palos-Sánchez, P., Cerdá Suárez, L.M.: Understanding the digital marketing environment with KPIs and web analytics. Future Internet **9**(4), 76 (2017)
2. (Alice) Li, H., Kannan, P.K.: Attributing conversions in a multichannel online marketing environment: an empirical model and a field experiment (2014). https://doi.org/10.1509/jmr.13.0050
3. Yuvaraj, C.B., Chandavarkar, B.R., Kumar, V.S., Sandeep, B.S.: Enhanced last-touch interaction attribution model in online advertising. In: 2018 IEEE Distributed Computing, VLSI, Electrical Circuits and Robotics (DISCOVER), Mangalore (Mangaluru), India, pp. 110–114 (2018)

4. Zhao, K., Mahboobi, S.H., Bagheri, S.R.: Revenue-based attribution modeling for online advertising. Int. J. Market Res. **61**(2), 195–209 (2019)
5. Du, R., Zhong, Y., Nair, H., Cui, B., Shou, R.: Causally driven incremental multi touch attribution using a recurrent neural network (2019). arXiv:1902.00215
6. Gordon, B.R., Jerath, K., Katona, Z., Narayanan, S., Shin, J., Wilbur, K.C.: Inefficiencies in digital advertising markets (2019). arXiv:1912.09012
7. Vaver, J., Koehler, J.: Measuring ad effectiveness using geo experiments (2011)
8. Johnson, G.A., Lewis, R.A., Nubbemeyer, E.I.: Ghost ads: improving the economics of measuring online ad effectiveness. J. Mark. Res. **54**(6), 867–884 (2017). https://doi.org/10.1509/jmr.15.0297
9. Gordon, B., Zettelmeyer, F., Bhargava, N., Chapsky, D.: A comparison of approaches to advertising measurement: evidence from big field experiments at Facebook (2016)
10. Liu, C.H., Bettaney, E.M., Chamberlain, B.P.: Designing experiments to measure incrementality on Facebook (2018). arXiv:1806.02588
11. Facebook for Business: Conversion Lift: Helping Marketers Better Understand the Impact of Facebook Ads, January 2015. www.facebook.com/business/news/conversion-lift-measurement
12. European Viewability Steering Group (EVSG): European Viewability Measurement Principles Version 1.1., October 2017. www.eaca.eu/industry/european-viewability-initiative/
13. Bounie, D., Morrisson, V., Quinn, M.: Do You See What I See? Ad Viewability and the Economics of Online Advertising (2017)
14. Kohavi, R., Longbotham, R.: Online Controlled Experiments and A/B Testing (2017). https://doi.org/10.1007/978-1-4899-7687-1_891
15. Davis, R.B., Mukamal, K.J.: Hypothesis testing: means. Circulation **114**(10), 1078–1082 (2006)
16. Kwak, S.G., Kim, J.H.: Central limit theorem - the cornerstone of modern statistics. Korean J. Anesthesiol. **70**(2), 144–156 (2017). https://doi.org/10.4097/kjae.2017.70.2.144

Explainable Artificial Intelligence for Falls Prediction

Leeanne Lindsay[1(✉)], Sonya Coleman[2], Dermot Kerr[2], Brian Taylor[1],
and Anne Moorhead[3]

[1] Institute for Research in Social Sciences, Ulster University, Londonderry,
Northern Ireland, UK
{lindsay-1,bj.taylor}@ulster.ac.uk
[2] Intelligent Systems Research Centre, School of Computing, Engineering and Intelligent
Systems, Ulster University, Londonderry, UK
{sa.coleman,d.kerr}@ulster.ac.uk
[3] School of Communication and Media, Institute for Nursing and Health Research,
Ulster University, Newtownabbey, Northern Ireland, UK
a.moorhead@ulster.ac.uk

Abstract. With a rapidly ageing population, it is likely that we will encounter an older adult falling. Falls can cause death, serious injury or harm, loss of confidence and loss of independence. Falling can happen to any of us, however those over 65 years of age can be classified as a group of adults who are more vulnerable and at increased risk of falling. This paper focuses on applying explainable artificial intelligence techniques, in the form of decision trees, to healthcare data in order to predict the risk of falling in older adults. These decision trees could potentially be introduced for health and social care professionals to help aid their judgements when making decisions.

Keywords: Decision tree · Explainable AI · Classification · Risks · Falls

1 Introduction

A major public health issue is falls and in particular falls in older adults. A third of the population over 65-years old and half the population over 80-years old are likely to fall at least once per year according to Public Health England [1]. Falling not only affects an older adult physically, it can also lead to an individual having unnecessary stress, loss of confidence and loss of their ability to live independently [2]. If an adult is currently living at home, falls can lead to distress for their families, caregivers and health and social care professionals as they then have to come to a decision if the individual can continue living independently or if other arrangements need to be considered [3]. The future of the individual and their safety and wellbeing is of utmost importance. This means everyone involved needs to communicate effectively with each other about choosing the best outcome for the patient. Health and social care professionals make decisions every day and are focusing more of their attention on risks [4]. In order to help

© Springer Nature Singapore Pte Ltd. 2020
M. Singh et al. (Eds.): ICACDS 2020, CCIS 1244, pp. 76–84, 2020.
https://doi.org/10.1007/978-981-15-6634-9_8

aid their judgements and decisions, decision trees are becoming popular for classifying or calculating risks in healthcare as they can be easily understood and interpreted [5].

The use of Artificial Intelligence (AI) has become popular in industries such as healthcare, education, manufacturing and finance. Explainable AI differs from commonly used opaque AI techniques in that it aims to provide an understanding into how AI decisions are made. Decision tree algorithms provide clarity within machine learning as it is possible to clearly interpret how decisions are reached and the attributes that are deemed important in reaching that decision. Prediction accuracy is used within decision trees to explain how conclusions are made at each stage of the decision-making process, providing a cognitive understanding and ultimately trust from humans.

The remainder of this paper is organized as follows: the motivation to study Explainable AI for Falls Prediction is introduced in Sect. 2. Section 3 outlines the decision tree results from the study. Finally, Sect. 4 summarises the work completed in this paper.

2 Explainable AI Methodology for Falls Prediction

Data mining involves extracting useful information from datasets and displaying this in an interpretable way [6]. Decision trees are commonly used for data mining purposes to develop prediction algorithms for a specific target variable. A decision tree can be described as an inverted tree which contains a root node, internal nodes and leaf nodes which are all split into branch-like segments [6]. A decision tree can also be identified as a prediction tree [7]. Decision trees are appealing to use due to their simplicity and their ability to handle mixed data [8]. A decision tree is simply a tree structure that defines a sequence of decisions and their consequences [7]. In this work we use four different types of decision trees to evaluate the effectiveness at measuring the risk of falling: Fast and Frugal Trees, Classification and Regression Trees, Conditional Inference Trees and the J48 decision tree.

Fast and Frugal Trees (FFTs) are a supervised learning algorithm used to create binary classification tasks [9]. This type of decision tree uses sequentially ordered cues, every cue breaks of into two branches, one of these being the exit point. The final cue in the tree will have two exit points for the decision. For the experiments presented here we use the R Studio implementation found under the package FFTrees.

The Classification and Regression Tree (CART) is a form of binary recursive partitioning. Each node in a decision tree can be split into two binary groups. Recursive refers to the binary process being applied over and over again. The partitioning refers to the dataset being split into training and testing sections. An advantage of using the CART decision tree is that it can identify the splitting variables based on searching through all possibilities from the input variables [10]. The building of a tree begins from the root node, this is the beginning of the dataset whereby the variables are split to find the best variable for the root node. The recursive nature of the algorithm ensures that all input variables are checked to find the best variable within the tree. When building the tree, CART recursively splits nodes. As each node is split it is assigned to a predicted class [10]. Branches are then split below each node in the tree and the decision tree becomes complete when a terminal node is in place as the stopping rule [11]. The CART method used here is implemented in R Studio using the Breiman algorithm [12]. CART uses the 'rpart' method to produce classification decision trees [7]. Rpart follows the simple process of:- 'rpart (formula, data=, method=, control=)' whereby the formula includes:- 'Outcome ~ predictor1 + predictor2' etc. 'Data=' specifies the data frame, 'method=' refers to 'class' if using a classification tree and 'anova' is used for a regression tree. Finally, 'control=' references the optional parameters used for controlling the growth of a tree.

Conditional Inference Tree (CTree) uses two steps to split the tree. CTree determines the variable to be split based on the outcome and the measure of association. After examining all variables, the variable determined to create the best split is then chosen as the root node. Instead of using the Rpart package, CTree uses the Party kit package. By default, the Party kit function uses a quadratic test statistic as it is found to produce more accurate splits [13].

The Waikato Environment for Knowledge Analysis (WEKA) learning environment was used to test the J48 decision tree approach. J48 is an open source Java implementation of the C4.5 algorithm. A J48 decision tree is constructed iteratively, one node at a time. Each lead in the tree represents a classification and the branches that connect the lead to the root node are the conditions that produce the classification. The different transparent decision trees can be compared and evaluated according to their individual predictive accuracies to ensure the model correctly predicts the class of either new or unseen data [14].

3 Experiments and Results

We use the dataset from the Irish Longitudinal Study on Ageing (TILDA) [15]. The dataset comprises information from over eight thousand adults whom are all over the age of 50 and living in the community. The dataset is split into three different waves. Wave 1 incorporates data that were collected between 2009 and 2011. Wave 2 data were collected between 2012 and 2013 and Wave 3 data were collected between 2015 and 2016. TILDA collects data from the community-dwelling participants in waves approximately every two years. In this paper we focus specifically on the use of Wave 1 and Wave 2 data only. The TILDA dataset has been previously used in studies such as predicting the likelihood of recurrent falls in older adults based on previous falls [16].

The four different decision tree methods outlined in Sect. 2 were used for this study, namely, Fast and Frugal Trees, Classification and Regression Trees, Conditional Inference Trees and a Decision Tree known as J48 in WEKA. The inputs into each of the decision tree algorithms remained the same in all cases. The data were split into a training set and a testing set using a 90:10 split for all four algorithms. The input variables used from the TILDA dataset were the same for all four decision tree algorithms: "Overall Health Description", "Emotional Mental Health", "Long-term Health Issues", "Previous Blackout or Fainting", "Afraid of Falling" and "Joint Replacements". The target output was defined as either *fall* or *no falls* using a binary classification represented by 0 (no falls) and 1 (falls).

Figure 1 presents the J48 decision tree produced in WEKA. The tree is significantly deeper than the other algorithms before the terminal nodes are defined. The J48 decision tree produced the best predictive classification accuracy out of all four trees. Each terminal node includes a final outcome and two predictive accuracies. Take for example:

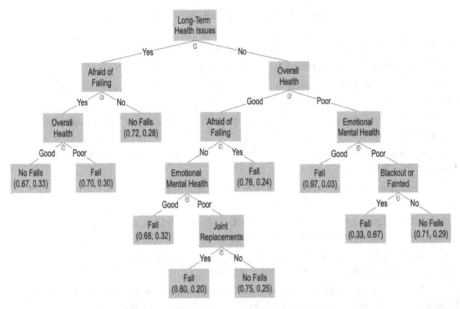

Fig. 1. J48 decision tree for falls and no falls

- Long-Term Health Issues – Yes (They do have a long-term health issue)
- Afraid of Falling – Yes (They are afraid of falling)
- Overall Health – Good (Their overall health is good)
- The terminal node concludes with No Falls, the first predictive accuracy is a 0.67 chance of no falls. The second predictive accuracy in the same terminal node is a 0.33 chance of falls.

However, if their long-term health issue is Yes, Afraid of Falling is Yes and their Overall Health is Poor then the terminal node is different from the above. The terminal node in this instance has a final outcome which is fall as there is a higher chance of a fall than no falls in this case. In each terminal node the outcome, either fall or no falls, relates to the first predictive accuracy in the node.

Table 1 explains the J48 rules. The J48 tree is the most complex tree generated in this study. There is a good range of outcomes for falling and not falling. There is a 28% chance of falling if you have a long-term health issue and you are not afraid of falling. If you do have a long-term health issue, you are afraid of falling and have a poor overall health description then you have a higher 70% chance of falling.

In Fig. 2 the Fast and Frugal tree is illustrated. In this tree it can be noted that if a subject has no long-term health issues, the tree branches off straight away into a terminal node. The Fast and Frugal Tree predicts that if you have no long-term health issues then you are not likely to have any falls. The J48 decision tree had an accuracy result of 69%, whereas the Fast and Frugal tree correctly classified 67% in the overall predicted accuracy (See Table 5). Both of these trees performed the highest out of the four and there is no significant differences between both trees.

Table 1. J48 rules for falls or no falls

0.28 Fall when Long-term Health Issues = Yes, Afraid of Falling = No
0.72 No Fall when Long-term Health Issues = Yes, Afraid of Falling = No
0.33 Fall when Long-term Health Issues = Yes, Afraid of Falling = Yes, Overall Health Description = Good
0.67 No Fall when Long-term Health Issues = Yes, Afraid of Falling = Yes, Overall Health Description = Good
0.70 Fall when Long-term Health Issues = Yes, Afraid of Falling = Yes, Overall Health Description = Poor
0.30 No Fall when Long-term Health Issues = Yes, Afraid of Falling = Yes, Overall Health Description = Poor
0.68 Fall when Long-term Health Issues = No, Overall Health Description = Good, Afraid of Falling = No & Emotional Mental Health = Good
0.32 No Fall when Long-term Health Issues = No, Overall Health Description = Good, Afraid of Falling = No & Emotional Mental Health = Good
0.80 Fall when Long-term Health Issues = No, Overall Health Description = Good, Afraid of Falling = No, Emotional Mental Health = Poor & Joint Replacements = Yes
0.20 No Fall when Long-term Health Issues = No, Overall Health Description = Good, Afraid of Falling = No, Emotional Mental Health = Poor & Joint Replacements = Yes
0.25 Fall when Long-term Health Issues = No, Overall Health Description = Good, Afraid of Falling = No, Emotional Mental Health = Poor & Joint Replacements = No
0.75 No Fall when Long-term Health Issues = No, Overall Health Description = Good, Afraid of Falling = No, Emotional Mental Health = Poor & Joint Replacements = No
0.76 Fall when Long-term Health Issues = No, Overall Health Description = Good & Afraid of Falling = Yes
0.24 No Fall when Long-term Health Issues = No, Overall Health Description = Good & Afraid of Falling = Yes
0.97 Fall when Long-term Health Issues = No, Overall Health Description = Poor & Emotional Mental Health = Good
0.03 No Fall when Long-term Health Issues = No, Overall Health Description = Poor & Emotional Mental Health = Good
0.29 Fall when Long-term Health Issues = No, Overall Health Description = Poor, Emotional Mental Health = Poor & Blackout/Fainted = No
0.71 No Fall when Long-term Health Issues = No, Overall Health Description = Poor, Emotional Mental Health = Poor & Blackout/Fainted = No
0.33 Fall when Long-term Health Issues = No, Overall Health Description = Poor, Emotional Mental Health = Poor & Blackout/Fainted = Yes
0.67 No Fall when Long-term Health Issues = No, Overall Health Description = Poor, Emotional Mental Health = Poor & Blackout/Fainted = Yes

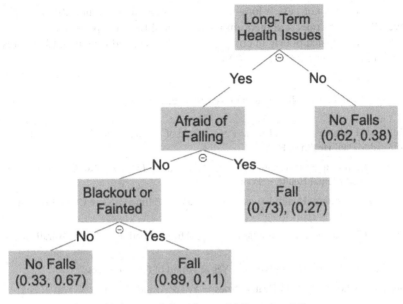

Fig. 2. Fast & frugal tree of falls and no falls

Presented in Table 2 are the rules that correspond to the Fast and Frugal tree in Fig. 2. The probabilities of falls are higher in two out of three of the condition statements. One of the condition statements has a considerably high 89% chance of falling if someone has a long-term health issue and if they are afraid of falling. The probabilities of each terminal node can be found in Table 2.

Table 2. Fast & Frugal Trees (FFTree)

0.33 Fall when Long-term Health Issues = Yes, Afraid of Falling = No & Blackout/Fainted = No
0.67 No Fall when Long-term Health Issues = Yes, Afraid of Falling = No & Blackout/Fainted = No
0.73 Fall when Long-term Health Issues = Yes & Afraid of Falling = Yes
0.27 No Fall when Long-term Health Issues = Yes & Afraid of Falling = Yes
0.89 Fall when Long-term Health Issues = Yes, Afraid of Falling = No & Blackout/Fainted = Yes
0.11 No Fall when Long-term Health Issues = Yes, Afraid of Falling = No & Blackout/Fainted = Yes

Presented in Table 3 are the rules that were generated by the Conditional Inference Tree. It can be seen from the input variables that "Joint Replacements" were discarded by the algorithm due to having no significance. The highest probability of falls is 59%

where someone does not have a long-term health issue, they are not afraid of falling, their overall health is poor and their emotional mental health is poor. The Conditional Inference tree performed poorly in comparison to the Fast and Frugal and J48 decision tree with a classification accuracy of 60%.

Table 3. Conditional inference tree (CTree)

0.59 Fall when Long-term Health Issues = No, Afraid of Falling = No, Overall Health = Poor & Emotional Mental Health = Poor
0.41 No Fall when Long-term Health Issues = No, Afraid of Falling = No, Overall Health = Poor & Emotional Mental Health = Poor
0.48 Fall when Long-term Health Issues = No, Afraid of Falling = No & Overall Health = Good
0.52 No Fall when Long-term Health Issues = No, Afraid of Falling = No & Overall Health = Good
0.46 Fall when Long-term Health Issues = Yes & Afraid of Falling = Yes
0.54 Don't Fall when Long-term Health Issues = Yes & Afraid of Falling = Yes
0.53 Fall when Long-term Health Issues = Yes, Afraid of Falling = No & Blackout/Fainted = Yes
0.47 No Fall when Long-term Health Issues = Yes, Afraid of Falling = No & Blackout/Fainted = Yes

The CART tree algorithm focused only on "Long-term Health issues" and "Afraid of Falling" and its output is presented in Table 4. The probability of falling when someone does have a long-term health issue and when someone is afraid of falling is 54%, this is significantly lower than the Fast and Frugal tree which had 89% for the same circumstances. In previous work, Brighton compared the predictive accuracy of fast and frugal trees with classification and regression trees and found that varying the size of the training sets made a difference to which tree outperformed the other which may explain the difference between the two predictive accuracies [17]. The CART algorithm was the poorest performing tree possibly because it only created a tree using a small number of the risk factors that were inputted.

Table 4. Classification & Regression Trees (CART)

0.57 Fall when Long-term Health Issues = Yes
0.43 No Fall when Long-term Health Issues = No
0.54 Fall when Long-term Health Issues = Yes, Afraid of Falling = Yes
0.46 No Fall when Long-term Health Issues = Yes, & Afraid of Falling = No

Using the testing data, the prediction accuracy for each of the decision trees is presented in Table 5. The overall differences between each of the algorithms are not significant however the best performing algorithm, Decision Tree (J48) obtained an overall accuracy of 69% correct classifications of fall or no fall.

Table 5. Results for each decision tree

Decision tree classifier	Correctly classified %
Fast & Frugal Trees (FFT)	0.67
Classification & Regression Trees (CART)	0.58
Conditional Inference Trees (CTree)	0.60
Decision Tree (J48)	0.69

Although the results are in favour of the J48 decision tree, Fast and Frugal trees may be preferred by Health and Social Care professionals as a Fast and Frugal tree has two branches at every node and each branch is the opposite of each other. This allows professionals to process the tree much quicker and use the process of elimination while interpreting the tree. Gerd Gigerenzer [18] states that the fast and frugal trees are still being used by physicians as they are easily adapted compared with complex machine learning algorithms [18].

4 Conclusion

This study has explored four decision trees algorithms using data from The Irish Longitudinal Study on Ageing. Each decision tree presents the relationship between each of the inputted different risk factors. The health and social care factors that were explored were "Overall Health Description", "Emotional Mental Health", "Long-term Health Issues", "Blackouts/Fainting", "Afraid of Falling" and "Joint Replacements". For all of the algorithms other than J48 decision trees, Joint Replacements were removed as they are considered to have no significance compared to the input factors towards the risk of falling. Considering the overall accuracies, although each of the trees were between 58% to 69% accurate, these results are based on self-declared qualitative data which would be typical of the accuracies obtained for these type of data. It is apparent from the classification results that the explainable decision trees are easily interpreted. The most important aspect of these models is to ensure health and social care professionals understand and accept the models that may help in their day-to-day work with the ability to help and provide the necessary knowledge that can help guide and support their decisions [3]. Further work will consider a visualization dashboard to compare how risks can be visualized in the real world when health and social care professionals are faced with risks every day in their work.

References

1. Fenton, K.: The human cost of falls - public health matters (2014). https://publichealthmat ters.blog.gov.uk/2014/07/17/the-human-cost-of-falls/. Accessed 29 Jan 2020
2. National Institute for Health and Care Excellence: Falls in older people: assessing risk and prevention. Clinical Guideline Nice 2020 (2013). www.nice.org.uk/guidance/cg161
3. Godolphin, W.: Shared decision-making. Healthc. Q. **12**(sp), e186–e190 (2009). https://doi. org/10.12927/hcq.2009.20947
4. Stevenson, M., McDowell, M., Taylor, B.: Concepts for communication about risk in dementia care: a review of the literature. Dementia (2017). https://doi.org/10.1177/1471301216647542
5. Dillibabu, R., Suresh, K.: Designing a machine learning based software risk assessment model using Naïve Bayes algorithm (2018). www.tagajournal.com
6. Song, Y.Y., Lu, Y.: Decision tree methods: applications for classification and prediction. Shanghai Arch. Psychiatry **27**(2), 130–135 (2015). https://doi.org/10.11919/j.issn.1002-0829. 215044
7. Sharma, A., Srivastava, A.: Understanding decision tree algorithms by using R programming language, pp. 177–182 (2016)
8. Su, J., Zhang, H. (n.d.): A fast decision tree learning algorithm introduction and related work. www.aaai.org. Accessed 20 Jan 2020
9. Phillips, N.D., Neth, H., Woike, J.K., Gaissmaier, W.: FFTrees: a toolbox to create, visualize, and evaluate fast-and-frugal decision trees. Judgm. Decis. Mak. **12**(4), 344–368 (2017)
10. Lewis, R.J. (n.d.): An introduction to classification and regression tree (CART) analysis. https://www.researchgate.net/publication/240719582. Accessed 9 Jan 2020
11. Venkatasubramaniam, A., Wolfson, J., Mitchell, N., Barnes, T., Jaka, M., French, S.: Decision trees in epidemiological research. Emerg. Themes Epidemiol. **14**, 11 (2017). https://doi.org/ 10.1186/s12982-017-0064-4
12. Speybroeck, N.: Classification and regression trees. Int. J. Public Health **57**(1), 243–246 (2012). https://doi.org/10.1007/s00038-011-0315-z
13. Hothorn, T., Hornik, K., Wien, W., Zeileis, A., (n.d.): CTree: conditional inference trees. Vignette R package partykit version 1.1-1 (2016). https://CRAN.R-project.org/web/packages/ partykit/vignettes/ctree.pdf. Accessed 28 Jan 2020
14. Zheng, Y., Peng, L., Lei, L., Junjie, Y.: R-C4.5 decision tree model and its applications to health care dataset. In: International Conference on Services Systems and Services Management, Proceedings of ICSSSM 2005, vol. 2, pp. 1099–1103 (2005). https://doi.org/10.1109/ICS SSM.2005.1500165
15. The Irish Longitudinal Study on Ageing. Trinity College Dublin, The University of Dublin. https://tilda.tcd.ie/
16. Lindsay, L., Coleman, S., Taylor, B., Kerr, D., Moorhead, A.: Using machine learning algo-rithms to predict the likelihood of recurrent falls in older adults. In: 15th International Conference on Machine Learning and Data Mining, pp. 1–5 (2019)
17. Guo, P., Pedrycz, W. (eds.): Human-Centric Decision-Making Models for Social Sciences. SCI, vol. 502. Springer, Heidelberg (2014). https://doi.org/10.1007/978-3-642-39307-5
18. Wasley, D., Araujo, L.S., Raab, M., Gigerenzer, G.: The power of simplicity: a fast-and-frugal heuristics approach to performance science. Front. Psychol. **6**, 1672 (2015). https://doi.org/ 10.3389/fpsyg.2015.01672

Enhanced UML Use Case Meta-model Semantics from Cognitive and Utility Perspectives

Mahesh R. Dube[✉]

Department of Computer Engineering, Vishwakarma Institute of Technology, Pune,
Maharashtra, India
mahesh.dube@vit.edu

Abstract. Unified Modeling Language (UML) formalized by Object Management Group (OMG) to express analysis and design models is a general-purpose graphical language for visualization and documentation of software system artifacts. UML diagrams are interdependent and hence a change in one diagram at a level would introduce changes in the entire related diagrams. Since UML divides the system model into functional requirement capture views modeled by use case diagrams, static structural views modeled by class diagrams, and dynamic behavior views modeled by interaction and state-machine diagrams. As domain consists of concepts, the higher-order views can be formed from the recognized concepts so that the structuring is visible at the initial development efforts. The models are required to be platform-independent so that they can be mapped to any available platform using migrations. From the model semantics, a metalanguage representing the model language can be created so that the model transformations can be applied vertically and horizontally. In this regard, an attempt to narrate enhanced semantics for use cases and its relationship has been made.

Keywords: Unified Modeling Language · Use cases · Meta-model · Non-Functional requirement · Semantics

1 Introduction

The rational exemplifications of knowledge about classifications of recognizable entities and know-hows are known as concepts. Concepts provide distinctive or abstract information of things associated with a suitable linguistic form to derive interpretations. A metaphor is a cognitive instrument that provides the structural mapping between abstract and concrete concepts based on domains. The source concepts are somatic, while the targets are more intangible and theoretical. The metaphor represents the symbolic language that expresses things. The linguistic semantic theory conceptual semantics signifies linguistic expressions relationship to human cognition [1]. Cognition follows language independence based on human aptitude dealing with sensitivity and system interactions as indicated in Fig. 1.

© Springer Nature Singapore Pte Ltd. 2020
M. Singh et al. (Eds.): ICACDS 2020, CCIS 1244, pp. 85–95, 2020.
https://doi.org/10.1007/978-981-15-6634-9_9

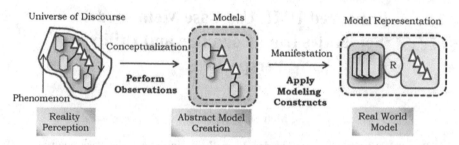

Fig. 1. Reality to real world mapping

A conceptual, phonological, and syntactic organization represents conceptual semantics that values mental representations rather than truth-centered philosophies. Conceptual semantics follows a predicate-argument pattern incorporated in a componential metalanguage based on universal semantic primitives. Context states expression interpretation drawn from the background pooled by the participants. Components carry natural language meanings used by semantic metalanguage to denote decompositional semantic theory. The set or series of expressions is known as discourse or the universe of discourse based on events or themes [2].

The sentence meaning investigations and interactions are the basis for Discourse Representation Theory (DRT). It is a Model-Theoretic approach that performs a series of Discourse Representation Structures (DRS) having discourse referents and predicates applicable. Denotational Semantic Methodology or Formal Semantics based on beliefs and scientific logic formulates expressions language and the world consisting the things or situations. Representational methodology or Natural Semantic Metalanguage models human mind representations.

A formal metalanguage consisting of logic that signifies natural language meanings results in formal semantic theory. The metalanguage has precise rules and meaningful to avoid vagueness. The lexicon is a set of expressions that language users can grasp rather than develop new sets repeatedly. Lexicon can relate to a classical language dictionary: language vocabulary or a particular language user's conceptual lexicon. A metalanguage should provide a comprehensive and explicit description of the object language. A primitive or atomic unit cannot be decomposed or defined at lower levels of hierarchy [3].

The attributes or characteristics are categorized and named as property. The lexicalization of properties represented by abstract nouns or adjectives and can be quantified in scalar or absolute magnitude. The concepts set or a vocabulary theoretic representation is recognized as a semantic field. Frame semantics is a hypothesis that relates etymological semantics to entire encompassing information. In intellectual phonetics, where undertone is called all-encompassing importance, the demonstrative significance is given a significantly focal job. Denotation is additionally called word reference importance in psychological etymology and appeared differently in relation to exhaustive significance [4]. Section 2 focuses on the Literature Review of semantics theory and approaches. Section 3 describes requirements to use case relationships. Section 4 represents generic metamodel semantics that can be applied to all UML diagrams. It conveys the layered form of model-to-metamodel terminology. Section 5 discusses use case semantics from cognitive and utility dimensions.

2 Literature Review

Efforts were taken to evaluate the similarity between Cognitive Semantics and Relevance Theory. The absence of harmony inside the directive is the situation for the most fundamental representation of expressive importance in lexical semantics, on which such a large number of consequent hypothetical investigations rest [5]. The semantic research was centered on representation of the semantic structure of languages with an assumption that etymological articulations work in phonetic cooperation by passing on fixed and invariant constituent, which suggests the disputable proposition of a reasonable limit among etymological and non-semantic data, and sidelines a hermeneutic point of view [6, 7].

Hermeneutics as the practice of analysis deals with questions that emerge while managing substantial human activities and the results of these activities, in particular contents of text. The differentiation between the ontological and epistemological level is generally recognized as it verifies whether it is to be sure productive to totally disregard the composition and architecture of the factual entities with the action of understanding [8]. Wierzbicka proposed Natural Semantic Metalanguage (NSM) hypothesis that attempts to lessen the semantics of all vocabularies down to a limited arrangement of semantic natives, or primes. Primes are widespread in that they have a similar interpretation in each language, and they are unsophisticated in that they can't be characterized utilizing different words. The semiotic origination of language uses words as overall signs for suggestions, things, or ideas [9].

Fundamental semantics expects relationship between semantic properties and increasingly key non-semantic properties. Chomsky broadly recognizes two originations of language. Firstly, a language as a computational framework, acknowledged in a person's mind, that creates the structures explicitly embroiled in phonetic conduct. Secondly, a language that addresses autonomously of the properties of the intellect [10]. A semantic hypothesis is one of the parts of a transformational generative language structure. The key semantic part was the 'semantic marker', which names an idea that any individual can think about; consequently, the hypothesis is material to every single common language. Katz tried to build up a hypothesis of importance (sense) that would do the entirety of the accompanying: characterize what it is; characterize the type of lexical sections; relate semantics to linguistic structure and phonology by hypothesizing semantic hypothesis as an indispensable segment of a hypothesis of language structure; set up a metalanguage in which semantic portrayals, properties, and relations are communicated; guarantee the metalanguage is general by corresponding it with the human capacity to conceptualize; recognize the segments of significance and show how they join to extend meaning onto basically complex articulations [11].

In Frame semantics, a semantic casing is characterized as an intelligent structure of ideas that are connected to such an extent that without information on every one of them, one doesn't have total information on one of them either, and are in that sense sorts of gestalt. Casings are evoked by words as the semantic reasonable substance of the word actuates the casing of the broad implying that is required for the comprehension of that word. Conventional semantics is driven by objectivism and fundamentally based on the supposition that will be that significance alludes legitimately to or indicates articles and relations in the outside world.

The psychological etymologists designate basic semantics by word implications or lexical implications that can be separated into atomic semantic highlights, which are in a way the particular properties of the significance of a word. Semantic highlights are accepted to allude to real properties, items or relations in the real-world. As a major aspect of the field of subjective phonetics, the psychological semantics approach dismisses the conventional custom modularization of etymology into phonology, linguistic structure, pragmatics, and so forth [12].

Intellectual semantic hypotheses are ordinarily based on the contention that lexical significance is theoretical. That is, the importance of a lexeme isn't referring to the element or connection in reality that the lexeme alludes to, however to an idea in the mind-dependent on encounters with that substance or connection. Semantic structure is the reasonable encapsulated structure which means representation is comprehensive and development is conceptualization. Categorization is the procedure wherein incremental knowledge and ideas are perceived and comprehended [13].

3 The Requirements and Use Cases

The requirement is an unambiguous, testable or quantifiable, and qualitatively acceptable description of process or products with constraints satisfaction. The concerns of stakeholders are about utilizing, profiting, underprivileged, and influenced by product realizations. Requirements engineering includes recognizable proof of client necessities, detailed investigation of the necessities to determine prerequisites, and approval of the reported necessities against client needs. The requirement dependency is represented as a matrix or version graph in order to avoid loose coupling and uncertainty [14].

Functional necessities describe the capability of the product. They describe precisely what tasks the software should perform. Functional necessities define the scope of the machine, the product boundaries, and its connections to adjoining systems. Non-Functional necessities describe the appearance and feel of the system. This includes the visual residences of the system, its usability, and the performance necessities. Non-Functional requirements also encompass cultural and political troubles as well as legal requirements that the software should conform to. It is critical to factor out that numerous sources, which include literature surveys, describe the different styles of software necessities the usage of other categories [15, 16]. The system types and formation is indicated in Fig. 2.

Use cases coordinate the requirements into a complete package that portrays the association of the client with the framework. Ivar Jacobson first presented the Use Case and the Use Case Diagram as a significant application investigation apparatus. Ivar Jacobson alongside Grady Booch and James Rumbaugh is credited with Unified Modeling Language (UML). Use Cases are content visual records that portray a situation that associates with the "Framework". The Actor can be an individual, another framework or another environment. The actors are not the elements of the use case but have access to

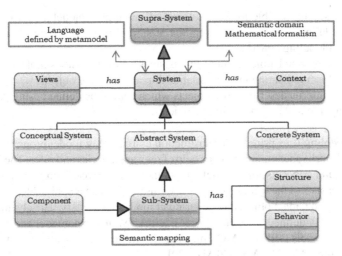

Fig. 2. System components

permission and are associated with the framework. Use Case depicts just how the system responds and static behavior. Use Cases are influenced by Domain Experts, Users, and Stakeholders [17].

4 The Metamodel Semantics

The real-world objects are expressed at multiple abstraction levels, granularities, and are based on multiple themes, context, and intents. Figure 3 shows that the metamodel is used to create the terminal model specific knowledge using which the modeler can create multiple platform independent models that represent the system under study or domain. The concepts are specified using metamodels having specific purpose, scope and viewpoints. The conceptual domain specification is defined by the metamodel. The concrete model creates the model level specifications [18]. A model profile consists of metamodel standard, patterns, and stereotypes and the terminal model conforms to the metamodel using this profile and backward traceability links. The stereotypes and patterns follow modeling constructs ensuring reusability required by object models supporting following features:

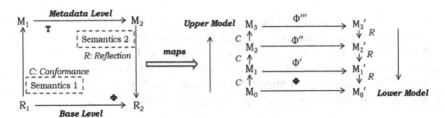

Fig. 3. Metamodel layers

- The model should be able to adapt to a set of predefined modeling constructs supporting modeling notations. The common atomic objects are expected to form the predefined modeling constructs.
- Common aggregated objects represent business entities with simple or complex definitions form indicating the interrelationship amongst the objects.
- The object patterns describe those objects which cannot be defined using predefined common aggregated objects or simple objects.

Meta-modeling uses principle of abstraction emphasizing model properties in relation with real world phenomena with the help of relevant theories and rules. The instances of properties and concepts for a modeling language are derived from meta-model. The metamodel consists of the syntax, semantics, terms used, context applicable, and number of participating metadata objects in the terminal model vocabulary. The abstract syntax of the metadata is defined by Meta Object Facility (MOF) metamodel and referenced in lower-level models [19]. The MOF metadata framework is typically depicted as four layer architecture (M_0, M_1, M_2, and M_3) [20].

There are following features on MOF-based metamodel and UML profile [21, 22]:

- UML notation subset usage constrained by metamodel semantics.
- Metamodel description are written using common style incorporating prescriptive and descriptive declarative languages.
- The abstract syntax for the metamodel level M_1 is defined at a higher level metamodel level M_2.
- The metamodel M_1 is used to describe the real world objects and M_1 is called as a set of platform-independent terminal models.
- The constraints for the metamodel are coming from UML semantics for things, namespaces, elements, relationships, tagged values, and stereotypes.
- UML profile and the metamodels are mapped with relationship preservation both in forwards and backward traceability link resolution which can be used to articulate the development tool.
- The tools incorporating the metamodel terminology are permitted to exchange model-metamodel-metadata-data information between reach other.

Let L_0, L_1, L_2, and L_3 be the languages corresponding to models M_0, M_1, M_2, and M_3 then transformation T represents the transformation of M_2 of L_2 to M_1 of L_1. The following features of vertical and horizontal model transformation from a source model M_0, M_1, M_2 to a target model M_1, M_2, and M_3 are significant:

- If M_1 is at the same, lower or higher abstraction level as M_0
- If M_1 is an semantic extension, stronger or weaker semantics of M_0
- If M_2 is at the same, lower or higher abstraction level as M_1
- If M_2 is an semantic extension, stronger or weaker semantics of M_1
- If M_3 is at the same, lower or higher abstraction level as M_2
- If M_3 is an semantic extension, stronger or weaker semantics of M_2.

Let ModelLangSet$_L$ addresses the plan of models used to assemble the metamodel language L. The ModelLangSet$_L$ parts express M advancements which have surmising M.D for each datum type D of M, including a lot of thing identifiers M.MEC for each metaclass MEC of L, and limits f_M : M.MEC → M.D for each metaobject f : D specified for instances of MEC in L. The structures M should contain no other additional segments not specified in L, and should satisfy any reasonable properties defined for L. M: L is supported than M: ModelLangSet$_L$. The property f_M (segment) by and large created as element.f_M and the model change for L_1 to L_2 is made as RelationT: ModelLangSetL$_1$ ↔ ModelLangSetL$_2$. Appearance in UML establishment addresses metainformation available at a base level as data at metadata level. Metarepresentation describes metalevel substances that can be controlled and changed [23, 24].

5   The Use Case Semantics

Classifiers, events, and behaviors represent model elements categories. The objects having a state linked with other objects is designates as a classifier. The set of possible occurrences as processing result designates event. The set of possible executions on the basis of algorithmic performance bounded by rules designates behavior. Objects belong to the model domain and models of objects can be imperfect, inaccurate, and abstract in relation with their scope and utility in the model. UML metaclasses has two distinct levels with metasemantic associations between the metaclasses; firstly, type level representing generic entities use cases, classes, states, and activities in models secondly, instance level representing things runtime that models represent [25, 26].

Let there be behavioral models p and q where $p, q \in UML$. The syntactic domain composition is indicated by $p \oplus q$ with $\oplus$ indicating composition. The semantic domain $SD(p \oplus p) = SD(p) \cap SD(q)$ expresses semantics $p \oplus q$. The multiple views semantics for UML constructs is indicated by $View \rightarrow SD(\{construct_1, ..., construct_n\}) = SD(\{construct_1\}) \cap \cdots \cap SD(\{construct_n\})$. If $SD(\{construct_n\}) \neq \emptyset$ is satisfied, the UML constructs holds consistent properties. The semantic domain consistency supports view integration and model consistency aspects. If $SD(q) \subseteq SD(p)$ then refinement of $p \in UML$ that belongs to $q \in UML$ is same.

The elements and its characteristics are represented by system model with abstract set of objects having selector functions which defines properties of set. Appropriate naming conventions are also incorporated to address the model terminologies preservation as far as possible. This means that the thing or element without specification does not carry any meaning. These underspecified closes are strong for framework model specialization since for a property, modeler can, (a) check whether property exists; (b) if coming up short on, a rising property can be found; (c) if not, check whether it is totally important; and (d), if indeed, an extra limitation can be included.

Use Case Relationship is a twofold relationship that shows the circumstances and logical results connection between two components: source component and target component. This is a statement that demonstrates that the impact component can be activated uniquely by the reason component accordingly indicating that impact component is executed simply after the reason component executes. Logical Use Case Relationship expresses functional association between at least two associates having a place with Functioning Model which can be modeled as conjunction (and), disjunction (or), and exclusive disjunction (xor).

Use Case Cycle expresses the coordinated utilitarian pattern of the framework; it comprises of components and connections between them. It shows primary usefulness that has a fundamental significance in the working of a framework, i.e., by interfering with the principle cycle the framework can never again capacity or its working is disfigured. Use Case Operation is a conduct highlight of classifier that indicates the name, type, parameters, and imperatives for conjuring a related behavior, and related useful highlights and associates for determining circumstances and logical results relations inside framework, subsequently permitting a circumstances and logical results relations to be demonstrated inside the framework by methods for collective highlights.

A functional component is a representation of atomic business activity (i.e., it can't be bound into various differing business exercises). Every workable component is a particular tuple associated with merge class and procedure reference (sections C and O). Remarkable tuple significance of utilitarian part XTid is as per the following:

XTid = <IDS, AP, OP, R, P, C, ST, PreCond, PostCond, E, S>, where
  IDS is an identifier of a useful component,
  AP is Action of Process P,
  Operation will be Operation which will give value described by association A,
  RS is the consequence of activity AP,
  P is a Process that gets the consequence of that is utilized in real-life AP,
  C is Class which will express to forms P in the static perspective of the framework,
  ST is the new condition of procedures P in the wake of performing activity AP,
  PreCond is a set of preconditions,
  PostCond is a set of postconditions,
  E is an entity responsible for performing activity AP,
  S is subservience of useful component (can be interior or outer).

The universes of segments for use case plots with properties and associations can be described from UML metamodel. The universe of type names is UCTYPE, the universe of characteristics UCVAL, association REL that accomplices type names and their characteristics, the universe of usage case names USECASE, the universe of use case identifiers UCID, the universe of on-screen characters with express occupations UACTOR, and the universe of speculations USTYPES. Type indicates sets of multiple categories for the system model and $T \in$ UCTYPE is used for naming purpose that refers to model elements.

Type names T of this universe UCTYPE are ordinarily not pointed by point further. In spite of the fact that models a sort, T really represents a name, and in short, we state type T for it. In that regard, we utilize a profound installing of the sort arrangement of UML, by speaking to it through sort names and a vast expanse of qualities as it were. By profound installing, we imply that we don't outline of the UML to a kind arrangement of the fundamental numerical structure, yet unequivocally model sorts as the top of the line components. typeOf assigns a type to values and used when there is absence of default type for values and the carrier sets are the same such that for type information (t_i, T), $t_i \in$ REL (T).

```
[Type]
UCTYPE UCVAL
REL: UCTYPE → P(UCVAL)
```

$\forall u \in$ UCTYPE: REL (u) $f = \emptyset$

[typeOf] use Type

typeOf : UCVAL ~ UCTYPE

$\forall_v \in$ UCVAL: v $\in$ domain (typeOf) $\Rightarrow$ v $\in$ REL (typeof (v))

Void $\in$ UCTYPE void $\in$ UCVAL REL (Void) = {void}

[Boolean-Integer] use Type

Boolean, Integer $\in$ UCTYPE

true, false $\in$ UCVAL

REL (Boolean) = {true, false} true f = false

REL (Integer) = X $\subseteq$ UCVAL

UCTYPE (UCVAL) at least contains Boolean and integer (values).

[Variable] use Type UCVAR

vartype : UCVAR $\rightarrow$ UCTYPE varsort : UCVAR $\rightarrow$ P(UCVAL)

VarAssign = RECORD (UCVAR, varsort)

Let a : T denotes a typed variable with name and type of variable stated explicitly.

vartype(a : T) = T $\forall_v \in$ UCVAR : varsort(v) = REL(vartype(v))

UCVAR is the set of all variable names in the system model.

VarAssign is set of all variable assignments for variables from UCVAR.

[RecordType] use Type, Variable

TRecord : P_f (UVAR) $\rightarrow$ UCTYPE

RECORD(UCVAR, varsort) $\subseteq$ UCVAL

Record {a1 : T1,. .., an : Tn} stands for type TRecord ({(a_1, T_1),. .., (a_n, T_n)})

$\forall_i$: vartype(a_i) = T_i $\Rightarrow$ REL (Record{a1 : T_1, …, a_n : T_n}) = SortRec({(a_1, …, a_n)})

[TupleType] use Type, Variable

Tuple : List(UCTYPE) $\rightarrow$ UCTYPE TUPLE(UCVAL) $\subseteq$ UCVAL

REL(Tuple[T_1,. .., T_n]) = UCTuple[REL(T_1),. .., REL(T_n)]

Tuple[.. .] acts as type constructor.

Usecase use Type

USECASE, UCID, INSTANCE

att = ribute: USECASE $\rightarrow$ P_f (UCVAR) ucids: USECASE $\rightarrow$ P(UCID)

ucinstance: USECASE $\rightarrow$ P(INSTANCE) ucinstance: UCID $\rightarrow$ P(INSTANCE)

usecaseOf : INSTANCE $\rightarrow$ USECASE usecaseOf: UCID $\rightarrow$ USECASE

$\forall c \in$ USECASE, ucid $\in$ UCID :

ucinstance (ucid) = {(ucid, r) | r $\in$ VarAssign $\wedge$ attribute(r) = attribute(C)}

$\forall$ ucid $\in$ ucids(C) : usecaseOf (ucid) = C

$\forall_o \in$ ucinstance (C) : usecaseOf (o) = C

UCID contains use case identifiers universe, USECASE classifier, INSTANCE.

attribute doles out attributes to every use case.

```
ucids dole out a stack of identifiers to a use case.
usecaseOf ensures every use case event and each identifier recog-
nizes its use case.
```

6 Conclusion

A model is an applied and improved outline of the present reality. It allows the visualizer to isolate and grasp the different properties of a given structure as showed by a specific perspective, which is associated with his unbending nature of interest. Model is used to address in a solid way the coordination of all sub-estimations in the general one and it can guarantee the association of various business points of view. A use case creates serviceable essentials, models the destinations of structure or coordinated efforts, and records circumstances from trigger events to goals. The proposed semantic model for use case obviously characterizes the use case and participants alongside the metamodel phraseologies.

References

1. Chandler, D.: Semiotics: The Basics, 2nd edn. Routledge, London (2007)
2. Croft, W., Cruse, D.A.: Cognitive Linguistics. Cambridge University Press, Cambridge (2004)
3. Jackendoff, R.: Foundations of Language: Brain, Meaning, Grammar, Evolution. Oxford University Press, Oxford (2002)
4. Jackendoff, R.: On conceptual semantics. Intercult. Pragmat. **3**, 353–358 (2006)
5. Wilson, D.: Parallels and differences in the treatment of metaphor in relevance theory and cognitive linguistics. Intercult. Pragmat. **8**, 177–196 (2011)
6. Riemer, N.: Word meaning. In: Taylor, J.R. (ed.) The Oxford Handbook of the Word. Oxford University Press, Oxford (2015)
7. Gadamer, H.: Truth and Method, 2nd edn. Continuum, London (2004)
8. Scholz, O.: On the very idea of a textual meaning. In: Daiber, J., Konrad, E.-M., Petraschka, T., Rott, H. (eds.) Understanding Fiction: Knowledge and Meaning in Literature, pp. 135–145. Mentis, Münster (2012)
9. Wierzbicka, A.: Imprisoned in English. The Hazards of English as a Default Language. Oxford University Press, Oxford (2014)
10. Chomsky, N.: New Horizons in the Study of Language and Mind. Cambridge University Press, Cambridge (2000)
11. Katz, J.J.: Recent issues in semantic theory. Found. Lang. **3**, 124–194 (1967)
12. Fillmore, C., Baker, C.: Frame semantics for text understanding. In: Proceedings of WordNet and Other Lexical Resources Workshop, NAACL (2001)
13. Cruse, A.: Meaning in Language. An Introduction to Semantics and Pragmatics, 2nd edn. Oxford University Press, New York (2004). ISBN 978-0-19-926306-6
14. Jarke, M., Pohl, K.: Requirements engineering in 2001: (virtually) managing a changing reality. Softw. Eng. J. **9**, 257–266 (1994)
15. Cysneiros, L., Leite, J.: Nonfunctional requirements: from elicitation to conceptual models. IEEE Trans. Software Eng. **30**(5), 328–350 (2004)
16. Gregoriades, A., Sutcliffe, A.: Scenario-based assessment of nonfunctional requirements. IEEE Trans. Software Eng. **31**(5), 392–409 (2005)

17. Jacobson, I.: Formalizing use-case modeling. J. Object-Oriented Program. JOOP **8**(3), 10–14 (1995)
18. Gargantini, A., Riccobene, E., Scandurra, P.: A semantic framework for metamodel-based languages. Autom. Softw. Eng. **16**(3–4), 415–454 (2009). https://doi.org/10.1007/s10515-009-0053-0
19. Meta Object Facility (MOF) Core Specification OMG Specification, Version 2.0 (2006)
20. Object Constraint Language, OMG Specification, Version 2.2 (2010)
21. OMG Unified Modeling Language (OMG UML) Superstructure, Version 2.3 (2010)
22. OMG Unified Modeling Language (OMG UML) Infrastructure, Version 2.3 (2010)
23. Zabawa, P.: Context-Driven Meta-Modeling Framework (CDMM-F) – context role. Tech. Trans. Fundam. Sci. **112**, 106–114 (2015)
24. Shan, L.J., Zhu, H.: Unifying the semantics of models and meta-models in the multi-layered UML meta-modelling hierarchy. Int. J. Softw. Inform. **6**(2), 163–200 (2012)
25. Rumpe, B.: Modeling with UML: Language, Concepts, Methods. Springer, Cham (2016). https://doi.org/10.1007/978-3-319-33933-7. ISBN 978-3-319-33933-7
26. Kraas, A.: On the automated derivation of domain-specific UML profiles. In: Anjorin, A., Espinoza, H. (eds.) Modelling Foundations and Applications. LNCS, vol. 10376, pp. 3–19. Springer, Cham (2017). https://doi.org/10.1007/978-3-319-61482-3_1

The Impact of Mobile Augmented Reality Design Implementation on User Engagement

Mervat Medhat Youssef[1,2] ⓘ, Sheren Ali Mousa[1,2(✉)] ⓘ, Mohamed Osman Baloola[3], and Basma Mortada Fouda[1,2] ⓘ

[1] College of Mass Communication, Ajman University, Ajman, UAE
D.mervatmedhat@gmail.com, {s.nawar,b.fouda}@ajman.ac.ae
[2] Art and Design Academy Higher Institution of Applied Art, Cairo, Egypt
[3] College of Engineering and Information Technology, Ajman, UAE
m.baloola@ajman.ac.ae

Abstract. With the rapid advances in mobile technologies, one of the main goals of Colleges and universities is to reach national and international recognition. Recently many Universities used new strategies to stay competitive, trying to adopt new technologies like Augmented Reality (AR) To capture students' attention. Augmented Reality (AR) had used as a useful tool in education, but besides teaching and learning, technology may play an important role in university and student engagement. The research aims to study the effect of using augmented reality (AR) on human interaction and engagement. The study focuses on designing an AR prototype framework and a pilot AR application to test its impact on the user's interaction engagement with the university information platform. The researcher assumes that by applying AR technology, it will improve two way of communication between universities and students, employees and visitors. Research findings show that AR UI and UX collaborate to achieve students engagement. Universities may consider thinking of the efficacy of AR implementations within their business strategy to enrich its recognition in recent intense competition.

Keywords: Augmented Reality (AR) · User Experience (UX) · User interface (UI) · Engagement

1 Introduction

Augmented reality (AR) has entered our life as a new interactive technology in its way of engagement. AR improve the user's view of the world by overlaying digital visuals view in their surrounding physical space. AR Crosse the gap between printed media new technology in digital, "adding another layer of (virtual) reality to every day (real) life" [1]. According to [2], AR has the potential to become the next popular branding people may like. [3] define AR as the augmentation of the real physical world with virtual objects evolving technology and applications delivered through mobile devices. Several works of literature empirical and experiment studies discussed AR technologies effectiveness on human interaction provide evidence that augmented reality systems improve human

M. Singh et al. (Eds.): ICACDS 2020, CCIS 1244, pp. 96–106, 2020.
https://doi.org/10.1007/978-981-15-6634-9_10

performance and engagements. Drawing from the literature related, Authors, propose that AR application UI and UX design have a significant effect on user engagement with AR application and their surroundings. To empirically test this proposition, Authors conducted a conceptual design framework to design AR application and test its design UI and UX principle on user engagements.

2 Literature Review

There are many AR implementation enables users to experience objects and spaces around them in novel ways. "the powerful significance of the information blended concept with the real world has pushed AR to business, entertainment, technology, branding, and education" [4]. For example, In the advertisement business, many brands used AR to create an interactive experience, packaging, and to develop interactive AR games [5]. "Design, Discovery, Details, Desire and Delivery" [6]. [7] even in the navigation system. For sure, Using AR in education institutions has significant consequences on Students. For instance, both the Massachusetts Institute of technology and Georgia institute of technology is working to enhance student learning through using AR gaming simulations [8]. According to [9], new technologies uniquely present visuals to improve the learning experience. AR can support students by using the 3D object to be seen from different perspectives to enhance their knowledge [10]. Squire and Klopfer (2007) also suggested that AR games can increase the student level of engagement in academic activities [11]. Moreover, AR also can enhance collaboration between students and each other, even with their instructors [12]. Most studies showed how AR environments could motivate students' and develop a better understanding of learning contents. Also, AR can promote social interaction among students [13]. Universities can benefit from applying AR platforms, to shape student's way of gaining knowledge and information and to reach them before they even arrive on campus. Simpson College placed advertisement banner that incorporated AR technology at a crowded mall in 2013. AR can enhance the value of engagement. For instance, [14] the study discovered that unique AR technique enrich experience but requires an attractive design to prevent user distraction by technology instead of the application itself. Researchers were focusing on the application design to sustain users' engagement levels with the app. keeping engagement to build or enhance users relationships [15]. For this purpose, Authors will introduce AR design through UI and UX principles, to study the effectiveness of mobile AR systems on user's engagement.

2.1 AR Graphic User Interface (GUI)

In AR graphic user interface GUI is a combination of virtual and physical objects. These elements work together to achieve a fundamental goal, mapping user physical input onto Mobile application output. GUI focus on three main factors; the physical design elements of the interface, the virtual visual display and the interaction metaphors between these [16]. The AR graphical design should facilitate usability, and provides a high level of user satisfaction, perceived usefulness and consistency, so the design could easily be recognized and understand by user [17]. Few studies discuss the AR graphic user interface GUI principle and its relation to ("user-centred design") and how to make using an appropriate interface action symbol ("metaphors") easy to use, easy to understand and easy to discover. Therefore, it is fundamental to know how users behave when they access an AR application to create a better interface design.

AR Graphic User Interface (GUI) Principles are a high-level concept that should be used to guide Application design creation [18]. The main of excellent AR application interface is to create a combination of design elements easy to recognize, to understand, and to interact with, so some design principles must be applied in design to improve the usability, utility, and usefulness. Jakob Nielsen and Rolf Molich's present Ten User Interface Guidelines that keep interface easy to understand and attractive within a good amount of time [17]. like Visibility, Low Physical Effort, Learnability, User Satisfaction, Error Tolerance, Flexibility and Simplicity [17]. (Theo, Mandel. 1997) Bring up three main areas of UI design: 1. users always be in control of the interface 2. Users' memory load reduction 3. consistently of the interface [18]. the paper will discuss them:

Place Users in Control. That's mean let users control their trip in an application. The big UI in AR challenge is the unpredictable contribution of the real world. UI technique will rely on a particular object being in the user's field of view. A lot of UI design principles can allow users to be in control or lead to it, like Visibility and Affordance of AR system. Using easily objects metaphors that users can identify it's use or function once they observe it [17]. Error prevention actions that may reduce errors like Offer users go back steps, including undoing and redoing previous actions [19]. Provide immediate and reversible responses and feedback. Using terms in UI design, users can understand to offer guidance through common tasks. Giving users alternative options—to navigate easily, cancel and to save or even return to where they left off [18].

Reduce Users' Memory Load. Human attention is selective and limited, Short-term memory usually keeps information as a pack up so you can restore it in a short period. Based on that, UI design shouldn't force users to keep information in their short-term memory while they navigate between task [18]. UI design should Low user Physical Effort by doing tasks for them or help them achieve tasks easily. Focus on Recognition rather than recall. UI AR design should Provide visual cues so users know where they are and what they can do next. Learnability will follow recognition as a cause when the user will be able to use the system [17]. Also, the design will be more flexible by reducing using Mistry abbreviations and too much function keys.

Make the Interface Consistent. Interface Consistency defined as keeping both the graphic elements and design behavior maintained similar everywhere in your application. However, Consistency in AR may be harder to continue because of the dynamics movements effects of the physical and virtual design elements. UI design needs to be flexible Because of these unpredictable dynamics. Otherwise, traditional UI design suggests keeping design simple to minimize dynamic changes in the UI composition [20]. Simplicity Promote Visual consistency. Simple UI design means that anyone can learn your application alone. Applying simplicity increase legibility, enjoyment and increase User satisfactory. Simple design catches attention and creates emotional engagement: the more aesthetically pleasing UI, the more likely to deliver a satisfying User Experience (UX).

2.2 AR User Experience UX

UX is a multidimensional phenomenon [21], and it has been around for many years [22]. ISO 9241-210 [23] defines user experience as "a person's perceptions and responses that result from the use of a product, system or service" [24] identify users, context and interaction as the critical elements of UX. User experience defined as how a service, a product or an application make users feel while using it [25]. Hassenzahl and Tractinsky [26] saw that interactive products and services become trendy and fashionable, with the advancement of technology. AR often implies the use of 3D graphics over the user's view of the world [27]. However, in the real user experience, it can be any an environment space surrounding like visuals, sound, etc. [28]. Also depends on the user's position, and ongoing interactions with the augmented world. "creating great user experiences, need to maintain three things: usability, utility and desirability" [29]. Usability and utility; both are important to make something useful. Usability has many components, such as efficiency, effectiveness, satisfaction, Learnability, and recall [30]. Usability focuses on how users can learn and use a product to achieve their goals. It also refers to how satisfied users are with that process. Usability evaluates Efficiency How quickly and effectively user can perform basic tasks. Learnability describes How easy is it for users to accomplish necessary tasks in the first time, even more, recall it: When users return to the design after a while. We can describe Application with Utility and desirability once it provides the features user need in a pleasing way [31].

2.3 The Nature of Engagement

It is necessary to understand what user engagement is and which key attributes make up engagement to determine how mobile augmented reality applications can simulate user engagement. Engaging is to occupy the attention of someone "to attract, hold fast or please" and "become involved". According to [32], "User engagement is the behavioural and emotional relationship that can exist, at any time between a user and a resource". Augmented Reality applications can engage the audience and attract them more than any other mediums [33] AR Engagement user experience can be visualized through the time spent, the intensity of effort, concentration and propensity to start and stay in the task. [34] classified user engagement to two types of user-brand engagement and user-user engagement. Users may share the same AR campaign in the same physical space even they aren't all engaged directly to the brand.

Engagement Key Factors. The research raised the understanding of how user engagement can be essential to a mobile augmented reality application's success, and also presented how it is necessary to company's looking for more user attraction. Not only in terms of getting more visibility, but it is also essential for flexibility of work by company employees [35]. Recently, academic institutions brought this concept into their core business strategy. By stimulating high levels of interaction and involvement, creating more meaningful and long-lasting relationships with their students. Based on literature review key factors that make up engagement: **1. Focused attention** – this factor is inherently related to human cognition and presents both a physical and social dimension. This concerns users' intensity of concentration in the application and losing the consciousness of the surroundings, both objects and people [36]. A highly engaged user is more likely not to be aware of the passage of time. **2. Positive affect** – engaged users frequently found to be affectively involved with the app [36]. The user creates a personal connection to the experience at hand [37]. Other authors, such as Vreede et al. 2013, consider this investment is seen as one of the major contributors to customer loyalty [38]. **3. Aesthetics** defined by [37] as Aesthetics represents and visual beauty. Some authors pinpoint to its strong contribution to users' first impressions of the platform [39], while other Studies found that general factors, such as symmetry and balance, as well as specific issues, like graphics and screen layout, were strong contributors of user engagement [36]. **4. Endurability** is strongly linked to the concept of user loyalty. This factor refers to users' likelihood of remembering and wanting to repeat an experience that brought them enjoyment, usefulness or any other type of value. Also referred to behavior that will entice them to continue using it [40]. Other studies linked endurability with a user's recommendation of their own experience to others [41, 42]. **5. Novelty** a new, unexpected or unfamiliar thing or experience can be perceived as valuable to users. Unexpected visual or changes may cause joyful and exciting to users [43]. Huang [44] also defined novelty as the introduction of features, in a given interface. However, as high degrees of novelty can lead users to "become lost" and confused when using the application. **6. Trust and expectation** trust are a requirement for attaining user engagement. The factor is built upon the technological- dependent assurances, such as encryption of data, and users' attitudes towards applying that technology [45]. In more recent studies, [46] also debated how online deception was shaping the trust of users in social media platforms, as fake content can produce negative effects on user engagement. **7. User context** takes into account how the user's specific state may shape how their engagement takes place. Context of the user states that real engagement is not measurable since the real-world state of the user can vary greatly [47]. Additional research on the matter of context has unveiled two sub-dimensions: incentives and motivations. Incentives such as users' interest (personal preferences), personality, social context, the influence of trends. Motivation, such as rewards in a game, can also act as additional triggers to engage users and keep them interested in the application.

3   Research Importance

Based on the Literature review, the research focuses on applying AR as a smart communication tool to solve a practical problem of student's engagements with a university campus. Authors will answer the proposed research questions:

1. How could the students evaluate the AR application UI?
2. How could the students evaluate the AR application UX?
3. How could the students evaluate the AR application Engagement?
4. Are there any correlations among the three factors UI, UX and Engagement?

4   Methodology

The study followed the Empirical approach in testing and analyzing the effect of UI and UX design of AR application on volunteer student's engagement with Ajman university campus. As a start to investigate the research problem. A pilot study survey was designed and conducted on a sample of 100 Ajman university newcomers' students asking them about their source of information about the university, their usage of the university application, and how they evaluate the university application. The finding shows that the highest source of information was 47% from friends, 41% Uni. Website, 38% direct contact with Uni. employee 35% media, 30% Uni. social media accounts. The survey found About 93% of the sample using the Ajman University Application. Between strongly agree, agree and disagree, the application has an attractive design 40% agree. Has enough information about university 54% agree. Easy to use 53% agree, but Need more data more interactivity, more photos and video to cover many subjects.

Due to this finding, Authors designed an Initial Experiment prototype to visualize the main concept of the AR application design. Users will be exposed in the university campus to a poster as a trigger marker to examine AR application. 2 different levels of interactive media (AR video stream promotes the university colleges and interactive buttons contacted to university's website and social media), were used to examine how students engaged with the university environment, as shown in Fig. 1.

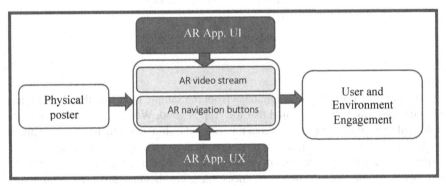

Fig. 1. AR application Conceptual Design Framework to enhance engagement with the surrounded environment.

5   AR Application Design

The AR application design process consists of two stages. The first stage, the off-line processing covers the selection and sittings of identification objects (poster) and the geometric model of 3D objects (video and buttons) in the virtual world. The physical model establishment, the generation of behavior model and three-dimensional scene, were created by Unity 3D. The process started with Creating a license key for our AR App in Vuforia, copy license key API to Unity, Create Database in Vuforia Target Manager, Add our poster as image target to Database, and Checking the rating of image's features. Secondary the on-line process includes identification of objects and producing the corresponding feedback information in the real-time and real view, finally, the video applied above image target in Unity.

6 Results

An online survey designed to ask about UI and UX design of the AR Application. The number of samples was 200 volunteer newcomers' students in Ajman University. They expressed how was their experience while using the APP and how was their engagement with the information presented.

Table 1 refers to the sample variables, the Gender: male and female students, Colleges: human sciences, and applied sciences, and the academic study level.

Table 1. Sample variables.

Sample		Frequency	Percent
Gender	Male	92	46
	Female	108	54
College	Humanities	109	54.5
	Sciences	91	45.5
Year	1 and 2	119	59.5
	3 and 4	81	40.5

N = 200

Participants respond to UI design, 65% from the sample indicated that the App has a good design, and 27.5% indicated that the App is good. While 7.5 0% indicated that it is natural (not good, not bad). Authors measured the student evaluation to the App. in three factors Table 2 shows the Mean of each sentence in the measurement scale.

Interface: according to the student evaluation of the App interface, three factors were above (4) in a scale from (5) these sentences related to the innovativeness, Simplicity, Visibility. While legibility and Recognition scored less than (4), experience: Usability and Recall were the most scored factors in the experience measurement with means above (4). In contrast, Learnability, Efficiency, Control scored less than (4) Engagement: most of the engagement factors scored above (4) except the factor related to the feeling of boring from the app. The main reason that push students to suggest the App. To others was because it is interesting to use, it is a new tool, a new source of information. And it is a good university service.

According to "Pearson" correlations in Table 3, There are strong positive correlations between "interface, experience, and engagement" (sig. = 0.000) that means each factor in a relationship and affected by others.

Table 2. Sample evaluation of the App UI, UX and engagement.

		N	Mean	Std. D.
AR application Design user interface				
1.	The application is very innovative and novel	200	4.26	.765
2.	Recognition: I can easily recognize the application if I visit it for the second time	200	2.69	1.278
3.	Simplicity: the AR application interface easy to understand so that I can get the information in a simple way	200	4.24	.851
4.	Visibility: there is a contrast between content and background in the designs, so it's legible	200	4.13	.690
5.	Legibility: the size of the text/buttons is legible easy to read and interact with it	200	3.87	1.001
AR application Design user experience UX				
6.	Usability: The AR application interface efficacy and efficiency to use	200	4.20	.634
7.	Learnability: I can quickly complete basic tasks in the first use time	200	3.99	.802
8.	Efficiency: I can quickly perform basic tasks	200	3.97	.904
9.	Control: I can go backward steps and undoing move forward in the Application	200	3.93	.905
10.	Recall: The AR application experience helps me in recalling information across parts of the dialogue	200	4.09	.728
AR application Design engagement				
11.	While using the application, I feel bored that makes me want to end the application	200	3.97	.994
12.	I think that using the application fits well with the way I like to use mobile devices	200	4.08	.956
13.	The information is presented in a unique, pleasant and logical order	200	4.04	1.022
14.	The AR application is an easy and good source of information about the university	200	4.45	.678
15.	The design elements and my experience in using the application influence my attention	200	4.15	.759

Table 3. Correlation among interface, experience, and engagement

Factors		Interface	Experience	Engagement
AR application design user interface	Pearson Correlation	1	.823[**]	.789[**]
	Sig. (2-tailed)		.000	.000
	N	200	200	200
AR application design user experience UX	Pearson correlation	.815[**]	1	.873[**]
	Sig. (2-tailed)	.000		.000
	N	200	200	200
AR application design engagement	Pearson correlation	.789[**]	.874[**]	1
	Sig. (2-tailed)	.000	.000	
	N	200	200	200

**Correlation is significant at the 0.01 level (2-tailed).

7 Discussion

The study presents a comprehensive framework for building an application using AR techniques. The result shows how students experience with AR application UI, and UX had influenced their engagement. Findings show that Students are looking for exciting and new technology to use. Simplicity and Visibility in the design attracted students. Animation in the video stream ease communicates university information with students. Using AR makes students enjoy what they are doing. Usability and Learnability used in application helped students to complete basic tasks and recall the app in the second use. Most of the students think the application fits well with the way they like to use smart devices, Novelty, good design and the logical order of information attracted their attention. They considered it as an easy and useful source of information about the university, that influences them to suggest it to others. Generally, the study address principles of UI and UX design as the key components of user engagement with their surroundings. On the other hand, authors faced challenges in implementation limitation in the way video appeared on the trigger poster, the content is very small, so the design showed a lack of control and efficiency in use. Dark colors in video design background reduced the Legibility and recognition of the UI design. Students face some challenges like response time delays of the AR system, or even hardware or software failures. Also, their ability to Interact with UI design buttons, due to the small display of the mobile device, but In general, AR experiment develops students' soft skills. It provides a better understanding of how using AR technology as a unique interactive tool. The experiment will increase AU reliance on innovative, Investments in smart and efficient technologies.

8 Conclusion

Study framework can be a guideline to others in the field of academic institutions, to think of applying AR technology not only in education but also in engagement strategies. Furthermore, findings show the direct relation between design and user engagement. The study is an on-going project consist of several levels toward enhancing student's engagement with Ajman University campus. Authors will suggest future studies may ably indoor AR navigation system and AR game filters to study their efficacy on student's engagement. There are several challenges and limitations with the technology that needs to be overcome by Authors. For example, designing AR game design with 3D models and high level of AR system tracking. Also, AR navigation system has to deal with a vast amount of information, in reality, to provide an accurate marker for helping in the navigation system.

Acknowledgements. Authors acknowledge the experiment design supports: Gaffer El Manasab for video and poster design, Saad Oriekat for photography and Shaza Mohammed in modelling and voice over.

References

1. Dholakia, N., Reyes, I.: Virtuality as place and process. J. Mark. Manage. **29**(13), 1580–1591 (2013)

2. Layar: Layar Augmented Reality Platform: Information for brands and publishers (2010)
3. Yuan, M., Ong, S.K., Nee, A.C.: Augmented reality for assembly guidance using a virtual interactive tool. Int. J. Prod. Res. **46**(7), 1745–1767 (2008). https://doi.org/10.1080/002075 40600972935
4. Johnson, L., Smith, R., Willis, H., Levine, A., Haywood, K.: The 2011 Horizon Report. The New Media Consortium. Austin, Texas (2011)
5. Scholz, J., Andrew, N.: Augmented reality: designing immersive experiences that maximize consumer engagement. Bus. Horiz. **59**(2), 149–161 (2016)
6. Young, O.: Augmented reality technology shapes the future of retail and commerce (2013)
7. Thomas, B., et al.: ARQuake: an outdoor/indoor augmented reality first person application. In: Proceedings of the 4th International Symposium on Wearable Computers, pp. 139–146 (2000)
8. Georgiatech. AR Spot: An augmented-reality programming environment for children (2014)
9. Liarokapis, F., Anderson, E.F.: Using augmented reality as a medium to assist teaching in higher education. In: Proceedings of the 31st Annual Conference of the European Association for Computer Graphics (Eurographics 2010), 4–7 May 2010, pp. 9–16. Education Program, EurographicsAssociation, Norrköping, Sweden (2010)
10. Chen, Y.C., Chi, H.L., Hung, W.H., Kang, S.C.: Use of tangible and augmented reality models in engineering graphics courses. J. Prof. Issues Eng. Educ. Prac. **137**(4), 267–276 (2011)
11. Squire, K., Klopfer, E.: Augmented reality simulations on handheld computers. J. Learn. Sci. **16**(3), 371–413 (2007). https://doi.org/10.1080/10508400701413435
12. Billinghurst, M.: Augmented Reality in Education. New Horizons for Learning, December 2002
13. Cobb, S., Heaney, R., Corcoran, O., Henderson-Begg, S.: Using mobile phones to increase classroom interaction. J. Educ. Multi. Hyper. **19**(2), 147–157 (2010)
14. Koh, R.K.C., Duh, H.B.L., Gu, J.: An integrated design flow in user interface and interaction for enhancing mobile AR gaming experiences. In: Proceedings of IEEE International Symposium on Mixed and Augmented Reality-Arts, Media, and Humanities (ISMAR-AMH), pp. 47–52. IEEE Computer Society, Washington (2010)
15. Roderick, J.B., Ilic, A., Juric, B.J., Hollebeek, L.: Consumer engagement in a virtual brand: an exploratory analysis. J. Bus. Res. **66**(1), 105–114 (2013)
16. Billinghurst, M., Grasset, R.: Designing augmented reality interfaces. ACM SIGGRAPH Comput. Graph. **39**(1), 17–22 (2005). https://doi.org/10.1145/1057792.1057803
17. Ejaz, A., Ali, S.A., Ejaz, M.Y., Siddiqui, F.A.: Graphic user interface design principles for designing augmented reality applications. Int. J. Adv. Comput. Sci. Appl. **10**(2), 209–216 (2019)
18. Mandel, T.: The Elements of User Interface Design. Wiley, New York (1997). ISBN 978-0-471-16267-4
19. Euphemia, W.: User Interface Design Guidelines. https://www.interaction-design.org/litera ture/article/user-interface-design-guidelines-10-rules-of-thumb. Accessed 10 Jan 2020
20. Shneiderman, B.: Designing the User Interface, 3rd edn. Addison-Wesley, Boston (1998)
21. Tokkonen, H., Saariluoma, P.: How user experience is understood?. In: Proceedings of the Science and Information Conference (SAI), London, UK, 7–9 October 2013 (2013)
22. Dirin, A.: From Usability to User Experience in Mobile Learning Application, p. 316. Aalto University, Helsinki (2016)
23. DIS, 9241-210: Ergonomics of human system interaction-part 210: human-centered design for interactive systems International Standardization Organization (ISO). Switzerland (2010)
24. Carlos, J., Nicolás, O., Aurisicchio, M.A.: Scenario of user experience. Design (2011)
25. Feiner, S., Terauchi, T., Rashid, G., Hallaway, D.: Exploring MARS: developing indoor and outdoor user interfaces to a mobile augmented reality system. Comput. Graph. **23**(6), 779–785 (1999)

26. Hassenzahl, M., Tractinsky, N.: User experience - a research agenda. Behav. Inf. Technol. **25**, 91–97 (2006)
27. Azuma, R., Baillot, Y., Behringer, R., Feiner, S., Julier, S., MacIntyre, B.: Recent advances in augmented reality. IEEE Comput. Graph. Appl. **21**(6), 34–47 (2001)
28. Shute, T.: Is it OMG finally for augmented reality? Interview with Robert Rice. Virtual Realities world 2.0., Ugotrade, August 2009
29. Molich, R., Nielsen, J.: Improving a human-computer dialogue. Commun. ACM **33**(3), 338–348 (1990)
30. Finstad, K.: The usability metric for user experience. Interact. Comput. **22**(5), 323–327 (2010)
31. https://www.nngroup.com/articles/usability-101-introduction-to-usability/
32. Attfield, S., Kazai, G., Lalmas, M.: Towards a science of user engagement (Position Paper). In: WSDM Workshop on User Modelling for Web Applications, pp. 9–12 (2011)
33. Leach, D.: Augmented reality and its potential for advertising (2013). http://blog.ad-tech.com/augmented-reality-and-its-potential-for-advertising/
34. Scholz, J., Andrew, N.S.: Augmented reality: designing immersive experiences that maximize consumer engagement. Bus. Horiz. **59**(2), 49–161 (2016)
35. Permadi, D., Rafi, A.: Developing a conceptual model of user engagement for mobile-based augmented reality games. J. Teknologi **77**(29), 9–13 (2015)
36. O'Brien, H.L., Toms, E.G.: What is user engagement? A conceptual framework for defining user engagement with technology. J. Am. Soc. Inf. Sci. Technol. **59**(6), 938–955 (2008)
37. Jennings, M.: Theory and models for creating engaging and immersive ecommerce Websites. In: Proceedings of the 2000 ACM SIGCPR Conference on Computer Personnel Research, pp. 77–85 (2000)
38. Vreede, T., Nguyen, C., Vreede, G.-J., Boughzala, I., Oh, O., Reiter-Palmon, R.: A theoretical model of user engagement in crowdsourcing. In: Collaboration and Technology, pp. 94–109 (2013)
39. Lindgaard, G., Fernandes, G., Dudek, C., Brown, J.: Attention web designers: you have 50 milliseconds to make a good first impression! Behav. Inf. Technol. **25**(2), 115–126 (2006)
40. Read, J., Macfarlane, S., Casey, C.: Endurability, engagement and expectations : measuring children's fun. Interact. Des. Child. **2**, 1–23 (2002)
41. Jung, S., Lee, S.: Developing a model for continuous user engagement in social media. In: Proceedings of the 10th International Conference on Ubiquitous Information Management and Communication, vol. 19, pp. 1–4 (2016)
42. O'Brien, H.L., Toms, E.G.: The development and evaluation of a survey to measure user engagement. J. Am. Soc. Inf. Sci. Technol. **61**(1), 50–69 (2010)
43. Aboulafia, A., Bannon, L.J.: Understanding affect in design: an outline conceptual framework. Theor. Issues Ergon. Sci. **5**(1), 4–15 (2004)
44. Huang, M.H.: Designing website attributes to induce experiential encounters. Comput. Hum. Behav. **19**(4), 425–442 (2003)
45. Khare, R., Rifkin, A.: Weaving a web of trust. World Wide Web J. **2**(3), 77–112 (1997)
46. Tsikerdekis, M., Zeadally, S.: Online deception in social media. Commun. ACM **57**(9), 72–80 (2014)
47. Law, E.L.-C., Roto, V., Hassenzahl, M., Vermeeren, A.P.O.S., Kort, J.: Understanding, scoping and defining user experience. In: Proceedings of the 27th International Conference on Human Factors in Computing Systems - CHI 2009, p. 719 (2009)

Intelligent Mobile Edge Computing: A Deep Learning Based Approach

Abhirup Khanna[✉], Anushree Sah, and Tanupriya Choudhury

University of Petroleum and Energy Studies, Dehradun, India
{akhanna,asah,tanupriya}@ddn.upes.ac.in

Abstract. In recent times researchers across the globe have shown keen interest towards advancements in the domain of edge computing. Mobile Edge Computing (MEC) is a new age computing paradigm wherein cloud services are made accessible at network edges via the use of mobile base stations. It is a promising technology that helps in overcoming the limitations of mobile cloud computing. MEC facilitates seamless integration of various application services, thereby proving cloud resources at the edge of the network, within the vicinity of the end-user's locality. It can effortlessly be integrated with the upcoming 5G architecture, hence supporting the execution of resource-rich applications that require low network latency. In order to enhance the levels of intelligence at mobile base stations, deep learning algorithms can be implemented over network edges for rendering optimized communication and workload balancing. The paper discusses a conceptual architecture for creating a mobile edge computing environment involving the applicability of deep learning algorithms. The paper discusses the fundamentals of MEC along with specific applications of reinforcement and continuous learning in an edge environment. We list the benefits of MEC along with a discussion on how its amalgamation with deep learning models can prove beneficial in case of a computation offloading scenario.

Keywords: Edge computing · Mobile Computation Offloading · Deep learning · Mobile computing

1 Introduction

Predominance of mobile terminals, like smartphones or tablet, computers and so on has a lot of effect on mobile and wireless networks, which in a way creates problems for the mobile system all over the world. Cellular networks experiences high power consumption, less storage capacity, high latency and lower bandwidth. Apart from this, it is believed that the exponential growth of new technologies, i.e., the Internet of Things (IoT), will lead to another convergence of the cellular and wireless networks. Mobile Cloud Computing (MCC), which is the amalgamation of the cloud computing in the mobile environments, provides significant opportunities for mobile devices, which enables them to store, compute, and power through the use of centralized cloud resources. However, while opening a multitude of mobile devices, MCC faces significant

© Springer Nature Singapore Pte Ltd. 2020
M. Singh et al. (Eds.): ICACDS 2020, CCIS 1244, pp. 107–116, 2020.
https://doi.org/10.1007/978-981-15-6634-9_11

problems [5], like high latency, low coverage, security vulnerabilities, and data transmission latency, which in a way can be cumbersome, especially for the next generation mobile networks (for example, 5G).

It has been seen that MCC is less applied to Quality of service and real life applications. As per a report from Cisco Visual networking index the use of mobile devices is increasing drastically and after 2020 it will further gain more and more usage [2]. The increase of mobile usage is because of increase mobile users and development and availability of mobile applications. In the era of Information technology, EDGE computing has become the leading mobile cellular computing (MEC) in cellular networks.

The term mobile edge computing was first introduced in 2013 when Nokia Siemens Networks and IBM developed the MEC platform, which allows applications to work directly. This platform only speeds up the local area, which does not support the application migration, compatibility. MEC is also recognized as the leading new technology for 5G networks by the European 5G PPP (Public-Private Partnership 5G Infrastructure). The Natural Language Processing helps the Edge Computing to be more robust and increase the rate of data processing so that the output can be generated at much faster rate. It's a type of program which delivers low latency. According to Karim Arabi, edge computing is a broadly network whatever is happening outside the cloud and in the applications where real time processing of data is required.

The world is seeing a constant rise in the numeral of smart cities. Edge computing looks like a viable solution for making the smart city environments as it facilitates the extension of the cloud resources to the network edge [20]. Henceforth, enhancing the service awareness, scalability and low latency. Characteristics high context awareness, single hop connectivity and geo distribution are the reasons for rapid adoption of mobile edge computing. The amalgamation of edge computing along with deep learning mechanisms can be used to create real life health applications. A similar application being, HealthFog, a heart disease analysis application [6]. It delivers healthcare as a service using IoT devices over a fog network.

Organization of rest of the paper is done as follows: Sect. 2 explains some of the contemporaries of edge computing. Section 3 discusses the applicability of deep learning algorithms in an edge environment. It talks through some of the most popular verticals wherein extensive research is happening at a global scale. Section 4 which suggests some of the benefits of mobile edge computing. Finally we have, Sect. 5 presenting the experimental analysis of MEC and deep learning models.

2 Related Technologies

Cloudlet is a small data center that usually happens as a wireless transition from public places such as mobile devices for the convenience of the hospital, mall, office building, etc. It is a convenient approach wherein many blocks of multi-core computers create a cloud that connects remotely to a cloud server [1]. Cloudlet is presented as a promising solution for remote extended area networks (WAN) and latency in cellular Energy consumption due to the use of mobile communication for data transmission. The primary goal of the cloud server cloud cover is to carry the cloud. The end user's technology which provides resource support more time-sensitive applications. Further studies confirmed that using resource wealthy machines close to cellular customers, called cloudlets, pro-vides offerings typically observed inside the cloud, furnished improvements in execution time while some of the responsibilities are offloaded to the edge node. On the alternative hand, offloading each task may also bring about a slowdown due to switch instances between device and nodes, so depending at the workload a foremost configuration may be defined [18].

Fog computing [17] which is also known as edge computing, which supports uni-versally connected equipment. The word fog computing was made CISCO system that brings cloud services to enterprise point of view Network like MEC [6]. In fog comput-ing, processing is mostly done at the end of the LAN on the IoT gateway or fog node. The advantage of allowing only processing equipment in fog computing is the benefit is to collect data from various sensors and act in accordance to it [19]. However, due to their dependence on the wireless connection, there are some limitations in ambiguous calculations, which must be active in order to perform the complicated actions. The estimate of fog and MECs are widely used interchangeably except in some cases where they differ [21, 22].

Blockchain and the use of smart contracts are the best known examples for creating distributed applications. The expression "smart contract" was first authored in mid-1990s by researcher and cryptographer Szabo, who characterized a smart contract as a lot of guarantees, indicated in computerized structure, including rules inside which the different groups perform on these guarantees. In his popular model, Szabo analogized smart agreements to vending machines: machines take in coins, and by means of a simple algorithmic system (e.g., finite automata), dispenses change and item as indicated by the showed cost. Smart contracts go past the candy machine by proposing to install contracts in a wide range of properties by advanced methods. As a rule, smart agreements or contracts can be characterized as the digital rules that digitally encourage, confirm, and uphold the agreements made between at least two gatherings on blockchain. As smart contracts are normally used and verified by blockchain, they have some novel attributes. In the first place, the program code of a smart contract will be recorded and confirmed on blockchain, subsequently making the agreement alter safe. Secondly, the execution of a smart contract is implemented among different, trustless individual hubs without incorporated control, and coordination of third part specialists [20, 23].

3 Deep Learning Applications in Edge

Fog or edge computing is a kind of program which delivers low latency. The following figure illustrates an edge environment. In both IoT and Mobile Computation Offload-ing tasks and data were communicated to the remote cloud for performing computa-tional operations [4]. Whereas, through edge computing code offloading and data trans-fer is performed at the network edge level, thus helping in quick response time and enhanced Quality of Service. Further in this section we talk about some of the fields wherein the deep learning algorithms [10] can be implemented in an edge environmen-tal setup. Similar to these areas, a use case for mobile code offloading is discussed in Sect. 5 (Fig. 1).

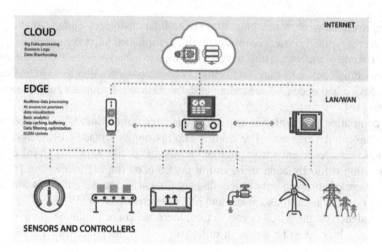

Fig. 1. Conceptual architecture for edge computing [16]

3.1 Internet of Things (IoT)

In today's world of IoT we need lots of storage and massive amount of data to data centers, due to which bandwidth is affected. Besides all these improvements in to-day's world, data centers still cannot guarantee the transfer rate and response times which is very important. Edge software services lessen the volumes of information that must be moved, the resultant site visitors, and the space that records need to move. It provides decrease latency and decreases transmission prices.

3.2 Micro Data Centers and Cloudlets

Computation offloading for actual-time programs, consisting of facial reputation algorithms, confirmed widespread enhancements in reaction times as established in early studies. Further studies confirmed that using resource wealthy machines close to cellular customers, called cloudlets, and providing offerings typically observed inside the cloud, furnishes improvements in the execution time while some of the responsibilities are offloaded to the edge node. On the alternative hand, offloading each task may also bring about a slowdown due to switch instances between device and nodes, so depending at the workload and foremost configuration may be defined.

3.3 Virtual Reality

Another use of the architecture is cloud gaming, in which some components of a sport may want to run within the cloud, even as the rendered video is transferred to lightweight customers together with mobile, VR glasses, and many others. Such sort of streaming is also referred to as pixel streaming. Conventional cloud video games might also suffer from excessive latency and inadequate bandwidth, due to the fact the amount of records transferred is huge because of the excessive resolutions required through some offerings.

In such cases Edge computing plays a vital role. An-other use of edge computing architecture is cloud gaming, in which some components of a sport may want to run within the cloud [8], even as the rendered video is transferred to lightweight customers together with mobile, VR glasses, and many others. Such sort of streaming is also referred to as pixel streaming. Conventional cloud video games might also suffer from excessive latency and inadequate bandwidth, due to the fact the amount of records transferred is big because of the excessive resolutions required through some offerings.

3.4 NLP

Natural Language Processing helps the Edge Computing to be more robust and increase the rate of data processing so that the output can be generated at much faster rate. Natural Language Processing, is an intelligent technology that helps to create an interaction between the machine and the humans. NLP helps to fill the gap between the machine and human by processing various codes, computation linguistics and also computer science to manipulate the human language and help machine to be more precise with the output. NLP helps to deliver the cognitive solutions by improving the output result. NLP helps to improve the latency by processing the data of the idle time a person spends over the service without sending any data packets. This helps to improve the route of the data transfer. NLP helps in real-time analytics to make the CPU workload more intensive. The secret is in knowing your field of interest and market, and by analyzing your activity and making a suitable design to support your system. The key here is develop an 'intimacy' with the customer through new technologies.

3.5 Computer Vision

An example of applicability of deep learning in an edge environment can be, a facial recognition system that verifies the face of the entering and exiting facilities has been used by Department of Defense. This technique requires machine learning and natural language processing, neural networks and other statistical computing for the process. This technology in particular has storage requirement to make the setting feasible. For this, edge computing provides crucial storage required for the processing of data.

In relation to the five research areas discussed previously, the following figure showcases a comparison between works having implemented deep learning models in an edge setup (Table 1).

Table 1. Selected works on union of edge computing and deep learning

Work	Deep learning model	Application	Performance parameters
DeepIoT [9]	VGGNet	Image recognition	Latency, memory
DeepMon [11]	Yolo [12]	Object detection	Latency
VideoEdge [13]	AlexNet	Image classification	Accuracy
DeepThings [14]	Yolo [12]	Object detection	Accuracy, latency
MCDNN [15]	AlexNet	Image classification	Energy, memory

4 Advantages of Mobile Edge Computing

MEC focuses on essential metrics like the concept delay. Moreover, high bandwidth, which is achieved by limiting data movement. Then for MEC servers, and then for a centralized server with a long delay cost. Apart from this, electricity consumption is also a significant problem. Computational work related to external resource systems lead to increased battery life of user devices.

- Speed: Edge computing helps to respond to data almost instantaneously by eliminating the lags. The internet bandwidth is reduced by providing the data processing computing near to the source. Since the resources are remotely available, the efficiency is increased and cost is reduced. The data is not required to be put up on cloud, and the security is ensured for the sensitive data. The most vital advantage of edge computing is their ability to recover the network's performance by combating latency.
- Security: The edge computing architecture's distributed nature makes it easy to execute the security protocol, which can separate the hacked parts without turning the whole network off. For example, a facial recognition system that verifies the face of the entering and exiting facilities has been used by Department of Defense. This technique requires machine learning and natural language processing, neural networks and other statistical computing for the process. This technology in particular has storage requirement to make the setting feasible [7].
- Scalability: Border computing provides a much less expensive way for scalability. The utility of edge computing devices with the ability to process also reduces the cost of development, because each added new device does not apply the critical requirements on the network core bandwidth. At this point, edge computing comes in handy as reduced form factor is critical for IoT implementation and edge computing provides crucial storage required for the processing of data.
- Reliability: With the security advantages of edge computing, it is not surprising that it also provide high reliability. Since IoT edge computing devices and edge data centers are located close to end users, the probability of problems with the net-work in remote locations affecting local customers is less likely. IoT edge computing devices will work fine even if there is some fault with the nearby data centers if the edge computing architecture is followed. When edge computing architecture is used then it is seen that complete service will never disable completely. The key here is develop an 'intimacy' with the customer through new technologies [3].

5 Implementation and Simulation

In this section we would be discussing an application for mobile computation offloading wherein resource intensive tasks of an application are offloaded in order to ensure reduced battery consumption and response time. The sample application comprises of three different categories of tasks that are I/O intensive, CPU intensive and Data intensive in nature. According to the proposed hypothesis "Mobile Computation Offloading performed in an Edge environment making use of a deep learning algorithm is most appropriate in terms of ensuring response time and mobile battery consumption".

The following are the three set of experiments that have been performed for validating the proposed hypothesis.

1. Task Offloading in an MCC Environment
2. Task Offloading in an Edge Environment
3. Task Offloading in an Edge Environment assisted by Deep Learning

Convolution Neural Network (CNN) is the deep learning model used in this experiment. The CNN is trained on the basis of task execution patterns. It identifies relationships between tasks in terms of a task being independent or dependent on other task executions.

Table 2. Task offloading

Experiment number	Number of tasks offloaded	Battery consumption (%)	Latency (ms)
1	9	12.64	32
2	5	12.22	21
3	5	11.87	16

The above-mentioned Table 2 has been further illustrated using the following graphs in form of Fig. 2 and Fig. 3

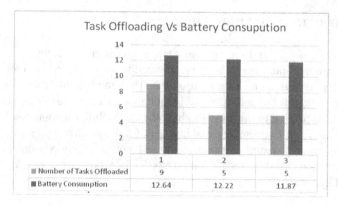

Fig. 2. Effect of task offloading on battery consumption

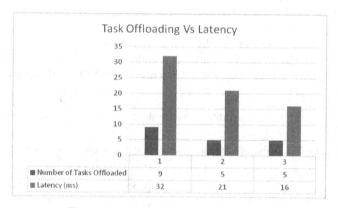

Fig. 3. Effect of task offloading on latency

6 Conclusion

Mobile edge computing (MEC) is one of the new age computing paradigm and is seeing rapid adoption among researchers across the globe. Characteristics such as enhanced scalability and performance are the reasons for it being a successful successor to Mobile Cloud Computing. Significant research is being conducted towards ensuring amalgamation of deep learning models and the edge environment. In recent times, areas like natural language processing and computer vision have seen works involving the use of deep learning algorithms and edge computing. Furthermore, the amalgamation can lead to enhanced performance of mobile computation offloading with respect to performance parameters of mobile battery consumption and latency.

References

1. Xu, X., Chen, Y., Yuan, Y., Huang, T., Zhang, X., Qi, L.: Blockchain-based cloudlet management for multimedia workflow in mobile cloud computing. Multimed. Tools Appl. **79**(15), 9819–9844 (2019). https://doi.org/10.1007/s11042-019-07900-x
2. http://www.cisco.com/c/en/us/solutions/collateral/service-provider/visual-networking-ind exvni/mobile-white-paper-c11-520862.html
3. Wang, S., et al.: Edge server placement in mobile edge computing. J. Parallel Distrib. Comput. **127**, 160–168 (2019)
4. Khanna, A. et al.: Adaptive mobile computation offloading for data stream applications. In: 2017 ICACCA (Fall), pp. 1–6. IEEE, September 2017
5. Khanna, A. et al.: Mobile cloud computing architecture for computation offloading. In: 2016 NGCT, pp. 639–643. IEEE, October 2016
6. Tuli, S., Basumatary, N., et al.: HealthFog: an ensemble deep learning based smart healthcare system for automatic diagnosis of heart diseases in integrated IoT and fog computing environments. Future Gener. Comput. Syst. **104**, 187–200 (2020)
7. Amanullah, M., et al.: Deep learning and big data technologies for IoT security. Comput. Commun. **151**, 495–517 (2020)
8. Tomar, R., Khanna, A., Bansal, A., Fore, V.: An architectural view towards autonomic cloud computing. In: Satapathy, S.C., Bhateja, V., Raju, K.Srujan, Janakiramaiah, B. (eds.) Data Engineering and Intelligent Computing. AISC, vol. 542, pp. 573–582. Springer, Singapore (2018). https://doi.org/10.1007/978-981-10-3223-3_55
9. Yao, S. et al.: DeepIot: compressing deep neural network structures for sensing systems with a compressor-critic framework. In: Proceedings of the 15th ACM Conference on Embedded Network Sensor Systems, p. 4. ACM, November 2017
10. Bahri, Y., Kadmon, J., Pennington, J., Schoenholz, S.S., Sohl-Dickstein, J., Ganguli, S.: Statistical mechanics of deep learning. Annu. Rev. Condens. Matter Phys. **11**, 501–528 (2020)
11. Huynh, L.N. et al.: DeepMon: mobile GPU-based deep learning framework for continuous vision applications. In: Proceedings of the 15th Annual International Conference on Mobile Systems, Applications, and Services, pp. 82–95. ACM, June 2017
12. Redmon, J., Farhadi, A.: YOLO9000: better, faster, stronger. In: Proceedings of the IEEE Conference on Computer Vision and Pattern Recognition, pp. 7263–7271 (2017)
13. Hung, C.C. et al.: VideoEdge: processing camera streams using hierarchical clusters. In: 2018 IEEE/ACM Symposium on Edge Computing (SEC), pp. 115–131. IEEE, October 2018
14. Zhao, Z., et al.: DeepThings: distributed adaptive deep learning inference on resource-constrained IoT edge clusters. IEEE Trans. Comput. Aided Des. Integr. Circuits Syst. **37**(11), 2348–2359 (2018)
15. Han, S. et al.: MCDNN: an approximation-based execution framework for deep stream processing under resource constraints. In: Proceedings of the 14th Annual International Conference on Mobile Systems, Applications, and Services, pp. 123–136. ACM, June 2016
16. IIoT Edge Computing vs. Cloud Computing. https://openautomationsoftware.com/blog/iiot-edge-computing-vs-cloud-computing. Accessed Dec 2018
17. Yang, Y., Luo, X., Chu, X., Zhou, M.T.: Fog computing architecture and technologies. In: Fog-Enabled Intelligent IoT Systems, pp. 39–60. Springer, Cham (2020)
18. Fore, V., Khanna, A., Tomar, R., Mishra, A., Intelligent supply chain management system. In: 2016 International Conference on Advances in Computing and Communication Engineering (ICACCE), pp. 296–302. IEEE, November 2016
19. Calheiros, R.N.: Fog and edge computing: challenges and emerging trends (invited talk). In: 2nd Workshop on Fog Computing and the IoT (Fog-IoT 2020). Schloss Dagstuhl-Leibniz-Zentrum für Informatik (2020)

20. Khan, L.U., Yaqoob, I. et al.: Edge computing enabled smart cities: a comprehensive survey. IEEE Internet Things J., 1 (2020)
21. Kumar, P., Choudhury, T. et al.: Fog computing: common security issues and proposed countermeasures. In: System Modeling Advancement in Research Trends (SMART) (2016)
22. Garg, V., Choudhury, T., et al.: Advance survey of mobile ad-hoc network. IJCST 2(4), 552–555 (2011)
23. Choudhary, T., Choudhury, V., et al.: An approach to improve task scheduling in a decentralized cloud computing environment. Int. J. Comput. Technol. Appl. 3(1), 312–316 (2012)

Analysis of Clustering Algorithms in Machine Learning for Healthcare Data

M. Ambigavathi$^{(\boxtimes)}$ and D. Sridharan

Anna University, CEG Campus, Chennai, India
ambigaindhu8@gmail.com

Abstract. Clustering algorithm is one of the most popular data analysis technique in machine learning to precisely evaluate the vast number of healthcare data from the body sensor networks, internet of things devices, hospitals, clinical, medical data repositories, and electronic health records etc. The clustering algorithms always play a crucial role to predict the diseases by partitioning the similar patient's data based on their relevant attributes. The vast number of clustering algorithms have been developed for analyzing several healthcare data sets so far. However, the algorithms presented in the literature may achieve a better result with a particular type of data set but may fail or provide poor results with the data set of other types. Many of the research studies considered specific or multiple data sets for clustering analysis. But there are only a few studies used mixed type of data for analyzing and verifying the optimal number of clusters. To alleviate these issues, this paper aims to inspect various clustering algorithms from the theoretical and experimental perspectives. The experimental results elucidate the best algorithm from each categories using a physiological data set. The efficiency of each clustering algorithm in machine learning is validated using a number of internal as well as stability measures. Finally, this paper highlights the future directions with a proper clustering algorithm for handling high dimensional healthcare data sets.

Keywords: Machine learning · Clustering algorithms · Unsupervised learning algorithms · Big data · Healthcare applications

1 Introduction

The numerous records of healthcare data generated every day are increasing astronomically in today's modern era [1]. The explosion of medical sensors, internet of things devices, and digitalization of medical records have created a flood of data typically landing in different medical storage repositories. Then, various kinds of operations such as analytical, process, and retrieval are performed to extract valuable insights from the raw data [2]. With the help of real-time alerts, doctors or medical practitioners will take perspective decisions about treatment at the right time [3, 4]. Therefore, big data analytics solutions can be used to save human lives, provide analysis much faster, ultimately save money and improve the efficiency of treatment [5].

© Springer Nature Singapore Pte Ltd. 2020
M. Singh et al. (Eds.): ICACDS 2020, CCIS 1244, pp. 117–128, 2020.
https://doi.org/10.1007/978-981-15-6634-9_12

The healthcare data is captured from various sources that include [6] hospitals, clinical, medical research, electronic records, and authorized websites respectively. They are stored in different formats such as text, video, audio, image, impala complex types, and sequence file respectively [7] and also make it very difficult to process and analyze all pieces of data effectively. One key strategy to solve this analytic issue is to group or cluster the big health data in a more compact format. In such a case, clustering algorithms contribute a major role to analyze the massive volume of healthcare data as small segments in a dispersed way and effectively aggregate all these data across different clusters to obtain the final processed medical data [8]. There are several clustering algorithms developed [9] to analyze the data but still, it is a challenging task which algorithm provides the best and the optimal number of clusters with respect to different data sets. Many authors have evaluated the clustering algorithms using different medical data sets with unique validation metrics [10–13]. Only a few authors [14] have been used synthetic data sets with real-time data sets to assess the variations and performance of three distinct clustering algorithms. Each data set is unique in its own way. No studies have been considered so far to estimate various clustering algorithms using the mixed type of physiological data. This type of analysis on vital parameters must require in the near future to identify the time-critical data than normal data. Therefore, this work considers only a synthetic data set instead of real-time data sets to evaluate the best number of clusters for healthcare data analysis. Moreover, the value of the raw healthcare data collected from hospitals or patients in real-time may be similar or slightly different from our synthetic data set. But the minimum and maximum values of vital data may only deviate from the considered ranges.

Despite the vast number of analysis for clustering algorithms using various healthcare data sets including heart rate [15], brain [16], body temperature [17], emotions [18], cancer [19], blood pressure, ambulatory, and emergency respectively, available in the literature. In such a case, it is very difficult for handlers to decide in advance which algorithm is most suitable one for identifying the abnormality in a given big health dataset. There are still many limitations exist in the literature that need to be addressed: (i) the unique attributes of various clustering algorithms especially for physiological data set are not analysed carefully, (ii) several clustering algorithms have been developed for healthcare domain but they were not deliberated any mixed type of vital information and (iii) only experimental analysis has been carried out to specific healthcare data set to study the significance of one algorithm over another. The aforementioned reasons are highly motivated us to inspect various clustering algorithms, especially for the mixed type physiological data set. The main contributions are outlined as follows:

- To study three distinct types of clustering algorithms based on the theoretical perspectives.
- To validate the different clustering algorithms using internal and stability metrics.
- To analyze the most optimal clustering algorithm with respect to clinical perspectives.

Therefore, this article provides readers with a sufficient analysis of particular clustering algorithms by theoretically and experimentally comparing them on the synthetic physiological data set. Other sections of this paper are described as follows: The theoretical details of clustering algorithms are summarized in Sect. 2. Section 3 describes the internal and stability validation measures for various clustering algorithms. The experimental and comparative analysis of different clustering algorithms are explained in Sect. 4. Finally, Sect. 5 concludes the paper with appropriate clustering algorithm with future scope.

2   Analysis of Clustering Algorithms

Clustering is one of the best known algorithm in machine learning domain, named as an unsupervised learning algorithm [20]. The significance of clustering algorithm is to divide the large volume of data into smaller groups of data when there is no class labels available to process the datasets. Each cluster contains a set of data points where clustering algorithm mainly used to classify and group each data point into a particular cluster. Besides, the data points within the same cluster should have similar properties, while data points in the different cluster should have highly dissimilar properties and/or features [21]. Many clustering algorithms for analyzing healthcare data sets have been introduced in the existing research works [22–26]: K-means, K-Medoids or Partitioning Around Medoids (PAM), and Hierarchical. The main procedures of these algorithms are classified as follows.

2.1   K-means Clustering Algorithm

K-means is a simple and most general clustering algorithms which is mainly used to classify the given dataset that is unlabeled. This algorithm mainly aims to find similar clusters represented by variable k. For this purpose, this algorithm uses the mean or centroid as a metric to characterize the cluster. A centroid is a data point that indicates the center of the cluster, and it might not necessarily be a member of the dataset. So, it divides n data points into k number of clusters and then each data point n belongs to appropriate cluster with the nearest possible centroid. Next, the Euclidean distance is accurately calculated from each data point n to the centroid in a given cluster. Always, the data points in a cluster are assigned to the centroid depending on the minimum euclidean distance from that centroid point. When there no data point is available to assign, an early grouping is considered. In such case, 'c' new centroids are re-calculated, thus new iteration continues until the 'c' centroids stop changing their position.

2.2   K-medoids Clustering Algorithm

K-medoid is a variant type of algorithm which is also termed as Partition Around Medoids (PAM). In this algorithm, data point act as a medoid within a cluster that are centrally located whose disparity over all data points in the cluster is minimal. Therefore, this medoid can be used as a representative of other data points within a cluster. The main core idea of PAM is to first calculate major data point as a medoid in a specific cluster, group the set of medoids, and then each data point is assigned to the nearest medoid in a given cluster. Moreover, this algorithm generally follows two phases: build and swap phase. The role of the first phase is to select the first medoid as the data point with the lowest mean dissimilarity with respect to the whole dataset. Likewise, in the second phase, given the current set of 'k' medoids, all the neighbor data points are evaluated. A new medoid is created by exchanging data points in the old medoid with the data points in a new non-medoid.

2.3 Hierarchical Clustering

Hierarchical is a special type of unsupervised machine learning algorithm, also referred as Hierarchical Cluster Analysis (HCA). The goal of hierarchical cluster analysis is to cluster similar unlabeled data points into number of clusters using tree based structure. The data points in the end of tree forms a set of clusters, where each and every cluster is distinct from other clusters. Besides, the data points within a specific cluster is mostly identical to other clusters in the data set. This algorithm uses a tree-type structure (dendrogram) based on the hierarchy. Basically, there are two types of hierarchical clustering algorithms include Agglomerative hierarchical clustering or AGNES (Agglomerative Nesting) and Divisive hierarchical clustering or DIANA (Divisive Analysis). Both this algorithm is exactly the reverse of each other. The summary of various algorithms with respect to various characteristics are listed in Table 1.

Table 1. Summary of clustering algorithms

Algorithm	Big data			Computation speed	Modifications corrections	Cluster shape	Results interpretation
	Size of data set	Type of data	Complexity				
K-means	Large	Numerical	$O(nkd)$	Fast	Flexible	Non convex	Easy
K-medoids	Small	Categorical	$O(n^2dt)$	Moderate	Difficult	Non convex	Difficult
Hierarchical	Large	Numerical	$O(n)$	Slow	Flexible	Non convex	Easy

3 Validation Measures

The performance of unsupervised learning algorithms is evaluated using different internal, and stability validation metrics. The internal measures are very important for evaluating the right number of clusters and computing the quality of the appropriate clustering algorithm. This measures consider only the internal information to calculate the quality of a clusters without using any external information. The basic internal validation measurements [27] are classified into three types: Connectivity, Silhouette and Dunn index. This section briefly presents the internal validation indices used for a physiological data set.

3.1 Internal Measures

Connectivity. This measure represents the total number of rows n (data points or observations) and columns m in a dataset. The values are always considered as numeric (e.g., a physiological parameter's values). Let $Y_{ni}(j)$ and $x_i Y_{ni}(j)$ be the jth nearest neighbor of data point i and zero, respectively, if both i and j are in the same cluster, and then $1/j$ otherwise. The connectivity is measured for a particular cluster $C = \{C_1, C_2 \ldots .C_k\}$ with n data points using the below equation

$$C = \sum_{i=1}^{n} \sum_{j=1}^{p} x_i Y_{ni}(j) \tag{1}$$

Where p represents a parameter value and if the connectivity measure has a value between 0 and ∞, it should always be decreased.

Silhouette Coefficient. This coefficient is a very useful metric for evaluating the performance of clustering results. This value measures how data points are grouped and computes the average distance available between the different clusters. The width of this coefficient always lies in the following interval $[-1, 1]$ that implies the super grouped data points with values near to 1 and lower grouped data points with values near to -1. Therefore, the coefficient for data point i is defined as

$$S(i) = \frac{(y_i - x_i)}{max(y_i, x_i)} \tag{2}$$

Where x_i and y_i denote the average distance between the data points in the same cluster and the average distance between the data points in the nearest neighboring clusters which can be expressed as

$$y_i = \min_{C_k \in \frac{C}{C_i}} \sum_{j \in C_k} \frac{dist(i, j)}{nC_k} \tag{3}$$

Where C_i indicate a cluster with data point i, $dist(i, j)$ presents the distance between the data points i and j, then nC_k implies cardinality of the cluster C.

Dunn Index. This is an important metric that presents the ratio of the lowest distance between the data points which is not available in the same cluster and the highest distance in the intra-cluster. The index value can be obtained as

$$D_C = \min_{C_k, C_l \in C, C_k \neq C_l} \frac{\left(\min_{i \in C_k, j \in C_l} dist(i, j) \right)}{\max_{C_m \in C} d(C_m)} \tag{4}$$

Where $d(C_m)$ indicates a cluster C_m with maximum distance and this index has a value between 0 and ∞, and it should always be increased.

3.2 Stability Measures

The stability measure is a special type of validation measure to individually evaluate the cluster results from the overall analysis by removing each column in the data set. This type of measure is very significant especially when the physiological raw data are highly correlated with others. For this purpose, this study uses stability measures to compare the consistency of raw data in the medical synthetic data set. Generally, the stability measures [28] are broadly classified into four different groups: (i) Average Proportion of Non-overlap (APN), (ii) Average Distance (AD), (iii) Average Distance between Means (ADM), and (iv) Figure of Merit (FOM).

Average Proportion of Non-overlap (APN). This measure is used to calculate the average proportion of data point that is not located in the same cluster with a particular or single column removed. Let consider $C^{i,0}$ be the cluster with data point i using the original cluster and $C^{i,l}$ be the cluster with column l removed in the data set. Then, APN value is always varied between the following interval [0, 1]. If the APN values close to 0 that indicates the highly consistent results. For the total number of cluster set K, the APN value is measured using given formula

$$APN(K) = \frac{1}{MN} \sum_{i=1}^{N} \sum_{l=1}^{M} \left(1 - \frac{n\left(C^{i,l} \cap C^{i,0}\right)}{n\left(C^{i,0}\right)} \right) \tag{5}$$

Average Distance (AD). The main function of AD measure is to predict the average distance between the data points that are placed in the same cluster by considering the aforementioned two cases. If the AD has a value between zero and ∞, and then the smaller values are always considered to evaluate the results. The following given expression is used to compute AD,

$$AD(K) = \frac{1}{MN} \sum_{i=1}^{N} \sum_{l=1}^{M} \frac{1}{n\left(C^{i,l} \cap C^{i,0}\right)} \left[\sum_{i \in C^{i,0}, j \in C^{i,l}} dist(i,j) \right] \tag{6}$$

Average Distance Between Means (AM). The main objective of this measure is to calculate the average distance between data points that are presented in the same cluster under the aforementioned two cases. However, only it uses the Euclidean distance with smaller values between 0 and ∞ is always preferred. Let $\bar{x}_{C^{i,0}}$ denote cluster contains average data points i and $\bar{x}_{C^{i,l}}$ indicate the cluster contains data point i with column l removed. Then, it is computed using the below formula,

$$ADM(K) = \frac{1}{MN} \sum_{i=1}^{N} \sum_{l=1}^{M} \frac{1}{n\left(C^{i,l} \cap C^{i,0}\right)} dist(\bar{x}_{C^{i,l}}, \bar{x}_{C^{i,0}}) \tag{7}$$

Figure of Merit (FOM). The decisive role of a FOM is to estimate the average variance of the deleted columns in different clusters and grouping is performed based on the remaining (undeleted) columns. The smaller values between 0 and ∞ are mostly preferred and also it computes the mean error rate using average number of clusters. Then, FOM predicts a particular left-out column l using the given formula

$$FOM(l, K) = \sqrt{\frac{1}{N} \sum_{k=1}^{k} \sum_{i \in C^{k,l}} dist\left(x_{i,l}, \bar{x}_{C_k(l)}\right)} \tag{8}$$

Where $x_{i,l}$ presents the value of ith observation in the lth column and $\bar{x}_{C_k(l)}$ denote the average of a cluster. Generally, FOM uses only Euclidean distance and also it is multiplied by the following adjustment factor $\sqrt{\frac{N}{N-K}}$, to decrease the amount of cluster expansions.

4   Experimental Results

The clustering algorithms are validated by including two packages defined in R programming tool. The two major packages used in this study are clValid [29] package and NbClust package [30], respectively. Both packages are very significant to determine the best optimal number of data clusters for a given data set and validate the effective results from the clustering analysis. This analysis study uses Euclidean distance as a parameter in NbClust function. The frequency of occurrence of time-critical data is measured with respect to the range of vital parameters, which are shown in Fig. 1.

4.1   Data Set

This experiment study uses statlog heartrate real-world data set (i.e., UCI machine learning repository) as a basic data set, which consists of 130 instances and 3 variables. To validate the advantages of the synthetic dataset, this work includes 5 additional variables by utilizing the same 130 instances. The data set contains only numerical values with different attributes. The vital ranges of each attribute are incorporated based on the conditions of the patient such as normal, moderate and extremely high. The various characteristics of both real world and synthetic healthcare data sets are mentioned in Table 2.

Table 2. Various characteristics of healthcare data sets

Name of data set	Type of dataset	Type of data	No of instances	No of attributes
Heart rate	Real world	Multivariate	130	3
Physiological data	Synthetic data	Numerical	130	8

4.2   Comparative Analysis

The aim of comparative analysis is to choose how accurately each and every algorithm can able to group similar health records from the mixed physiological data set. Further, to analyze the optimal number of the cluster's size for every algorithm and predict which algorithm performs better than others. The analysis results of three different clustering algorithms are validated using both internal and stability measures.

Evaluating Validity. The analysis results of various clustering algorithms based on the internal validity measures are presented in Table 3. Initially, algorithms are validated with the varying cluster size from k = 2 to k = 10.

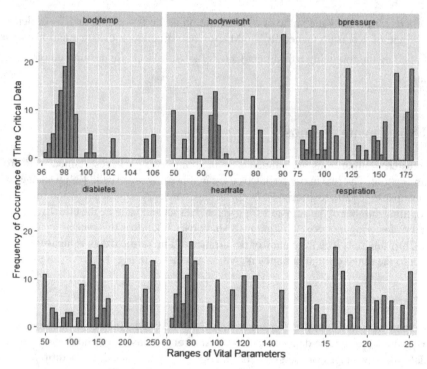

Fig. 1. Frequency of occurrence of vital data

From the cluster analysis, it is observed that the K-means algorithm with two clusters provides better results using connectivity and Dunn index measures as compared to the hierarchical clustering algorithm. However, the hierarchical algorithm achieves better output according to the silhouette validity measure. Therefore, it is the second best known clustering algorithm in terms of internal validity. Moreover, the comparative analysis suggested that the K-medoids yield no clustering results in comparison to K-means and hierarchical algorithms.

Evaluating Stability. The stability of three different clustering algorithms is validated to predict any variations in the clustering outputs based on the removal of one column in a given data set. The achieved results of stability for each clustering algorithm are displayed in Table 3. From the assessments, it is noticed that the hierarchical algorithm almost approaches the lower stability values based on the APN, ADM, and FOM respectively. Though it achieved better stability values with all three measures it is failed to provide the best result for AD measure. Further, the maximum stability value of K-means algorithm indicates that the algorithm is not able to yield better values. Likewise, the Pam algorithm is ineffective to give stable outputs in terms of stability measures. Hence the hierarchical clustering algorithm contributes the highest stability results in every aspects as compared with K-means and Pam algorithms.

Evaluating Optimal Scores. The optimal number of clusters and their scores are evaluated using two important measures such as internal and stability. The best optimal

Table 3. Internal validation of clustering algorithms

Type of measures	Clustering method	Validity measures	Cluster size			
			2	3	4	5
Internal validation metrics	Hierarchical	Connectivity	7.5556	10.4845	10.7845	14.5425
		Dunn	0.2943	0.2971	0.3140	0.3140
		Silhouette	0.3075	0.2093	0.2390	0.2261
	K-means	Connectivity	2.1940	41.1071	25.3369	35.9258
		Dunn	0.3450	0.1761	0.2356	0.1950
		Silhouette	0.2470	0.1743	0.2496	0.2345
	Pam	Connectivity	19.5016	45.1821	67.7214	67.7167
		Dunn	0.0763	0.0508	0.0330	0.0429
		Silhouette	0.2147	0.2094	0.1867	0.2152

scores of every algorithms are depicted in Table 4. Based on the observations, it is clearly shown that the K-means algorithm with two optimal clusters can provide the best results in terms of connectivity, Dunn index and silhouette, respectively. In contrast, the hierarchical algorithm with different clusters often yields the highest stability values for APN, ADM, and FOM except for AD among all considered clustering algorithms.

Table 4. Stability validation of clustering algorithms

Type of measures	Clustering method	Validity measures	Maximum cluster size			
			2	3	4	5
Stability validation metrics	Hierarchical	APN	0.0648	0.3074	0.0389	0.1035
		AD	3.5190	3.5114	3.0791	2.9625
		ADM	0.2221	0.9328	0.4820	0.4237
		FOM	0.9695	0.9520	0.9304	0.8599
	K-means	APN	0.1824	0.3675	0.1987	0.1940
		AD	3.6036	3.4353	2.9575	2.7903
		ADM	1.2480	1.2474	0.7353	0.7205
		FOM	0.9779	0.9509	0.9023	0.8712
	Pam	APN	0.1932	0.2424	0.3472	0.3391
		AD	3.4350	3.2164	3.1585	2.9340
		ADM	0.6666	0.8308	1.1647	1.0535
		FOM	0.9535	0.9196	0.9141	0.8984

Table 5. Optimal scores for various clustering algorithms

Type of validity measure	Name of validity metric	Optimal score	Clustering method	Optimal number of clusters
Internal	Connectivity	2.1940	k-means	2
	Dunn	0.3450	k-means	2
	Silhouette	0.3075	Hierarchical	2
Stability	APN	0.0389	Hierarchical	4
	AD	2.7903	K-means	5
	ADM	0.2221	Hierarchical	2
	FOM	0.8599	Hierarchical	5

However, the best optimal cluster size for a physiological data set is 2 and also it is significantly confirmed that the suitability for dealing with high-dimensional physiological datasets. Finally, this analysis suggested that the Pam algorithm failed to produce the optimal number of clusters on synthetic data set with high problem dimensionality, as mentioned in Table 5.

5 Conclusion and Future Work

This study provided a detailed theoretical view on clustering algorithms especially for healthcare data analysis from both theoretical and experimental perspectives. There are numerous clustering algorithms deliberated in the existing studies for analyzing healthcare data sets and also validated with different metrics. However, it is very hard to decide in advance which clustering algorithm would be the most suitable for a particular data set and what would be the best optimal number of clusters from a given a set. Based on these perceptions, this study analysed various clustering algorithms in clinical point of view and validated using internal and stability measures. The observed results reported a better solution to develop novel clustering algorithm and to recommend a specific algorithm for huge volume of physiological data set. The grouping of abnormal variations from different columns of data sets is the most significant requirement rather than grouping the normal variations when using the mixed or complicated vital data sets. In future, this study will further extend the analysis for big pandemic healthcare data sets with respect to similarity score, condition-specific, and then generic preference-based measures.

References

1. Dash, S., Shakyawar, S.K., Sharma, M., Kaushik, S.: Big data in healthcare: management, analysis and future prospects. J. Big Data 6(54), 1–25 (2019). https://doi.org/10.1186/s40 537-019-0217-0

2. Thasni, K.M., Haroon, R.P.: Application of big data in health care with patient monitoring and future health prediction. In: Smys, S., Senjyu, T., Lafata, P. (eds.) ICCNCT 2019. LNDECT, vol. 44, pp. 49–59. Springer, Cham (2020). https://doi.org/10.1007/978-3-030-37051-0_6

3. Dautov, R., Distefano, S., Buyya, R.: Hierarchical data fusion for smart healthcare. J. Big Data 6(19), 1–23 (2019). https://doi.org/10.1186/s40537-019-0183-6

4. Prosperi, M., Min, J.S., Bian, J., Modave, F.: Big data hurdles in precision medicine and precision public health. BMC Med. Inform. Decis. Mak. 18(139), 1–15 (2018)

5. Zillner, S., Neururer, S.: Big data in the health sector. In: Cavanillas, J.M., Curry, E., Wahlster, W. (eds.) New Horizons for a Data-Driven Economy, pp. 179–194. Springer, Cham (2016). https://doi.org/10.1007/978-3-319-21569-3_10

6. Ambigavathi, M., Sridharan, D.: A survey on big data in healthcare applications. In: Choudhury, S., Mishra, R., Mishra, R.G., Kumar, A. (eds.) Intelligent Communication, Control and Devices. AISC, vol. 989, pp. 755–763. Springer, Singapore (2020). https://doi.org/10.1007/978-981-13-8618-3_77

7. Ambigavathi, M., Sridharan, D.: Big data analytics in healthcare. In: 2018 Tenth International Conference on Advanced Computing (ICoAC), India, pp. 269–276. IEEE (2018)

8. Van Hieu, D., Meesad, P.: Fast K-means clustering for very large datasets based on MapReduce combined with a new cutting method. In: Nguyen, V.-H., Le, A.-C., Huynh, V.-N. (eds.) Knowledge and Systems Engineering. AISC, vol. 326, pp. 287–298. Springer, Cham (2015). https://doi.org/10.1007/978-3-319-11680-8_23

9. Fahad, A., et al.: A survey of clustering algorithms for big data: taxonomy and empirical analysis. IEEE Trans. Emerg. Top. Comput. 2(3), 267–279 (2014)

10. Hatamlou, A.: Heart: a novel optimization algorithm for cluster analysis. Prog. Artif. Intell. 2(3), 167–173 (2014). https://doi.org/10.1007/s13748-014-0046-5

11. Khalid, S., Prieto-Alhambra, D.: Machine learning for feature selection and cluster analysis in drug utilization research. Curr. Epidemiol. Rep. 6, 364–372 (2019). https://doi.org/10.1007/s40471-019-00211-7

12. Zhao, W., Zou, W., Chen, J.J.: Topic modeling for cluster analysis of large biological and medical datasets. BMC Bioinform. 15, 1–11 (2014)

13. Wei, P., He, F., Li, L., Shang, C., Li, J.: Research on large data set clustering method based on MapReduce. Neural Comput. Appl. 32, 93–99 (2020). https://doi.org/10.1007/s00521-018-3780-y

14. Patil, C., Baidari, I.: Estimating the optimal number of clusters k in a dataset using data depth. Data Sci. Eng. 4, 132–140 (2019). https://doi.org/10.1007/s41019-019-0091-y

15. Asril, H., Mousannif, H., Al Moatassime, H.: Reality mining and predictive analytics for building smart applications. J. Big Data 6(66), 1–25 (2019). https://doi.org/10.1186/s40537-019-0227-y

16. Durieux, J., Wilderjans, T.F.: Partitioning subjects based on high-dimensional fMRI data: comparison of several clustering methods and studying the influence of ICA data reduction in big data. Behaviormetrika 46, 271–311 (2019). https://doi.org/10.1007/s41237-019-00086-4

17. Obermeyer, Z., Samra, J.K., Mullainathan, S.: Individual differences in normal body temperature: longitudinal big data analysis of patient records. Bio Med. J. 359, 1–9 (2017)

18. Sharma, K., Castellini, C., van den Broek, E.L., Albu-Schaeffer, A., Schwenker, F.: A dataset of continuous affect annotations and physiological signals for emotion analysis. Nat. Sci. Data 6(196), 1–13 (2019)

19. Papachristou, N., Miaskowski, C., Barnaghi, P., Maguire, R., Farajidavar, N.: Comparing machine learning clustering with latent class analysis on cancer symptoms' data. In: IEEE Healthcare Innovation Point-Of-Care Technologies Conference (HI-POCT), UK, pp. 1–5. IEEE (2016)

20. Nerurkara, P., Shirkeb, A., Chandanec, M., Bhirudd, S.: Empirical analysis of data clustering algorithms. In: 6th International Conference on Smart Computing and Communications, ICSCC 2017, India, pp. 770–779. Elsevier (2018)

21. Tambe, S.B., Gajre, S.S.: Cluster-based real-time analysis of mobile healthcare application for prediction of physiological data. J. Ambient Intell. Hum. Comput. 9(429), 1–17 (2017)

22. Praveen Kumar, D., Amgoth, T., Annavarapu, C.S.R.: Machine learning algorithms for wireless sensor networks: a survey. Inf. Fusion 49, 1–25 (2019)

23. Rokach, L.: A survey of clustering algorithms. In: Maimon, O., Rokach, L. (eds.) Data Mining and Knowledge Discovery Handbook. Springer, Boston (2009). https://doi.org/10.1007/978-0-387-09823-4_14

24. Pérez-Suárez, A., Martínez-Trinidad, J.F., Carrasco-Ochoa, J.A.: A review of conceptual clustering algorithms. Artif. Intell. Rev. 52(2), 1267–1296 (2018). https://doi.org/10.1007/s10462-018-9627-1

25. Xu, D., Tian, Y.: A comprehensive survey of clustering algorithms. Ann. Data Sci. 2(2), 165–193 (2015). https://doi.org/10.1007/s40745-015-0040-1

26. Barbakh, W.A., Wu, Y., Fyfe, C.: Review of clustering algorithms. In: Non-standard Parameter Adaptation for Exploratory Data Analysis. Studies in Computational Intelligence, vol. 249. Springer, Heidelberg (2009). https://doi.org/10.1007/978-3-642-04005-4_2

27. Palacio-Nino, J.-F., Berzal, F.: Evaluation metrics for unsupervised learning algorithms 1, 1–9 (2019)

28. von Luxburg, U.: Clustering Stability: An Overview, pp. 1–41. Now Publishers Inc., Hanover (2010)

29. Brock, G., Pihur, V., Datta, S., Datta, S.: clValid: an R package for cluster validation. J. Stat. Softw. 25(4), 1–22 (2008)

30. Charrad, M., Ghazzali, N., Boiteau, V., Niknafs, A.: NbClust: an R package for determining the relevant number of clusters in a data set. J. Stat. Softw. 61(6), 1–36 (2014)

Securing Mobile Agents Migration Using Tree Parity Machine with New Tiny Encryption Algorithm

Pradeep Kumar[1]($\boxtimes$), Niraj Singhal[1]($\boxtimes$), and K. M. Chaitra[2]($\boxtimes$)

[1] Shobhit Institute of Engineering and Technology (Deemed to Be University), Meerut, India
pradeep8984@gmail.com
[2] JSS Academy of Technical Education, Noida, India

Abstract. A Mobile agents are combination of software programs which works automatically in homogeneous and non-homogeneous environment from one host to another for sharing information among users. Mobile agents migrate in unsecure network, so mobile agent's security is a major concern during the communication and sharing of data & information. Mobile agent's migration has major security issues i.e. data integrity, data confidentiality & authentication, on-repudiation, denial of service and access control. In this paper neural network based synchronization key exchange is proposed for Encryption and Decryption.

Keywords: Mobile agent · Neural networks · Tree parity machine · New Tiny Encryption Algorithm

1 Introduction

Mobile agents are self-dependent program, works on the basis of host. Mobile agent created by user and migrates from one host to another and works automatically. There are three main parts of mobile agent i.e. state, agent code and Agent Function. Mobile agent program executes on every host. A movable agents are software process automatically move through a heterogeneous network under self-control.

Mobile agents are migrating from one user to another user and contacting with other agents. It decides when and where to migrate. Mobile agents has specific life cycle from starting to end suspend its execution. Characteristic of Movable agents are autonomous, dynamic Behavior, intelligence, goal oriented, intelligence etc.

A mobile (or migrating agent) agent can do task on the behalf of host is host not connected to network. After reconnection of host to network mobile agent returns all the result to host. There are many advantages of using mobile agent like, requirement of less network bandwidth, less network delay (latency), synchronization of protocol, working for homogeneous and non-homogeneous network, automatic and dynamic behavior, robustness etc. Mobile Agents are used in different applications like, E-Commerce, Banking system, Security, data retrieval (centralized and distributed), Monitoring System etc.

© Springer Nature Singapore Pte Ltd. 2020
M. Singh et al. (Eds.): ICACDS 2020, CCIS 1244, pp. 129–138, 2020.
https://doi.org/10.1007/978-981-15-6634-9_13

1.1 Security of Mobile Agents

Mobile agent are collection of small code or software, which works on network to perform some tasks in place of a host. To develop and implement new features for mobile agents, security and application are important research area for researchers. The prime concern is attack on mobile agents. For providing security of agent's platforms (Host or User) by various threat attacks (Active as well as passive). Because of dynamic behaviour of mobile agents many types of attacks like as DOS, Masquerading, Hacking of agent and host, repudiation, eavesdropping, Modification of data and information etc.

Various types of attacks are as follows:-

- Agent to Agent's host,
- Agent to another agent,
- Agent's host to an agent.

Challenges in implementing Mobile Agents are Security, for securing agent platform and agents from misbehavior. Portability and Standardization, Performance and scalability.

1.2 Movable Agent's Process

The complete process of the agents as per Fig. 1 ensures that they are able to adapt the environment i.e. either home or foreign environment. They are able to switch among the positions of one node to other, and focused towards the final output.

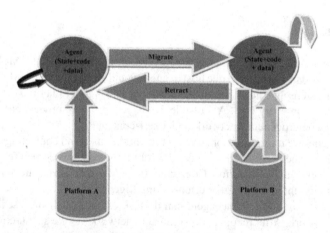

Fig. 1. Mobile agent's process

i. Generation: A newly agent is created and parameter of agent initialized.
ii. Reproduction: A duplicate movable agent is created and the parameter of the real is copy to cloning agent.

iii. Sending: An agent transfer to another host.

iv. Inactive: An agent is in sleep mode and its state is store in memory.

v. Activation: A deactivated agent is activated to lifecycle and its state is restored from memory.

vi. Retraction: A movable agent is ready for interaction with another agent and platform.

vii. Disposal: A movable agent is terminated from life cycle.

viii. Communication: Communication between an agent and host.

1.3 Tree Parity Key Generation Machine

Tree Parity key generation Machine is based on Neural Network having m input Neuron, n hidden neuron with single layer and one output neuron as shown in Fig. 2.

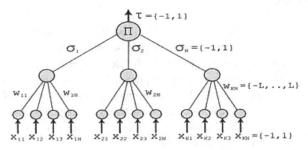

Fig. 2. Tree Parity key Generation Machine

Here, taking random input vector x generated by random no generator K*N the output of a tree parity machine is calculated by given formula:

$$x_{ij} \in \{-1, 0, +1\}$$

$$w_{ij} \in \{-L, \ldots, 0, \ldots, +L\}$$

$$\sigma_i = \text{sgn}\left(\sum_{j=1}^{N} w_{ij} x_{ij}\right)$$

$$\text{sgn}(x) = \begin{cases} -1 & \text{if } x < 0, \\ 0 & \text{if } x = 0, \\ 1 & \text{if } x > 0, \end{cases}$$

$$\tau = \prod_{i=1}^{K} \sigma_i$$

and if the outputs of Platform A and Platform B are equal, weights are update by using following learning rule:

- **Hebbian rule:**

$$w_i^+ = w_i + \sigma_i x_i \Theta(\sigma_i \tau) \Theta\left(\tau^A \tau^B\right)$$

- **Anti-hebbian rule:**

$$w_i^+ = w_i - \sigma_i x_i \Theta(\sigma_i \tau) \Theta\left(\tau^A \tau^B\right)$$

- **Random-walk:**

$$w_i^+ = w_i + x_i \Theta(\sigma_i \tau) \Theta\left(\tau^A \tau^B\right)$$

Tree parity machine have m*n weight matrix. Each weight are bound between lower bound and upper bound $\{-L\ldots\ldots, -5, -4, -3, -2, -1, 0, 1, 2, 3, 4, 5\ldots\ldots\}$.

2 Related Work

The work done by various researchers done in the field is as follows. Vogler *et al.* (1997) give framework for migrating agents provide security for agent's host and security for the movable agent. Framework provide fault tolerance at high level. To overcome these problem for mobile agents various cryptographic technique have been applied and a new frame work has been used for securing systems by applying OMG Object Transaction Service in the Framework. Jansen and Karayiannis (1999) discussed, variety of techniques and tools are available for providing security to agents. Not All the available techniques are suitable for many applications.

Alves-Foss and Harrison (2004) provides security for mobile agent using cryptographic function. Mukherjee *et al.* (2005) proposed private key exchange process by Lagrange polynomials using Shamir's threshold technique based on modular arithmetic.. This technique is highly secured. Kumar and Vatsa (2011) discussed a threshold cryptography based on Identity of users in MANET using Certification Authority. For authentication purpose used langrage's polynomial.

Mishra and Chaudhary (2012) provide method for securing migrating mobile agents and host. By this technique provide security at some extend. Homood (2012) discussed method for securing migrating agent behavior during transmission to share information among agents and hosts in the decentralized network. Proposed model based on the two state of AES and Lagrange key generation.

Ahila and Shunmuganathan (2014) discussed the security issues regarding mobile agent senario. These issues are security attacks, method how can we secure mobile and platform. Zrari *et al.* (2015) discussed to provide security is a most challenging issue in migrating mobile agent. So proposed the model for protecting mobile agents and agent's platform. Dhingra *et al.* (2016) focuses on security of information using Laplace transformation cryptographic technique. Karim (2018) proposed a new framework for betterment of mobile agent and platform security. Martinez Padilla (2018) described neural network-based Tree Parity key Generation Machines for securing encryption and decryption. Gang Wang (2010) proposed new intrusion detection Method based on artificial intelligence fuzzy clustering network known as FC-ANN. Niraj Singhal *et al.* (2017) proposed reliability based method to solve the issue of security. Provide a security model for migrating crawler. Platform oriented based approach used to provide security.

3   Problem Formulation

Studying the literature available for securing migrating agents shows that various techniques are available to provide security for migrating agents but none of the approach not provides a model that an effective security. So the main problem is how to provide high level security of agent's paradigm. New Frame work of secure migration agent is shown in Fig. 3.

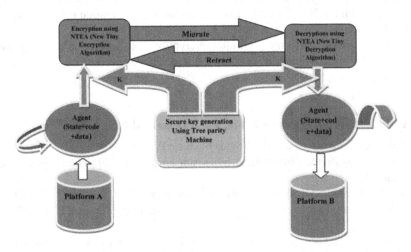

Fig. 3.   Secure agent migration framework

Secure Key Generation Using Tree Parity Machine

- Two parity machines used for Platform A and Platform B.
- Each parity machine for Platform A and Platform B start with random initial weight using random number generator.

Secure Key Generation Algorithm is as Follows

- Assign Common value for both platforms K, N, L.
- Using random number generator create weight matrix m*n for platform A and platform B.
- Assign common input Vector for platform A and platform B,
- Calculate the Output for both parity machine assign for platform A as well as platform B.
- Check output for platform A and platform B are same update the weight according to learning rule.
- Otherwise build random input Vector.
- If weight is same then generate session key.

Encryption Algorithm. Tiny Encryption Algorithm (NTEA).

NTEA Encryption consists of 8 bytes input and key size is 16 bytes. A unique global constant called magic constant is used for protecting attacks based on different number of rounds. Magic number is selected by $2^{32}/\varnothing$ where $\varnothing$ is the golden ratio.

Its properties are,

- It provides security through several rounds.
- Very high speed encryption Algorithm and provides security through several rounds.
- One single round of NTEA having 2 feistal operations, summation operation and bit wise XOR.
- Delta is calculated each time algorithm encryption and decryption methods were called and key size 16 bytes and two 8 bytes used. This is accessed each time the key is needed.

NTEA Block Diagram is Shown in Fig. 4 and Described As

- 8 bytes input is divided into two, 4 bytes blocks. (Y_i & Z_i).
- 16 bytes key is divided into four 4 bytes sub keys K [0...3]. These blocks were generated through following steps.
- 4 Bytes key generated in above is convert in to binary form. This binary number of 4 bytes is divided into four equal parts each size 1 byte part P1, part P2, part P3, part P4.
- Part P1 and Part P2 are used as least significant bit of key K1 & key K2 and Part P3 and Part P4 as most significant byte of KeyK3 & Keys K4.
- The resultant blocks were used in each round as shown in Fig. 4.
- The first round Z with the left shift of four is added to key Ko, Z is also added to delta$_i$, and Z also with the right shift of five is added to key K1.
- Addition and XORed operations were performed on input y and Z in each round.
- The resultant is the encrypted data known as cipher text.

NTEA Decryption

Decryption process of NTEA is similar to encryption process of NTEA but function are used in reverse order

- In decryption cipher text generated by encrypted process consider as input text which is final Y is written as input_Y_i and final_Z is written as the input_Z_i.
- key P(1...4) is generated through the process of key generation by using the formula of Lagrange interpolation explained above.
- The generation of blocks K(0...3) is same as of encryption process.
- The resultant blocks were used in each round as shown in Fig. 5.
- In the first round Z with the left shift of four is added to key K2, Z is also added to delta$_i$, and Z also with the right shift of five is added to key K3.
- Addition and XORed operations were performed on input Y and Z in each round.

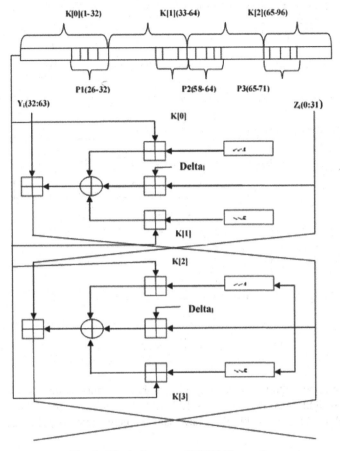

Fig. 4. Block diagram of NTEA Encryption

A full cycle of NTEA process repeated for 32 times to achieve the desired result of a full NTEA decryption. Finally result of decryption algorithm matched with the input of the encryption algorithm for verification. The NTEA decryption algorithm illustrated in Fig. 5.

Implementation and Results: Tree parity machine result based on the value of hidden layer (K), Input neuron (N) and weight (L). Key size of every transaction based on the no of hidden layers. Large no of hidden layer key size will increase as proportion of hidden layers provides higher security.

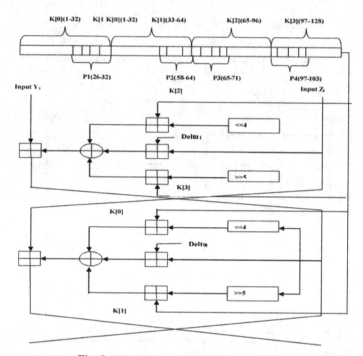

Fig. 5. Block diagram of NTEA Decryption

Initially, number of hidden layer are four, number of input neuron in parity machine are four and weight range is $\{-3, -2, -1, 0, 1, 2, 3\}$. After execution neural machine A and Neural machine B synchronized have same value generate key EXY as shown in Fig. 6.

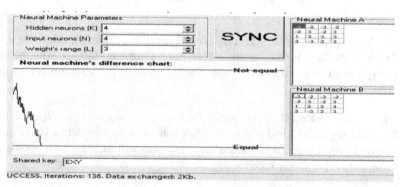

Fig. 6. Parity machine ($K = 4, N = 4, L = 3$)

In second experiment, number of hidden layer are five, number of input neuron in parity machine are four and weight range is $\{-4, -3, -2, -1, 0, 1, 2, 3, 4\}$. After execution neural machine A and Neural machine B synchronized have same value generate key XKOT as shown in Fig. 7.

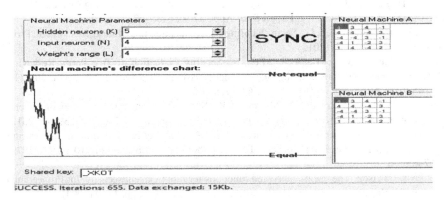

Fig. 7. Parity machine (K = 5, N = 4, L = 4)

4 Conclusion

Secure Mobile agent migration is of prime concern during the communication between two platforms. For secure communication tree parity machine based on neural network having K-input, N-hidden and one output neuron is used. Tree Parity Machine is highly secure algorithm for sharing of shared key(session key) which is used for encryption and decryption using New Tiny Encryption Algorithm. Here key length is larger if increase the number of hidden layers in Tree Parity key generation Machine. More no of Hidden layer generates large key length that provide higher security.

References

Ahila, S.S., Shunmuganathan, K.L.: Overview of mobile agent security issues. In: International Conference on Information Communication and Embedded Systems, pp. 1–6. S. A. Engineering College, Chennai (2014)

Dhingra, S., Savalgi, A.A., Jain, S.: Laplace transformation based cryptographic technique in network security. Int. J. Comput. Appl. **136**(7), 6–10 (2016)

Wayne, J., Tom, K.: NIST Special Publication 800-19–Mobile Agent Security, pp. 1–52. National Institute of Standards and Technology, Computer Security Division, Gaithersburg (1999)

Karim, M.R.: Security for mobile agents and platforms: securing the code and protecting its integrity. J. Inf. Technol. Softw. Eng. **8**, 1–3 (2018)

Kumar, P., Vatsa, A.K.: Novel security architecture and mechanism for identity based information retrieval system in MANET. Int. J. Mob. Adhoc Netw. **1**(3), 397–404 (2011)

Amit, M., Anamika, C.: Mobile agent: security issues and solution. Int. J. Comput. Technol. Electron. Eng. **2**(6), 30–32 (2012)

Vogler, H., Kunkelmann, T., Moschgath, M.L.: An approach for mobile agent security and fault tolerance using distributed transactions. In: Proceeding of the 1997 International Conference on Parallel and Distributed Systems, Germany, pp. 268–274 (1997)

Wang, G., Hao, J., Ma, J., Huang, L.: A new approach to intrusion detection using Artificial Neural Networks and fuzzy clustering. Expert Syst. Appl. **37**(9), 6225–6232 (2010)

Chadha, Z., Hela, H., Khaled, G.: Security during communication in mobile agents system. Proc. Comput. Sci. **60**, 17–26 (2015)

Martínez Padilla, J., Meyer-Baese, U., Foo, S.: Security evaluation of Tree Parity Re-keying Machine implementations utilizing side-channel emissions. EURASIP J. Inf. Secur. **2018**(1), 1–16 (2018). https://doi.org/10.1186/s13635-018-0073-z

Singhal, N., Dixit, A., Agarwal, R.P., Sharma, A.K.: Need of securing migrating crawling agent, remote platform and the data collection. Int. J. Electron. Comput. Sci. Eng. **2**(1), 39–45 (2012)

Singhal, N., Dixit, A., Agarwal, R.P., Sharma, A.K.: A reliability based approach for securing migrating crawlers. Int. J. Inf. Technol. **10**(1), 91–98 (2017). https://doi.org/10.1007/s41870-017-0065-0

Alves-Foss, Harrison: The use of encrypted functions for mobile agent security. In: Proceedings of the 37th Hawaii International Conference on System Sciences, pp. 1–10 (2004)

Mukherjee, A., Deng, H., Agrawal, D.P.: Distributed pair wise key generation using shared polynomials for wireless ad hoc networks. In: Belding-Royer, E.M., Al Agha, K., Pujolle, G. (eds.) MWCN 2004. IFIPAICT, vol. 162, pp. 215–226. Springer, Boston (2005). https://doi.org/10.1007/0-387-23150-1_19

Homood, K.: Lagrange interpolation for mobile agent connection encryption. Ibn Al-Haitham J. Pure Appl. Sci., 1–12 (2012)

An Approach to Waste Segregation and Management Using Convolutional Neural Networks

Deveshi Thanawala[✉], Aditya Sarin, and Priyanka Verma

Mukesh Patel School of Technology, Management and Engineering, Electronics and Telecommunication, Mumbai 400054, India
deveshi.m.thanawala@gmail.com, aditya28sarin@gmail.com, priyanka.verma@nmims.edu

Abstract. Population explosion in India has led to an outburst of some major concerns, one of which is the waste generation and disposal system. India is accountable for producing 12% of the global municipal waste. As a result, waste is collectively dumped irrespective of the type leading to drainage blockages, pollution and diseases. In this paper, we have proposed a model for waste segregation that uses neural networks to classify waste images into three categories namely recyclable, non-recyclable and organic. A training accuracy of 83.77% and testing accuracy of 81.25% was obtained. Along with the proposed neural network, five standard CNN architectures – VGG-16, Dense-Net, Inception-Net, Mobile-Net and Res-Net are also tested on the given dataset. The highest test accuracy of 92.65% was obtained from the Mobile-Net classifier. The paper also proposes an on-site waste management system using 8051 micro-controller and GSM technology.

Keywords: Convolutional Neural Network · Classification of waste images · GSM technology · 8051 micro-controller

1 Introduction

It is a known fact that Asia generated more than 3 billion tons of solid waste in 2000, which might rise to 9 billion tons by 2050 [1]. During the 2019 monsoon season, the Arabian sea gifted the city of Mumbai with 2,15,000 kg [2] of waste indicating that it's time for effective waste management.

The waste collection process is a critical aspect of efficient waste disposal methods. The traditional way of manually monitoring the wastes in bins is a complex, cumbersome process and utilizes more human effort, time and cost which can be made easier with present-day technologies. Irregular management of waste typically domestic, industrial and environmental is the root cause for many problems such as pollution, diseases and has adverse effects on the hygiene of living beings. To overcome such problems, this paper proposes the idea of waste segregation and management system which works without human interaction in order to maintain a clean environment. This concept is

© Springer Nature Singapore Pte Ltd. 2020
M. Singh et al. (Eds.): ICACDS 2020, CCIS 1244, pp. 139–150, 2020.
https://doi.org/10.1007/978-981-15-6634-9_14

implementable in cities where waste production is domestically high but the effort to control it is relatively very low. The model proposed mainly avoids congested collection of waste generated domestically whose disposal is difficult to manage.

The proposed model is designed not only for segregating waste into different categories but also to ensure efficient management of waste disposal on-site. The model is developed in 3 stages. In the first stage, we have designed our own CNN model and tested it on the WCD dataset. In the second stage, Convolutional Neural Network-based transfer learning models like VGG16, DenseNet, MobileNet, InceptionNet and ResNet are tested on the same dataset. In the third stage, an on-site mechanism has been proposed which will track the level of the garbage bin on a real-time basis and would send a notification to the concerned authority as soon as the bin reaches a certain limit along with the location of the bin.

The structure of the paper is as follows: Sect. 2 explains the literature survey. Section 3 explains System Description. Section 4 proposes the waste management model. Section 5 lists the results. Section 6 includes the conclusion.

2 Literature Survey

Paper [3] has an auto-clean feature that can be used to keep the dustbin clean. This model also analyses the amount and type of garbage disposed of by the user (Daily, weekly, monthly). However, the bin has an automatic lid that can open even when someone is walking close to the bin. Also, the segregation of waste is done using capacitive plates and copper plates. Capacitive sensors are sensitive to temperature and humidity change which could affect it's working. Paper [4] uses a cloud system that is set up using Long-range (LoRa) technology and Message queuing telemetry transport (MQTT) protocol that updates real-time data. The Speed up robust features (SURF) algorithm is used for the conversion of a three-dimensional image into a two-dimensional image. Paper [5] integrates an 8051 microcontroller and uses 'Caffe' which is a standard deep learning framework. Since it is a self-learning algorithm, the system can learn and train by itself thus does not require manual intervention.

Paper [6] Uses Capsule Neural network which is a type of Artificial Neural Network which mimics biological neural system in humans. The experimentation is done on two datasets-private household images and public waste images.

Paper [7] develops a transfer learning model based on Convolutional Neural Network (CNN) using classifiers like VGG-16, ResNet-50, Mobile-NET V2, DenseNET-121. From 4 architectures 8 learning approaches are used i.e. 4 waste-item classifiers and 4 waste-type classifiers. There are 20 waste-item classes and four waste-type classes in the dataset. Paper [8] uses CNN and support vector machines (SVM) using nonlinear mapping functions: polynomial, sigmoid and gaussian RBF. The image classification was performed quickly with an average classification time of 0.1 s and a standard deviation of 0.005 s. Paper [9] uses Recycle-Net optimized deep CNN architectures such as Res-NET, Mobile-NETs, Inception Res-NET and Dense-NETs. Paper [10] trained on the ImageNet Large Visual Recognition Challenge dataset attaining a final test accuracy of 87.2%.

3   System Description

The objective of the research is:

- To design a convolutional neural network that can classify waste images into recyclable, non- recyclable and organic categories.
- To compare classification performance of the designed neural network with five standard CNN architectures.

A. Data Collection

The dataset used in the transfer learning model, as well as the designed CNN model, has been taken from Kaggle [11]. It is waste classification dataset which consists of about 25,000 images of waste, divided into 3 categories namely: Organic, Non-recyclable and recyclable. Table 1 depicts the number of images in each class and Fig. 1 shows a pie-chart representation of the same.

Table 1. Number of waste images in each class

Waste type	Number of images
Organic waste	13,966
Non-recyclable waste	3,243
Recyclable waste	8,664
Total	25,873

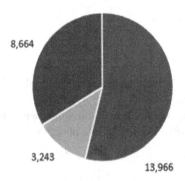

Fig. 1. Number of images in each class

The proportion in which the training and testing images are divided is 80:20. The training dataset is labeled into 3 classes. However, the testing images are not labeled and are from one single test folder.

The organic class consists of images of wasted fruits, vegetables, meat, pulses, etc. The non-recyclable class consists of images of bottles, plastic bags, chemical substances of mineral origin, etc. Recyclable class consists of images of paper, cardboard, metal, electronics, etc. Table 2 depicts some of the example images in our dataset.

Table 2. Example images in our dataset

B. Classification Model

A CNN based classification model has been designed for waste segregation. The neural network consists of four Convolutional layers, four Maxpooling layers, one flatten layer and two fully connected layers.

The Neurons are the basic computational unit which when stacked in a single line forms layers of a neural network. The layers in a neural network are broadly classified into 3 types:

- Input layer
- Hidden layer
- Output layer

Figure 2 illustrates the proposed Convolutional Neural Network model. It is explained in detail as follows:

$$z = wx + b \tag{1}$$

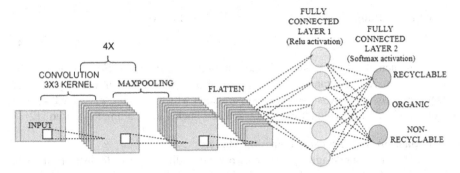

Fig. 2. Layers in the designed neural network

In Eq. (1) each variable takes an input 'x' and multiplies it with a variable 'w'. Some bias variable 'b' is added to give an output 'z' [12].

The output 'z' is then passed to an activation function. The activation function will introduce non-linearity in the system. Non-linearity of the system is required as linear functions won't be able to learn and analyze complex functions. Such activation functions also provide differentiability which is useful during backpropagation. The activation function used in this model is the 'Rectified Linear Unit' (ReLu) as the function and derivative, both are monotonic.

ReLu is represented by Eq. (2)

$$R(z) = \max(0, z) \tag{2}$$

The final layer of the model uses the SoftMax activation function as it limits the output between 0 to 1. Since it is a squashing function it allows the output to be interpreted directly as a probability, hence it is used in the final layers of the model [12].

The convolutional layer is the first layer in the proposed model which learns the image features using small kernels of input data. Kernel size of 3×3 is used in the convolutional layer. A small kernel size adds a benefit to the model in the reduction of computational power and balances weight sharing. The number of output filters(F) is the number of neurons in that convolutional layer. Starting with 32 it goes on increasing till 128 so that the convolutional layer can extract larger depths of the image characteristics. No padding(P) is used in the convolutional layer so the output has the same length as the input matrix. Output size is given by Eq. (3).

$$Output\ size = N - F + 2P/S + 1 \tag{3}$$

Where S is the stride, which is the movement of the kernel from pixel to pixel. A default stride of '1' was given. Max pooling layer is used after the convolutional layer to reduce the spatial size of the output. A maximum operation is performed on a matrix of size F $\times$ F. Two dense layers are used, these dense layers are fully connected i.e. each input is connected to every other output with a specified weight. Dense layers are also followed by a non-linear activation function [12].

The input image size is 224 $\times$ 224. Adaptive moment estimation (Adam) was used to optimize the model. It uses estimations of first and second moments of the gradient to adapt the learning rate for each weight of the neural network.

$$m_n = E[X^n] \tag{4}$$

The mathematical expression of Adam optimizer is shown in Eq. (4), where 'm' is the moment, X is the random variable and 'n' represents the number of gradients. Categorical cross-entropy is used as a loss parameter which is a multi-class log loss. The model is divided into 3 categorical targets. Hence, such a loss parameter is used [12].

C. Comparison with Transfer Learning Models

We have compared our designed neural network with 5 different transfer learning models which are briefly discussed below. Figure 3 explains the general flow of the following transfer learning networks.

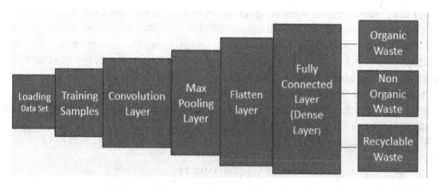

Fig. 3. Generic block diagram of standard CNN architectures

VGG-16

Vgg-16 is a standard CNN architecture. It consists of Convolutional layers, Max Pooling layers, some Activation functions and finally fully connected layers or Dense layers. There are 13 convolutional layers, 5 Max Pooling layers and 3 Dense layers which sums up to 21 layers but have only 16 weight layers.

Res-Net

Res-Net architecture has a fundamental building block (Identity) where a previous layer is merged into a future layer. By adding additive merges, the network can learn residuals (errors i.e. diff between some previous layer and the current one).

Mobile-Net

Mobile-net is a lightweight architecture. It uses depth-wise separable convolutions. Thus, it performs a single convolution on each colour channel instead of combining all three and then applying flatten layer on it. Thus, a filter is applied to the input channels.

Inception-Net

The 3rd version of the InceptionNet is used which consists of 9 stacked inception modules. The version 1 of the network proposed the model which was 27 layers deep. The overview of such a model was to replace full connected network architectures with sparsely connected network layers.

Dense-Net
The dense-net consists of very narrow layers that add a small set of new features maps. However, this model provides strong gradient flow and increased parameter efficiency and also maintains low key complexity features.

4 Proposed Waste Management Model

The proposed waste management model focuses on on-site waste management. Even after the segregation of waste into the designated categories, the method might not work if authorities fail to arrive in time for treating the waste. Waste thrown in bins accumulates leading to overfilling and ill-treatment. As a result, waste is squandered on roads and never treated leading to foul smell hence disrupting the eco-system.

A. Model Specifications
8051 Micro-controller: The 8051 micro-controller consists of 4 ports with 2 internal timers, 128 bytes of RAM and also supports full-duplex UART protocol.
Ultrasonic sensor: The ultrasonic sensor works on the SONAR principle and consists of a transmitter and receiver, thus providing time for the ultrasonic waves to strike an obstacle and return.
GSM module: The GSM technology uses AT commands which transmits data using radio waves. GSM is used to establish a connection with the electronic application. Tn this case, it interfaces with the micro-controller.
GPS module: The global positioning system is based on satellite navigation technology and provides information about location in real time.
Servo motor: A servo motor is either digital or continuous based on the application and is used to provide a rotatory mechanism.
IR sensor: An IR sensor uses an infrared signal to detect the presence of any obstacle in its path. The range of an IR sensor is generally very small.

B. Flow of the Model
The segregated waste enters the respective bins as recyclable, organic and non-recyclable. Each bin consists of a micro-controller which forms the foundation of the proposed system. The ultrasonic sensor connected to the microcontroller is programmed to calculate the distance from its transmitter to the obstacle. The sensor is attached below the bin lid, in such a way that the transmitter and receiver of the sensor is facing towards the bottom of the garbage bin (Fig. 4).

The ultrasonic sensor continuously measures real-time data and sends it to the microcontroller. The microcontroller is programmed in such a way that as soon as the threshold reaches 75% of the distance from the surface of the bin, it sends a trigger signal to the GSM module which acts as the base station subsystem, forming the base transceiver station, this station then sends the signal to the concerned authority which has a mobile

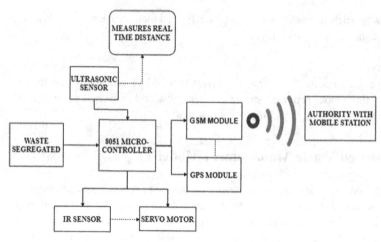

Fig. 4. Flow of the hardware model

station (mobile phone) and would receive an alert message along with the expected time until which the dustbin is about to be full. The GPS module is used to get the latitude and longitude of the current location which is then converted and the exact location of the garbage bin is sent to the concerned authority.

We have also used an IR sensor that works with the servo. A high torque servo motor is used to rotate the lid of the garbage bins. A high torque servo motor is preferred in compliance with varying lid sizes and weights. An IR sensor is placed just in front of the lid, which is set to a range of 15 cm. Such a range is chosen so that it is not too little for the user to nearly touch the bin and not large enough such that the bin lid opens even if a person walks by.

5   Results

The highest testing accuracy of each of the standard CNN models was obtained with 30 epochs and a batch size of 50. The model has been trained and tested with different architectures, epochs, batch sizes and augmentation.

A history of training and testing accuracies for all five classifiers namely VGG-16, Dense-Net, Inception-Net, Mobile-Net and Res-Net are listed in Table 3.

Table 3. The accuracy obtained by using standard CNN architectures

Method	Epochs	Batch size	Augmentation	Train accuracy	Test accuracy
Vgg-16	10	50	None	82	72
	30	50	None	96	75
	30	10	None	87	72
	10	10	Yes	86	85
	30	**50**	**None**	**96.04**	**87.41**
Dense-Net	**30**	**50**	**None**	**97.4**	**91.36**
	30	20	None	89.56	89.25
	30	20	Yes	87.12	90.17
	10	50	None	85.69	90.12
Inception-Net	30	10	None	85.25	80.12
	10	10	None	83.97	81.01
	30	**50**	**None**	**96.17**	**86.71**
	30	50	Yes	96.55	83.15
Mobile Net	30	10	None	87.63	84.19
	10	10	None	82.12	81.15
	30	**50**	**None**	**99.52**	**92.65**
	30	50	Yes	98.12	92.61
Res-Net	**30**	**50**	**None**	**98.14**	**91.38**
	30	20	None	87.12	87.10
	30	20	Yes	86.15	88.12
	10	50	None	90.17	88.66

Our designed model was trained with 10 epochs and a batch size of 50 attaining a training accuracy of 83.77% and a testing accuracy of 81.25%. The history of training and testing accuracies of the designed neural network is listed in Table 4 (Table 5).

Table 6 above compares the designed model with standard CNN architectures on the basis of their highest test accuracies. Figure 5 depicts bar chart representation for the same.

Table 4. Accuracy of proposed CNN architecture

Epochs	Batch size	Augmentation	Train accuracy	Test accuracy
5	10	None	72.18%	68.23%
7	10	None	75.67%	68.35%
10	10	None	78.19%	70.02%
10	10	Yes	78.56%	70.00%
10	30	None	79.22%	76.12%
10	50	None	78.12%	78.10%
10	40	None	80.48%	78.18%
10	35	None	83.72%	81.22%
10	**50**	**None**	**83.77%**	**81.25%**
10	35	Yes	84.00%	80.98%

Table 5. Comparison of model with standard CNN architectures (with 10 epochs)

Models	Accuracy (with 10 epochs)
Proposed neural network	81.25
Dense-Net	90.12
Inception-Net	81.01
Mobile-Net	84.19
Res-Net	88.66
Vgg-16	85.00

Table 6. Comparison of proposed model with standard CNN architectures

Architecture	Highest accuracies
Proposed neural network	81.25
Dense-Net	91.36
Inception-Net	86.71
Mobile-Net	92.65
Res-Net	91.38
Vgg-16	87.41

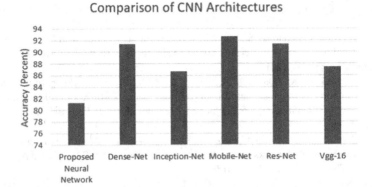

Fig. 5. Bar chart representation of accuracies of various models

6   Conclusion

Waste segregation and management has been a problem for a long period of time and has affected a large portion of the eco-system. By using present day technologies, it is easier to manage waste, if efficiently used. The proposed model segregates waste images with the highest accuracy of 92.65% among the transfer learning models and an accuracy of 81.25% with our designed neural network. The model also proposes on-site management of waste.

However, we were not able to train the designed model up to 30 epochs due to large requirement for computational power.

In the future, we plan to integrate the on-site mechanism with waste segregation using computer vision and design a complete model with a robotic segregator.

References

1. Amit, R.: Waste management in developing asia. J. Environ. Dev. **17**(1), 3–25 (2008). https://doi.org/10.1177/1070496507310742
2. https://www.cntraveller.in/story/no-thanks-says-sea-spits-9-tonnes-trash-back-mumbai/
3. Pereira, W., Parulekar, S., Phaltankar, S., Kamble, V.: Smart bin (waste segregation and optimisation). In: Amity International Conference on Artificial Intelligence (AICAI), Dubai, United Arab Emirates, pp. 274–279 (2019)
4. Shamin, N., Mohamed Fathimal, M.P., Raghavendran, R., Kamalesh, P.: Smart garbage segregation & management system using Internet of Things (IoT) & Machine Learning (ML). Department of Computer Science & Engineering, SRM Institute of Science & Technology, Chennai (2019)
5. Sudha, S., Vidhyalakshmi, M., Pavithra, K., Sangeetha, K., Swaathi, V.: An automatic classification method for environment: friendly waste segregation using deep learning. In: IEEE Technological Innovations in ICT for Agriculture and Rural Development (TIAR), Chennai, pp. 65–70 (2016)
6. Sreelakshmi, K., Akarsh, S., Vinayakumar, R., Soman, K.P.: Capsule neural networks and visualization for segregation of plastic and non-plastic wastes. In: 5th International Conference

on Advanced Computing & Communication Systems (ICACCS), Coimbatore, India, pp. 631–636 (2019)

7. Srinilta, C., Kanharattanachai, S.: Municipal solid waste segregation with CNN. In: 5th International Conference on Engineering, Applied Sciences and Technology (ICEAST), Luang Prabang, Laos, pp. 1–4 (2019)

8. Sakr, G.E., Mokbel, M., Darwich, A., Khneisser, M.N., Hadi, A.: Comparing deep learning and support vector machines for autonomous waste sorting. In: IEEE International Multidisciplinary Conference on Engineering Technology (IMCET), Beirut, pp. 207–212 (2016)

9. Bircanoğlu, C., Atay, M., Beşer, F., Genç, O., Kızrak, M.A.: RecycleNet: intelligent waste sorting using deep neural networks. In: 2018 Innovations in Intelligent Systems and Applications (INISTA), Thessaloniki, pp. 1–7 (2018)

10. Rabano, S.L., Cabatuan, M.K., Sybingco, E., Dadios, E.P., Calilung, E.J.: 2018 IEEE 10th International Conference on Humanoid, Nanotechnology, Information Technology, Communication and Control, Environment and Management (HNICEM). 978-1-5386-7767-4

11. https://www.kaggle.com/sapal6/waste-classification-data-v2

12. https://cv-tricks.com/tensorflow-tutorial/training-convolutional-neural-network-for-image-classification/amp/

Open Source Intelligence Initiating Efficient Investigation and Reliable Web Searching

Shiva Tiwari$^{(\boxtimes)}$, Ravi Verma$^{(\boxtimes)}$, Janvi Jaiswal$^{(\boxtimes)}$, and Bipin Kumar Rai$^{(\boxtimes)}$

Department of IT, ABES Institute of Technology, Ghaziabad 201009, Uttar Pradesh, India
shivatiwari757@gmail.com, ravipcm2000@gmail.com,
janvijaiswal104@gmail.com, bipin.rai@abesit.in

Abstract. Open Source Intelligence (OSINT) is the collection and processing of information collected from publicly available or open-source web portals or sites. OSINT has been around for hundreds of years, under one name or another. With the emergence of instantaneous communication and rapid knowledge transfer, a great deal of actionable and analytical data can now be collected from unclassified, public sources. Using OSINT as the base concept, we have attempted to provide solutions for two different use cases i.e. the first is an investigation platform that would help in avoiding manual information gathering saving time and resources of information gatherers providing only the relevant data in an understandable template format rather than in graphical structure and focuses on demanding minimal input data. The second is a business intelligence solution that allows users to find details about an individual or themselves for business growth, brand establishment, and client tracking further elaborated in the paper.

Keywords: Open Source Intelligence (OSINT) · Machine learning · AI · Investigation · Security · Automation · Web crawling

1 Introduction

Security researcher Mark M. Lowenthal defines OSINT as "any and all information that can be obtained from the overt collection: all media types, government reports, and other files, scientific research and reports, business information providers, the Internet, etc." [1].

The major help that Open Source Intelligence does is the wide variety of information it can give which is not restricted to only a single format such as text or image but the entire possible and available format of data can be extracted from a publicly accessible domain such as audio, video, etc.

In this paper, we are aiming to provide digital solutions that would help in collecting information about the targeted entity through a single platform, saving most importantly time.

For the first one, we have proposed a solution named OSINTEI (Open Source Intelligence for Efficient Investigation) that helps in the investigation and data extraction of a

© Springer Nature Singapore Pte Ltd. 2020
M. Singh et al. (Eds.): ICACDS 2020, CCIS 1244, pp. 151–163, 2020.
https://doi.org/10.1007/978-981-15-6634-9_15

target host, particularly administrative officers as they are responsible for all the administrative duties and root development of the nation. Government officials are responsible for the development of the nation and its citizens. But what if the officials who are looked upon for carrying out administrative responsibilities, involve in owning illegal assets, show abnormality in expenses, have eye-catching work behavior, etc. For such officials, information gathering is started by looking into various different records and files available in public records present on governmental and web portals. The process takes a lot of time and it all goes to nothing when nothing suspicious is found. The time consumed in such cases could have been used for other tasks too.

Investigation being the most crucial, time and effort consuming, cost absorbing phase which is done manually and is being done the same way for ages. This traditional way costs a lot than just money [2]. Even the risk of life for most of the investigation team officials. This has risen the need for an automated investigation platform that is the product of a cohesive technological advancement which reduces all the above-mentioned investments and results in a digital age information-gathering protocol that may ensure efficient investigation.

OSINT has been used for various other tools that fulfil searching, data extraction and compilation goals earning fame and wide user demand due to the compatibility, efficiency, and ease of use they provide [3].

For the next one, we have proposed a solution named OSINTSF (Open Source Intelligence as Social Finder) that aims to provide a tool that makes searching and finding details/information about an entity easy and accessible. Also, to benefit businesses with the search capability and would allow content to be searched present on social networking sites in real-time and provide profound analytical data.

2 Related Work

J. Pastor-Galindo, P. Nespoli, F. Gómez Mármol and G. Martínez Pérez, [4] in "The Not Yet Exploited Goldmine of OSINT: Opportunities, Open Challenges and Future Trends," – 2020, described the existing state of OSINT and gave a detailed review of the system, concentrating on the methods and strategies that strengthen the area of cybersecurity. They shared their views and notes on the problems that need to be solved in the future. Also, they researched the role of OSINT in the governmental public domain, which is an ideal place to exploit open data. The paper lacked in providing an approach that formulated the development of a solution that could help validate their idea to save open-source data. Unlikely, our paper concentrates on the development of platforms that solve applicable use-cases mentioned ahead.

In "The Evolution of Open Source Intelligence" - 2010, Florian Schaurer and Jan Störger [5] very clearly discuss the reason, history, and challenges behind the evolution of OSINT. The adoption of Open Source Intelligence by the private sector or agencies such as CIA and its partnerships for development in the field has also been discussed. It fulfills the theoretical details of evolution but lacks in the information related to the possible practical application.

3 Proposed Solution

The Investigation [6] and policing methodologies have evolved with time. But neither has been embedded totally with the technology.

The first proposed solution aims to provide a complete one destination platform for the entire information extraction investigation, focusing on the targeted individual and delivering miscellaneous as well as sorted data to the researcher. Data being either an age ago news coverage or the current financial status, this product would search each and every single module present in the entire webspace and find out the relevant information about the host needed for the investigation. Once complete, it would help to attain all the necessary information or data that are needed to find a direction for investigation, saving time that manual information gathering consumes. Instead of manually searching into different online sources, making a report on the same and taking days for a single task, they can use our solution to find relevant data about the target in few seconds and would get a data-filled report via a template on the same. Data is extracted from various public domains that can be legally used, government websites such as Supremo where data gets updated each year and is accurate. So, only the rich data would be extracted and would help in the investigation or information gathering. Making the primary stage of Investigation an easy task with efficient and reliable results.

The second solution aims to help businesses that keep records of their clients and seek new interested hosts. Would help individuals keep track of their social and web presence. Social media is becoming a crucial part of digital communication strategies. It is now an effective tool not only to improve brand loyalty and win new clients but also to strengthen the customer service by allowing businesses to access the social media networks to establish relationships and expand the span of their interactions.

It would also help businesses that work in insurance, banking or any other invest-ments industry to build peer-to-peer networks to meet non-contactable consumers whose renewal or incentive or maturity programs rest unclaimed. They can use this solution to find details about the customers and contact them.

Both of these solutions ask for minimal resource requirements for usage and also minimal input data about the target. With only a single laptop/device having a connection to it and you get the results immediately.

Being the technology, which has not been used commonly, OSINT [7] has a lot more in its treasure of usability that can be scraped out to create a software product with higher usability strength [8]. [9] The solution, being a software product would use Machine learning and Artificial Intelligence, classification and regression algorithms such as the Naïve Bayes algorithm. Selenium [10], which is an efficient and portable framework used for crawling and testing web applications and ensuring quality would also be used. The solution uses JWT (Java Web Tokens) for session management bringing security and uses microservices to upgrade the scalability of the product. A detailed description of the technologies used and their roles are present ahead.

Both the solutions use a similar technological stack but are different when it comes to their functionalities and use cases. Facial/pictorial data or image can also be used as an alternative input for investigation but only if name (being the primary input) is unavailable. This feature increases the ease of use and broadens functionalities of the solution (Fig. 1).

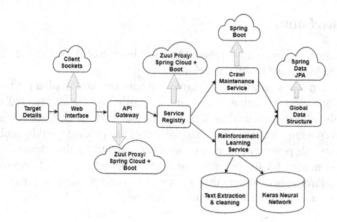

Fig. 1. Technological design of the solution.

The proposed solution works by following a particular algorithm Global Search (GS). This GS is used as a Global data structure that contains all the various details of a particular person, a foreign agent or a group.

3.1 Global Search (GS)

This GS Algorithm further gets divided into four smaller sub algorithms,

1. GS-Crawl
2. GS-Extraction
3. GS-Reinforce
4. GS-Template.

3.1.1 OSIGS-Crawl

In this part, a single keyword is considered as an input parameter which is generally supposed to be the name of that individual or group to be search on the basis of that input the crawler goes into the web and find digital footprints of the targeted individual or group. This algorithm returns a JSON of a Global Data structure which can be new if the target is unavailable on public domains.

- **GS-Crawl(Pi)**

```
driver unit.
Search_t ← Pi.
If(! Search_t)
        cw←driver instance
        for each i in S:
                temp←S[i] □ i←S
                cw □ temp.
                1st ← cw
```

```
            call insertIntoGS.
    db_init(JSON_F).
```

- **insertintoGS(cwi)**

```
    retrieve cw_i.
    Push into Global Stack(GS).
        if (GS_count < 0)
            Return null
        else
            Return GS(r1,r2,r3.........rn)
    G_Stack←GS(r1,r2,.........rn)
    If G_Stack is null:
        goto  1.
    else
        item_i ← pop(G_Stack).
        serialize (item_i).
        JSONs ← serialized(item_i)
        goto 4.
    return JSONs.
```

3.1.2 OSIGS-Extraction

OSIGS-Extraction algorithm takes input of JSONs and provides output that as keywords of the JSONs file in JSONF file using the following formula:

$$GS\text{-}TF(t,d) = ft,d/(\text{no. of words in } d)$$

$$GS\text{-}IDF(t,d) = \log(N \ /\{d \ E \ D: t \ E \ d\}$$

$$GS\text{-}E(t,d) = GS\text{-}TF(t,d) * GS\text{-}IDF(t,d).$$

where,

f = frequency of letter in the document.
d = JSONF document.
D = total no. of JSONF documents.
N = no. of d in which t occurs.

3.1.3 OSIGS-Reinforce

OSIGS-Reinforce algorithm takes the input from the extraction maintenance service as a JSON and builds a date set after deserializing the response. The OSIGSR behaves as a Rest end point consumer for the processing of consuming JSONF in order to train the model.

The various results that have been gathered from the public domains against the target are used as different parameters in order to train the learning model.

- **OSIGS-Reinforce(JSONF)**

```
init φ (JSONF, t).
Ji E JSONf, Ji □ JSONF:
        temp = Ji
        for each (i in j):
                Select t from JSONf
                Do trigger t,
                watch output 0 and next Ji+1
φ(Ji, t)←Q(Ji, t) + ß [ O + ρ . maxß, φ(Ji', t') - φ(Ji, t)].
        Ji ← Ji'
Push JSON[Ji1', Ji2'.........Jin'] in db.
```

Explanation of this algo

Initialize the φ value i.e., φ(JSON, trigger) then watch the current state JSONi choose a trigger it, only belonging to the Ji. Now provide the output 0 and watch out for new or next state Ji + 1. Update φ values until all values of JSON are exhausted. After this the new and approved results of the JSON[Ji1', Ji2', Ji3', Ji4'....... Jn'] will be reduced by the reinforcement learning service, which will go into the DB service.

3.2 Concepts and Technologies Used

3.2.1 High Level Design

It is an architecture that is used for software application development. The architecture diagram provides an overview of a system as a whole, defining the key components that would be developed for the product and its interfaces (Figs. 2 and 3).

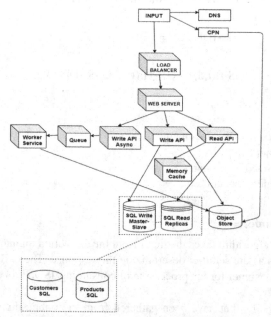

Fig. 2. Design for handling scalability.

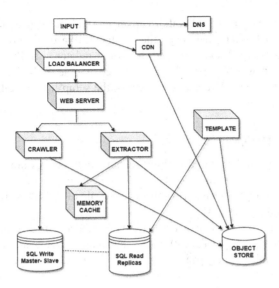

Fig. 3. Crawler flow and extractor flow.

3.2.2 Kafka

Kafka architecture is being used here as it provides higher throughput, speed, scalability, reliability and replication characteristics for any real-time streaming data architectures, big data collection or can provide real-time analytics [11].

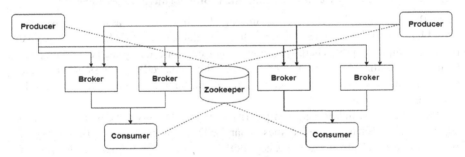

Fig. 4. Kafka architecture

Kafka is made up of Records, Topics, Consumers, Producers, Brokers, Logs and Clusters. Records can have base, meaning and timestamp. Kafka archives are the unchangeable. A Kafka Subject is a database stream ("/orders," "/user-signups"). On many servers the Kafka Cluster consists of many Kafka Brokers. Broker also refers to more of a logical system, or to the entire Kafka system.

To handle the cluster Kafka uses ZooKeeper. ZooKeeper is used to organize the topology for the brokers/clusters. ZooKeeper is a reliable, configuration information file system. ZooKeeper is used by Broker Subject Partition Members for leadership election [12].

Kafka uses Zookeeper to hold Kafka Broker and Subject Partition pair leadership elections. To Kafka Brokers that shape the cluster, Kafka uses Zookeeper to handle the application discovery. Zookeeper sends topology updates to Kafka, so every cluster node learns when a new broker joined, a Broker died, a topic was deleted or a topic was introduced, etc. Zookeeper offers a sync view of the setup for the Kafka Cluster [13] (Fig. 4).

3.2.3   Consistent Hashing

Consistent Hashing is a distributed hashing scheme that operates in a distributed hash table independently of the number of servers or objects by assigning them a position on an abstract circle, or hash ring. This allows for scale of servers and objects without affecting the overall system [14].

Consistent hashing is based on mapping each object to a point in a circle (or mapping each object to a real angle, equivalently). The system maps each machine (or other storage bucket) that is available to many pseudo-randomly distributed points on the same circle.

3.2.4   Data Set and Training of the Model

The data being extracted from the public and authentic government sources would be converted into data sets that would be used to train our model which further would help us to relevantly classify between the most relevant new link or information which would be added to the template (Fig. 5).

3.2.5   Relevancy Factor and Data Classification via Naïve Bayes Theorem

Naive Bayes is a simple technique for building classifiers: models assigning class labels to problem instances defined as vectors of feature values, where the class labels are taken from some finite set [15]. There is no single algorithm for training such classifiers, but a family of algorithms based on a common principle: all naive Bayes classifiers conclude that, given the class variable, the value of a particular feature is independent of the value of any other attribute.

To check which link to be given priority of being shown via the template over the other, the Naïve Bayes theorem comes in handy. The crawler extracts various links that are yet to be checked for the relevancy. For the check, all the links are checked and compared with the trained data searches for the probability of relevancy for each of the query link.

The link with the highest probability is then chosen to be showcased in the template.

After all the links are compared, as being shown in all above graphs we can notice a link showing higher relevancy probability ratio than the others in each graph and so they'll be used in the template that would ensure higher efficiency and information relevancy.

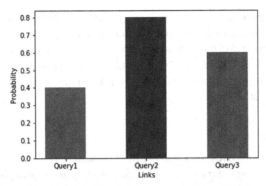

Fig. 5. Relevancy factor identification (*For the above graph, it is easily clear that Query 1 has the highest probability of relevancy than Query 1 or 3 and so, Query 2 will be added to the template.*)

3.2.6 Using Spring Boot to Create Micro Service

Spring Boot, being a Java-based open source framework, helps create micro Service [16]. It would make our solution more scalable.

It offers a flexible way to configure Java Beans, XML, and Database Transactions. This offers efficient batch processing and REST endpoints management. Everything is auto-configured in Spring Boot; no manual settings are required. It offers Spring application based on annotation. Managing reliance eases. This requires Embedded Servlet Container [17].

Micro Service is an architecture that allows the developers to independently develop and deploy services. Every program running has its own mechanism and this enables the lightweight business application support model.

Spring Boot provides Java developers with a good platform to develop a stand-alone and production-grade spring application that they can just run. With minimal configurations, you can get started without having to set up a whole Spring configuration [18].

3.2.7 Selenium for Crawling

Selenium is an open source tool designed to automate web browsers. This provides a single interface that allows you write test scripts in various programming languages such as Ruby, Java, NodeJS, PHP, Perl, Python, and C#, and more.

Versatility of Selenium is part of the reason why selenium is so popular. Anyone who codes for the web may use Selenium to check their code/app–from individual freelance developers running a short series of debugging tests to UI engineers conducting visual regression tests after a new integration process [19].

3.2.8 Spring Reactive

Reactive programming is indeed a programming model that advocates a data processing approach which is asynchronous, non-blocking, event-driven. Reactive programming includes modeling data and events as measurable sources of data and applying routines for data processing to respond to changes in those streams [20].

We make a request for the resource in the reactive programming style, and start performing other things. When the data is available, we receive the notification in the form of a call back function along with the data. We manage the response in callback feature as per application/user needs.

3.2.9   Load Balancer

Load balancing [21] improves workload distribution across multiple computer resources, such as computers, a computer cluster, network connections, central processing units, or disk drives. Load balancing aims at optimizing resource utilization, maximizing through-put, minimizing response time and avoiding overloading of any resource. Use through load balancing modules instead of a single system will improve the efficiency and availability by redundancy. Load balancing [22] typically involves specialized software or hardware, such as a multilayer switch or a cloud Domain Name Network.

3.2.10   JWT

JWT (JSON Web Token) is an Internet standard for creating JSON access tokens which assert a number of claims. A server could, for example, generate a token that has the claim "logged in as an admin" and provide it to a client. Then the client could use the token to show it's signed in as admin. The tokens are signed by the private key of one party (usually the server's), so that both parties (the other being already in control of the respective public key by some appropriate and trustworthy means) can check that the token is valid. The tokens are designed to be lightweight, URL-safe and especially usable in a single-sign-on (SSO) web browser setting.

Usually, JWT [23] statements can be used to transfer identification of authenticated users between an identity provider and a service provider, or any other form of assertion that business processes require.

3.2.11   Face Recognition Using Amazon's ReKognition boto3 API

Amazon Rekognition offers fast and accurate face recognition, enabling us to use our private face picture repository to identify individuals on a photo or video. We can also check identity by evaluating a facial picture for contrast to photographs that you have kept [24].

We can easily detect when faces appear in images and videos with Amazon Rekognition, and get attributes such as gender, age range, eyes open, glasses, facial hair for each. We can also calculate how these facial features change over time in film, such as creating a timeline of an actor's articulated emotions.

So, no matter how old the image is, it can efficiently recognize the individual and help our software to search and find out the related data about that particular entity.

4   Future Work

Based on the proposed solution we can create a platform that can make investigation stress and hassle-free process and providing a business-friendly solution that solves crucial business use cases. With the adoption of such a high-end digital platform, national security and efficient investigation won't be a dream anymore. The success rate of any investigation would be maintained by making efficient use of time and resources. Tracking presence and staying updated on one's or the client's web/social presence would be much easy. Estimating the buzz of your brand would be handy. OSINT can prove to be a wonderful choice when putting in terms of data extraction and processing. It truly helps in getting and availing the best of the best data from the abundantly available data stocks available on legal and accessible public domains. This solution when ready can bring a new wave in the field of investigation and real-time searching with huge pros to the nation.

5   Conclusion

To establish efficient and reliable law solutions that solve real-world problems related to the use of data, records, and information, OSINT can be considered as a better option and can be considered as the time, money and resource saver making it an efficient option. In the digital age, to investigate manually and traditionally by visiting record rooms and searching through the document files can be called foolish and so bringing out a software product can brief these efforts and make investigation a lot easier and also broader in perspective. The application of Open Source Intelligence in the investigation process can truly help in receiving better, optimized, accurate and reliable results all in one single platform with only minimal input information and no physical effort at all. It can also help track the presence and online fame of an individual gathering a lot of commercial project ideas in the future using the same technology.

Appendix

f = frequency of letter in the document.
d = $JSON_F$ document.
D = total number of $JSON_F$ documents.
N = number of d in which t occurs
pi = ith person
cw = crawler
S = set of links to be searched
lst = local storage of each crawl result
G_Stack = Global stack
$JSON_F$ = final JSON
t = triggers
$ß$ = learning rate.

References

1. George, R.Z., Kline, R.D., Lowenthal, M.M.: Intelligence and the National Security Strategist: Enduring Issues and Challenges, vol. 58, pp. 273–284. Rowman and Littlefield (2005). ISBN 9780742540392
2. Byrne, J., Marx, G.: Technological innovations in crime prevention and policing. A review of the research on implementation and impact. J. Police Stud. **20**(3), 17–40 (2011). ISBN 978-90-466-0412-0
3. Rico, R.A.P., Medina, M.J.H., Hernández, C.C.P., López, D.O.D., Ruíz, J.C.C.G.: Open source intelligence (OSINT) as support of cybersecurity operations. use of OSINT in a colombian context and sentiment Analysis. Revista Vínculos: Ciencia, Tecnología y Sociedad **15**, 195–214 (2018)
4. Pastor-Galindo, J., Nespoli, P., Mármol, F.G., Pérez, G.M.: The not yet exploited goldmine of OSINT: opportunities, open challenges and future trends. IEEE Access **8**, 10282–10304 (2020)
5. Schaurer, F., Störger, J.: Guide to the Study of intelligence. The evolution of open source intelligence (OSINT) intelligencer. J. U.S. Intell. Stud. **19**(3), 53–56 (2010)
6. Adderley, R., Musgrove, P.: Police crime recording and investigation systems – a user's view. Polic. Int. J. Police Strat. Manag. Emerald **24**(1), 100–114 (2001)
7. Clive, B.: Web mining for open source intelligence. In: IEEE. 12th International Conference Information Visualisation, London, pp. 321–325 (2008)
8. Nacci, G.: The general theory for open source intelligence in brief. A proposal, pp. 1–3. Intellilsfèra (2019)
9. Hassan, N.A., Hijazi, R.: Open Source Intelligence Methods and Tools, pp. 15–18. Apress Media LLC, New York (2018). ISBN-13 (pbk): 978-1-4842-3212-5. ISBN-13 (electronic): 978-1-4842-3213-2
10. Satheesh, A., Singh, M.: Comparative study of open source automated web testing tools: selenium and sahi. Indian J. Sci. Technol. **10**(13), 1–9 (2017). ISSN (Print): 0974-6846. ISSN (Online): 0974-5645
11. Dobbelaere, P., Esmaili, K.S.: Kafka versus RabbitMQ: a comparative study of two industry reference publish/subscribe implementations. Industry Paper, pp. 227–238 (2017)
12. Jason, B.: Machine Learning Streaming with Kafka, pp. 239–303. O'Reilly, Sebastopol (2020)
13. Shree, R., Choudhury, T., Gupta, S.C., Kumar, P.: KAFKA: the modern platform for data management and analysis in big data domain. In: 2nd International Conference on Telecommunication and Networks (TEL-NET), pp. 1–5 (2017)
14. Wang, X., Loguinov, D.: Load-balancing performance of consistent hashing: asymptotic analysis of random node join. IEEE/ACM Trans. Netw. **15**(4), 892–905 (2007)
15. Zhang, Z., Qiang, Y., Li, Y.: Using Naïve Bayes classifier to distinguish reviews from non-review documents in Chinese. In: 2007 International Conference on Management Science and Engineering, Harbin, pp. 115–121 (2007)
16. Francesco, P.D., Malavolta, I., Lago, P.: Research on architecting microservices: trends, focus, and potential for industrial adoption. In: 2017 IEEE International Conference on Software Architecture (ICSA), Gothenburg, pp. 21–30 (2017)
17. Christudas, B.: Spring Boot, Practical Microservices Architectural Patterns, pp. 147–182. Apress, New York (2019)
18. Reddy, K.: Web Applications with Spring Boot - Beginning Spring Boot 2: Applications and Microservices with the Spring Framework, pp. 107–132. Apress, New York (2017)
19. Chen, R., Miao, H.: A selenium based approach to automatic test script generation for refactoring JavaScript code. In: 2013 IEEE/ACIS 12th International Conference on Computer and Information Science (ICIS), Niigata, pp. 341–346 (2013)

20. Cosmina, I.: Building Reactive Applications Using Spring. Pivotal Certified Professional Core Spring 5 Developer Exam, pp. 903–955 (2020)
21. Abdhullah, S.S., Jyoti, K., Sharma, S., Pandey, U.S.: Review of recent load balancing techniques in cloud computing and BAT algorithm variants. In: 2016 3rd International Conference on Computing for Sustainable Global Development (INDIACom), New Delhi, pp. 2428–2431 (2016)
22. Prakash, S.W., Deepalakshmi, P.: Server-based dynamic load balancing. In: 2017 International Conference on Networks & Advances in Computational Technologies (NetACT), Thiruvanthapuram, pp. 25–28 (2017)
23. Wehner, P., Piberger, C., Göhringer, D.: Using JSON to manage communication between services in the Internet of Things. In: 9th International Symposium on Reconfigurable and Communication-Centric Systems-on-Chip (ReCoSoC), Montpellier, pp. 1–4 (2014)
24. Mishra, A.: Amazon Rekognition - Machine Learning in the AWS Cloud, pp. 421–444. Wiley, New York (2019). Chap. 18

A Neural Network Based Hybrid Model for Depression Detection in Twitter

Bhanu Verma$^{(\boxtimes)}$, Sonam Gupta, and Lipika Goel

Department of Computer Science, Ajay Kumar Garg Engineering College, Ghaziabad, India
{mebhanuverma96,guptasonam6,lipika.bose}@google.com

Abstract. Depression is a serious mental illness that leads to social disengagement, affects an individual's professional and personal life. Several studies and research programs are conducted for understanding the main causes of depression and an indication of psychological problems through speech and text-based data generated by human beings. The language is considered to be directly related to the current mental state of an individual, that's why social media network is utilized by researchers in detecting depression and helps in the implementation of the intervention program. We proposed a hybrid model using CNN & LSTM models for detecting depressed individuals through normal conversation-based text data, that retrieved from twitter. We employ machine-learning classifiers and proposed method on twitter dataset to compare their performance for depression detection. The proposed model provides an accuracy of 92% in comparison with the machine learning technique that gives a maximum accuracy of 83%.

Keywords: Deep learning · Machine learning · Depression · Twitter · Text data processing

1 Introduction

Mental health issues are in a peak state, most of the world population are affected by severe to major level psychiatric illness. Depression is one of the common mental illness and it is characterized as a mood disorder in which, suffering individuals experience sudden mood variations that are much different from non-depressed person mood variation. Research shows that depression is the result of childhood traumas, social inferiority, and loss of a beloved one. According to a survey conducted by the WHO (World Health Organization) [1], there are 322 million people worldwide suffering from depression. WHO organized world mental health survey in 17 countries that provide results that every 1 in 20 people have experienced an episode of depression in past years.

Depression is recognized as leading cause of disability for both males & females, whereas females are most prone to get depressed with a 50% higher rate as compared to males. According to NIMH [2], major depression was 8-7% higher in females and 5.3% for males. The study conducted by WHO shows China is the most depressed country in the world followed by India. Both the countries are most populated in the world and highly rich in recourses and level of income are sufficient for living, instead of this

© Springer Nature Singapore Pte Ltd. 2020
M. Singh et al. (Eds.): ICACDS 2020, CCIS 1244, pp. 164–175, 2020.
https://doi.org/10.1007/978-981-15-6634-9_16

90% of depression survivors doesn't receive any form of help or treatment [3]. In India, every 1 in 5suffering from depression throughout their life that is equal to 200 million people. The mental health problems have stigmatized image in India that's why only 10–12% of the affected people will get any type of treatment. Depression leads to various health-related issues as well, such as heart disease, thyroid problem and diabetes [4]. Moreover, Depression is the second leading cause of suicide among 15-29 years old & 788000 people died due to suicide in 2015. Mainly, 78% of suicide happens in countries that have low income, an estimate by WHO in 2017 [1]. A report by the World Health Organization reported that on average every 40-s suicide occurs worldwide [5, 6].

WHO study in 2017, states that 18.4% of depression survivors are increased from 2005 to 2015. For the treatment and screening of depression sufferers, intervention programs are conducted. There is a vast majority of sufferers who never receive any form of care due to cultural biases, social inferiority, discrimination fear, and improper knowledge. The intervention programs are useful when individuals participate in them [7]. The Health Activity Program (HAP) is an intervention program organized in India [8]. Intervention programs are participation-based to make it available and reachable to every individual, social media platforms are useful. Social media networks are the rich source of vast data generated by millions of users worldwide. Famous social media applications such as Twitter, Facebook, Instagram, and Reddit are accessed by millions of users. The social media network provides access to a broader part of the population for screening of social media users for depression [7].

The work presented in this paper is using neural network models for depression detection using textual data retrieve from twitter. The paper is organized as follows: Sect. 2 presents previous work on language for depression, social media as a tool for various health researches and twitter usage in the field of depression detection. Section 3 gives details about dataset and data preprocessing. In Sect. 4, there is a discussion about CNN & LSTM. Section 5 consists of the main experiment and Sect. 6 show experiment findings.

2   Related Work

Mental illness changes sufferer perception towards people, society and even for themselves. It was found that depression patients are mainly self -focused. They tend to isolate themselves and show a low social activity rate. Moreover, utilizing social networking platforms is increased and people who have social inferiority or suffering depression & anxiety issues tend to use social media for sharing their sentiments and experiences with others [9, 10]. There are mental health forms available and social networking sites are used many people to communicate their thoughts and feelings without disclosing their real identity and people discuss their mental health issues such as depression [11], BPSD (Behavioral and Psychological Symptoms of Dementia) [12], Bipolar disorder [13]. This stigmatized illness made people feel ashamed of themselves and increase the chances of social media use for help.

2.1 Twitter: A Tool for Mental Health Research and Indication of Depression Through Language Use

Twitter is a famous microblogging platform, it is estimated that 330 million active monthly users on twitter. There are about 500 million tweets were posted in a day, that's a huge amount of data generated by Twitter. Being a rich source of data, twitter supports various research work such as population-based surveys, mental health or consumer sentiment analysis [14]. The 38% of twitter user belongs to an age group of 18 to 29 years. Thus, social media networks are suitable for studying depression based behaviour and measuring the severity of depression. Some of the early works on twitter are [15, 16].

Tweets are textual data written by users that consist of maximum characters limits to 280. So, a user expresses their thought in a more concise manner due to the character limit. Researchers confirm that language and social behaviour has a direct relation to mental health issues. Several indications of psychiatric problems can be recorded through a proper examination of the language used by an individual. The language use by depressed individuals is different from individuals who never get depressed. In [17] suggests that depression survivors use more negative polarity words and the text written by depressed person mainly consists of the first-person pronoun such as (I, me and myself) and less first plural pronoun [18]. In [19] there is a discussion about the self-focus trait of depression survivors. They conduct a study for investigation of words used by suicidal and non-suicidal poets; the study shows that depressed poets use distinctive language as compare to non-suicidal poets. The suicidal poet uses more first person pronoun and put less focus on social integration. A self-focus model put by [20] that depicts "self-focus" as an initial trait of getting depression illness and people who are more concerned about their role in every situation of life and looking for fulfilling self-centric desires are closer to get depressed.

The depressive language was first studied by [21, 18] in textual data and in [19] states that the study of language for psychological health indications provides remarkable traits of problems faced by an individual in life. LIWC (Linguistic Inquiry Word Count) software is developed for natural language processing [21, 22]. LIWC is a text analysis tool used for the study of cognitive and emotional components in individual written and speech data. The LIWC dictionary categorizes words into language variables, description categories, 21 linguistic dimensions, psychological words categories & informal language markers [23]. In [24] utilizes LIWC software for distinguishing between the language used by the control group (a group of people who indulge in normal life activities such as singing, dancing, art, and sports) and clinical group (a group consisting mental health survivors and initial state depressed person). For the study, they use LIWC psycholinguistic word categories for understanding message communication between the control and clinical group on a social network. Social networking site Facebook used by [25] for identification of depressed individuals by categorizing text data using LIWC software that includes (present, future & past focus) and utilizes 9 linguistic dimensions for categorizing comments retrieve from Facebook and applying machine learning classifiers. The LIWC dictionary is outdated now, as it doesn't process slang and sarcastic comments or text. To use LIWC, the user must update the dictionary according to the latest trend and needs for the analysis of textual data. Another senti-ment analysis tool is Vader (Valence Aware Dictionary for Sentiment Reasoning) [26]. Vader outperforms LIWC in social media content analysis, Vader is more sensible and consist 25 feature that provides five "golden items" with realistic sentiment rate. It is freely available for use and it composed all the human emotions and language variation biomarkers that make it more useful for research and analysis.

2.2 Machine Learning-Based Framework for Depression Detection

The task of effective depression detection is performed using machine learning tech-niques. Machine learning algorithms require a well-defined feature set for classification of data. Researchers apply machine learning techniques for conducting various studies and surveys for health information, commercial and mental health. A remarkable work by [27] states that social media gives clues about user's mental health by showing some behavioral symptoms such as high social network activity from 10 pm to 6 am, small social network and discussion about treatment & medication related to psychiatric prob-lems. The supervised learning algorithm (SVM) is trained in behavioral attributes and predicts the chances of getting depressed. Another work [28] using KNN for depression identification among Facebook users. LIWC software is utilized for categorizing data using 5 emotional variable & 9 standard linguistic dimensions. A large size corpus of 2.5 million tweets are used for classification and apply a bag of words approach with machine learning techniques that provides good results [29].

Deep learning is applied for text classification in [30]. Researchers apply deep learn-ing models for population-level monitoring and data collection for health research & population-based surveys. Social media have a billion users from around the world; twitter is useful for depression & anxiety monitoring [31] and for PTSD (Post Trau-matic Stress Disorder) [32] in social media users. In [33] shows about early depression detection using the eRisk 2017 dataset. The early detection task is based on sequence word prediction in textual data, The CNN model is applied for early detection of depres-sion and a new modified version of the ERDE evaluation metric is proposed. This work states ERDE is better to use for the early detection task. Language is an indicator of men-tal health that's why CLPsych works on data collected from social media & hackathon as unshared and shared data respectively. The linguistic characteristics show a more distinct understanding of mental health signals without applying any machine learn-ing technique. In [34] proposed a DK-LSTM model for depression detection in social media content. This shows the increasing use of deep learning techniques over machine learning.

3   Neural Network and Baseline Model

In this section, we discussed about neural network-based hybrid model utilized in this paper. We use CNN and LSTM for this task of depression detection and discuss about some early work using these deep learning models.

3.1 Convolutional Neural Network (CNN)

CNN are popular for image classification work as well as for pattern recognition task [35]. Researchers apply a convolution neural network for sentence classification and shows significant results [36, 37]. CNN extracts features from textual data; firstly, it converts data into vector form using word2vector. It is a word-embedding model created in 2013 at Google [38, 39]. Word2Vector is a 2-layer neural network architecture that consists of CBOW (Continuous of Words) & Skiw gram model. It trained on 3 million words and has 300 dimensions. After Word2Vector by Google, Stanford and Facebook developed their word embedding models that are Glove by Stanford [40] & Fast text by Facebook in the year 2017 [41]. A C-LSTM model proposed by [42] is a text classification model that aims to configure local features & global features from the small text. The Stanford Sentiment Treebank (SST) benchmark [43] dataset is used and CNN extracts feature through the convolution matrix & fed it to LSTM to learn features for long -term dependency. Instead of textual data, speech data are also used for depression detection [44] using CNN are proposed for detecting the severity of depression using speech data on the basis of advantages of deep learning features extraction. The work provides significant results and use AVEC2013 & AVEC2014 depression based dataset. In [45] proposed word-level CNN architecture (deep-pyramid CNN). The paper shows shallow level CNN provide better categorization then character level CNN. This model outperforms the previous deep CNN model for text classification and categorization. CNN and SVM based hybrid model is used for sentiment analysis using NLPCC2014 dataset [46]. The CNN-SVM model provides a recall of 88% and compare with baseline model NLPCC_SCDL_best that score recall of 86%.

3.2 Long Short Term Memory (LSTM)

LSTM is a modified version of RNN (Recurrent Neural Network). RNN can learn sequences of data and produce time-series data. The feedback loop transfers data to one state to another and the basis of RNN is recursive formula. Some applications of RNN are; Chatbots, machine translation, image categorization, and text classification. The problem of vanishing gradient makes RNN provide low-level results. RNN cannot remember data pattern for long period means as the gap increases the pattern of data cannot formed. LSTM introduced in 1997 by (Hochreites and Schmidhuber) [47] and utilizes in tasks like natural language processing, speech recognition. LSTM overcome the problem of remembering data pattern as long as user requires and exploding gradient problem through cell state that consist "tanh" squashing function [-1, +1]. LSTM can easily excess 1000 discrete time steps by using cell state whereas RNN fails to learn when time step increases from 5 discrete time steps [48]. The architecture of LSTM consist of three gates are input, forget and output gate, the input gate performs the "selection" task means it decides what information should be kept and what to ignore. The forget gate consist of logistic function and output gate update cell state by providing what value should be the output. These three states and one Cell state mechanism provide support for long-term dependency of data patterns.

Long Short Term Memory model used in various text classification tasks such as; a work by MIT to detect depression using speech and textual data [49]. The model trained on data consists of 142 Interview from DAIC that includes text and speech data of these 142 mental health patients. A questionnaire is prepared and question asked to the patients, the score is given between 0 to 27 to each patient in terms of severity of depression. So, patient with score range (10–14) is moderate depressed and 15–19 are severely depressed patients. The LSTM model gives a recall of 83% for speech data and 0.67 f1-score for text data.

3.3 Baseline Model

We compare our model results with machine learning classifiers that applied on the same dataset of 15,000 tweets consisting of depressed and non-depressed text. The machine learning classifier used in this work is one that previously used for text classification, depression detection, and sentiment analysis task. We use Logistic regression, Decision tree, and Naïve Bayes, SVM, and KNN classifier. These classifiers utilized in various sentiments analysis task using social network data and provide significant results in classifying text document polarity and internet reviews about product, movie or any social media post.

4 Experimental Setup

In this paper, two experiments performed, firstly classification of depression through machine learning classifiers using twitter data and second, the experiment is using deep learning models for depression detection on 15000 tweets dataset. The machine learning classifiers act as baseline work to verify and compare with deep learning technique results.

4.1 Baseline Model (Experiment 1)

Firstly, the machine-learning classifier used for classification task and results are verified using the evaluation metric. There are a total 5 classifiers used and trained on the 15000-tweet dataset, there is no constraint posed on dataset for a better understanding of human psychology. The classifiers are Logistic regression- a classifier used to decide the class of object for binary classification using a sigmoid function. The output of logistic regression is either 0 or 1. Decision tree- is a supervised learning algorithm, commonly used for a classification task. Decision tree consist of node and branches and perform decision-making. Naïve Bayes- is based on Bayesian network and mainly works on large datasets. Support Vector Machine- a supervised machine learning algorithm, famous for handling linear and non-linear datasets with a higher dimension and form hyperplane for binary classification. K-Nearest Neighbor- used for regression and classification task, trained on a labeled dataset. KNN algorithm decides the class of objects by finding the minimum distance using Euclidean distance. Evaluation metric utilized in the task of verifying machine learning classifier accuracy and it includes F1-Score, Precision and Recall.

In Table 1 the results of the baseline model shown, the best classification task performed by the decision tree algorithm, giving a recall of 85%. Naïve Bayes classifier does the worst performance by giving a recall of 54% only. KNN provides a recall of 78%, second-best performance between these five models. Logistic regression and SVM recall value are 73% and 75% respectively. The table also includes precision and FI-score of each classifier. The evaluation metric is a common method for checking and validating machine learning classifier results. Now, these results used for measuring the classification accuracy of the CNN-LSTM hybrid model.

Table 1. The results of baseline model

Machine learning algorithm (Baseline approach)	Evaluation metric scores		
	F1-Score	Recall	Precision
LR	0.74	0.73	0.76
DT	0.85	0.85	0.85
NB	0.46	0.54	0.70
SVM	0.75	0.75	0.75
KNN	0.78	0.78	0.79

4.2 Proposed Method (CNN-LSTM Based Hybrid Model)

In this work, we use a small dataset according to our system computational power. The dataset covers natural interaction between people on twitter, to make the dataset more useful for deep learning models to extract features and the way a human can interact with another human. The internal mechanism of the deep learning model, that how the model perceives these findings and on what basis it relates to human behavior presented through text data. The Fig. 1 provides an overview of CNN-LSTM model. We employ CNN (Convolutional Neural Network) with the ReLu (Rectified Linear unit) activation function for the task of extracting the features from the dataset, through the convolution layer (Conv1D). The dimension set to 300; we apply 32 filters to the text document with a maximum sequence length of 140 characters. The kernel size set to 3 to achieve the best results after kernel the processed text padded with 'same'. The feature map (trainable feature) is generated passes to the pooling layer (max pooling) with a pool size of 3 which moves the filter and reduces the dimensionality of feature set. For the regularization of output, we apply a dropout of 0.3 to the output of the pooling layer then again it fed to the convolution layer, the same process will follow, after dropout of 0.3 apply to the output the dataset feed to next layer (LSTM model). The second deep learning model utilized is LSTM (Long Short Term Memory), the advantage of LSTM to remember data pattern for long period is useful in human psychiatric research because it requires remembering sequences of characteristics retrieve from text data and apply learned features on new data for retrieving new features and forming a sequence with them. The LSTM layer processes the output CNN layer and using popular optimizer 'Adam'. For loss function, 'Binary cross-entropy' is used and a dropout of 0.4 applied to the output of this layer. For a binary classification task, the sigmoid activation function [0,1] is used on dense layer. The batch size of 100 and 15 epochs and patience value set to 3 for conducting this experiment. The numbers of epochs are decided by first applying a set of 5 and 10 epochs for accuracy and then using 15 epochs to gain better results.

5   Results

5.1   Performance of CNN-LSTM Model (Proposed Method)

The CNN-LSTM model gives a significant accuracy of 92% that verifies using the evaluation metric and later compare the results with machine learning classifiers. These results obtained by using the same dataset. For comparison, we use the accuracy level of each model and recall value. It is clear from Table 2 that our proposed model performs well on small datasets with diverse emotional traits. The proposed model outperforms these machine learning models for depression detection task and it clearly indicates that machine learning algorithm has drawbacks of feature feature extraction task and differentiating between multiple sequences of text data for prediction. This work utilized the Keras library that runs on TensorFlow for implementing neural network models. The python framework with natural language processing based tools employed in this work.

Table 2. Results of proposed model

Label = 1 (Non-Depressed) cases		
Evaluation Metric Score	Recall	0.96
	Precision	0.90
	F1-Score	0.93
Label = 0 (Depressed) cases		
Evaluation Metric Score	Recall	0.85
	Precision	0.94
	F1-Score	0.89

Our proposed model provides significant results for conducting another study using a much larger size dataset for the depression detection task.

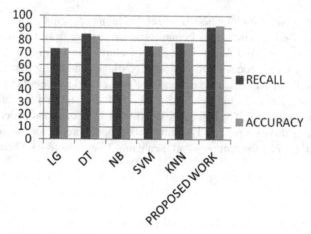

Fig. 1. Comparison graph for CNN-LSTM vs. baseline model.

5.2 Comparison of Proposed Model with Baseline Approach

The performance of Baseline model on depression detection task using twitter data shown in Table 1. We first perform a depression classification task using a machine-learning classifier that shows classification accuracy of 83% highest from all five classifiers employed. The Table 3. elaborates about performance of both model used in this paper and Fig. 1 is used for showing an graph representation of comparison between two models. We use baseline model to compare with deep learning model results and CNN-LSTM performance is more significant as compared to traditional machine learning techniques. The ability of deep learning to extract features and use it for classification gives better results. In Table 2 the results of CNN-LSTM are described, the precision value for depressed and non-depressed tweets is quite up to the level and recall for non-depressed tweets is 0.85.

Table 3. Comparison between baseline model and proposed model

Model name	Classifier name	Recall	Accuracy
Baseline model	LG	73	73%
	DT	85	83%
	NB	54	53%
	SVM	75	74%
	KNN	78	78%
Proposed model	CNN-LSTM	91	92%

6 Conclusion and Future Scope

In this paper, we proposed our CNN-LSTM model and verify its performance by comparing its results with machine learning classifiers. The performance of deep learning models is better in comparison with the machine learning model for textual data. The mental health research field requires a better dataset that can help models to provide diverse features and knowledge to understand and pick human mental characteristics. Twitter is a good source of social networking data for the number of previous research and it helps depressed people to share their views and sentiment with other people. The results of this proposed work help in organizing better intervention programs for screening and treatment of depressed people. For future work, we consider both Facebook and twitter data because the risk of increased depression issues is high among teenagers and around millions of young kids use Facebook for sharing their emotions and experience. Capturing images and text data provide a better understanding of user's mental state. By utilizing data from different social media platform better results can be obtained and use for helping the sufferer.

References

1. Depression and Other Common Mental Disorders: Global Health Estimates. World Health Organization (2017)
2. Key Substance Use and Mental Health Indicators in the United States: Results from the 2016 National Survey on Drug Use and Health (2017). https://www.samhsa.gov/data/
3. Patel, V.: Talking sensibly about depression. PLoS Med. **14**, e1002257 (2017)
4. Dhar, A.K., Barton, D.A.: Depression and the link with cardiovascular disease. Front. Psychiatry **7**, 33 (2016)
5. O'Dea, B., Wan, S., et. al.: Internet intervention-detecting suicidality on Twitter. Elsevier (2015)
6. Depression and Other Common Mental Disorders: Global Health Estimates. World Health Organization (2014)
7. Park, M., Cha, C., et al.: Depressive moods of users portrayed in Twitter (2012)
8. Patel, V., Weobong, B., et al.: The Healthy Activity Program (HAP), a lay counsellor delivered brief psychological treatment for secure depression, in primary care in India: a randomized control trial. Lancet **10065**, 176–185 (2017)
9. Gowen, K., Deschaine, M., et al.: Young adults with mental health conditions and social networking websites: seeking tools to build community. Psychiartic Rehabil. J. **35**, 245 (2012)
10. Naslund, J.A., Grande, S.W., et al.: Naturally occuring peer support through social media: the experiences of individual with severe mental illness using youtube. PLoS One **9**, e110171 (2014)
11. Berger, M., Wagner, T.H., Baker, L.C.: Internet use and stigmatized illness. Soc. Sci. Med. **61**, 1821–1827 (2005)
12. Cerajeira, J., et al.: Behavioural and psychological symptoms of demantia (2012)
13. Hilty, D.M.: A review on bipolar disorder in adults. https://www.ncbi.nlm.nih.gov/ (2006)
14. Clement, J.: Twitter: number of monthly active user 2010–2019 (2019)
15. Gui, T., Zhu, L., et al.: Cooperative multimodal approach to depression detection in Twitter. In: The Thirty-Third AAAI Conference on Artificial Intelligence (AAAI-2019) (2019)
16. McClellan, C., Ali, M.M., et al.: Using social media to monitor mental health discussions-evidence from Twitter. J. Am. Inform. Assoc. **24**, 496–502 (2017)

17. Rude, S.S., Gortner, E.M.: Language use of depressed and depression-vulnerable college students. Cogn. Emot. **18**(8), 1121–1133 (2004)
18. Stirman, S.W.: Word use in the poetry of suicidal and nonsuicidal poets. ncbi.nlm.nih.gov (2001)
19. Ramirez-Esparza, N., et. al.: The Psychology of word use in depression forums in English and in Spanish: testing two text analytic approaches (2008)
20. Pyszczynski, T.: Self-regulatory perseveration and the depressive self-focusing style: a self-awareness theory of reactive depression. Psychol. Bull. **102**(1), 122 (1987)
21. Tausczik, Y.R., et al.: The psychological meaning of words: LIWC and computerized text analysis methods. J. Lang. Soc. Psychol. **29**(1), 24–54 (2010)
22. Tauscizk Pennebaker, J.W., Francis, M.E., Booth, R.J.: Linguistic Inquiry and Word Count: LIWC 2001, vol.71, p. 2001. Lawrence Erlbaum Associates, Mahway (2001)
23. Pennebaker, J.W.: The development and psychometric properties of LIWC (2015)
24. Nguyen, T., Phung, D., et al.: Affective and content analysis of online depression communities. IEEE Trans. Affect. Comput. **5**(3), 217–226 (2015)
25. Rafiqul Islam, M.D., Wang, H., et al.: Depresssion detection from social network data using machine learning techniques. Health Inf. Sci. Syst. **6**(1), 8 (2018)
26. Hotto, C.J., Gilbert, E.: VADER: A Parsimonious rule-based model for sentiment analysis of social media text. In: AAAI-2014 (2014)
27. De Choudhary, M., et al.: Predicting depression via social media. In: AAAI Conference on Weblogs and Social Media (2013)
28. Islam, M.R.: Detecting depression using K-nearest neighbors (KNN). IEEE (2018)
29. Nadeem, M., Horn, M., et al.: Identifying depression on Twitter. https://arxiv.org/ (2016)
30. Ganda, R., et al.: Efficient deep learning model for text classification based on recurrent and convolutional layers. In: 16th ICMLA (2017)
31. De Choudhury, M., et al.: Social media as a measurement tool of depression in population. In: Proceedings of the 5th Annual ACM Web Science Conference (2013)
32. Coppersmith, G., et al.: CLPsych 2015 shared task: depression and PTSD on Twitter. In: ACL Anthology (2015)
33. Trotzek, M., Koitka, S., et al.: Utilizing neural networks and linguistic metadata for early detection of depression indications in text sequences. IEEE (2018)
34. Li, W., et al.: Applying deep learning in depression detection. In: PACIS (2018)
35. Goodfellow, I., Bengio, Y., et al.: Deep Learning. MIT Press, Cambridge (2016). http://www.deeplearningbook.org
36. Kim, Y: Convolutional neural network for sentence classification (2014)
37. Zhang, Y., Wallace, B.C.: A sensitivity analysis of convolutional neural networks for sentence classification. https://arxiv.org/ (2015)
38. Mikolov, T., et al.: Efficient estimation of word representation in vector space. https://arxiv.org/ (2013)
39. Joulin, A, et al.: Bag of tricks for efficient text classification. https://arxiv.org/ (2016)
40. Pennington, J., et al.: Glove: global vectors for word representation. In: Emperical Methods in Natural Language Processing (EMNLP) (2014)
41. Bojanowski, P., et al.: Enriching word vectors with subword information. Trans. Assoc. Comput. Linguist. **5**, 135–146 (2017)
42. Liu, Z., Lau, F.C.M..: A C-LSTM neural network for text classification (2015)
43. Koutnik, J: A clockwork RNN. https://arxiv.org/ (2014)
44. Lang, H.E., Cao, C.: Automated depression analysis using convolutional neural networks from speech. J. Biomed. Inf. **83**, 103–111 (2018)
45. Johnson, R., Zhang, T.: Deep pyramid convolutional neural networks for text categorization (2017)

46. Chen, Y., Zhang, Z.: Research on text sentiment analysis based on CNN and SVM. IEEE (2018)
47. Hochreiter, S.: The vanishing gradient problem during learning recurrent neural nets and problem solutions. Int. J. Uncertainity Knowl.-Based Syst. **6**(02), 107–116 (1998)
48. Gers, F.A., et al.: Learning to forget: continual prediction with LSTM. IEEE (2016)
49. Alhanai, T., Ghassemi, M.: Detecting depression with audio/text sequence modeling of interviews. In: Interspeech (2018)

Unleashing the VEP Triplet Count of Virtually Created 3D Bangla Alphabet to Integrate with Augmented Reality Application

Apurba Ghosh[1（✉）], Anindya Ghosh[1（✉）], Arif Ahmed[1（✉）], Md Salah Uddin[1（✉）], Mizanur Rahman[1（✉）], Md Samaun Hasan[1（✉）], and Jia Uddin[2（✉）]

[1] Department of Multimedia and Creative Technology, Daffodil International University, Dhaka, Bangladesh
{apurba.mct,salah.mct,mizan.mct,hasan.mct}@diu.edu.bd,
anindya.ghosh835@gmail.com, arifahmed@daffodilvarsity.edu.bd
[2] Woosong University, Daejeon, Korea
jia.uddin@wsu.ac.kr

Abstract. This paper demonstrates the process of calculating VEP triplet count of Bangla Alphabet which are unlike traditional hand written or printed letters, but created entirely in 3 dimensional virtual environment to effectively use in Augmented Reality Based applications. This count can be a potential support for 3d modelers, Augmented Reality based application developers and academic researchers. The challenges that we faced and the limitations of our contribution are also mentioned here. It is expected that future academic researchers and commercial application developers will get a great support on developing 3D Bangla alphabet from our research.

Keywords: 3D model of Bangla Alphabet · Augmented Reality technology · Euler's Polyhedron formula · Optimized 3D model for AR based application

1 Introduction

Every aspect of education is now continuously being modernized by the magical touch of advanced computing technologies. Elementary education is not an exception. People now a days have access to a computer or at least a smart phone which is also a powerful computer indeed. As the overall number of smart phone users is much higher than regular computers particularly in Bangladesh, targeting this particular segment to figure out the count of- vertex, edge and polygon to create virtual assets for elementary educational applications based on Bangla language that uses augmented reality technology is more logical. It should be mentioned that numerous researches have done on this segment based on many emerging technologies. For instance, previous researches on 3D as a medium of strong multimedia tool have focused on many important aspects. Research on classification of 3D models has already done for 3D animation environments [1]. Specific 3D tool based research for film and television [2, 3], rapid 3D human modeling

© Springer Nature Singapore Pte Ltd. 2020
M. Singh et al. (Eds.): ICACDS 2020, CCIS 1244, pp. 176–186, 2020.
https://doi.org/10.1007/978-981-15-6634-9_17

and animation based on sketch and motion database [4], real water simulation for 3D animation [5] - researches on all those areas have done. Even researches on- emotion based facial animation [6, 7], real-time speech driven facial animation using neural network [8], 3D measurement technologies for computer animation [9], 3D cartoon character animation engine [10] have come up with new findings. The area of interactive 3D animation system [11] and experimental teaching of 3D animation [12] have also explored. But none of those research works have published any specific count of the crucial virtual parameters that act as the building blocks of any 3D asset inside virtual environment. Particularly for this research we targeted that specific domain so that new findings can be brought.

2 Background Study

The sector of multimedia is a very dynamic field which is becoming enriched by the continuous evolution of technologies like- Augmented Reality (AR). Like many other sectors, AR has made a huge impact on education. In today's world of engineering education- AR based lab system exists which enables teachers and students to work remotely via internet/intranet in current classroom labs, including virtual elements which is capable to interact with real ones [13]. Impact of AR is clearly visible in the sector of medical education as well where training in real-life context is not always possible due to safety, costs and many other factors. In those cases, AR can potentially offer a highly realistic suited learning experience supportive to complex medical learning and transfer [14]. Apart from engineering and medical education, AR is also putting its bold footprint in skill based trainings [15].At present AR is not only conquering the different domains of higher education where the target group is adults [16] but also knocking the door of elementary education. Researchers have begun to study AR for teaching colors and shapes to kids [17]. AR as a technology of elementary education- most of the previous studies targeted languages like English [18], Japanese [19], Kanji [20]. Another notable thing is most of the previous works are tightly coupled with internet connectivity. Majority of them requires internet connection during runtime [21] which sometimes come out as a big issue particularly for those places where internet connection is not available or worst case scenarios like natural disasters. Previous researches on AR as a medium of learning also did not pull out the importance of Bengali language. They had serious lack of necessary details in the modeling process of Bangla alphabet [22], and mainly focused on supplementary elements [23]. Bangla alphabet were largely represented in the form of 2D images.

To overcome those limitations our effort of calculating the VEP count of every 3D modeled Bangla letters which has tested through an augmented reality based mobile application makes a significant number of contributions. Firstly, we are focusing on Bengali language so that the young learners who are the native speaker of this language are getting the opportunity to have a smart learning aid at their fingertips. This application can also be a great help for those learners who are not native speaker of Bengali. Secondly, our developed work does not require internet connection during operation so that this application can run effectively in those places where high speed internet connection is not available. Thirdly, we are adding a new dimension in the way of teaching kids in comparison with the traditional teaching approach where they can experience the Bangla alphabet in 3D. When a kid can perceive a letter at its fullest extent, acquisition is simply a joyful consequence and undoubtedly it is an effortless approach of learning. To be mentioned, our proposed model does not use VR technology and this is because-VR based applications require gadgets like google card board or VR box. And it is troublesome for those users who use spectacles.

Important to mention- this research work mainly focuses on the 3D modeling mechanism. Development process of the application that we created as a prototype is not in the main focus as creating this kind of application has some discrete methods. However a keen focus has given in describing the process of calculating the count of vertex, polygon and edge as these crucial parameters of 3d assets have never been published before. At the initial stage of this research, it was a real challenge to identify the most appropriate mathematical model that can compute the counts of our interest. After so many trails and errors we finally found that Euler's polyhedron theorem is our key to move forward. Next section of this paper clearly discuss how this theorem comes into action while calculating the counts of desired parameters. As a limitation of this work- we clearly would like to say that the counts we are unleashing may change depending on the modeling mechanism. Continuous upgrade on 3d modeling tools may also reflect a change. Thus it can be regarded as a scale or threshold point of standard. If the VEP triplet count of any virtually created Bangla Alphabet does not exceed our proposed value, it is expected that the developed asset will work perfectly fine as part of an augmented reality application.

3   Proposed Model with Detailed Workflow

Figure 1 represents a block diagram which indicates different stages that were involved in the implementation of our proposed learning application for elementary learners. As we can see, there are six major steps involved in this system implementation process. 3D modeling and poly optimization is described in details in this paper as the VEP triplet counting process took place in this stage and is the major focus of this paper. As we know Bengali language has vowel and consonant just like any other languages. Our determination was to make a one stop solution. So at this stage we decided to include both vowels and consonants in our project. Thus we ended up with the inclusion of 50 letters at the selection of characters stage while developing the prototype that is able to simulate. As our intention was to integrate 3D letters with our learning application, so we had to perform modeling tasks. Here creating the 3d models of Bengali alphabet were the major concern. Inside the virtual environment we created the line art of every single letter. Later on we converted them to poly objects. At that point there was no depth on those letters. So we had to apply "Shell" modifier on them just to make them a proper three dimensional model as demonstrated in Fig. 2. The process ends up with the conversion of those shapes into poly objects.

Table 1. Vertex, polygon and edge count during the formation of every single 3D letter

Sl. No.	Name of 3D Model	Vertex Count (V)	Poly-face Count (F)	Edge Count (E = V + F - 2)
1	অ	544	274	816
2	আ	772	390	1160
3	ই	1080	542	1620
4	ঈ	1212	608	1818
5	উ	1168	586	1752
6	ঊ	1156	580	1734
7	ঋ	624	314	936
8	এ	800	402	1200
9	ঐ	1120	562	1680
10	ও	900	452	1350
11	ঔ	1140	572	1710
12	ক	556	280	834
13	খ	628	316	942
14	গ	632	318	948
15	ঘ	468	236	702
16	ঙ	888	446	1332
17	চ	400	202	600
18	ছ	768	386	1152
19	জ	1132	568	1698
20	ঝ	352	178	528
21	ঞ	1512	758	2268
22	ট	876	440	1314
23	ঠ	732	368	1098
24	ড	512	258	768
25	ঢ	452	228	678
26	ণ	676	340	1014
27	ত	580	294	872
28	থ	684	344	1026
29	দ	324	164	486
30	ধ	512	258	768
31	ন	288	146	432
32	প	664	334	996
33	ফ	656	330	984
34	ব	208	106	312

(*continued*)

Table 1. (*continued*)

35	ভ	672	340	1010
36	ম	544	274	816
37	য	316	160	474
38	র	432	220	650
39	ল	744	374	1116
40	শ	712	358	1068
41	ষ	312	158	468
42	স	576	290	864
43	হ	708	356	1062
44	ড়	680	344	1022
45	ঢ়	592	300	890
46	য়	564	286	848
47	ৎ	656	330	984
48	where	512	260	770
49	where	840	424	1262
50	ঃ	428	218	644

Being poly objects, all the letters that we created for our learning application had three crucial parameters. These parameters are – edge, vertex and polygon. Table 1 shows the polygon count and vertex count for every single Bengali letter that we modeled for the implementation of our learning application. These two parameters are very easy to figure out inside the virtual 3D environment of almost any 3D development tool. But our Table has an additional column in it that shows the edge count of every single 3D letter. And this is a unique contribution of our endeavor indeed as the complete VEP triplet count of 3D models (of Bangla alphabet) have not been revealed by any enthusiasts or researchers so far even though these values have crucial impact on the development of any 3D model. So we would like to mention the process of how we figured out this VEP triplet count.

We applied Euler's Polyhedron formula at this stage. But before diving into that formula, it would be great to explore the core concepts of polyhedron a bit more. Polyhedron is basically a solid object and a handful of flat faces make the surface of it. Those flat faces are bordered or surrounded by straight lines. In a closer look each face turns out to be a polygon. Those polygons are closed shapes in 2D plane formed by points and connected by straight lines. Under any circumstances polygons can never have holes in them as illustrated in Fig. 4. Here the left-hand shape has every reasons to be claimed as a polygon while the right hand shape is not, as a hole exists in it.

```
┌─────────────────────────────────────────────────┐
│              Selection of Characters              │
└─────────────────────────────────────────────────┘
                        ↓
┌─────────────────────────────────────────────────┐
│          3D Modeling & Poly Optimization          │
└─────────────────────────────────────────────────┘
                        ↓
┌─────────────────────────────────────────────────┐
│             Exporting Optimized Models            │
└─────────────────────────────────────────────────┘
                        ↓
┌─────────────────────────────────────────────────┐
│           Preparation of Triggering Input         │
└─────────────────────────────────────────────────┘
                        ↓
┌─────────────────────────────────────────────────┐
│   Integration of Triggering Input with 3D Models  │
└─────────────────────────────────────────────────┘
                        ↓
┌─────────────────────────────────────────────────┐
│                      Deploy                       │
└─────────────────────────────────────────────────┘
```

Fig. 1. Different stages of implementing the proposed learning model

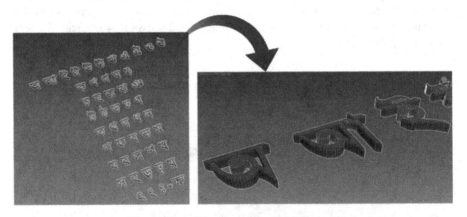

Fig. 2. Bengali letters getting their 3D shape

In a state where all the sides are of same length and their in between angles are also equal, we can refer a polygon as regular. The triangle and square stated in Fig. 3 replicates regular polygons. A regular polygon many have shapes other than triangle or square, for instance- the shape of a pentagon or hexagon or any ideal form of n-gon are also valid.

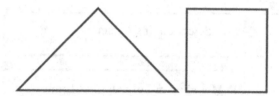

Fig. 3. Illustration of polygonal shapes

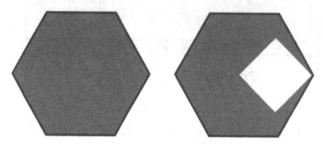

Fig. 4. Illustration of hole on polygonal surface

From those 2D shapes if we move one dimension up then we will end up finding a polyhedron. An ideal polyhedron is absolutely closed and its surface is formed by a number of polygonal faces. Important to mention that polyhedron is solid in nature. In the context of polyhedron- the sides of polygonal faces are identified as edges while vertices are the corners of a poly-face. Thus, when two faces meet along, any vertex lies on at least three different faces. Figure 5 illustrates this discussion by the help of a very well-known polyhedron.

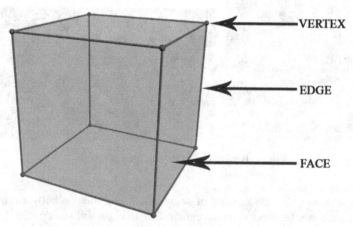

Fig. 5. Pointing vertex, edge and face on a cubic polyhedron.

A polyhedron should always have to be in one piece. For instance, two or more individual parts linked by only an edge or a vertex cannot build it. This means the shape stated in Fig. 6 is not a polyhedron even though two separate shapes are connected by edges.

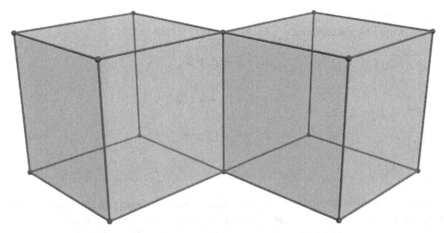

Fig. 6. Illustration of not forming a polyhedron even after edge connection

For the exact same reason mentioned earlier, the shape in Fig. 7 is also not a polyhedron even after having vertex connection.

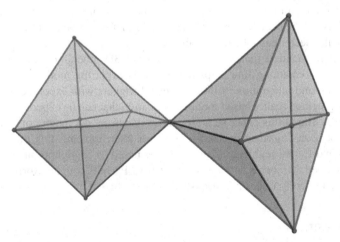

Fig. 7. Illustration of not forming a polyhedron even after vertex connection

After all these discussion, now we are ready to integrated Euler's formula with our work. Let us consider the polyhedron stated in Fig. 5 which is a cube. This cube has 8 vertices and variable V holds this value. On the other hand it has 12 edges and variable

E holds this value. Lastly variable F holds the number of faces. For the case of cube it has 6 faces. Euler formula suggests that, $V - E + F = 2$.

If we articulate this, the number of vertices minus the number of edges plus the number of face is equal to two. Getting back to our example here we have-

$$V = 8; \; E = 12; \; F = 6$$

Following Euler's Formula,

$$V - E + F = 2$$

$$=> 8 - 12 + 6 = 2$$

$$=> 14 - 12 = 2$$

$$\therefore 2 = 2$$

And the result is totally as expected. Now if we see closely at the equation, we already have two values in our hand- Vertex and Face, and we need the value of Edge. So if we simple plot the values of V and F for every 3D alphabet we inevitable end up by getting the value of E that is the edge count. And that is exactly what we did to calculate the edge count for every alphabet. If a 3D modeler knows all these three values, it certainly would be a great help to get an idea to predict the formation of a particular 3D model. We would like to mention that many groups of enthusiasts have done the 3D modeling of alphabet from many languages including Bangla. But none of them have ever published the entire polygon, vertex and edge count of the alphabet. Unlike them we have published the detailed count of each Bengali letter so that the future researchers in this particular sector can take this finding as a start.

Important to mention that after creating all the 3D models of Bangla alphabets we followed the remaining four steps that involves exporting models, preparation of triggering input and integration of those prepared triggers with exported 3d models to deploy an augmented reality based mobile application that can be experienced by smart phones. As the title of this paper suggests, those stages are not major concerns for this paper and not been described in details. However after implementing the system and deploying it in different segments of smart phones based on configuration, we found out that our developed 3d models with the mentioned VEP triplet count performs absolutely fine and faces no faulty issues like back-facing or so.

4   Conclusion

The process of calculating vertex, edge and polygon (VEP triplet count in short) count of Bangla Alphabet which are developed in 3D virtual environment to effectively use in Augmented Reality Based mobile applications is discussed in this paper. The count is represented in the form of a table and this can be a great anchor for 3d modelers, Augmented Reality based application developers and needless to mention- academic researchers. We would like to conclude by saying that there are even larger scopes to do research in this particular sector and researchers of coming times can start their journey right from where we are concluding as the continuous advancement of technology is going to offer more convenient development tools and advanced hardware.

References

1. Yuesheng, H., Tang, Y.Y.: Classification of 3D models for the 3D animation environments. In: 2009 IEEE International Conference on Systems, Man and Cybernetics (2009). https://doi.org/10.1109/icsmc.2009.5346642
2. Huang, L., Pei, Y.: Film and television animation design based on Maya and AE. In: 2010 3rd International Congress on Image and Signal Processing (2010). https://doi.org/10.1109/cisp.2010.5646354
3. San-ao, X.: Application of Maya in film 3D animation design. In: 2011 3rd International Conference on Computer Research and Development (2011). https://doi.org/10.1109/iccrd.2011.5764150
4. Xu, X., Leng, C., Wu, Z.: Rapid 3D human modeling and animation based on sketch and motion database. In: 2011 Workshop on Digital Media and Digital Content Management (2011). https://doi.org/10.1109/dmdcm.2011.52
5. Feng, Y., Zhan, S.: Simulation of real water in 3D animation. In: 2011 International Conference on Multimedia Technology (2011). https://doi.org/10.1109/icmt.2011.6001656
6. Zhou, W., Xiang, N., Zhou, X.: Towards 3D communications: real time emotion driven 3D virtual facial animation. In: 2011 Workshop on Digital Media and Digital Content Management (2011). https://doi.org/10.1109/dmdcm.2011.76
7. Mohammadi, S., Gervei, O.: 3D Res. **10**, 22 (2019). https://doi.org/10.1007/s13319-019-0232-0
8. Hong, P., Wen, Z., Huang, T.S., Shum, H.-Y.: Real-time speech-driven 3D face animation. In: Proceedings First International Symposium on 3D Data Processing Visualization and Transmission (2002). https://doi.org/10.1109/tdpvt.2002.1024147
9. Suenaga, Y.: 3D measurement technologies for computer animation. In: Proceedings Computer Animation 1996. https://doi.org/10.1109/ca.1996.540499
10. Ying, G., Xuqing, L., Xiuliang, W., Yi, F., Shuxia, G.: Design and realization of 3D character animation engine. In: 2009 2nd IEEE International Conference on Broadband Network & Multimedia Technology (2009). https://doi.org/10.1109/icbnmt.2009.5347860
11. Furukawa, M., Fukumoto, S., Kawasaki, H., Kawai, Y.: Interactive 3D animation system for Web3D. In: 2012 IEEE International Conference on Multimedia and Expo Workshops (2012). https://doi.org/10.1109/icmew.2012.122
12. Lam, A.D.K.-T., Su, Y.-Y.: A study on experimental teaching of 3D animation. In: 2016 International Conference on Applied System Innovation (ICASI) (2016). https://doi.org/10.1109/icasi.2016.7539903
13. Mejías Borrero, A., Andújar Márquez, J.M.: J. Sci. Educ. Technol. **21**, 540 (2012). https://doi.org/10.1007/s10956-011-9345-9
14. Kamphuis, C., Barsom, E., Schijven, M., Christoph, N.: Augmented reality in medical education? Perspect. Med. Educ. **3**(4), 300–311 (2014). https://doi.org/10.1007/s40037-013-0107-7
15. Lee, K.: Techtrends tech trends **56**, 13 (2012). https://doi.org/10.1007/s11528-012-0559-3
16. Miyosawa, T., Akahane, M., Hara, K., Shinohara, K.: Applying augmented reality to e-learning for foreign language study and its evaluation. In: Proceeding of the 2012 International Conference on e-Learning, e-Business, Enterprise Information Systems, & e-Government, pp. 310–316, July 2012

17. Dalim, C.S.C., Piumsomboon, T., Dey, A., Billinghurst, M., Sunar, S.: TeachAR: an interactive augmented reality tool for teaching basic english to non-native children. In: ISMAR 2016 (2016)
18. Li, S., Chen, Y., Lafayette, W.: A pilot study exploring augmented reality to increase motivation of Chinese college students learning English. Comput. Educ. **25**(1) (2015)
19. Rose, H., Billinghurst, M., Zengo, S.: An immersive educational environment for learning japanese, Report No. r-95-4. University of Washington, Human Interface Technology Laboratory (1995)
20. Wagner, D., Barakonyi, I.: Augmented reality kanji learning. In: 2nd IEEE and ACM International Symposium on Mixed and Augmented Reality, ISMAR 2003, pp. 335–336 (2003)
21. Tsai, C.W.: Univ Access Inf. Soc. (2017). https://doi.org/10.1007/s10209-017-0589-x
22. https://play.google.com/store/apps/details?id=com.FeenixLab.BLK_AR_Bangla. Accessed 31 Jan 2019
23. https://play.google.com/store/apps/details?id=com.microtech.neelimarbioscope. Accessed 10 Feb 2019

A Hybrid Machine Learning Framework for Prediction of Software Effort at the Initial Phase of Software Development

Prerana Rai[✉], Shishir Kumar, and Dinesh Kumar Verma

Department of Computer Science and Engineering, Jaypee University of Engineering and Technology, Raghogarh-Vijaypur, India
prerana.rai99@gmail.com, dr.shishir@yahoo.com,
dinesh.hpp@gmail.com

Abstract. In the era of software application, the prediction of the effort of software plays an essential role in the success of project software. The inconsistent, inaccurate, and unreliable prediction of software leads to failure. As the requirement and specification changes as per the software needs, accurate prediction of effort is a difficult task for developing software. This prediction of effort must be calculated accurately to avoid unpredicted results. At the early stages of development, these inaccurate, unreliability, and uncertainty are the drawback of previously developed models. The main aim of the study is to overcome the drawbacks and develop a model for the prediction of the effort of software. A combination of regression analysis and genetic algorithm has been used to develop the model. The model is trained and validated using the ISBSG dataset. The proposed model is compared for performance with a few baseline models. The results show that the proposed model outperforms most of the baseline models against different performance metrics.

Keywords: Software effort estimation · Regression analysis · Deep learning · Genetic algorithm · Evaluation metrics

1 Introduction

Effort prediction of software is a prominent way to know the cost, time and man-power used to design the software product at the starting phase of development [1]. Unlike other projects, software projects are having more uncertainty, challenges, technical complexity, etc. Several studies proposed various models for parameter's prediction in software. Software developers still facing the challenges in prediction of software parameter's early in the software development life cycle. Early prediction of software parameter's help project manager in taking decision whether to accept or reject the project. Software effort estimation helps developers in proper project planning, scheduling of different project task, resource and people management, and controlling cost and progress of project. There may be three faces of outcome of an effort estimation process; underestimation, accurate estimation, and overestimation. There would be serious consequences

© Springer Nature Singapore Pte Ltd. 2020
M. Singh et al. (Eds.): ICACDS 2020, CCIS 1244, pp. 187–200, 2020.
https://doi.org/10.1007/978-981-15-6634-9_18

of underestimates and overestimates, such as, the former leads to understaffing, over-budgeting and late delivery, whereas the later leads to resource and budget wastage. These two faces of effort prediction process increases the project failure with respect to time and budget [2]. There are several reasons for the failure of the software project uncertainty of the requirements that occur in the software and the system and estimation of cost, duration, staff, and size of the project not done properly [3].

In the past three decades, several researchers focused on designing a model for accurately predicting the effort of the software, i.e., the third face of effort prediction. Numerous effort estimation techniques have been proposed which can be described broadly into three categories [4]: expert judgment [5], parametric techniques [6] and machine learning techniques [7]. Jorgensen et al. [8] did a literature review on designing effort estimation model using several techniques and concluded that 11 estimating techniques have been used vastly; among which regression techniques was the dominant on the other techniques. Nowadays Machine Learning (ML) techniques are being used by many researchers to design the model for effort estimation. The complexity between the dependent variable and other software attributes can be decreased by designing a model with the latest techniques, particularly with respect to the ML.

After reviewing different techniques for effort estimation, analogy-based techniques have manifest to be promising methods for effort prediction. The analogy-based estimation method has two advantages: 1) ease of understanding because of human problem-solving approach and, 2) the approach can model the complex relation between variables. Despite their advantages, classical analogy-based estimation approaches are limited by their inability in handling linguistic values and managing imprecision and uncertainty [9]. Although there are large number of prediction models in Software Development Effort Estimation (SDEE) [10], very few of them achieved good result under all circumstances, as their performance changes according to dataset, which makes them unstable [11]. It has been observed that the performance of multiple techniques for designing the effort estimation model was better than any single technique used to design the model [12]. The earlier studies shows that the combination of more than one technique into an ensemble form leads to good accuracy of the model in comparison to the model designed with the single techniques [13].

In SDEE, more than one technique is ensembled with the help of specific combination rule that is Ensemble Effort Estimation (EEE) technique. The SDEE defines two types of EEE techniques 1) Homogeneous EEE [1] is an ensemble that consists of members having a single-type base learning algorithm [14]. 2) Heterogeneous EEE is an ensemble that consists of having different base learning algorithms [12]. Since each single estimation techniques have its strengths and weaknesses [15], motive of the EEE is to alleviate the weaknesses of the techniques and combine the advantages of the techniques by combining one or more solo techniques through an ensemble. This helps in finding more accurate effort estimation at the initial level of the development phase. With the consideration of this rule, this study proposed a hybrid model using Genetic Algorithm (GA) and Deep Neural Network Regressor (DNNR). The proposed model has been compared with the single techniques (GA and DNNR) to know the accuracy of prediction in respect with proposed hybrid model.

2   Background

Software effort prediction is a complex and most critical, but inevitable task in software development. In last 4 decades, drastic upward changes have been observed in the techniques for software effort prediction. There are several researchers who have proposed various models for predicting effort with reference to the different datasets and techniques. These techniques would be having some pros and cons according to the situation and the input criteria. The first idea of effort prediction of software was using manual rule of thumb [16], in which each required element of the model was used to prepare the test cases and the sum of the test cases was taken to find out the defects in the software. Finding out the defect helps in overall effort estimation of software. In late 1970s, Barry Boehm proposed a new method known as COCOMO model in which mathematical models were formulated on the large historical dataset. Later, he continued to worked on the estimation and proposed many other algorithms described in his book "Software Engineering Economics" [6].

The algorithmic models were used to predict the effort of software by establishing the relationship with other characteristics of software project such as between software size, design metrics, project metrics, etc. These models are parametric in nature which represents the formula of standard form which is parameterized and derived from historical data. Few examples of such models are function points analysis [17] and software lifecycle management [18]. The algorithmic model was unable to capture the complex set of relationships which gave a failure to the model designed using it. In recent years, computational intelligence models are used to predict effort of software. This includes fuzzy logic [19], neural networks [4], regression trees [20], genetic programming [21], case-based reasoning [22] and support vector regression [23], etc. The advantage of these models include the ability to learn from historical projects data [24]. Although the field of effort estimation appears to be drenched with large number of predictive models, researchers are working to find a new approach in formulation of predictive model. In this process "Ensemble" approach plays a vital role in designing the model with the help of exiting models. This approach helps in measuring the accuracies and drawback of existing methodologies.

The conclusion of the above studies is that considering a greater number of factors or metrics and the use of different parameters improve the accuracy of effort prediction in the developing software. The set of selected metrics for checking the performance of model can be termed as the evaluation metrics. Further, the type of hybrid method in designing the model also effects the result of the model. In this paper, a systematic approach has been followed from selection of appropriate parameters, collection of data for selected parameters, pre-processing of collected data, formulation of hybrid model, analyzing the model with data set, to presenting and validating the result of analysis.

3   Proposed Approach

This section highlights the approach used in formulation of effort prediction model. The task has been divided in four steps. In the first step, the raw input data set needs to pre-process for eliminating anomalies from it. The pre-processed data set contains many numbers of features, the relevance feature(s) are selected in the second step. In third step, the proposed model has been developed using genetic algorithm and deep neural network regressor. Finally, in the fourth step, evaluation metrics values have been generated for performance evaluation of the proposed H-GADNNR model with the existing models. The development of proposed model has been graphically represented in Fig. 2 and 3.

3.1 Dataset Used

The use of real-life projects data in effort prediction model has been the most reliable approach. The major concern of selecting a data set is that the constant results should repeatedly occur during testing of the model [7]. The industrial dataset, used in this study, was published by ISBSG Release R12 [25] which has been extensively used by researchers and organization's for analyzing and validating their respective models. The dataset comprises of 8000+ cross company projects data, 251 parameters grouped into 30 categories, and with a combination of numerical and string dataset. The only limitation of this data set is its heterogeneity because of collection of data from large number of organization's [43].

Table 1. Project selection for effort

Attribute	Selected values
Data quality	A, B
Functional sizing methods	1 IFPUG V4+
Count approach	IFPUG
Unadjusted function point	A, B
Development type	New development

This large data sets needs to filter according to the appropriate projects and variables for the proposed approach. The attributes selected from the filter dataset has been done on project basis which has been showed in Table 1. In this study, appropriate categories have been selected on chronological manner for testing and training of proposed H-GADNNR model. Data pre-processing has been performed to eliminate missing values and to avoid biased estimates of coefficients. The qualitative variables were balanced and aggregated to reduce the number of classes. The variables with less than five observations have been eliminated.

3.2 Proposed H-GADNNR Model for Effort Prediction

3.2.1 Genetic Algorithm

Genetic Algorithm (GA) has been a search methodology which uses probability transform rule and fitness function for search direction. In GA, a population used is a set of candidate solutions which is composed of chromosomes. Each individual chromosome is a candidate solution. GA works on Darwinian principle of "survival of the fittest", after a series of iterative computation, GA obtains the optimal solution. The fitness function reflects each chromo's adaptive potential for its living environment. In the evolutionary process a fitness function estimates the efficiency of a solution. The fitness value is affected randomly by the genetic operation, which includes the crossover operator, the mutation operator, and the selection operator.

The chromosome solution which obtains the best fitness value is the optimal solution after many generations. The crossover is the most important genetic operator, a random mechanism for exchanging genes between two chromosomes using one crossover point, two crossover point, or crossover homologue. Mutation is the genes in the chromosome of which may be changed periodically, i.e. genes code from 1 to 0 or vice versa. The process flow of crossover and mutation has been shown in Fig. 1.

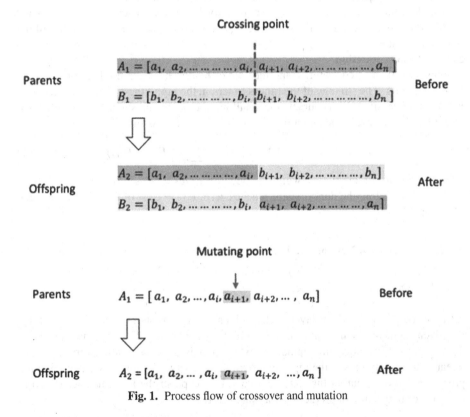

Fig. 1. Process flow of crossover and mutation

In selection, the highest fitness value chromosomes are more likely to be selected by the roulette wheel or tournament selection methods into the recombination pool. GA is using evolutionary strategy, which includes elitist GA selection, father-offspring combined GA selection, and so on, to ensure GA convergence.

The detailed workflow of genetic algorithm for proposed model is as follows:

Step 1: Selection of initial population

Initial population $P_o(0)$, which consists of n pairs of parent, is generated randomly. If the population size is too large, then the algorithm complexity is too high, and the quantity of the computation is too large. If the population size is too small then the algorithm's optimal performance is reduced, and the algorithm is easily plunged into the optimal local solution. Size of the training samples should be considered from the range of 30 to 200.

Step 2: Crossover operation

Crossover process produces intermediate population $Co(t)$ by independently performing crossover for parent having n pairs in current population $Px(t)$. Operator crossover C_t: $P_x(t) \rightarrow C_0(t)$ acts upon the individual space's subspace.

Step 3: Mutation operation

Mutation activity produces population $Mu(t)$ by independently conducting mutation for intermediate individuals in intermediate population $Po(t)$. The switch controller M_t: $P_0(t) \rightarrow Mu(t)$ constantly changes the subspace which has full space search capability of the individual space.

Step 4: Fitness function

The number of selected features, and the cost of feature has been used to construct the fitness function. The strategy of the fitness function, the high fitness value is determined by the high classification accuracy, the small number of features and the low total cost of the feature. To avoid the denominator to reach zero, fitness function is calculated as:

$$ff = We_a \times Acc + We_f \times (C + (\sum_{i=1}^{n} Fe_c \times Fe_i))^{-1} \tag{1}$$

where Acc = accuracy, We_a = weight of accuracy, We_f = weight feature, C = setting constant of avoiding the denominator reaches zero, Fe_c = Cost of feature, Fe_i = value of feature.

3.2.2 Deep Neural Network Regressor

The Deep Neural Network (DNN) technique is artificial neural network which consists of multiple hidden layers between the input and output layers. This approach is useful for studying the effectual and discriminative features in nature. This network consists of input, output and hidden layers. The hidden layer consists of units that transform the given input into something that the output layer can use. The deep network finds the correct mathematical manipulation, whether it is a linear relationship or a non-linear relationship, to turn the input into output. The network moves by calculating the probability of each output through the layers. In this paper the DNN has been used in the form of regression in predicting software effort.

3.2.3 Proposed H-GADNNR

The overall architecture of the proposed H-GADNNR (Hybrid – Genetic Algorithm Deep Neural Network Regressor) model has been shown in Fig. 2 which comprises of the entire workflow of developing the model.

The proposed model has been designed as follows: Genetic algorithm has been used to find out the best feature. These best features will then be used as an input

Fig. 2. Overall architecture of proposed H-GADNNR

parameter to train and test the model and find out the best effort prediction for effort. The random population of n chromosomes has been generated. The fitness function of each chromosome (ch) in the population is evaluated. After finding the fitness function a new population has to be generated. This has been done by selection, crossover and mutation. The selection process helps to select the best fit individuals. It chooses two pairs of individuals (parents) based on their fitness ratings. High-fitness individuals have a higher chance of being selected for reproduction. In the crossover process each pair of parents are mated to form a new offspring. In the mutation phase the new offspring formed is subjected to a mutation with a low random probability. The process goes on until it reaches to the termination point and the new population has been formed. The considered parameters are as follows: population size $= 1100$, generation $= 50$, crossover probability $= 0.7$, mutation probability $= 0.4$, crossover technique $=$ random single point. This new population has been used as an input parameter in the deep neural network regressor. Seven input parameters, two hidden layers and one output layer has been used in designing the model. Sigmoid activation function has been used in this model $\emptyset(z) = \frac{1}{1+e^{-z}}$. The graphical design architecture of the proposed model has been shown in Fig. 3. The main steps of the workflow are as follows:

Step 1: Input dataset:
It contains the training and testing dataset.

Step 2: Data pre-process:
The data pre-process has been done by linear scaling. Each feature present in the dataset can be linearly scaled to the range [0,1]. The advantage of linear scaling is to avoid attributes in large numerical ranges that dominate those in smaller numerical ranges, and to avoid numerical difficulties during the value of the calculation feature, and to help achieve higher accuracy.

Step 3: Select feature subset:
The feature subsets which are relational has been chosen and the unrelated feature subsets has been discarded. After doing the feature subset the training and testing dataset has been divided.

Step 4: Train DNNR:
The DNNR has been trained by the training dataset.

Step 5: Calculate evaluation metrics:
The proposed model has been evaluated on four different evaluation metrics (viz. Accuracy, MAE, MSE, RMSE).

Step 6: Fitness evaluation:
The number of selected features and cost of feature has been used to construct a fitness function shown in Eq. 1.

Step 7: Ending condition:

The ending condition is to achieve good accuracy and minimum error. When the ending condition has been satisfied, the operation ends; otherwise the next generation operation is proceed until the condition is not satisfied.

In the later stage, proposed H-GADNNR model has been compared with GA and DNNR. The single models selected for comparison have been widely used by researchers as a single model for prediction of effort [26]. For comparing the performance of proposed H-GADNNR model against the performance of GA and DNNR model, few evaluation metrics have been identified. The proposed H-GADNNR model algorithm has been describe below in Algorithm 1.

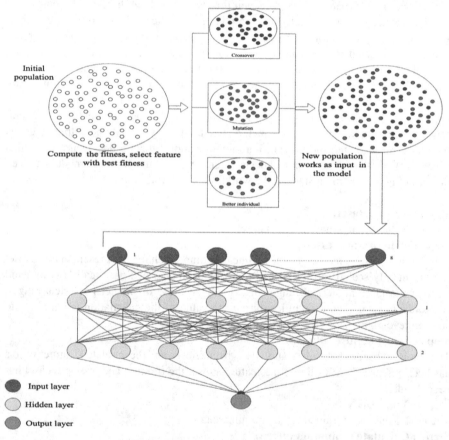

Fig. 3. Process of building effort model

Algorithm 1. H-GADNNR model

Step 1: Read data D ← Input

Step 2: Apply data pre-processing
Step 3: Feature analysis i.e. Total Number of Feature ← D(1: end − 1, end)
Step 4: Initialize the population
Step 5: Generate initial configuration of deep neural network regressor and genetic algorithm parameters popsize, co_{prob} mu_{prob}
Step 6: Random weight (w) and fitness function assignment is initialized
Step 7: Apply feature selection

Begin ch = total number of chromosomes
Initial population generation pop = {pop1, pop2, pop3, pop4, ..., pop_{ch}}
Fitness computation

Step 8: Generate new population pop_{new}
Step 9: Randomly select two new chromosomes as ch1 and ch2 from pop_{new}
Step 10: Apply crossover functionality to obtain new chromosomes with the help of ch as cross probability
Step 11: Apply mutation functionally with mutation probability (muprob)
Step 12: Compute fitness of ch1 and ch2
Step 13: Insert ch1 and ch2 in the pop_{new}
Step 14: Pick best chromosomes from popn-1 and pop_{new} to formulate pop_n
Step 15: Obtain optimal weights and best fit values as feature selection fs_{best}
Step 16: Obtain fs_{best} is used as an input parameter in DNNR algorithm
Step 17: Apply hidden layer computation and activation function to find accurate effort estimation
Step 18: Performance evaluation in terms of evaluation metrics (MSE, MAE, RMSE).

3.3 Performance Evaluation

The major challenge in estimating effort of by a model is its effectiveness to produce an accurate prediction of effort. So many studies have been performed to identified the reasons for inaccuracy of estimates by a model. Jorgensen et al. highlighted the three major reasons; unexpected events and change done by the client, problem with resource allocation, and less amount of time spent on effort estimation work [27]. Magnitude of Error Relative (MER) and Mean Magnitude of Relative Error (MMRE) have been widely used by the researchers as evaluation metrics for their prediction model. But, Shepperd et al. suggest not to use these metric because of their biased nature [28]. Due to this reason the unbiased estimation criteria (viz. Mean Squared Error (MSE), Mean Absolute Error (MAE), and Root Mean Squared Error (RMSE)) described in the equations below have been used to validate the accuracy of the model. The evaluation metrics are defined as follows:

$$MSE = \frac{1}{n}\sum (y_i - \bar{y}_i)^2 \tag{2}$$

$$MAE = \frac{1}{n}\sum_{j=1}^{n} |y_i - \bar{y}_i| \tag{3}$$

$$RMSE = \sqrt{\frac{1}{n} \sum_{j=1}^{n} (y_i - \bar{y}_t)^2} \qquad (4)$$

where,

y_t = actual value, yt' = estimated value, n = number of samples.

3.4 Results and Discussion

This section explains the result obtained based on the evaluation metrics used for all three models. The proposed H-GADNNR model shows good accuracy, capability of handling outliners and noises in the input dataset in comparison to two other models. The obtained values of different evaluation metrics for all the models have been shown in Figs. 4, 5, 6 and 7.

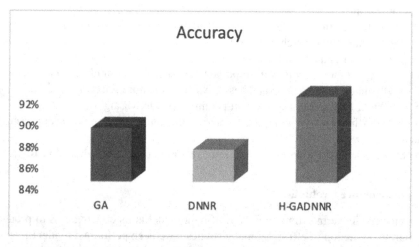

Fig. 4. Variation of accuracy from H-GADNNR Model, GA and DNNR

The lower MSE, MAE, and RMSE value indicates higher accuracy of prediction it is inferred from Table 2 that DNNR has the lowest MSE in comparison to GA and proposed H-GADNNR model. In the case of MAE, proposed H-GADNNR model has the lowest value in comparison to GA and DNNR. The RMSE value for GA and proposed H-GADNNR model are better than DNNR model. In terms of accuracy of the model in effort prediction, proposed H-GADNNR model shows 92% whereas for GA, it is 89% and for DNNR it is 87%.

Finally, it is concluded from the results that proposed hybrid approach is helpful in accurate effort prediction with very less error rate in comparison to the GA model and DNNR model. This H-GADNNR model helps the developers to make accurate effort prediction for their project before starting the development processes. It also shows that when the software has been developed with the help of different parameters shows good effort. Ultimately, the accurate effort prediction helps in proper resource allocation in

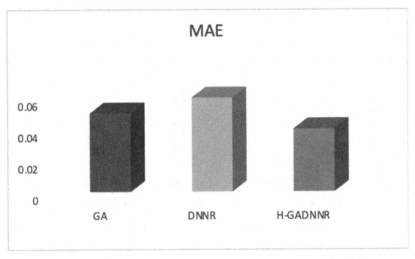

Fig. 5. Variation of MSE from H-GADNNR Model, GA and DNNR

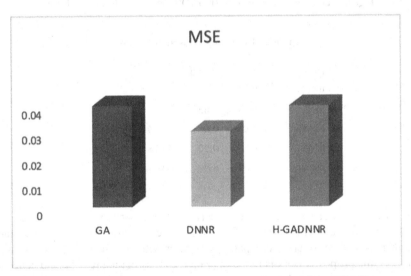

Fig. 6. Variation of MAE from H-GADNNR Model, GA and DNNR

different development activities throughout the development life cycle as well as help to design better software.

3.5 Threats to Validity

Threats that might affect the validity of the model is described in this section. This includes:

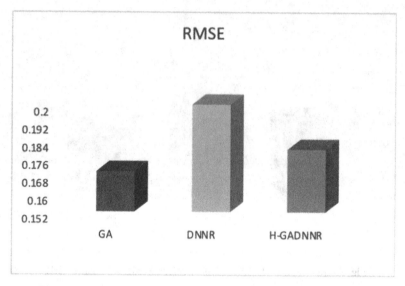

Fig. 7. Variation of RMSE from H-GADNNR Model, GA and DNNR

Table 2. Evaluation measure for effort

Evaluation metrics	GA	DNNR	H-GADNNR
MSE	0.04	0.03	0.04
MAE	0.05	0.06	0.04
RMSE	0.17	0.20	0.18
Accuracy	89%	87%	92%

- The ISBSG dataset has been used to perform the experiments. The dataset contains cross-company data collected from different companies. Though it's debatable whether the data from cross-company projects or within-company projects is superior [29], the proposed work is relied only on the cross-company data, and hence the validity can be questioned on this aspect.
- As recommended by Shepperd et al. [28] we did not use the evaluation metrics MMRE and MMER. As described by Shepperd et al., these evaluation criteria are biased which favours the underestimate models. Though MMRE have successfully been used in various studies for comparison, the absentia of the same in the proposed work might pose a threat to validity.

4   Conclusion

This paper investigates the key approach on examining the prediction of software effort. Several researchers had tried to formulate a model for accurate prediction of software effort at the initial phase of software development. GA and DNNR are the widely used approaches for classification, regression, and prediction etc. These two algorithms have been used in this study to formulate a new model for prediction of effort using some known parameters. The proposed H-GADNNR model has been trained and tested on the dataset collected from ISBSG data repository. The performance of the proposed H-GADNNR model has been compared with existing GA and DNNR models. It is concluded from the results that proposed H-GADNNR model is helpful in accurate effort prediction with very less error rate in comparison to the GA model and DNNR model. This proposed model helps the developers to make accurate effort prediction of software before starting the development processes. Ultimately, the proposed H-GADNNR model provides support in making decisions in developing software by resource planning, managing and controlling the activities. The organizations can be benefitted by utilizing this early effort estimates for proper budgeting, duration, and distribution of resources among different activities during software development.

References

1. Wen, J., Li, S., Lin, Z., Hu, Y., Huang, C.: Systematic literature review of machine learning based software development effort estimation models. Inf. Softw. Technol. **54**, 41–59 (2012). https://doi.org/10.1016/j.infsof.2011.09.002
2. Eck, D., Brundick, B., Fettig, T., Dechoretz, J., Ugljesa, J.: Parametric Estimating Handbook, 4th edn. The International Society of Parametric Analysis (ISPA), Chandler (2009)
3. Fedotova, O., Teixeira, L., Alvelos, A.H.: Software effort estimation with multiple linear regression: review and practical application. J. Inf. Sci. Eng. **29**, 925–945 (2013). https://doi.org/10.1016/S0950-5849(96)00006-7
4. de Barcelos Tronto, I.F., da Silva, J.D.S., Sant'Anna, N.: An investigation of artificial neural networks based prediction systems in software project management. J. Syst. Softw. **81**, 356–367 (2008). https://doi.org/10.1016/j.jss.2007.05.011
5. Hughes, R.T.: Expert judgement as an estimating method. Inf. Softw. Technol. **38**, 67–75 (1996)
6. Boehm, B.W.: Software Engineering Economics. Prentice-Hall, Englewood Cliffs (1981)
7. Idri, A., Khoshgoftaar, T.M., Abran, A.: Can neural networks be easily interpreted in software cost estimation? In: IEEE International Conference Fuzzy System, pp. 1162–1167 (2002)
8. Jorgensen, M., Shepperd, M.: A systematic review of software development cost estimation studies. IEEE Trans. Softw. Eng. **33**, 33–53 (2007). https://doi.org/10.1109/TSE.2007.256943
9. Pandey, P., Kumar, S., Shrivastava, S.: A fuzzy decision-making approach for analogy detection in new product forecasting. J. Intell. Fuzzy Syst. **28**, 2047–2057 (2015)
10. Aslam, W., Ijaz, F., Lali, M.I., Mehmood, W.: Risk aware and quality enriched effort estimation for mobile applications in distributed agile software development. J. Inf. Sci. Eng. **33**, 1481–1500 (2017). https://doi.org/10.6688/JISE.2017.33.6.6
11. Shepperd, M., Kadoda, G.: Comparing software prediction techniques using simulation. IEEE Trans. Softw. Eng. **27**, 1014–1022 (2001)
12. Kocaguneli, E., Menzies, T., Keung, J.W.: On the value of ensemble effort estimation. IEEE Trans. Softw. Eng. **38**, 1403–1416 (2011)
13. Elish, M.O.: Assessment of voting ensemble for estimating software development effort. In: 2013 IEEE Symposium on Computational Intelligence and Data Mining, pp. 316–321 (2013)
14. Braga, P.L., Oliveira, A.L.I., Meira, S.R.L.: Software effort estimation using machine learning techniques with robust confidence intervals. In: 19th IEEE International Conference on Hybrid Intelligent Systems, pp. 181–185 (2007)

15. Elish, M.O., Helmy, T., Hussain, M.I.: Empirical study of homogeneous and heterogeneous ensemble models for software development effort estimation. Probl. Eng. (2013). https://doi.org/10.1155/2013/312067

16. Jones, C.: Estimating Software Costs: Bringing Realism to Estimating, 2nd edn. McGraw-Hill, New York (2007)

17. Jeng, B., Yeh, D., Wang, D., Lhu, S.L., Chen, C.M.: A specific effort estimation method using function point. J. Inf. Sci. Eng. **27**, 1363–1376 (2011)

18. Putnam, L.H.: A general empirical solution to the macro sizing and estimating problem. IEEE Trans. Softw. Eng. **4**, 345–361 (1978)

19. Lopez-Martin, C., Yanez-Marquez, C., Gutierrez-Tornes, A.: Predictive accuracy comparison of fuzzy models for software development effort of small programs. J. Syst. Softw. **81**, 949–960 (2008)

20. Baskeles, B., Turhan, B., Bener, A.: Software effort estimation using machine learning methods. In: Proceedings of 22nd International Symposium on Computer Science, pp. 209–214 (2007)

21. Shukla, K.K.: Neuro-genetic prediction of software development effort. Inf. Softw. Technol. **42**, 701–713 (2000)

22. Chiu, N.H., Huang, S.J.: The adjusted analogy-based software effort estimation based on similarity distances. J. Syst. Softw. **80**, 628–640 (2007)

23. Oliveira, A.L.I.: Estimation of software project effort with support vector regression. Neurocomputing **69**, 1749–1753 (2006)

24. Idri, A., Abran, A., Khoshgoftaar, T.: Estimating software project effort by analogy based on linguistic values. In: Proceedings of 8th IEEE Symposium on Software Metrics, pp. 21–30 (2002)

25. ISBSG (2018)

26. Nassif, A.B., Azzeh, M., Capretz, L.F., Ho, D.: Neural network models for software development effort estimation: a comparative study. Neural Comput. Appl. **27**(8), 2369–2381 (2015). https://doi.org/10.1007/s00521-015-2127-1

27. Jorgensen, M., Molokken-Ostvold, K.: Reasons for software effort estimation error: impact of respondent role, information collection approach, and data analysis method. IEEE Trans. Softw. Eng. **30**, 993–1007 (2004)

28. Shepperd, M., MacDonell, S.: Evaluating prediction systems in software project estimation. Inf. Softw. Technol. **54**, 820–827 (2012)

29. Kitchenham, B.A., Mendes, E., Travassos, G.H.: Cross versus within-company cost estimation studies: a systematic review. IEEE Trans. Softw. Eng. **33**, 316–330 (2007)

Chronic Disease Prediction Using Deep Learning

Jyoti Mishra[✉] and Sandhya Tarar[✉]

School of ICT, Gautam Buddha University, Greater Noida, India
Jyotidotmishra27@gmail.com, tarar.sandhya@gmail.com

Abstract. Nowadays data is growing rapidly in bioscience and health protection, in clinical information, an exact investigation can benefit early infection identification, patients' social insurance, and community services. Prediction is an significant aspect in the health care domain. In this paper, we establish ML and deep learning algorithms for Prediction of patients' chronic diseases. Experiment with the refitted prediction model from the standard dataset available. Objective of this paper is to forecast chronic diseases in the individual patient by using the machine learning method, K-nearest neighbor, decision tree and deep learning using (RELU or Rectified linear activation function, sigmoid activation function, deep sequential network) and Adam as an optimizer. Examine to several ordinary algorithms, the accuracy of the proposed system is enhanced. With the comparison of other algorithms, deep learning algorithms will give better accuracy it's about 98.3%. These techniques are applied to predict heart, breast cancer, and diabetes chronic diseases.

Keywords: Healthcare · Deep learning · Machine learning · Chronic disease · RELU activation function

1 Introduction

Artificial intelligence is the study of getting personal computer to Gain or acquire knowledge by studying and behave like a human being, and improve their learning by experience after some time in self-sufficient design, by providing them data and information in the form of observations and real-word connections [1, 2]. In the domain of ML, we have our dataset which has to be trained by the machine learning model that is the chosen algorithm. As machine learning has some limitations like ML are not useful when no. inputs are large amounts of data and cannot solve crucial Artificial intelligence problems like NPL, image recognition, etc. to overcome these limitations of machine learning, deep learning has come into existence. Deep learning is the newest field under machine learning which is gaining huge popularity as the amount of data is increasing exponentially [3]. Chronic diseases are a significant reason for death and inability around the world. In India, chronic sicknesses are anticipated to represent 53% of all things considered. Absolute death in India, 2005 is 10,362,000, Total anticipated chronic infection-related death in India, and 2005 is 5,466,000 [4]. WHO extends that throughout the following 10 years in India, Over 60 million individuals will die due to chronic diseases. Death from irresistible sicknesses, maternal and perinatal conditions, and dietary

© Springer Nature Singapore Pte Ltd. 2020
M. Singh et al. (Eds.): ICACDS 2020, CCIS 1244, pp. 201–211, 2020.
https://doi.org/10.1007/978-981-15-6634-9_19

insufficiencies consolidated will diminish by 15% [5]. Death from incessant ailments will increment by 18% most uniquely, death from diabetes will increment by 35%. 15% of death in India was because of heart sicknesses in 1990; presently up to 28%. The heart sicknesses sped up more than 2.1 million death in India in 2015 at all ages, or over a fourth all things considered. The age between 30 to 69 years, 1.3 million deaths are caused by cardiovascular [6]. Grown-ups brought into the world after the 1970s are significantly more defenseless against such death than those brought into the world before, the examination appears [7]. As per the union ministry of health, breast cancer disease positions as the principle danger among the Indian females as high as 25.9 per 100,000 women and mortality of 12.8 per 100,000 women. As indicated by gauges, in any event, 18, 97,901 ladies in India may have breast cancer by 2020 [8]. Raised weight record (overweight and underweight) is a significant reason for chronic disease. The commonness of overweight in India is relied upon to increment in the men and women throughout the following 10 years. The anticipated commonness of overweight, India, guys, and females matured 30 years or increasingly, 2005 and 2015 in 2005 22% of men were overweight, and in 2015 it expanded by 31%. Similarly, 21% women were over-weight in 2005, and it expanded by 29% in 2015. Economic effect chronic ailments make enormous unfriendly – and overlooked – financial impacts on families, networks and nations [9]. In 2005 alone, it is evaluated that India will lose 9 billion dollars in national pay from unexpected losses because of heart illness, stroke and diabetes (announced in worldwide dollars to represent contrasts in buying power between countries [10]. These misfortunes are anticipated to keep on expanding in total, India stands to lose 237 billion dollars throughout the following 10 years from unexpected losses because of heat illness, stroke, and diabetes. In any event, 80% of untimely heart illness and type 2 diabetes, and 40% of cancer could be forestalled through solid eating routine, normal physical action, and evasion of tobacco items [11]. Financially savvy intercessions exist, and have worked in numerous nations: the best methodologies have utilized a scope of populace wide methodologies joined with mediations for people. WHO assesses that a 2% yearly decrease in national-level chronic disease demise rates in India would bring about a monetary addition of 15 billion dollars for the nation throughout the following 10 years [12]. Different methodologies have been acquainted with prevent chronic disease, and the vast majority of them center on the way of life. Notwithstanding, it is really difficult for people to change their habits to forestall chronic infections, in light of the fact that numerous individuals don't know which interminable ailments they might be vulnerable to dependent on their physical illness and medicinal history[13]. It examined that smoking, drinking, and elevated cholesterol levels cause incessant ailments. With the upgrading of big data investigation innovation, more consideration has been paid to illness forecast. However, to the best of our knowledge. None of the previous work used RELU or rectified activation function and it has been demonstrated to function admirably in a neural network, it has been gotten exceptionally famous in the previous year. It's just $R(x) = \max(0, x)$ i.e. if $x < 0$, $R(x) = 0$ and if $x \geq 0$, $R(x) = x$ and Adam is used as an optimizer, the objective of this study is to forecast illness accurately with high accuracy by using ML, deep learning, decision tree and KNN algorithms by Appling these algorithms on a standard dataset [14].

The suggested work started with the Sect. 1, as brief Introduction of chronic diseases, machine learning and deep learning i.e. RELU and Sigmoid activation function. Section 2, literature survey, which defines the related work of this paper by different researcher. Section 3, describes the detail description of dataset used in this study. Section 4, describes the architecture of proposed chronic disease prediction. Section 5, describes method evaluation. Section 6, discuss about result analysis in detail. Section 7, conclude the whole research paper.

2 Literature Review

M. Chen proposed [15] another CNN based multimodal ailment chance forecast calculation by using structured information and unstructured information about the medical clinic. Wang imagined ailment expectation framework for the various districts. They performed ailment expectation on three infections like diabetics, cerebral infarction, and coronary illness. The malady expectation is done on organized information. The expectation of coronary illness, diabetes and cerebral infarction is completed by utilizing diverse AI calculations like gullible Naïve-Bayes, Decision tree and K-Nearest Neighbor calculation [16]. The aftereffect of Choice tree calculation is superior to K-Nearest Neighbor and Naïve Bayes calculation. Likewise, they anticipate that whether a patient encounters from the more danger of intellectual localized necrosis or less the danger of cerebral localized necrosis. For the hazard expectation of cerebral infraction, they used Convolutional Neural Network based multimodal illness chance forecast on the content information [17]. The precision examination happens between CNN based unimodal infection hazard expectations against CNN based multimodal infection hazard expectation calculation. The precision of ailment expectation comes to up to 94.7% with quicker speed than CNN based unimodal sickness hazard forecast calculation. The CNN based multimodal infection chance expectation calculation steps are comparable CNN of the CNN-UDRP calculation is the step of testing comprise of "2" extra advances [18]. In the paper takes a shot at both the kind of dataset like organized and unstructured information. Creator took a shot at unstructured information. While past work is just dependent on organized information, none of the creators took a shot at unstructured and semi-organized information. Be that as it may, this paper relies upon organized just as unstructured information. IM. Chen, C. Youn, D. Wu, Y. Mama, Y. Zhang and Y. Li [19] they proposed a wearable 2.0 framework in which configuration savvy launder able dress that modify the QoE and QoS for next generation human services framework. Chen planned a updated version of internet of things (IOT) related information assortment framework. In the current sensor-based savvy launder able material concocted. By the utilization of this fabric, the specialist caught the patient's physiological condition [20]. What's more, with the help of the physiological information further examination. Right now a reversal of launder able brilliant material, for the most part, comprises various sensors, cables and terminal with the assistance of this segment client can ready to gather the biological state of the patient just as enthusiastic wellbeing status data by the utilized of the cloud-based framework. With the assistance of this material, it caught the biological state of the patient [21]. Furthermore, for examination reason, this information is utilized. Examined the matters which are confronting. Negative mental impacts, against remote for body region organizing and Sustainable huge physiological information assortment and so forth. The different activities performed on documents like investigation on information, observing and expectation [22]. Hang Lai, Assemble a successful prescient model with high affectability and selectivity to all the more likely recognize Canadian patients in danger of having diabetes mellitus dependent on understanding demographical information and the lab result during their visits to clinical offices. Utilizing the latest record of 13,308 patient matured somewhere in the range of 18 and 90 years, alongside their research center data such as age, sex, body mass index, blood pressure, etc. Fabricate prescient model utilizing logistic regression and gradient boosting machine techniques. The area under the receiver operating characteristic curve (AROC) was utilized to assess the unfair ability of these model. In GBM model AROC is about 84.6% and the sensitivity of this model us about 71.5%. In Logistic model AROC is about 84.1% and sensitivity is 73.5% [23].

3   Dataset and Model Description

The dataset that used in this study contains a patient's details. Our information center on inpatient division information which remembered 1643 records for all out and 128 items, for example, circulatory strain, glucose, pregnancy, BMI, insulin, and so forth. The inpatient division information is mostly made out of structured and unstructured content information. The data information incorporates lab information and the patient's essential data, for example, Age of patient, sexual orientation and lifestyle, and his/her illness, the specialist's cross-examination records and analysis, and so forth. The dataset contains heart, diabetics, and cancer-related data. Using these datasets we are trying to predict these chronic diseases [24].

Right now, diabetic information records were gained from various sources, a modified electronic record device, and a paper record. In the program, contraption had an internal clock to timestamp event. While in paper records, it gives "authentic time' space (breakfast, lunch, dinner, rest time). Also, for paper record fixed events were given out to break-fast (08:00 AM), lunch (12:00), dinner (6:00 PM), and rest time (10:00 PM). In this way, paper records have imaginary uniform account times while electronic records have increasingly practical time stamps. The characteristics of a dataset are multi-variant and time series, data contains 20 attributes, the characteristics of the attribute is categorical and integer. Data contains no missing value. Diabetes Attributes comprise of four fields for every record. Each field is isolated by a tab and each record is isolated by a newline. File Names and formats are data in MM-DD-YYYY, Time in XX: YY format, code, and value.

Highlights are calculated from a digital images of a fine needle suction (FNS) of breast lump, they depict qualities of the cell cores from the picture of breast mass, characteristics of a dataset is multi-variant, data contains 32 attributes, attribute characteristics are real, the associated test is classification, the no. instances are 569, breast cancer attribute comprises of the identity document (ID). And Diagnosis (malignant (M), (benign (B)). The real value features for every cell nuclei are: Radius, texture as standard deviation of gray-scale values, area, perimeter, smoothness as local variation in radius lengths.

The dataset of heart disease contains 77 attributes, but experiments refer to using only 14 subsets of 76 attributes. In heart disease data set column describes age, sex (male = 1& female = 0), chest pain type (cp), chol, testing food slugger (TFS) TFS > 120 mg/dl (TRUE = 1, FALSE = 0), etc. As shown in Table 1.

Table 1. The detail description dataset.

Item	Description
Demographics of the patient	Gender, Age, etc.
Living habits	The history of patients drinking habits, Smoking habits, A genetic history, etc.
Examination items and results	Includes 128 items, such as blood pressure, glucose, pregnancy, BMI, insulin, etc.
Diseases	Patient's disease such as, heart disease, diabetics, and Wisconsin breast cancer
Patients readme sickness	Patients disease, and medical back ground detail
Doctors records	Doctors cross-examinations record
Number of attributes	1643

4 Architecture of the Proposed Work

See Fig. 1.

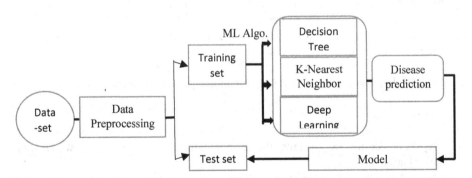

Fig. 1. The architecture for our proposed work to predict chronic disease with high accuracy.

Steps to Implement the Proposed Work:

Step 1: load the disease-dataset, at beginning we take illness datasets from UCI machine learning web site and dataset contain a list of disease and its symptoms.
Step 2: Necessary libraries and packages are loading.

Step 3: preprocessing is performed for cleaning. That is expelling comma, accentuations and white spots etc.

Step 4: After preprocessing data is spat into the training set and testing set. What's more, that is utilized as preparing datasets. After that component separated and chose.

Step 5: Than model classify that information using classification technique or grouping procedures, using machine learning algorithms (KNN and Decision tree), deep learning (ReLU or Rectified linear activation function and sigmoid activation function), and the sequential model is used. Now the model is compiled by taking two parameters: optimizer and lose, the optimizer controls the learning. To control the learning rete Adam is used as an optimizer, the learning rate determines how fast the optimal weight for the model is calculated (up to the certain point) but the time it takes to compute the weight will be longer by using standard datasets to predict the possibility of patients with and without chronic disease.

Step 6: If the prediction of the patient with or without the chronic disease is not an accurate, model test the model again and repeat the process until the model gets an accurate prediction of chronic disease.

(A). Algorithms

1. **The KNN**

   Step 1: Collect the data and load it.
   Step 2: Initialize the value of K to your picked number of neighbors.
   Step 3: For each and every instance in a data.
   Step 4: measure the distance between the query instance and the present instance from the data.
   Step 5: Include the distance and the file of the instance to an arranged collection.
   Step 6: Sort the arranged collection of distance and records from littlest to biggest (in increasing order) by the distance.
   Step 7: Pick the primary K entries from the arranged collection.
   Step 8: Get the marks of the chose K entries.
   Step 9: If the regression, it will return the mean of the K labels.
   Step 10: If classification, it will return the method of the K names.

2. **Decision Tree**

   Step 1: Select the best attribute (the attribute which said to be the best which split and separate the data).
   Step 2: Provide relevant question as input.
   Step 3: Follow the appropriate response way.
   Step 4: Go to stage 1 until you get the appropriate answer.

3. **Deep Learning**

   Step 1: Input data is changed into vector structure.
   Step 2: The word installation did with adopt "0" qualities to full-fill the information.

Step 3: In the pooling layer, the convolutional layer is taken as input and perform max poling on it.

Step 4: The data is converted into a fixed-length vector form known as max pooling. This layer is fully connected with neural network.

Step 5: The full association layer associated with the classifier that is a softmax classifier. In softmax classifier, full connection layer is connected to classifier.

5 Method Evaluation

To measure the performance of calculation in this trial, initially given are TP, FP, TN and FN as TP is "true positive" i.e. number of cases forecast accurately as needed, TN is "true negative" i.e. number of cases forecast accurately as not required, FP is "false positive" i.e. number of cases can't forecast accurately, FN is "false negative" i.e. number of cases can't forecast accurately as not required. Now we try to obtain calculations using the following formulas explained in the Fig. 2;

$$\text{Accuracy} = \frac{(TP) + (TN)}{(TP) + (FP) + (TN) + (FN)}$$

$$\text{Precision} = \frac{(TP)}{(TP) + (FP)}$$

$$\text{Recall} = \frac{(TP)}{(TP) + (FN)}$$

$$\text{F1 Measure} = \frac{2(\text{Precision} \times \text{Recall})}{\text{Precision} + \text{Recall}}$$

Fig. 2. Formulas for accuracy, precision, Recall, f1-meaure

Where, F1-Measure is the weighted consonant mean of the accuracy Recall speaks to the general execution.

6 Analysis of Result and Discussion

In this part, we are going to portray the general outcome. Python language is used to implement all the experimental cases, using anaconda software (Spyder). The battling classification approach alongside various component extraction techniques, and run in an environment with System having a design of Intel CORE i5, 2.30 GHz, Windows 10 with the 64-bit machine and 8 GB RAM. We are going to present the prediction of the disease using decision tree, K-Nearest Neighbor, and deep learning algorithms. In terms of the possibility of having a disease or not to a particular patient. Accuracy of each method for the data used. The accuracy is shown for each disease and each algorithm separately. In Fig. 3, shows the possibility of a patient with and without heart disease. Whereas '0' represents the without heart problem and '1' represents with a heart problem, the Percentage of patients without heart problems is 45.54% and the Percentage of patients with heart problems is 54.46%. Figure 4, shows the distribution of various attributes.

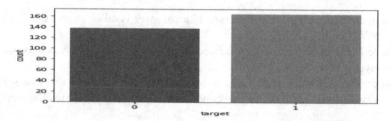

Fig. 3. The possibility of a patient with and without heart disease.

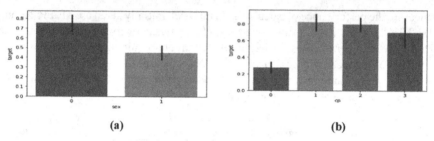

(a)   (b)

Fig. 4. Distribution of various attributes.

Now splitting the data into testing and training to achieve the accurate prediction of heart disease. The accuracy score of heart diseases prediction using KNN is 67.21%, the accuracy score of heart chronic disease using Decision Tree is 81.97%. And the accuracy score of heart disease prediction using Neural Network is 94.26%. In a comparison of other algorithms, accuracy is improved by using neural network. The accuracy of cancer prediction using Deep Network is 98.2%. The accuracy score achieved for cancer using Decision Tree is 92.98%. And the accuracy score achieved for cancer using KNN is 98.25%. The accuracy of diabetes prediction using Deep Network is 99.2%. The accuracy score achieved for diabetes using Decision Tree is 71.356%. And the level of accuracy score achieved for diabetes using KNN is 79.17%. In Fig. 5, shows the possibility of a patient with and without diabetic's disease. Whereas '0' represents the without diabetics problem and '1' represents with a diabetics problem. Dataset is splitting into test and train and it shows the accuracy of data after training and testing the data. The accuracy level get increased after training the data as shown in Fig. 6. Figure 7, Accuracy level of

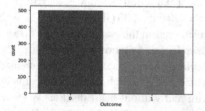

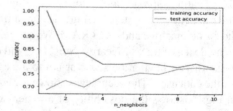

Fig. 5. The possibility of a patient with or without Diabetic chronic disease.

Fig. 6. The accuracy of training and test dataset.

heart, breast cancer and diabetics disease by using Decision tree, K-Nearest Neighbor, and Deep learning (Table 2).

Table 2. The accuracy of chronic diseases by using KNN, Decision tree and Deep learning techniques.

Chronic diseases	K-nearest-neighbors technique	Decision tree technique	Deep learning
Heart disease	67.21%	81.97%	94.26%
Diabetics	79.17%	71.3	99.2%
Breast cancer	98.25%	92.98%	98.3%

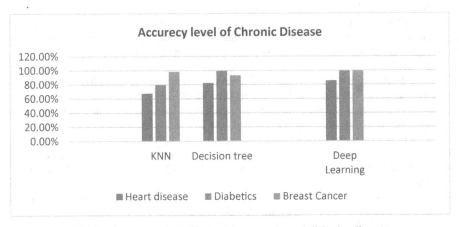

Fig. 7. Accuracy level of heart, breast cancer and diabetics disease.

7 Conclusion

In this proposed work, a chronic disease prediction system dependent on a ML algorithm has been developed. The Decision tree and deep learning algorithms are being utilized to characterize patient's information. In this paper, KNN, decision tree and deep learning algorithm are being utilize i.e. RELU or Rectified linear activation function and sigmoid activation function and Adam as an optimizer by using standard datasets to predict the possibility of patients with and without chronic disease. To the best of studies, there is no existing work done by using these techniques. Compared to several ordinary prediction algorithms, the accuracy level of the proposed system is enhanced. The level of accuracy on the training set is 98.3% and on the test set is 71.354.

References

1. Groves, P, Kayyali, B., Knott B, Kuiken, S.V.: The'big data'revolution in healthcare: accelerating value and innovation (2016)
2. Chen, M., Mao, S., Liu, Y.: Big data: a survey. Mobile Netw. Appl. **19**(2), 171–209 (2014)
3. Jensen, P.B., Jensen, L.J., Brunak, S.: Mining electronic health records: towards better research applications and clinical care. Nat. Rev. Genet. **13**(6), 395–405 (2012)
4. Tian, D., Zhou, J., Wang, Y., Lu, Y., Xia, H., Yi, Z.: A dynamic and self-adaptive network selection method for multimode communications in heterogeneous vehicular telematics. IEEE Trans. Intell. Transp. Syst. **16**(6), 3033–3049 (2015)
5. Chen, M., Ma, Y., Li, Y., Wu, D., Zhang, Y., Youn, C.: Wearable 2.0: enable human-cloud integration in next generation healthcare system. IEEE Commun. **55**(1), 54–61 (2017)
6. Chen, M., Ma, Y., Song, J., Lai, C., Hu, B.: Smart clothing: connecting human with clouds and big data for sustainable health monitoring. Mobile Netw. Appl. **21**(5), 825–845 (2016). https://doi.org/10.1007/s11036-016-0745-1
7. Chen, M., Zhou, P., Fortino, G.: Emotion communication system. IEEE Access (2016). https://doi.org/10.1109/ACCESS.2016.2641480
8. Qiu, M., Sha, E.H.-M.: Cost minimization while satisfying hard/soft timing constraints for heterogeneous embedded systems. ACM Trans. Des. Autom. Electron. Syst. (TODAES) **14**(2), 25 (2009)
9. Wang, J., Qiu, M., Guo, B.: Enabling real-time information service on telehealth system over cloud-based big data platform. J. Syst. Architect. **72**, 69–79 (2017)
10. Bates, D.W., Saria, S., Ohno-Machado, L., Shah, A., Escobar, G.: Big data in health care: using analytics to identify and manage high-risk and high-cost patients. Health Aff. **33**(7), 1123–1131 (2014)
11. Qiu, L., Gai, K., Qiu, M.: Optimal big data sharing approach for tele-health in cloud computing. In: IEEE International Conference on Smart Cloud (SmartCloud), pp. 184–189. IEEE (2016)
12. Zhang, Y., Qiu, M., Tsai, C.-W., Hassan, M.M., Alamri, A.: Healthcps: healthcare cyber-physical system assisted by cloud and big data. IEEE Syst. J. **11**, 88–95 (2015)
13. Lin, K., Luo, J., Hu, L., Hossain, M.S., Ghoneim, A.: Localization based on social big data analysis in the vehicular networks. IEEE Trans. Industr. Inf. **13**, 1932–1940 (2016)
14. Lin, K., Chen, M., Deng, J., Hassan, M.M., Fortino, G.: Enhanced fingerprinting and trajectory prediction for IoT localization in smart buildings. IEEE Trans. Autom. Sci. Eng. **13**(3), 1294–1307 (2016)
15. Oliver, D., Daly, F., Martin, F.C., McMurdo, M.E.: Risk factors and risk assessment tools for falls in hospital in-patients: a systematic review. Age Ageing **33**(2), 122–130 (2004)
16. Marcoon, S., Chang, A.M., Lee, B., Salhi, R., Hollander, J.E.: Heart score to further risk stratify patients with low timi scores. Crit. Pathways Cardiol. **12**(1), 1–5 (2013)
17. Bandyopadhyay, S., et al.: Data mining for censored time-to-event data: a bayesian network model for predicting cardiovascular risk from electronic health record data. Data Min. Knowl. Discov. **29**(4), 1033–1069 (2015)
18. Qian, B., Wang, X., Cao, N., Li, H., Jiang, Y.-G.: A relative similarity based method for interactive patient risk prediction. Data Min. Knowl. Discov. **29**(4), 1070–1093 (2015)
19. Singh, A., Nadkarni, G., Gottesman, O., Ellis, S.B., Bottinger, E.P., Guttag, J.V.: Incorporating temporal EHR data in predictive models for risk stratification of renal function deterioration. J. Biomed. Inform. **53**, 220–228 (2015)
20. Wan, J., et al.: A manufacturing big data solution for active preventive maintenance. IEEE Trans. Ind. Inf. (2017). https://doi.org/10.1109/tii.2017.2670505

21. Yin, W., Schütze, H.: Convolutional neural network for paraphrase identification. In: HLT-NAACL, pp. 901–911 (2015)
22. Kaur, G., Chhabra, A.: Improved J48 classification algorithm for the prediction of diabetes. Int. J. Comput. Appl. **98**, 13–17 (2014)
23. Zhai, S., Chang, K.-h., Zhang, R., Zhang, Z.M.: Deepintent: learning attentions for online advertising with recurrent neural networks. In: Proceedings of the 22nd ACM SIGKDD International Conference on Knowledge Discovery and Data Mining, pp. 1295–1304. ACM (2016)
24. Hwang, K., Chen, M.: Big Data Analytics for Cloud/IoT and Cognitive Computing. Wiley, Hoboken (2017). ISBN 9781119247029

A Deep Learning Based Method to Discriminate Between Photorealistic Computer Generated Images and Photographic Images

Kunj Bihari Meena and Vipin Tyagi[✉]

Jaypee University of Engineering and Technology, Raghogarh, Guna, MP, India
`dr.vipin.tyagi@gmail.com`

Abstract. The rapid development of multimedia tools has changed the digital world drastically. Consequently, several new technologies like virtual reality, 3D gaming, and VFX (Visual Effects) have emerged from the concept of computer graphics. These technologies have created a revolution in the entertainment world. However, photorealistic computer generated images can also play damaging roles in several ways. This paper proposes a deep learning based technique to differentiate computer generated images from photographic images. The idea of transfer learning is applied in which the weights of pre-trained deep convolutional neural network DenseNet-201 are transferred to train the SVM to classify the computer generated images and photographic images. The experimental results performed on the DSTok dataset show that the proposed technique outperforms other existing techniques.

Keywords: Digital forensics · DenseNet-201 · Deep convolutional neural network · Computer generated images · SVM · Photographic image · Transfer learning

1 Introduction

In today's era of digital technology, digital images have become a primary carrier of information. At the same time, an unprecedented involvement of the digital images with misinformation or fake news can be seen on the social media platforms [33]. When an image with false or misleading information goes viral on social media it can disrupt social harmony. Moreover, political parties are using social media for their election campaigning purpose frequently. From the study conducted by Machado et al. [16], it has been observed that more fake posts are shared on these platforms during such political campaigning. This study revealed that 13.1% of Whatsapp posts were fake during the Brazilian presidential elections. Recently, uncountable fake posts were shared globally on social media related to the novel coronavirus and COVID-19 (coronavirus disease 2019) pandemic [11]. There are various ways to tamper with an image. The most common types of forgeries are copy-move forgery [17] and image splicing forgery [18]. In the last decades, several methods [1, 19, 20], were developed to detect these forgeries.

© Springer Nature Singapore Pte Ltd. 2020
M. Singh et al. (Eds.): ICACDS 2020, CCIS 1244, pp. 212–223, 2020.
https://doi.org/10.1007/978-981-15-6634-9_20

The invention of Computer Generated (CG) imagery has enabled various new technologies like virtual reality, 3D gaming, and VFX (Visual Effects). These technologies are widely used in the fields of the film industry, education, and medicine. Though, there are so many good applications of CG images, the computer generated images that were created with malicious intent may create the problems. The problem becomes worse when the CG image is highly photorealistic, as human eyes cannot differentiate between the CG image and actual photographic (PG) image. GAN (Generative Adversarial Network) tool can generate CG images with high photorealism. A good collection of CG images that are generated using GAN is available on the website www.thisartworkdoes notexist.com. CG detection has become an open area for research. In past, a good number of techniques were presented to distinguish the CG and PG images. Recently, Meena and Tyagi [21] have surveyed the existing methods that were developed to distinguish the CG images from PG images. This survey paper discussed various methods from the literature, and then these methods were grouped into four classes: acquisition process based, visual feature based, statistical feature based, and deep learning based.

The methods based on the deep Convolutional Neural Network (CNN) have gained unprecedented success for image classification. A series of deep convolutional neural networks were proposed in the last few years to solve the different challenging problems. This paper proposes a deep learning based technique to discriminate between CG and PG images. The contributions of this paper are twofold; first, a fully automated model based on the deep CNN DenseNet-201 and transfer learning is proposed to mitigate the laborious task of designing the hand-crafted features; second, to the best of our knowledge, first-time DenseNet-201 network is used to solve this task. The proposed technique shows comparatively better detection accuracy and lower time complexity.

2 Related Works

The recent survey paper [21] has discussed a total of 52 state-of-the-art techniques available in the literature, therefore, a brief summary of the related works is presented in this section. The existing techniques to identify PG and CG images can be categorized as traditional or hand-crafted feature based, and deep learning based. The existing hand-crafted feature based techniques have two basic steps, feature extraction, and classification. In the past, the authors have explored various feature extraction mechanisms and classifiers to improve the detection accuracy of their proposed methods. Conversely, in deep learning based techniques, the image features are leaned using a specific neural network. Generally, in deep learning based techniques feature extraction and classification steps are performed in a single step by CNN.

Lyu et al. [15] designed a statistical model based on the first-order and higher-order wavelet statistics to identify CG and PG images. Two supervised machine learning methods, linear discrimination analysis and Support Vector Machine (SVM) were used for the classification task. A low CG detection rate (71%) was the main drawback of this method. Wu et al. [35] put forward a technique based on histogram features to solve this problem. This method achieved a good detection accuracy of up to 95.3%. However, this method was evaluated on a comparatively small image dataset. Fan et al. [6] employed contourlet transform to propose an approach to discriminate between the CG and PG images. This method follows a statistical model similar to [15], but in place of wavelet transform the authors have used contourlet transform. The authors have recommended that the HSV color model can improve the detection accuracy. Wang et al. [34] designed a technique based on color quaternion wavelet transform. Recently, Meena and Tyagi [22] developed an approach to detect CG and PG images based on Tetrolet transform and Neuro-fuzzy classifier.

Cui et al. [4] proposed a deep learning based approach to distinguish the CG and PG images. This approach first applies high-pass filters to pre-process all the images in the dataset. After that, this model was trained on the pre-processed images. He et al. [9] combined CNN and recurrent neural network to propose a model to detect CG and PG images. The detection accuracy of this method was 93.87% on the image dataset comprising 6,800 CG and 6,800 PG images. A CNN based framework to classify CG and PG images was introduced by He [8]. In this method, the author has explored two different networks VGG-19 and ResNet-50. Rezende et al. [30] have proposed a deep learning based model where they have used ResNet-50 network as a feature extractor. In this method for classification purposes, several classifiers such as softmax, SVM, k-nearest-neighbor were investigated. Meanwhile, Quan et al. [29] developed a CNN model to classify CG and PG images. This model was trained from scratch on the Columbia image dataset [25]. Recently, Ni et al. [26] have presented a comprehensive survey of the deep learning based CG detection methods.

3 The Proposed Technique

The overview of the proposed technique is presented in Fig. 1. The following subsections will describe the major steps of the proposed technique.

3.1 Pre-processing

Generally, the image datasets comprise the images of various pixel resolutions. However, the proposed technique can work on the images of size 224×224 pixels. The only reason for selecting image size as 224×224 pixels is that the DenseNet-201 is trained on the images of size 224×224 pixels. Therefore, as pre-processing we have resized all the images in the DSTok dataset [32] before training the network. Similar to [30], the mean RGB value that was computed over the ImageNet dataset is subtracted pixel-wise from each image in the DSTok dataset.

3.2 DenseNet-201 Network

Several numbers of deep CNNs are available online with the pre-trained models. Some of the popular deep CNNs are AlexNet (2012) [29], VGG-16 (2014) [8], VGG-19 (2014) [8], ResNet-50 (2015) [30], Inception-v3 (2015) [10], Xception (2016) [10], DenseNet-121 (2017) [10], and DenseNet-201 (2017) [10]. A large number of applications of these networks can be seen in areas like data science, image processing, computer vision, and digital image forensics [21]. More specifically, VGG-19 and ResNet-50 have been used for CG detection in [8], and [30] respectively. Recently, Cui et al. [4] also developed a

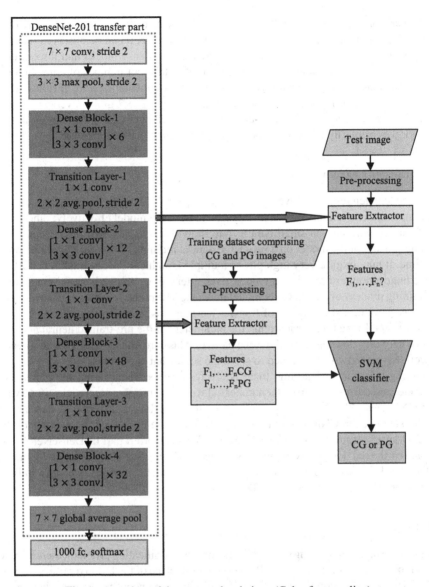

Fig. 1. Overview of the proposed technique (Color figure online)

CNN model for CG and PG image classification purposes. In this paper, many experiments were performed with a varying number of hidden layers in CNN. Based on these experiments the authors have suggested that the detection accuracy can be improved if a CNN with more number of layers is used. Therefore, in this paper, we have tried to solve the problem of CG detection by utilizing the recently proposed very deep CNN DenseNet-201.

The complete layer-wise architecture of the DenseNet-201 is shown as a gray rectangular box in the left part of Fig. 1. From the architecture, it can be observed that this network comprises a total of 201 layers. There are four dense blocks and three transition layers. Each dense block is a combination of a different number of convolutional layers. Whereas, each transition layer contains a 1 × 1 convolutional layer followed by a 2 × 2 average pooling layer. The second last layer is a 7 × 7 global average pooling layer. Finally, the last layer is a fully-connected softmax layer with 1000 neurons as this network was primarily designed to classify the images into 1000 categories.

3.3 Transfer Learning

Two main challenges arise if we try to train a deep CNN based technique from scratch. First, an enormous amount of data is required to train the model effectively; otherwise, if the model is trained on small data it may show the sign of overfitting; second, the training process of the model on a very large dataset requires a huge computation power and time. It may require very high power multiple Graphical Processing Units (GPUs) with a huge amount of physical memory. Even after using such high power computers, the training process of a deep CNN model may take several hours or days. To overcome these two limitations, a concept of transfer learning has gained much attention in the last few years. In the transfer learning, the parameters of the pre-trained neural network (source network) that was trained for one particular task are transferred to the new neural network (target network) designed to solve the somewhat similar tasks.

During the transfer learning, there is always a scope of adjusting the number of parameters used from the particular number of layers. The proposed technique uses a very deep CNN DenseNet-201 that comprises a total of 201 layers. It becomes impractical to train such a deep CNN from the beginning, hence we have used the weights of the first 200 layers of pre-trained DenseNet-201. The transferred part of DenseNet-201 is denoted by the red dotted box in Fig. 1. Note that the DenseNet-201 network was trained on the very large image dataset namely ImageNet [5] that comprises over 1.28 million images for object classification in 1000 classes.

3.4 Classifier

The problem of distinguishing CG and PG is a binary classification problem; therefore we have employed a non-linear binary SVM classifier in place of the last layer of DenseNet-201 which is a fully connected softmax layer with 1000 neurons. Though there exist several variants of an SVM, we have used a non-linear binary SVM with Radial Basis Function (RBF) kernel. In an SVM there are two parameters Cost (C) and gamma (γ) that need to be set appropriately to manage the trade-off between variance and bias. A high value of γ may give better accuracy but results may be more biased, the reverse is also true. Similarly, a large value of C shows poor accuracy but low bias. The optimum values of these two parameters can be found using the grid-search method. Based on the experiments we have set the value of C as 10.0 and the value of γ as 0.001.

4   Experimental Results

4.1   Datasets

For evaluating and analyzing the performance of any technique the availability of the image dataset plays a crucial role. There are very few image datasets available to asses the effectiveness of the methods that are developed to discriminate between the CG and PG images. Ng et al. created the Columbia image dataset [25] to evaluate their method of CG detection in 2004. Most of the early works were evaluated only on this image dataset; this is because no other image dataset was available until 2013. There are two main drawbacks of this dataset: first, less number of images (800 CG and 800 PG images) is available in the dataset, and second, the CG images are less photorealistic. As deep learning based methods need more image data to train the model effectively, hence Columbia dataset was less relevant for evaluating the proposed technique. Due to this reason, the proposed approach is assessed on the well-designed image dataset that was created by Tokuda et al. [32]. This dataset is commonly referred to as the 'DSTok' dataset in the literature.

There are 4,850 CG images and 4,850 PG images in the DSTok dataset. The computer graphics images in this dataset are having high photorealism as compared to Columbia dataset. The CG images were collected from various sources such as gaming websites and screenshots of the latest 3D computer games. All the images are stored in JPEG format, and the physical size of these images varies from 12 KB to 1.8 MB. Figure 2 shows some of the example images from the DSTok dataset. The top row in Fig. 2 illustrates CG images whereas the bottom row shows the PG images.

Fig. 2. Sample images from DSTok dataset [32], top row: computer generated images; bottom row: photographic images

4.2   Validation Protocol and Evaluation Metrics

Due to the hardware limitation, it becomes impractical to evaluate the proposed technique on the images in the actual size in the DSTok dataset. Thus, all the images in the DSTok dataset are resized to 224 × 224 pixels for all the experiments. The 5-fold cross-validation approach, similar to [32] and [30], is considered to analyze the proposed technique. Note that, the authors in [32] and [30] have also used the resized images of size 224 × 224 pixels. All 9,700 images (4,850 CG and 4,850 PG) are partitioned into five folds of equal size. Hence, each fold comprises 1940 images. In each cross-validation step, any four folds are used to train the model whereas the remaining other fold is used to test the model.

The proposed technique is assessed based on three metrics [3, 9]; True Positive Rate (TPR), True Negative Rate (TNR), and detection accuracy. These metrics are defined in Eq. 1–3.

$$TPR = \frac{total\ number\ of\ correctly\ detected\ CG\ test\ images}{total\ number\ of\ CG\ test\ images} \tag{1}$$

$$TNR = \frac{total\ number\ of\ correctly\ detected\ PG\ test\ images}{total\ number\ of\ PG\ test\ images} \tag{2}$$

$$Detection\ accuracy\ (Acc) = \frac{TPR + TNR}{2} \tag{3}$$

The TPR represents the detection rate of computer generated images, and TNR represents the detection rate of photographic images. Whereas, the detection accuracy is a simple mean of TPR and TNR.

The ROC (Receiver Operating Characteristics) curve provides important visual information of a binary-classification model. The ROC curve is drawn between two metrics true-positive rates and false-positive rates. The value of AUC (Area Under Curve) is also used as an evaluation metric, and this metric is used to determine the effectiveness of the binary classification model.

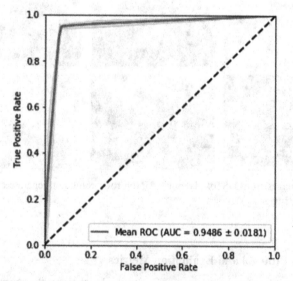

Fig. 3. ROC curve of the proposed technique on DSTok dataset

4.3 Implementation Details

The proposed technique has been implemented using the Python deep learning library Keras v2.2.4 with Python v3.6.10. The TensorFlow-GPU v1.13.1 is used as a backend. A computer system with a configuration of 16 GB RAM and Quadro RTX 4000 GPU from NVIDIA is used for all the experiments.

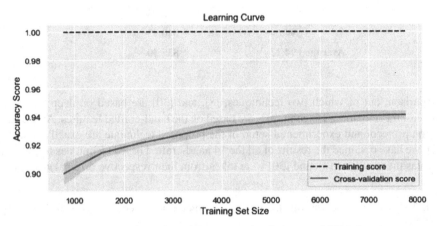

Fig. 4. The learning curve of the proposed technique on DSTok dataset

4.4 Results of the Proposed Technique

The detection accuracy and training time of the proposed technique is reported in Table 1. It can be observed that the average detection accuracy is 94.12%, and the average training time of the model is 835.30 s when the model is trained on the DSTok dataset. As there are a total of 9700 images in the DSTok dataset the average time to process an image of size 224 × 224 pixels is only 0.0861 s. Therefore the proposed technique can be used to distinguish the CG images from PG images in real-time. The ROC curve is shown in Fig. 3, it also shows the encouraging performance of our method as the obtained value of AUC is 0.9486 with a very small standard deviation of ±0.0181. The small value of the standard deviation is the indication of the better stability of our approach. Moreover, the learning curve of the proposed technique is shown in Fig. 4, from where it can be observed that the cross-validation score is increasing with the size of the training dataset. Hence, it can be believed that the accuracy of the proposed technique can further be enhanced if the size of the training dataset is increased.

4.5 Comparison and Analysis of Results

The proposed technique can distinguish between the CG and PG images with the detection accuracy of 94.12%. The comparative results of our technique with the existing techniques are reported in Table 2. A total of 16 techniques were considered for this

Table 1. Detection accuracy and training time of the proposed technique

Fold	Detection accuracy (%)	Time in seconds
1	94.48	835.12
2	93.92	838.37
3	94.23	832.01
4	93.35	830.68
5	94.64	840.33
Average	**94.12**	**835.30**

comparison, out of which two techniques, [8], and [30] are based on deep learning; whereas, the remaining 14 techniques are based on the hand-crafted features. As the validation protocol and experimental setup of the proposed technique are exactly same as [32], we have obtained the results of all the 14 hand-crafted based techniques from [32], whereas the results of [8], and [30] were taken from their respective original articles.

Table 2. Result comparison with the existing techniques that were proposed to distinguish between CG and PG images

Technique	Main concept	TPR(%)	TNR(%)	Acc(%)
SOB [32]	Sobel operator	54.4	55.2	55.3
BOX [14]	Boxes counting	54.1	56.8	55.4
POP [28]	Interpolator predictor	57.0	57.5	57.3
GLC [7]	Co-occurrence matrix	64.0	63.0	63.5
LSB [24]	Camera noise	67.2	65.1	66.2
SHE [12]	Shearlet transform	74.8	67.7	71.3
HOG [23]	Histogram of oriented gradients	75.4	72.0	74.0
HSC [31]	Histogram of shearlet coefficients	81.8	78.7	80.2
CUR [2]	Curvelet transform	80.6	80.5	80.5
LBP [27]	Local binary patterns	90.4	83.8	87.1
CON [6]	Contourlet transform	91.8	88.7	90.2
He [8]	VGG-19	92.2	91.8	92.0
LYU [15]	Wavelet transform	94.2	89.9	92.0
FUS1 [32]	Concatenation	93.4	92.2	92.8
LI [13]	Second order difference	94.8	91.1	93.0
Rezende et al. [30]	ResNet-50	–	–	94.0
Proposed	**DenseNet-201**	**93.6**	**94.6**	**94.12**

The rows in Table 2 are sorted according to the values of *Acc* in increasing order. The TPR and TNR values corresponding to the technique proposed by Rezende et al. [30] were not provided in the paper, therefore we have reported only the detection accuracy for this technique. It can be noticed that this technique shows the second best detection accuracy among all the referenced techniques. It can also be noticed that the detection accuracy of the proposed technique is greater than the detection accuracies of all the reported techniques. The proposed technique obtains the values of TPR and TNR as 93.6% and 94.6% respectively. The simultaneous higher values of these two parameters indicate that the proposed technique has shown balanced behavior while correctly predicting the more accurate results in each category. Additionally, the proposed technique can be used for real-time classification of CG and PG images.

5 Conclusion

The challenges to solve the problem of differentiating between computer generated images and photographic images are growing with the development of multimedia tools. Therefore, the techniques proposed so far have become less powerful to address this problem. This paper has introduced a technique to address this problem using the concept of deep learning. The very deep convolutional neural network DenseNet-201 was used as a feature extractor and then the support vector machine is applied as a classifier. The proposed technique achieved a detection accuracy of 94.12% on the DSTok dataset, which is higher than the detection accuracies of the existing techniques in the literature.

Additionally, the proposed technique can be used for real-time applications as it can process an image of size 224×224 pixels in 0.0861 s. In the future, the detection accuracy of the proposed technique can be improved further if the model is trained on the large training dataset. Furthermore, the proposed technique can also be modified to classify the computer generated images and photographic images, when the images are post-processed by various operations such as noise addition, image blurring, and contrast enhancement.

References

1. Ansari, M.D., Ghrera, S.P., Tyagi, V.: Pixel-based image forgery detection: a review. IETE J. Educ. **55**(1), 40–46 (2014)
2. Candes, E., Donoho, D.L., Candès, E.J., Donoho, D.L.: Curvelets: a surprisingly effective nonadaptive representation of objects with edges. Curves Surf. Fitting **C**(2), 1–10 (2000)
3. Chen, W., Shi, Y.Q.: Identifying computer graphics using HSV color model and statistical moments of characteristic functions. In: IEEE International Conference on Multimedia, pp. 1123–1126 (2007)
4. Cui, Q., McIntosh, S., Sun, H.: Identifying materials of photographic images and photorealistic computer generated graphics based on deep CNNs. Comput. Mater. Continua **55**(2), 229–241 (2018)
5. Deng, J., Dong, W., Socher, R., Li, L., Li, K., Fei-Fei, L.: ImageNet: a large-scale hierarchical image database. In: IEEE Conference on Computer Vision and Pattern Recognition, pp. 248–255 (2009)

6. Fan, S., Wang, R., Zhang, Y., Guo, K.: Classifying computer generated graphics and natural images based on image contour information. J. Inf. Comput. Sci. **10**(2010), 2877–2895 (2012)
7. Haralick, R.M., Shanmugam, K., Dinstein, I.: Textural features for image classification. IEEE Trans. Syst. Man Cybern. **3**(6), 610–621 (1973)
8. He, M.: Distinguish computer generated and digital images: a cnn solution. Concurr. Comput. Pract. Exp. **4788**, 1–10 (2018)
9. He, P., Jiang, X., Sun, T., Member, S., Li, H.: Computer graphics identification combining convolutional and recurrent neural network. IEEE Signal Process. Lett. **25**(9), 1369–1373 (2018)
10. Huang, G., Liu, Z., van der Maaten, L.: Densely connected convolutional networks. In: IEEE Conference on Computer Vision and Pattern Recognition (CVPR), pp. 4700–4708 (2017)
11. McDonald, J.: Social Media Posts Spread Bogus Coronavirus Conspiracy Theory (2020). https://www.factcheck.org/2020/01/social-media-posts-spread-bogus-coronavirus-conspiracy-theory/. Accessed 24 Feb 2020
12. Kutyniok, G., Lim, W.-Q.: Compactly supported shearlets are optimally sparse. J. Approx. Theory **11**, 1564–1589 (2011)
13. Li, W., Zhang, T., Zheng, E., Ping, X.: Identifying photorealistic computer graphics using second-order difference statistics. In: International Conference on Fuzzy Systems and Knowledge Discovery, pp. 2316–2319 (2010)
14. Liebovitch, L.S., Toth, T.: A fast algorithm to determine fractal dimensions by box counting. Phys. Lett. A **141**(8), 386–390 (1989)
15. Lyu, S., Farid, H.: How realistic is photorealistic? IEEE Trans. Signal Process. **53**(2), 845–850 (2005)
16. Machado, C., Kira, B., Howard, P.N.: A study of misinformation in WhatsApp groups with a focus on the Brazilian presidential elections. In: WWW 2019: Companion Proceedings of the 2019 World Wide Web Conference, pp. 1013–1019 (2019)
17. Meena, K.B., Tyagi, V.: A copy-move image forgery detection technique based on Gaussian-Hermite moments. Multimed. Tools Appl. **78**, 33505–33526 (2019)
18. Meena, K.B., Tyagi, V.: Image forgery detection : survey and future directions. In: Data, Engineering and applications, pp. 163–195 (2019)
19. Meena, K.B., Tyagi, V.: A copy-move image forgery detection technique based on tetrolet transform. J. Inf. Secur. Appl. **52**, 102481–102490 (2020)
20. Meena, K.B., Tyagi, V.: A hybrid copy-move image forgery detection technique based on Fourier-Mellin and scale invariant feature transforms. Multimed. Tools Appl. **79**(11), 8197–8212 (2020)
21. Meena, K.B., Tyagi, V.: Methods to distinguish photorealistic computer generated images from photographic images: a review. In: Singh, M., Gupta, P.K., Tyagi, V., Flusser, J., Ören, T., Kashyap, R. (eds.) ICACDS 2019. CCIS, vol. 1045, pp. 64–82. Springer, Singapore (2019). https://doi.org/10.1007/978-981-13-9939-8_7
22. Meena, K.B., Tyagi, V.: A novel method to distinguish photorealistic computer generated images from photographic images. In: 2019 Fifth International Conference on Image Information Processing (ICIIP), pp. 385–390 (2019)
23. Dalal, N., Triggs, B.: Histograms of oriented gradients for human detection. In: IEEE International Conference on Computer Vision and Pattern Recognition (CVPR), pp. 886–893 (2005)
24. Ng, T., Chang, S.: Distinguishing between natural photography and photorealistic computer graphics. IEEE Signal Process. Mag. **26**(2), 49–58 (2009)
25. Ng, T., Chang, S., Hsu, J., Pepeljugoski, M.: Columbia Photographic Images and Photorealistic Computer Graphics Dataset. ADVENT Technical Report #205-2004-5, Columbia University (2005)

26. Ni, X., Chen, L., Yuan, L., Wu, G., Yao, Y.E.: An evaluation of deep learning-based computer generated image detection approaches. IEEE Access **7**, 130830–130840 (2019)

27. Ojala, T., Pietikäinen, M., Mäenpää, T.: A generalized local binary pattern operator for multiresolution gray scale and rotation invariant texture classification. In: International Conference on Advances in Pattern Recognition (ICAPR), Brazil, pp. 399–408 (2001)

28. Popescu, A.C., Farid, H.: Exposing digital forgeries in color filter array interpolated images. IEEE Trans. Signal Process. **53**(10), 3948–3959 (2005)

29. Quan, W., Wang, K., Yan, D.M., Zhang, X.: Distinguishing between natural and computer-generated images using convolutional neural networks. IEEE Trans. Inf. Forensics Secur. **13**(11), 2772–2787 (2018)

30. De Rezende, E.R.S., Ruppert, G.C.S., Archer, C.T.I.R.: Exposing computer generated images by using deep convolutional neural networks. In: 30th SIBGRAPI Conference on Graphics, Patterns and Images, pp. 71–78 (2017)

31. Schwartz, W.R., da Silva, R.D., Davis, L.S., Pedrini, H.: A novel feature descriptor based on the shearlet transform. In: IEEE International Conference on Image Processing (ICIP), Belgium, pp. 1053–1056 (2011)

32. Tokuda, E., Pedrini, H., Rocha, A.: Computer generated images vs. digital photographs: a synergetic feature and classifier combination approach. J. Vis. Commun. Image Represent. **24**(8), 1276–1292 (2013)

33. Tyagi, V.: Understanding Digital Image Processing. CRC Press, Boca Raton (2018)

34. Wang, J., Li, T., Luo, X., Shi, Y., Liu, R., Jha, S.K.: Identifying computer generated images based on quaternion central moments in color quaternion. IEEE Trans. Circ. Syst. Video Technol. **29**(9), 2775–2785 (2018)

35. Wu, R., Li, X., Bin, Y.: Identifying computer generated graphics via histogram features. In: 18th IEEE International Conference on Image Processing, pp. 1973–1976 (2011)

Load Balancing Algorithm in Cloud Computing Using Mutation Based PSO Algorithm

Saurabh Singhal[✉] and Ashish Sharma

GLA University, Mathura, India
{saurabh.singhal,ashish.shamra}@gla.ac.in

Abstract. Cloud computing is a prominent technology that uses dynamic allocation technique to assigns tasks to virtual machines (VM). As per the usage, users are charged by the cloud service provider. There are various challenges that a Cloud service provider (CSP) faces, out of which load balancing being one of the significant challenge. Many algorithms have been proposed till now for load balancing algorithm, where each one focuses on the different parameters. However these proposed approached in the literature experiences various issues such as poor speed for convergence, untimely convergence, the first random chosen solution. None of the algorithms has proven to be completely sufficient. To solve the problems associated with existing meta-heuristic techniques, paper discuss a load balancing approach that is based on mutative Particle Swarm optimization. The load on the data centers are balanced by the help of proposed algorithm and parameters such as makespan are minimized while improving the overall fitness function of algorithm.

Keywords: MPSO · PSO · Makespan · Energy efficiency

1 Introduction

In recent years in the area of distributed computing and high performance computing, Cloud computing has seen a tremendous growth [14]. The users can access resources with on-demand services to shared pool of resources over Internet as a self service. It is dynamically scalable. Cloud computing is in its early ages, so as to procure its full benefits, much research has been done and is yet to be done over a broad area of topics. Users submit their jobs to the cloud which are then dynamically allocated on virtual machine for execution [13]. Load balancing is one of the challenge that the cloud computing faces. The term load balancing can be defined as the process of load sharing among systems in such a way that each system is equally loaded. There could be a variety of workload which can be classified into different classifications based on the resources that they use. In cloud computing, the term load balancing means sharing of the jobs submitted by the user, to virtual machine using virtual machine manager. The

M. Singh et al. (Eds.): ICACDS 2020, CCIS 1244, pp. 224–233, 2020.
https://doi.org/10.1007/978-981-15-6634-9_21

data centers of cloud computing are designed in such a way that they can handle large number of users job. For running these jobs a large amount of energy is required that lead to heat and wastage of energy. As resource utilization directly impacts the energy consumption, therefore resource utilization has to be optimized so as to avoid wasting energy [3].

The load balancing techniques in cloud computing are required for optimizing workload shared among data centers to prevent the overloading of nodes while other nodes remain in the state of under loaded or maybe idle. Considering various constraints of energy conservation and optimal resource utilization, the workloads are to be allocated to available resources. Load balancer distributes submitted jobs of the client efficiently across multiple servers available to them. Load balancing are a NP hard problems [5]. With the help of virtual machine manager, the load balancing algorithms in cloud are able to utilize the idea of virtualization.

In cloud, the load balancer can be static or dynamic [1]. The static algorithms do not depend on the system state and have all information such as task details and system resources in prior. These algorithms are not able to handle a sudden spike in load or failure of any resource. In dynamic algorithm the decision is taken as per system's current state and does not require any information beforehand. The dynamic algorithms are complex in nature but are fault tolerant and have better performance than static algorithm [4].

1.1   Metrics of Load Balancing

Throughput. Number of the tasks which have gone through the execution within a given time limit. The time should have maximum throughput for maximising the efficiency of system [8,14].

Overhead. The overhead needs to be minimum for better execution of the algorithm. It is a measure of the operating cost of the given load balancing algorithm at the time of execution.

Fault Tolerance. A fault tolerance parameter manages the breakdown occurred in the node factually and continuous manner. It changes the nodes in case of faulty nodes.

Transfer Time. It is the unit of time for the movement and reallocation of the resources from one node to another node. Transfer time should be minimized for achieving better efficacy of system.

Resource Utilization. This parameter assures the utilization of the resources in load balancing. This parameter should be as maximum as possible in load balancing environment to achieve cost minimization.

Makespan. Makespan is time required for executing the entire submitted tasks to the system. Makespan is taken as the maximum time required for executing cloudlets running on the data centers. The Above metrics needs to be fulfilled for the productive execution of the system.

The rest section of this research is organized as follows: Sect. 1.2 provides related work in load balancing in cloud computing, proposed algorithm is discussed in Sect. 1.3, Simulation and result are discussed in Sect. 1.4. Section 1.5 discusses conclusion and future work of the proposed work.

1.2   Related Work

For balancing load in cloud, a fuzzy row penalty method is given by [11]. In this approach fuzzy method are used to solve the uncertainty of response time in the cloud environment. For solving both balanced and unbalanced fuzzy load balancing a fuzzy row penalty method is used. The authors in [7] have combined osmotic behavior to propose a hybrid metaheuristic approach that balance the load in bio-inspired algorithms. The migrated VMs are automatically deployed using the osmotic behavior. Combing the advantages of bio-inspired algorithm the approach balance load in cloud environment.

[6] gave a load balancing algorithm based on genetic algorithm including ACO, ABC and PSO. To balance the load in cloud with economic rule in consideration, Ant Lion Optimizer (ALO) was introduced by the authors. The algorithms gave a better result than in terms of latent period, and makespan time.

The authors in [10] contemplated load-adjusting calculations in circulated frameworks. They grouped the critically utilized load-adjusting calculations in circulated frameworks, including cloud processing, bunch figuring, and network registering. They displayed a similar investigation of different burden adjusting systems on different effectiveness markers roundtrip time, throughput, movement time, reaction time, and so forth. They likewise exhibited a portrayal of the principle highlights of burden adjusting calculations. They broke down points of interest of different burden adjusting calculation. Pattern of difficulties and issues of different calculations is absent in their work.

In [15] a optimization technique for energy consumption based on the flock of birds known as cuckoos is discussed by authors for balancing the load in cloud computing. The cuckoo species winged animals do not assemble their homes for themselves so they lay their eggs in the homes of different fowls who have the comparable eggs so that these feathered creatures can raise their young. The heap adjusting technique pro- posed in the paper comprises of unique advances. In the initial step, to distinguish used hosts the COA calculation is done. In next step, few VMs are chosen to relocate from the overused host to different hosts Till this procedure is not completed, the under used hosts are kept dynamic. Thus, for selecting under used VMs from overused hosts Minimum Migration Time approach is used. The Simulation results yields that the proposed methodology diminished the vitality utilization.

In [12], the authors proposed HBB-LB algorithm for balancing load. The algorithm is based on Honey bee foraging behavior. This technique minimize the waiting time of the tasks in waiting queue. This technique uses the behavior of honeybees in searching and reaping food. Inspired by the behavior of honey bees, the tasks s removed from overloaded nodes. Whenever a new job task arrives to a VM, the priority of that job is updated by virtual machine and it seeks the help of other tasks in choosing a underloaded virtual machine. In the proposed work, the authors have picturized honeybees as tasks and virtual machines as food sources. This algorithm minimizes waiting time of tasks in waiting queue and job execution time.

In [2], the authors have used ant colony optimization (ACO) based technique to balance the load on virtual machine in cloud computing. They have developed a multi agent system to mimic the behavior of ants. The proposed algorithm distributes workload in the environment and thus tries to optimize the response time and load on the virtual machine.

In [9], the authors have proposed an algorithm based on the honey bee foraging strategy is proposed to balance the load in cloud. The proposed algorithm tries to minimize the processing time and overall response time of the system as it shares the workload on different virtual machine by looking into the load and availability of each virtual machine. If the processing time of virtual machine becomes equal or more than a predefined value from average value, the algorithm limits the allocation of workload on the virtual machine to improve the response time.

1.3   Proposed Work

The architecture of Load Balancing Algorithm is given in Fig. 1. The user submits its job to the cloud broker via internet. Broker uses the load balancer to distributes the job taking the load on each server into consideration. If the load on every server is high, then the balancer selects the node that has recently executed a similar kind of job.

In this section, a new variant of particle swarm optimization (PSO), called Mutation Based PSO (MPSO) is described. The proposed algorithm emphasizes the searching process of the under-utilized servers in data centers. In cloud environment, the users are able to use VMs via Internet. CSP has to take care of resources while allocating them to the end users. For the execution, the task submitted by users propagates among multiple cloud data centers. Each DCs in turn divides task into sub-tasks known as jobs. The jobs are allocated to the DCs by selecting available under-utilized machine in each DCs. The proposed algorithm reduce the makespan time by assigning the user jobs to the available DCs.

Particle Swarm Optimization. For each particle, fitness is calculated after updating the particle's position using 1 and 2.

The flow of the algorithm is shown in Fig. 2.

If the value of t_{best} gets better, the value of entire population improves. The mutation applied to the system will randomly change the global best value and

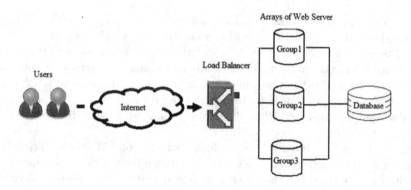

Fig. 1. Architecture of load balancing algorithm

as a result of it the global best value can improve. If the value improves, then it is accepted. To improve the search space, a Gaussian based mutation technique is used on G_best value.

1.4   Simulation and Result

Implementation Environment. CloudSim 3.0.3 s used to simulate the proposed algorithm. The tool has been configured with eclipse-mars. Using CloudSim, researchers can simulate massive scale cloud application demand with respect to the distribution of servers over a geographical location and the workload submitted by the users. The proposed algorithm improves the fitness function and minimizes the makespan time of jobs.

Parameters Setting. For comparing the performance of proposed algorithm, makespan time is considered. The value of makespan time of proposed algorithm is compared with existing nature inspired-algorithms for different values of data centers (DCs). The range of data centers is taken from 10–30. The performance With data center equal to 10 is shown in Fig. 3 and later for 20 and 30 DCs ares shown in Fig. 4 and 5 respectively. The range of cloudlets is taken from 50–300 for each number of DCs (Table 1).

Experimental Result. The proposed algorithm MPSO is compared with Artificial Bee Colony, Ant Colony Optimization, Particle Swarm Optimization. It is observed that MPSO algorithm gives reduce makespan time than the existing algorithm in literature for different values of data centers. The change varies from 5% to 10% for different sets of cloudlets and datacenters. In Fig. 6, the data centers are set kept constant and the number of cloudlets are varied. The figure shows that the performance is better in that case also.

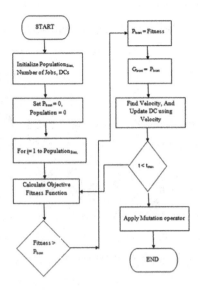

Fig. 2. Flowchart of MPSO Algorithm

Table 1. Experimental parameters for CloudSim

Entity type	Parameters	Values
User	Number of Cloudlets	50–300
Cloudlets	Length	1000–9000
Host	Number of Hosts	2
	RAM	4 GB
	Storage	20 GB
	Bandwidth	1000
Virtual Machine	Numbers of VMs	4
	RAM	1 GB
	Storage	4 GB
	Operating System	Windows
	Type of policy	Time sharing
	Numbers of CPUs	2
Data centers	Numbers of Data Centers	10–20

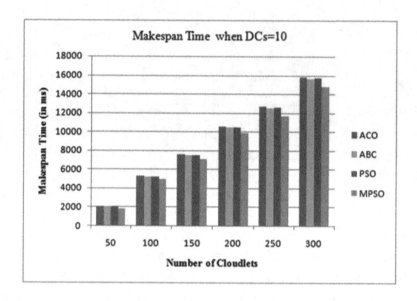

Fig. 3. Makespan Time comparison when DC is 10

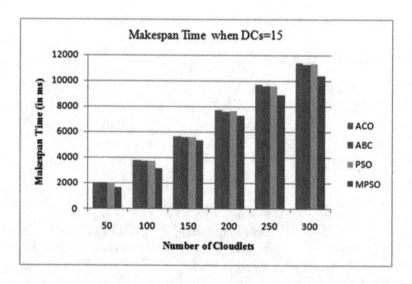

Fig. 4. Makespan Time comparison when DC is 15

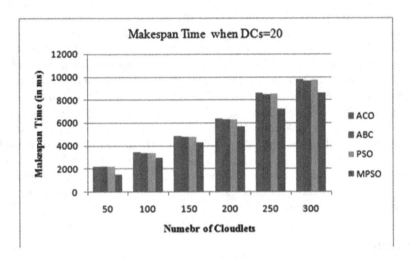

Fig. 5. Makespan Time comparison when DC is 20

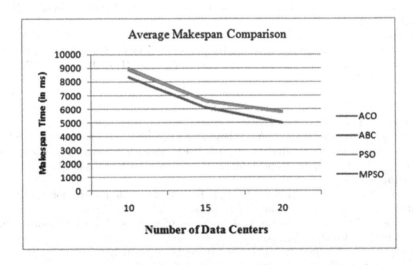

Fig. 6. Average Makesapn Time Comparison

1.5 Conclusions and Future Work

n cloud computing, balancing the load over data centers is one of the greatest challenge. The proposed algorithm betters fitness function while minimizing the makespan time. To achieve the objectives MPSO algorithm is discussed. Swarm based algorithm is a nature inspired technique and we have applied a genetic operator i.e. mutation to it. We are able to improve fitness function by applying the mutation operator to existing PSO algorithm and thus are able to minimize the makespan time. The proposed algorithm gives minimal makespan time as compared to other nature inspired algorithms. The parameter used for comparing different algorithm is makespan time. To balance the load on data centers, a preemptive scheduling on virtual machine is used. In future we can consider various other parameters like, resource utilization, throughput, waiting time etc. for comparing the performance. Also, we can also use other genetic operators i.e crossover and also we can apply successive mutation i.e Gaussian mutation followed by Cauchy mutation.

References

1. Aditya, A., Chatterjee, U., Gupta, S.: A comparative study of different static and dynamic load balancing algorithm in cloud computing with special emphasis on time factor. Int. J. Curr. Eng. Technol. 3(5), 64–78 (2015)
2. Dam, S., Mandal, G., Dasgupta, K., Dutta, P.: An ant-colony-based meta-heuristic approach for load balancing in cloud computing. In: Applied Computational Intelligence and Soft Computing in Engineering, pp. 204–232. IGI Global (2018)
3. Dasgupta, K., Mandal, B., Dutta, P., Mandal, J.K., Dam, S.: A genetic algorithm (GA) based load balancing strategy for cloud computing. Procedia Technol. 10(2), 340–347 (2013)
4. Deepa, T., Cheelu, D.: A comparative study of static and dynamic load balancing algorithms in cloud computing. In: 2017 International Conference on Energy, Communication, Data Analytics and Soft Computing (ICECDS), pp. 3375–3378. IEEE (2017)
5. Ebadifard, F., Babamir, S.M.: A PSO-based task scheduling algorithm improved using a load-balancing technique for the cloud computing environment. Concurr. Comput.: Pract. Exp. 30(12), e4368 (2018)
6. Farrag, A.A.S., Mahmoud, S.A., El Sayed, M.: Intelligent cloud algorithms for load balancing problems: a survey. In: 2015 IEEE Seventh International Conference on Intelligent Computing and Information Systems (ICICIS), pp. 210–216. IEEE (2015)
7. Gamal, M., Rizk, R., Mahdi, H., Elnaghi, B.E.: Osmotic bio-inspired load balancing algorithm in cloud computing. IEEE Access 7, 42735–42744 (2019)
8. Gupta, R.: Review on existing load balancing techniques of cloud computing. Int. J. Adv. Res. Comput. Sci. Softw. Eng. 4(2), 168–171 (2014)
9. Hashem, W., Nashaat, H., Rizk, R.: Honey bee based load balancing in cloud computing. KSII Trans. Internet Inf. Syst. 11(12) (2017)
10. Kirichenko, L., Ivanisenko, I., Radivilova, T.: Dynamic load balancing algorithm of distributed systems. In: 2016 13th International Conference on Modern Problems of Radio Engineering, Telecommunications and Computer Science (TCSET), pp. 515–518. IEEE (2016)

11. Kumar, N., Shukla, D.: Load balancing mechanism using fuzzy row penalty method in cloud computing environment. In: Mishra, D., Nayak, M., Joshi, A. (eds.) Information and Communication Technology for Sustainable Development. LNCS, vol. 9, pp. 365–373. Springer, Singapore (2018). https://doi.org/10.1007/978-981-10-3932-4_38

12. Dhinesh Babu, L.D., Krishna, P.V.: Honey bee behavior inspired load balancing of tasks in cloud computing environments. Appl. Soft Comput. **13**(5), 2292–2303 (2013)

13. Mousavi, S., Mosavi, A., Varkonyi-Koczy, A.R.: A load balancing algorithm for resource allocation in cloud computing. In: Luca, D., Sirghi, L., Costin, C. (eds.) INTER-ACADEMIA 2017. AISC, vol. 660, pp. 289–296. Springer, Cham (2018). https://doi.org/10.1007/978-3-319-67459-9_36

14. Nwobodo, I.: Cloud computing: a detailed relationship to grid and cluster computing. Int. J. Future Comput. Commun. **4**(2), 82 (2015)

15. Yakhchi, M., Ghafari, S.M., Yakhchi, S., Fazeli, M., Patooghi, A.: Proposing a load balancing method based on cuckoo optimization algorithm for energy management in cloud computing infrastructures. In: 2015 6th International Conference on Modeling, Simulation, and Applied Optimization (ICMSAO), pp. 1–5. IEEE (2015)

Statistical Model for Qualitative Grading of Milled Rice

Medha Wyawahare, Pooja Kulkarni, Abha Dixit, and Pradyumna Marathe[✉]

Department of Electronics and Telecommunication, Vishwakarma Institute of Technology, Pune, India
{medha.wyawahare,pooja.kulkarni,abha.dixit,
pradyumna.marathe16}@vit.edu

Abstract. Rice is a principal dietary component for most of the world's population and comprises many breeds and qualities. Qualitative grading of rice is essential to determine its cost. This grading process is currently conducted via physical and chemical invasive processes. Automated, non-invasive grading processes are pivotal in reducing the drawbacks of invasive processes. This research involves the construction of a method for automated, non-invasive, qualitative grading of milled rice. Percentage of broken rice is one of the factors which governs the grading of rice. The method developed is useful in predicting the percentage of broken rice from the image of a given sample of rice. Color images of rice were acquired using cellphone camera. The images were processed by a foreground detector program. Statistical analysis was implemented to extract features for the formation of a regression model. The entire method is executed in MATLAB. The method involves simple regression models and hence requires lesser runtime (4.0567 s) than existing methods of calculating percentage of broken rice. The process produces low root mean square error (0.69 and 0.977) and high r squared (0.999 and 0.999) values for overlapping and non-overlapping grains' dataset respectively.

Keywords: Rice grading · Quality of rice · Foreground detector · Standardization · Supervised learning · Regression models

1 Introduction

Food is a necessity for human beings. Assessment of food quality determines the cost incurred by the consumer. Rice (Oryza sativa) is the staple food for more than half the world's population. Rice is grown in various parts of the world and is affected by the climate and nutrients provided. The climate and agricultural researches determine the breed of rice that can be grown naturally in a region and the nutrients provided determine the quality of cultivated rice. This ensues, that the breed and quality shall affect the cost of rice. It therefore becomes significant to qualitatively grade every sample of cultivated rice.

Rice is initially obtained as paddy from the fields. It then undergoes a rigorous process called milling which provides us with a clean and polished product termed as

© Springer Nature Singapore Pte Ltd. 2020
M. Singh et al. (Eds.): ICACDS 2020, CCIS 1244, pp. 234–246, 2020.
https://doi.org/10.1007/978-981-15-6634-9_22

milled rice. The quality of milled rice can be assessed using factors specified by experts and validated by Government certified Food Authorities such as the Food Corporation of India [1]. The factors specified by the Food Corporation of India that affect the quality of rice are as follows:

• Broken Rice Percentage (Brokens)	• Damaged/Slightly Damaged Grains
• Foreign Matter	• halky Grains
• Discolored Grains	• Red Grains
• Moisture Content	• Dehusked Grains

The factors specified are identical for every breed of rice. This qualitative grading can be realized in two methods:

I. Invasive physical and chemical processes.
II. Non-invasive, automatic process.

The merits and demerits of the two methods are summarized in Table 1. Method I and Method II correspond to the invasive process requiring physical and chemical analysis and our non-invasive automated process respectively.

Table 1. Merits and demerits of grading processes

S No	Property	Method I	Method II
1	Automated	✓	✓
2	Non - invasive	✗	✓
3	Economical	✗	✓
4	Multiple sensors	✓	✗
5	Efficient	✓	✓
6	Breed invariant	✗	✓
7	Saves time	✗	✓
8	Samples remain fit for consumption after grading	✗	✓

The objective of our research is to develop a method that can predict the percentage of broken rice in a sample image while minimizing the computation time and the cost of the process. The method must possess a low Root Mean Square Error (RMSE) and a large R Squared Value. Statistical analysis will be used to create a dataset of significant features and utilize a simple regression model to predict the percentage of broken rice.

2 Literature Review

The method developed encompasses the domains of machine learning, image processing and statistics. The processing of images, extraction of features, manipulation of data and implementation of a regression-based predictor must be understood and implemented appropriately. Hence, it is important to review existing research to understand the minutiae of the process and build on the solid and ambitious research available. To implement image processing, it is important to understand the fundamentals of image processing [2] which are crucial in the construction of a valuable algorithm. To create an algorithm which would be light invariant, adaptive thresholding techniques are beneficial [3]. To understand the various methods available in qualitative grading of rice, researches employing different techniques were referred. To employ a classification process, acquirement of data is essential. Many researches employ edge detection techniques in their proposed methods [4–6]. Principal Component Analysis has been implemented in a research, but it provides low classification accuracy [6]. Reference [7] however, extracts region properties of the foreground as they would provide the exact physical properties of the rice grains and hence, we decided to utilize the region properties in our prediction technique. The next step would be to explore the establishment of thresholds for classification from the properties obtained. Many methods involved observation-based establishment of thresholds for grading [8–11]. From these researches, we identified that a process which employs computationally adaptive thresholding can be used. This was further explored in a research that increased the accuracy of grading via histogram-based thresholding [12]. The next stage of the method would involve a machine learning component which can be used to grade the rice sample. Most researches utilize Neural Networks for classification of milled rice [13–15]. Neural Networks provide good accuracy but processing numerous grains in the foreground is computationally expensive. Multiclass Support Vector Machines (Multiclass SVM) have also been successfully implemented but the technique is comparatively less accurate in grading rice [16]. Another research implemented Regression for determining percentage of broken rice [17]. As the first and foremost stage in our research was predicting the percentage of broken rice, this technique appealed to the aspect of being computationally inexpensive and considerably accurate in prediction of broken rice percentages. The most ambitious research for the grading of milled rice implemented Neural Networks, Support Vector Machines, Bayesian Network and Decision Trees which provided accuracies of 92.31%, 90.38%, 82.69% and 59.615% respectively [18]. However, it requires the creation of a large feature vector and the method was unsatisfactory with respect to fluctuating light conditions. Finally, we decided to utilize regression as we hypothesized that it would be computationally less expensive and using the region properties of rice would be apt for predicting the percentage of broken rice. To implement regression, the values of all the physical properties had to be converted into statistically relevant format to ensure that the range of place values for every property did not differ vastly from one another. [19] was referred to for standardizing all the properties into Z-scores for manipulation and analysis of the properties as a normal distribution about their respective means.

3 Materials and Method

The method requires creation of an extensive database to obtain the data matrix which is used to predict the percentage of broken grains from an input image. The block diagram for the method is illustrated in Fig. 1.

A. *Acquiring Rice Breeds*

To incorporate samples of various breeds and qualities, 18 breeds of rice were collected:

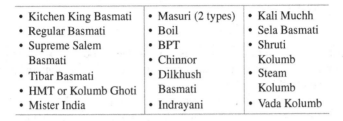

• Kitchen King Basmati	• Masuri (2 types)	• Kali Muchh
• Regular Basmati	• Boil	• Sela Basmati
• Supreme Salem Basmati	• BPT	• Shruti Kolumb
• Tibar Basmati	• Chinnor	• Steam Kolumb
• HMT or Kolumb Ghoti	• Dilkhush Basmati	• Vada Kolumb
• Mister India	• Indrayani	

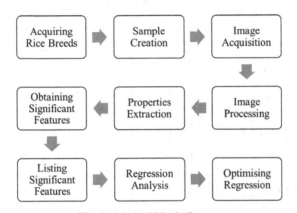

Fig. 1. Method block diagram

The above specific breeds are selected as they are readily available in the state of Maharashtra, India.

B. *Sample Creation*

To analyse sample images for the creation of statistically significant data, we have attempted to incorporate every combination of intrinsic variation probable in the creation of such samples. Such variations have been incorporated in the sample images through the following means:

1) *Number of Grains:* The number of grains was varied across each subsequent image of every breed to achieve a variation in the number of grains.
2) *Percentage of Broken Rice:* The first sample image of every breed of rice contains an arbitrary basic number of grains. Two to three broken rice grains were added in every subsequent image of each breed to increase the percentage of broken rice.
3) Overlapping of Grains: The grains present in any image for any breed of rice were overlapped or separated randomly across the entire dataset. A different dataset was also created for individually separated rice grains.

C. *Image Acqusition*

In the implementation of this method, we have used mobile phone camera of following specifications:

- 13 Megapixels
- Dimensions: 3096 * 4128

It is vital to note that if the resolution is too low, the properties extracted may not be suitable for further use. On the other hand, if the resolution is too high, the computation time of the process will increase.

Figure 2 describes the relationship between the continuities detected for increase in Megapixel specifications/settings of the mobile camera. It is inferred that low resolutions (below 8 Megapixels) will result in continuities being skipped and increasing resolution beyond 16 Megapixels is futile.

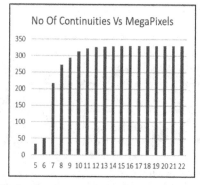

Fig. 2. Continuities detected in varying camera resolutions

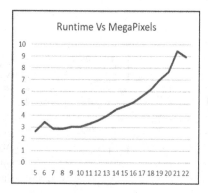

Fig. 3. Runtime for various camera resolutions

Figure 3 describes the relationship between runtime and Megapixels. The runtime increases somewhat exponentially therefore, increasing the resolution can be detrimental to the speed of the method.

D. *Image Processing*

The method designed to process the image has no shortcomings with respect to ambient light conditions. Processing the image involves the following stages:

1) *Conversion of Colorspace:* The images are converted to CIE XYZ color space because it is device independent, has no negative values and can be used for accurate color mixing as well as communicating and mapping a representation.
2) *Binarization*: Use Otsu's Adaptive threshold to improve the efficiency of the foreground detector program.
3) *Filter and Segmentation:* Median Filter is used to eliminate salt and pepper noise and the image is eroded.

4) *Boundary Labeling and Removal of Unusable Foreground:* Region properties are acquired, and undesirable light specks are removed using area threshold.

E. *Properties Extraction*
Physical properties of the foreground are extracted using available functions. This offers us several properties of which, five properties are useful. The properties were selected after observing their histograms for test images. These properties are as listed:

• Area	• Convex Area	• Minor Axis Length
• Major Axis Length	• Eccentricity	

F. Obtaining Significant features
Significant features for use in the regression classifiers are obtained after standardizing the properties extracted for each labeled continuity in the foreground. Stages for acquiring the standardized features are:

Table 2. Significant features matrix

Statistical property vs feature selected	Area	Major axis length	Minor axis length	Convex area	Eccentricity
μ/σ	✓	✓	✓	✓	✓
$Max\,(X_i)$	✓	✓	✓	✓	✓
$Min\,(X_i)$	✓	✓	✓	✓	✓
$Median\,(X_i)$	✓	✓	✓	✓	✓

• Compute population mean (μ).

$$\mu = \left(\sum x_i\right)/n \,|\, i \in [1,\, n] \tag{1}$$

• Compute population standard deviation (σ).

$$\sigma = \left[\sum (x_i)^2/n\right]^{(1/2)} |\, i \in [1,\, n] \tag{2}$$

• Standardize the values of each property (xi) for each labelled continuity using the standardization formula:

Table 3. Variations incorporated in imaging

Sr No	Variation	Total number of grains	Number of broken grains	Percentage of broken rice	Overlapping of grains
1	Image 1 Boil: Fig. 4	314	372	1.9108	Present
2	Image 30 Boil: Fig. 5	6	64	17.2043	Present

$$X_i = (x_i - \mu)/\sigma \,|\, i \in [1, n] \tag{3}$$

'X_i' denotes the standardized value of a property x_i. 'n' denotes the number of continuities

G. *Listing Significant Features*

The 20 features which are acquired through standardization and are used in regression are illustrated as a matrix of interactions (Table 2) between general statistical properties of a distribution and the features selected:

Output p-values for each of these features are observed against a typical significance value of 0.05 for their relevance in regression.

H. Regression Analysis

The 20 features thus obtained are used for regression modelling. The dataset is using inbuilt functions to check the fit of the dataset for regression and prediction.

I. Optimising Regression

The model specifications or types of regression available in the function are tested with the dataset to determine the appropriate regression model that must be used.

4   Data and Results

A. *Database Creation*

The database was created by incorporating all inherent variations probable in the construction of such a database. To exemplify the manner of imaging, observe Fig. 4 and Fig. 5. Each of the images (Fig. 4 and Fig. 5) represents the variations which must be incorporated. Each of these variations have been explained in Table 3.

B. *Features Obtained*

The data that is obtained from the images can be enlisted in tabular format and the features which are used for regression can be observed. Each of these features possess a general trend of increase or decrease with respect to increase in percentage of broken rice across 30 images. Consider the rice breed 'Boil'; the variation in the features for the breed as the percentage increases is represented in graphical format (Fig. 6 to Fig. 9).

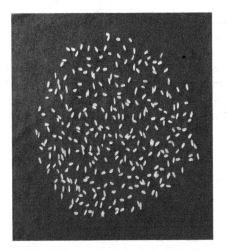

Fig. 4. Image 1 for rice breed 'Boil'

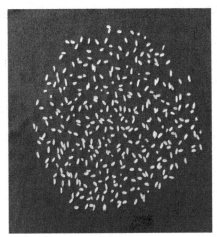

Fig. 5. Image 30 for rice breed 'Boil'

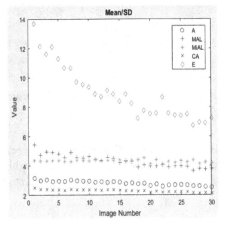

Fig. 6. Ratio of mean and standard deviation for all properties

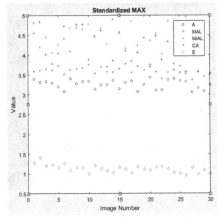

Fig. 7. Standardized Maximum for all properties

The legends of the graphs represent the properties as acronyms. 'A' represents Area, 'MAL' represents Major Axis Length, 'MiAL' represents Minor Axis Length, 'CA' represents Convex Area and 'E' represents Eccentricity. The Vertical axis indicates the values of the features and the horizontal axis indicates the serial number of the image.

Trends described from Fig. 6 through to Fig. 9 for the rice breed 'Boil' are:

- Fig. 6 illustrates the change in ratio of mean and standard deviation for all properties. Values remain constant for area, major axis length, Minor axis length and convex area. Whereas, the values decrease exponentially for eccentricity.

- Figure 7 illustrates the change in standardized maximum for all properties. It is observed that the pattern of change for area, convex area and eccentricity and the change for major axis length and minor axis length are roughly proportional.
- Figure 8 illustrates the change in standardized minimum for all properties. A gradual increase at varying intensities is observed for area, major axis length, minor axis length and convex area. An incongruous variation in the value for eccentricity is observed which results in an exceedingly independent vector.

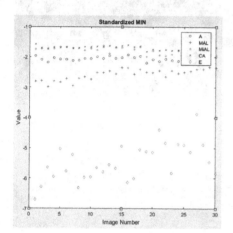

Fig. 8. Standardized minimum for all properties

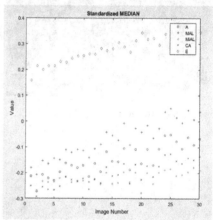

Fig. 9. Standardized median for all properties

- Figure 9 illustrates the change in standardized median for all properties. The area, major axis length, minor axis length and convex area report a gradual but roughly proportional increase in the value. The values for eccentricity, however, increase steadily.

C. *Fitting Regression Models*

The inbuilt MATLAB regression function provides output values regarding the fit of the data. These values must be analysed to determine the regression model which can be used in the final process. The function provides outputs for different kinds of fit for each the datasets i.e. the overlapping grains dataset and the non-overlapping grains' dataset.

D. *Selection of Regression Model*

The factors that govern the selection of a regression model and the criteria used to select a model are illustrated in Table 4. From the criteria listed in Table 4 and the models which were trained for both datasets, we have selected the appropriate regression models and listed them in Table 5.

E. *Output Matrix for Predictor*

The final process implemented contains a method to create a data matrix (X) from the image which contains the features required for prediction in a Quadratic Regression Model. The data matrix contains 1 row and 231 columns. The data matrix is

Table 4. Factors affecting regression models

Sr No	Factor	Criterion
1	RMSE	Lowest value of all iterations
2	R squared	Value closest to 1
3	Adjusted R squared	Value closest to 1
4	P-Value	Lowest value of all iterations
5	Error degrees of freedom	Lowest value of all iterations

Table 5. Selected regression models

Dataset -	Overlapping grains	Non-overlapping grains
Regression model	Quadratic	Quadratic
RMSE	0.69	0.977
R Squared	0.999	0.999
Adjusted R squared	0.994	0.991
P-Value	2.67e–37	6.48e–08
Error degrees of freedom	39	8

multiplied with the regression matrix (θ) attained as the part of the model provided by the inbuilt function. The regression matrix contains 231 rows and 1 column.

$$Op = X * \theta \tag{4}$$

After multiplication in the form represented in Eq. 4, a matrix of one row and one column *(Op)* is obtained which contains the percentage of broken grains predicted from the input image.

5 Discussion

The proposed method includes various stages which have been implemented through continuous testing of the process. The advantages and intricacies of the process are:

A. *Prominent Advantage:* Every existing research employs processing images containing rice grains which are non-overlapping or have been separated manually. Even with such segregation of rice grains, the accuracy of the algorithms is comparatively low. Analysis of overlapping grains by means of such impractical algorithms is impossible. Our proposed method provides output of a higher accuracy whilst also processing all overlapping rice grains present in the image.

B. *Comparison with Existing Researches:* The proposed method displays several advantages over the existing research currently conducted for grading. The similarities and advantages have been highlighted in Table 6.

Table 6. Advantages of proposed method

Sr No	Property	Existing research	Proposed method
1	Automated	✓	✓
2	Non-invasive	✓	✓
3	Economical	✗	✓
4	Minimum RMSE	✗, $1 < RMSE < 2$	✓, $RMSE = 0.69$ and 0.977
5	Maximum R squared	✗, R Squared $= 0.92$	✓, R Squared $= 0.999$ and 0.999
6	Breed invariant	✗	✓
7	Considers overlapping grains	✗	✓
8	Low runtime	Undefined; Neural Networks require higher Runtime	✓, Average runtime $= 4.0567$ S

C. *Use of CIE XYZ Color space:* It is also called the 1931 XYZ color space. The X, Y and Z values are crudely related to Red, Green and Blue respectively. The CIE XYZ color space defines a large extent of possible colors that we may not be able to create or observe. It is also device independent, has no negative values and can be used to convey image maps accurately. We have used it in this process because prior conversion to CIE XYZ color space followed by conversion to grayscale and binarization yields significantly improved results.

D. *Image Erosion:* The image is eroded using an inbuilt function available in MATLAB. As expected, the erosion operation reduces the area of the foreground along the discontinuities. We have used a line structuring element along with a small kernel size (2 by 2) for image erosion as these attributes allow the separation of sparsely overlapped grains but do not reduce the effective area of the grains.

E. *Adaptive Thresholding:* The early objective of the method comprised precise processing of the image using a foreground detector program. The main challenge at this stage, was to create a light invariant program because of the small size of the grains which would be affected by fluctuating light conditions. This would enable the visualization of the grains as foreground continuities. Binarization of the image would be most successful if and only if a sharp contrast between the rice grains and the background is created.

F. *Selecting Regression Model:* Each of the regression models in the inbuilt function of MATLAB were tested. For a Regression model to possess low error, the RMSE value must be as low as possible. For an accurate regression model which explains the variation in output as a function of input features, the R squared value must be closer to 1. A good regression model always possesses a low p-value, especially lower than the selected significance level. The error degrees of freedom are the number of observations in the dataset which are an exception to the estimated regression model. Hence, a quadratic regression model has been used which has a low RMSE value and p-value along with an R squared value closest to 1 and, a low number of error degrees of freedom with respect to the total number of observations in the dataset.

6 Conclusion

Qualitative grading of rice is a significant process that determines the cost incurred by the consumer and allows assessment of milled rice as per Government certified Food Authorities or the ISO. Qualitative grading may be used to determine the breeds of rice that can be purchased at a lower cost whilst safeguarding the dietary requirement of the consumer. For example, Indrayani and varieties of Kolumb are breeds of rice which are inexpensive and can be consumed daily whereas varieties of Basmati are expensive and are used occasionally in delicacies and special food products.

Existing researches regarding automated non-invasive grading of rice employ computationally expensive neural networks and, in some cases, expensive imaging setups which are not economical. It is therefore vital to develop a process which can serve the purpose of grading but is computationally less expensive and economical.

The proposed method in our research provides precise output with a very low root mean square error value (0.69 and 0.977) and a high R Squared value (0.999 and 0.999). Also, as simple regression models are used, the runtime for the method is low (4.0567 s). As simple cellphone camera is used for imaging, the cost of a dedicated imaging setup is also circumvented thus making it economical. The method is breed invariant and hence negates the requirement of a dedicated analytical technique for every existing breed of rice. The most prominent advantage is the incorporation of overlapping grains which establishes its practicality over existing researches which necessitate the separation of grains.

Statistical analysis by means of standardization creates versatility in terms of camera used, resolution of the image, overlapping of grains, depth of image and programming environment used. As the functions used in the implementation of the process are available in MATLAB, they can be reproduced in a lower level programming language to achieve further improved runtime and a ready to use process in the agriculture sector.

Acknowledgement. This research was supported by Vishwakarma Institute of Technology, Pune, India.

References

1. Uniform Specifications of Paddy, Rice And Coarse Grains For Kharif Marketing Season 2017–18, 16 August 2017. http://fci.gov.in/qualities.php?view=9
2. Jain, A.K.: Fundamentals of Digital Image Processing. Prentice-Hall, Englewood Cliffs (1989)
3. Otsu, N.: A threshold selection method from gray-level histograms. IEEE Trans. Syst. Man Cybern. **SMC-9**(1), 62–66 (1979)
4. Mahale, B., Korde, S.: Rice quality analysis using image processing techniques. In: International Conference for Convergence of Technology (2014)
5. Anami, B.S., et al.: Analysis of rice granules using image processing and neural network pattern recognition tool (2016)
6. Maheshwari, C.V., Jain, K.R., Modi, C.K.: Non-destructive quality analysis of Indian Gujarat-17 Oryza Sativa SSP Indica (Rice) using image processing. Int. J. Comput. Eng. Sci. (IICES) **2**(3), 48–54 (2012)
7. Kambo, R., Verpude, A.: Classification of basmati rice grain variety using image processing and principal component analysis. Corr Abs/1405.7626 (2014). N. Pag
8. Aulakh, J.S., Banga, V.K.: Grading of rice grains by image processing. Int. J. Eng. Res. Technol. (IIERT) **1**(4) (2012)
9. Ajay, G., Suneel, M., Kiran Kumar, K., Siva Prasad, P.: Quality evaluation of rice grains using morphological methods. Int. J. Soft Comput. Eng. (IISCE) **2**(6), 35–37 (2013)
10. Kuchekar, N.A., Yerigeri, V.V.: Rice grain quality grading using digital image processing techniques. IOSR J. Electron. Commun. Eng. (IOSR-JECE) **13**(3), 84–88 (2018). E-ISSN 2278-2834, P-ISSN 2278-8735, Ver. I
11. Mahajan, S., Kaur, S.: Quality Analysis of Indian Basmati Rice Grains Using Digital Image Processing-A Review (2014)
12. Sansomboonsuk, S., Afzulpurkar, N.: Machine vision for rice quality evaluation. In: Technology and Innovation for Sustainable Development Conference (TISD), pp: 343–346 (2008)
13. Siddagangappa, M.R., Kulkarni, A.H.: Classification And quality analysis of food grains. IOSR J. Comput. Eng. (IOSR-JCE) **16**(4), 01–10 (2014). E-ISSN 2278-0661, P-ISSN 2278-8727, Ver. Iii
14. Verma, B.: Image processing techniques for grading & classification of rice. In: 2010 International Conference on Computer and Communication Technology (ICCCT), pp. 220–223 (2010)
15. Gujjar, H.S., Siddappa, M.: A method for identification of basmati rice grain of India and its quality using pattern classification (2012)
16. Kaur, H., Singh, B.: Classification and grading rice using multi-class SVM (2013)
17. Aghayeghazvini, H., Afzal, A., Heidarisoltanabadi, M., Malek, S., Mollabashi, L.: Determining percentage of broken rice by using image analysis. In: Li, D., Zhao, C. (eds.) CCTA 2008. IAICT, vol. 294, pp. 1019–1027. Springer, Boston, MA (2009). https://doi.org/10.1007/978-1-4419-0211-5_27
18. Rexce, J., Usha Kingsly Devi, K.: Classification of milled rice using image processing. Int. J. Sci. Eng. Res. **8**(2), 10 (2017). ISSN 2229-5518
19. Montgomery, D.C., Runger, G.C.: Applied Statistics and Probability for Engineers, 3rd edn. Wiley, New Delhi (2007)

Measuring the Effectiveness of Software Code Review Comments

Syeda Sumbul Hossain$^{(\boxtimes)}$, Yeasir Arafat, Md. Ekram Hossain,
Md. Shohel Arman, and Anik Islam

Daffodil International University, Dhaka, Bangladesh
{syeda.swe,yeasir35-1501,ekram35-1936,arman.swe,anik35-1376}@diu.edu.bd

Abstract. Code reviewing becomes a more popular technique to find out early defects in source code. Nowadays practitioners are going for peer reviewing their codes by their co-developers to make the source code clean. Working on a distributed or dispersed team, code review is mandatory to check the patches to merge. Code reviewing can also be a form of validating functional and non-functional requirements. Sometimes reviewers do not put structured comments, which becomes a bottle neck to developers for solving the findings or suggestions commented by the reviewers. For making the code review participation more effective, structured and efficient review comments is mandatory. Mining the repositories of five commercialized projects, we have extracted 15223 review comments and labelled them. We have used 8 different machine learning and deep learning classifiers to train our model. Among those Stochastic Gradient Descent (SGD) technique achieves higher accuracy of 80.32%. This study will help the practitioners to build up structured and effective code review culture among global software developers.

Keywords: Empirical software engineering · Modern code review · Sentiment Analysis · Machine learning · Mining software repositories

1 Introduction

In the global software engineering era, code reviewing becomes a more popular technique to find out early defects in source code. Nowadays practitioners are going for peer reviewing their codes by their co-developers to make the source code clean. Clean source code increase the legibility of the code which is crucial for global software development. Nowadays, software industries are adopting agile software development methodology (Hossain 2019), and agile software development is done across the distributed and dispersed teams when the project in large scale. While working on a distributed or dispersed project, every patch need to be reviewed by other developers for reviewing not only the source code as well as to check if the patch meet the requirements or not. Sometimes reviewers do not put structured comments, which becomes a bottle neck to developers for solving the findings or suggestions commented by the reviewers. To make the

© Springer Nature Singapore Pte Ltd. 2020
M. Singh et al. (Eds.): ICACDS 2020, CCIS 1244, pp. 247–257, 2020.
https://doi.org/10.1007/978-981-15-6634-9_23

review participation more effective, structured and efficient review comments is mandatory.

In modern code review (Bacchelli and Bird 2013), tool-based review is becoming more popular. Different tools (Holzmann 2010; Ahmed et al. 2017) are being used by the practitioners for tool-based code review. An automated code review tool (Moskowitz et al. 2005) is developed by Google. There are many tools for reviewing source code, however, peer reviewing is one of the code quality assurance activity other than tool-based reviewing (Sadowski et al. 2018). CFar (Henley et al. 2018) is an automated review tool, while comments in any changeset automatically, tracks if any changes is occurred according to the comment, and also label any comment as "useful", "not useful" or "not understandable". Therefore, source code need to be reviewed by human reviewers. However, most of the times review comments do not make any sense to the developers while development is done in distributed manner. Therefore, review comments by the reviewers need to be well structured for the developers for further changing in the patch set.

The purpose of this study is to find the effectiveness of modern code review comments using different machine learning and deep learning techniques. Mining the software repositories of five commercialized projects, we have extracted 15223 review comments. In order to identify the sentiment polarity (or sentiment orientation) of review comments, we train our model with 8 machine learning and deep learning algorithms and compare those results to find the better ones. This research can be extended to develop a code reviewing tool, which can suggest structured and effective review comments. That will help the practitioners to build up structured and effective code review culture in global software development.

This paper is structured as follows: Related discussion is done at Sect. 2 that followed by Research Methodology and Result & Discussion at Sect. 3 and Sect. 4 respectively. The final Sect. 5 summarizes our contribution and furnishes the conclusion.

2   Related Work

Plenty of breakdowns of scientists working procedures with Sentiment Analysis are present today, from organized, general procedure disintegration like the sentiment analysis instruments, created for breaking down online life content or item reviews, work ineffectively on a Software Engineering (SE) data-set (Ahmed et al. 2017). Recently Sentiment Analysis has as of late assumed a critical job for researchers since the analysis of online content is valuable for the statistical surveying political issue, business insight, web-based shopping, and logical overview from psychological and so on (Vinodhini and Chandrasekaran 2012). Some of works (Rahman et al. 2019) also done on news headlines for finding the polarity of text.

There is an enormous assemblage of literature on code reviews and investigations. Basically, Code review is a key tool for quality affirmation in software development (Madera and Tomoń 2017). During the procedure of statically assessing code, developers of a code base can cooperatively identify conceivable code defects, just as use code reviews as a method for transferring information to improve the general comprehension of a system (Kalyan et al. 2016). Using machine learning for learning the source code revision, an automated review tool is suggested (Shi et al. 2019). In (Lal and Pahwa 2017) a machine learning approach is proposed to help in faster and cleaner code review. Sentiment analysis is done on code review comments to identify positive comments (Ahmed et al. 2017).

Deep learning (DL) is also becoming popular in software repositories mining research. For detecting redundant comments in code, (Louis et al. 2018) developed a tool using DL. They introduced a framework which can help the developers in writing better and informative programming comments. Using topic model and n-grams, (Movshovitz-Attias and Cohen 2013) presented a technique to predict programming comments. In (White et al. 2016), they used DL in detecting code clone by mining software repositories. Automated code review systems are proposed in (Gupta and Sundaresan 2018; Li et al. 2019) using deep learning techniques.

3   Research Methodology

In this section, we briefly describe the methodology we have adopted for the analysis. We are motivated to design our research by our prior work (Rahman et al. 2019). In order to analyze the review comments, first we have extracted code review comments from five commercialized iOS projects. Then we labeled the data for determining sentiments of those data. After preprocessing the labeled data, we trained our models. Figure 1 shows the overall process of our research work.

3.1   Data Extraction

We have selected five commercialized projects of a renowned company. All projects are iOS App development projects and are maintained in Gerrit[1]. Using the Gerrit API, we crawled the repositories and extracted the raw data. For extracting those data, we set the time line from January, 2018 to December, 2019. Table 1 is showing the details of data set extracted from different data source.

3.2   Data Labeling

One of our researchers had worked on labeling review comments. The raw data had been labeled as Efficient, Somewhat Efficient, Not Efficient and System

---

[1] https://www.gerritcodereview.com/.

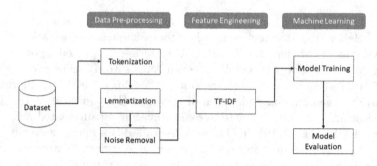

Fig. 1. Research methodology

Table 1. Summary of data set

Project name	#No of members	#No of review comments
Alpha	25	5223
Beta	17	3050
Gamma	11	2500
Delta	4	2350
Epsilon	3	2100

Generated. Then we have selected two reviewers and developers of respective projects to cross-check the sentiment of labeled data.

3.3 Data Preprocessing

For analyzing our data set, we preprocess all the code review comments. First of all, we tokenized the comments. Then we lemmatized the texts and removed the noise words. The overall preprocessing and training code is available at GitHub[2].

3.3.1 Tokenization

We segment our text data into words using word-tokenize which is from NLTK library. We make sure all the short forms like *don't, she'll* will remain as one word.

3.3.2 Lemmatization

Lemmatization reduces the inflected words ensuring that the root word belongs to the language. In lemmatization, root word is called lemma. Using Wordnet lemmatizer we lookup the lemmas of words. The parts of speech of a word is determined in lemmatization like 'V' as verb, 'A' as adjective, 'N' as noun. Table 2 is showing some sample lemma words from our dataset.

---

[2] https://github.com/AronnoDIU/Measuring-the-Effectiveness-of-Software-Code-Re view-Comments.

Table 2. Lemmatization process

Word	Lemma
Merged	Merge
Abandoned	Abandon
Done	Do
Refactored	Refactor

3.3.3   Noise Removal

We have removed all the punctuation and regular expression from the text as these cannot be used for analyzing sentiment.

3.4   Feature Engineering

After preprocessing, we used sklearn's Tf-Idf word vectorizer to convert code review comments into trainable features for classifier by making word vectors. For doing so, we have used N-gram (Brown et al. 1992) to make the features more trainable. Table 3 shows the N-gram words.

Table 3. N-gram words

Uni gram	Bi gram	Tri gram
"Activity"	"Assign variable"	"Assign new variable"
"Refactor"	"Authentication token"	"Band setting common"

3.5   Sentiment Determining

We quantify the polarity of review comments in a scale of 1 (Efficient), 0 (Somewhat Efficient), −1 (Not Efficient) and −2 (System Generated). The overall code reviews comment is inferred as four sentiment form in the sign of the polarity score. Table 4 shows how we consider polarity score of review comments.

Table 4. Polarity score of review comments

Scale	Polarity	Code review comments
1	Efficient	Need to release memory to avoid memory overflow
0	Somewhat Efficient	I will submit later. Please wait
−1	Not Efficient	Will correct in next patch
−2	System Generated	CL Validation Started: It will take a few seconds to validate

3.6   Model Training

We have used 8 different Machine Learning (ML) and Deep Learning (DL) algorithms for training our model. The used algorithms are: Logistic Regression (Ho et al. 1994), Multinomial Naïve Bayes (MNB) (Rennie et al. 2003), Linear Support Vector Classifier (SVC) (Gunn et al. 1998), Bernoulli Naïve Bayes (McCallum et al. 1998), Stochastic Gradient Descent (SGD) (Bottou 2010), XGBoost (Chen and Guestrin 2016), Perceptron (Stephen 1990) and Passive Aggressive (Crammer et al. 2006). Table 5 shows the training accuracy of different classifiers.

Table 5. Training accuracy of different classifiers

Algorithms	Accuracy (%)
Logistic Regression	95.36
Multi nomial Naïve Bayes (MNB)	91.43
Support Vector Machine (SVM)	95.23
Bernoulli Naïve Bayes	86.98
Stochastic Gradient Descent (SGD)	95.33
XGBoost	92.52
Perceptron	91.59
Passive Aggressive	92.33

3.7   Validation

Our corpus datasets contains total 15223 text data. In which System Generated is 4408, SomeWhat Efficient is 3785, Efficient and Not Efficient are 3680 and 3650 respectively. We evaluate our model using k-fold cross validation where K-Fold value 5 is used. Figure 2 and 3 shows the overall cross validation process and learning curve of Stochastic Gradient Descent (SGD).

Section 4 presents the details of Validation accuracy of different classifier based on N-gram. In this study, we have used Uni-gram, Bi-gram and Tri-gram for training our model and considered the average of those accuracy for final training.

4   Result and Discussion

In this section, we have explained the accuracy of different classifiers and the polarity estimation of our research.

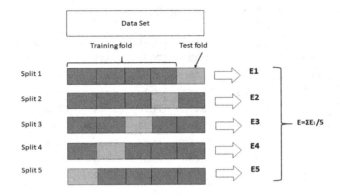

Fig. 2. Cross validation process

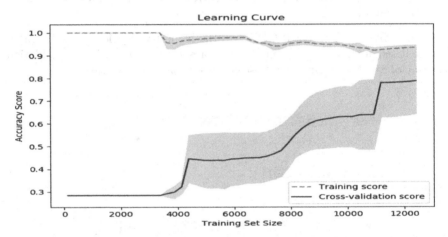

Fig. 3. Learning curve of SGD

4.1   Classification Results

In order to classify the review comments as per our polarity score, we used different classification algorithms. Table 6 shows the validation accuracy of different classifiers we used in this research. Here we have validated using uni-gram, bi-gram and tri-gram and have considered the average accuracy for finding the most accurate one for training our model.

We can see from the accuracy results that Bernoulli Naïve Bayes shows the lowest accuracy of 77.26% where Stochastic Gradient Descent (SGD) shows highest accuracy of 80.32% among the machine learning algorithms. An average accuracy of around 80.23% has shown by Multi nomial Naïve Bayes (MNB) and Support Vector Machine (SVM).

Table 6. Accuracy of different classifiers

Algorithms	Uni-gram	Bi-gram	Tri-gram	Average accuracy (%)
Logistic Regression	79.43	78.81	78.68	78.97
Multi nomial Naïve Bayes (MNB)	80.48	80.04	80.17	80.23
Support Vector Machine (SVM)	80.44	80.15	80.15	80.24
Bernoulli Naïve Bayes	81.15	77.70	72.94	77.26
Stochastic Gradient Descent (SGD)	80.30	80.27	80.39	80.32
XGBoost	76.37	76.88	76.47	76.57
Perceptron	77.76	79.35	78.60	78.57
Passive Aggressive	79.95	78.34	78.22	78.83

4.2 Precision, Recall and F-Measure for Different Labels of Code Reviews Data Sets

Stochastic Gradient Descent (SGD) gives the highest validation accuracy on an average based on different N-gram. Figure 4 presents the comparison of validation accuracy of Stochastic Gradient Descent (SGD) based on N-gram.

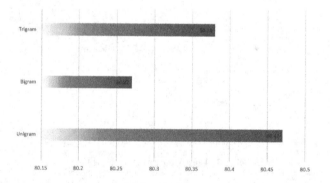

Fig. 4. Accuracy of SGD based on N-gram

Then we have trained our model with Stochastic Gradient Descent (SGD). The precision, recall, and F-measure of the Stochastic Gradient Descent (SGD) classifier is shown in Table 7 as per the polarity.

We can state from the above that-

– Every Efficient review comments and System Generated comments are correctly identified with 88% and 86% recall respectively. That means Efficient comments and System Generated comments are correctly identified with few false negative.
– Both Non-efficient and Some What Efficient is 84% likely to be correct (as high precision) which shows fewer false positives.
– By identifying the efficient comment correctly, this result can help the reviewers by suggesting comments polarities.

Table 7. Confusion matrix of reviews data sets

Sentiment	Precision	Recall	F-measure
Efficient	0.62	0.88	0.72
Not-Efficient	0.84	0.77	0.81
Some What Efficient	0.84	0.69	0.76
System Generated	0.99	0.86	0.92

4.3 Comparison of Machine Learning and Deep Learning Classifier

In this study, we have used both machine learning algorithms and deep learning algorithms. Among those Stochastic Gradient Descent (SGD) and XGBoost
shows the better accuracy of 80.32% and 76.57% from ML and DL algorithms
respectively. Other than Stochastic Gradient Descent (SGD), Multi Nominal
NB and Support Vector Machine (SVM) have given better accuracy. This study
shows that, we can get better accuracy from machine learning algorithms than
other deep learning algorithms.

5 Conclusion

The intention of this study to measure the code review comments done by the
peer reviewers. Here we have used both machine learning and deep learning
algorithms to train our model. From this study, we can state machine learning
algorithms outperforms over deep learning techniques. By measuring the efficiency of code review comments, development teams can save their time and
cost in reviewing and fixing the code as per the review comments. In this study,
we only extracted review comments only from Gerrit. However, in future work we
can build our corpus data set by extracting from different code review tools. This
research can also be extended to develop an automated code reviewing feedback
tool, which can suggest structured and effective review comments while com-
menting at the practise of code reviewing. That will help the practitioners to
build up structured and effective code review culture.

References

Hossain, S.S.: Challenges and mitigation strategies in reusing requirements in large-
scale distributed agile software development: a survey result. In: Arai, K., Bhatia,
R., Kapoor, S. (eds.) CompCom 2019. AISC, vol. 998, pp. 920–935. Springer, Cham
(2019). https://doi.org/10.1007/978-3-030-22868-2_63

Bacchelli, A., Bird, C.: Expectations, outcomes, and challenges of modern code review.
In: Proceedings of the 2013 International Conference on Software Engineering, pp.
712–721. IEEE Press (2013)

Holzmann, G.J.: SCRUB: a tool for code reviews. Innov. Syst. Softw. Eng. 6(4), 311–
318 (2010). https://doi.org/10.1007/s11334-010-0136-x

Ahmed, T., Bosu, A., Iqbal, A., Rahimi, S.: SentiCR: a customized sentiment analysis tool for code review interactions. In: Proceedings of the 32nd IEEE/ACM International Conference on Automated Software Engineering, pp. 106–111. IEEE Press (2017)

Moskowitz, M., Potter, W., King, W.: Automatic computer code review tool, 26 May 2005. US Patent App. 10/769,535

Sadowski, C., Söderberg, E., Church, L., Sipko, M., Bacchelli, A.: Modern code review: a case study at Google. In: Proceedings of the International Conference on Software Engineering: Software Engineering in Practice, pp. 181–190. Association for Computing Machinery (2018)

Henley, A.Z., Muçlu, K., Christakis, M., Fleming, S.D., Bird, C.: Cfar: a tool to increase communication, productivity, and review quality in collaborative code reviews. In: Proceedings of the 2018 CHI Conference on Human Factors in Computing Systems, pp. 1–13 (2018)

Vinodhini, G., Chandrasekaran, R.M.: Sentiment analysis and opinion mining: a survey. Int. J. **2**(6), 282–292 (2012)

Rahman, S., Hossain, S.S., Islam, S., Chowdhury, M.I., Rafiq, F.B., Badruzzaman, K.B.M.: Context-based news headlines analysis using machine learning approach. In: Nguyen, N.T., Chbeir, R., Exposito, E., Aniorté, P., Trawiński, B. (eds.) ICCCI 2019. LNCS (LNAI), vol. 11684, pp. 167–178. Springer, Cham (2019). https://doi.org/10.1007/978-3-030-28374-2_15

Madera, M., ł Tomoń, R.: A case study on machine learning model for code review expert system in software engineering. In: 2017 Federated Conference on Computer Science and Information Systems (FedCSIS), pp. 1357–1363. IEEE (2017)

Kalyan, A., Chiam, M., Sun, J., Manoharan, S.: A collaborative code review platform for GitHub. In 2016 21st International Conference on Engineering of Complex Computer Systems (ICECCS), pp. 191–196. IEEE (2016)

Shi, S.-T., Li, M., Lo, D., Thung, F., Huo, X.: Automatic code review by learning the revision of source code. In: Proceedings of the AAAI Conference on Artificial Intelligence, vol. 33, pp. 4910–4917 (2019)

Lal, H., Pahwa, G.: Code review analysis of software system using machine learning techniques. In: 2017 11th International Conference on Intelligent Systems and Control (ISCO), pp. 8–13. IEEE (2017)

Louis, A., Dash, S.K., Barr, E.T., Sutton, C.: Deep learning to detect redundant method comments. arXiv preprint arXiv:1806.04616 (2018)

Movshovitz-Attias, D., Cohen, W.W.: Natural language models for predicting programming comments. In: Proceedings of the 51st Annual Meeting of the Association for Computational Linguistics (Volume 2: Short Papers), vol. 2, pp. 35–40 (2013)

White, M., Tufano, M., Vendome, C., Poshyvanyk, D.: Deep learning code fragments for code clone detection. In: Proceedings of the 31st IEEE/ACM International Conference on Automated Software Engineering, pp. 87–98. ACM (2016)

Gupta, A., Sundaresan, N.: Intelligent code reviews using deep learning (2018)

Li, H.-Y., et al.: DeepReview: automatic code review using deep multi-instance learning. In: Yang, Q., Zhou, Z.-H., Gong, Z., Zhang, M.-L., Huang, S.-J. (eds.) PAKDD 2019. LNCS (LNAI), vol. 11440, pp. 318–330. Springer, Cham (2019). https://doi.org/10.1007/978-3-030-16145-3_25

Brown, P.F., Desouza, P.V., Mercer, R.L., Della Pietra, V.J., Lai, J.C.: Class-based n-gram models of natural language. Comput. Linguist. **18**(4), 467–479 (1992)

Ho, T.K., Hull, J.J., Srihari, S.N.: Decision combination in multiple classifier systems. IEEE Trans. Pattern Anal. Mach. Intell. **16**(1), 66–75 (1994)

Rennie, J.D., Shih, L., Teevan, J., Karger, D.R.: Tackling the poor assumptions of Naive Bayes text classifiers. In: Proceedings of the 20th International Conference on Machine Learning (ICML-03), pp. 616–623 (2003)

Gunn, S.R., et al.: Support vector machines for classification and regression. ISIS Tech. Rep. **14**(1), 5–16 (1998)

McCallum, A., Nigam, K., et al.: A comparison of event models for Naive Bayes text classification. In: AAAI-98 Workshop on Learning for Text Categorization, vol. 752, pp. 41–48. Citeseer (1998)

Bottou, L.: Large-scale machine learning with stochastic gradient descent. In: Lechevallier, Y., Saporta, G. (eds.) Proceedings of COMPSTAT 2010, pp. 177–186. Springer, Heidelberg (2010). https://doi.org/10.1007/978-3-7908-2604-3_16

Chen, T., Guestrin, C.: XGBoost: a scalable tree boosting system. In: Proceedings of the 22nd ACM SIGKDD International Conference on Knowledge Discovery and Data Mining, pp. 785–794 (2016)

Stephen, I.: Perceptron-based learning algorithms. IEEE Trans. Neural Netw. **50**(2), 179 (1990)

Crammer, K., Dekel, O., Keshet, J., Shalev-Shwartz, S., Singer, Y.: Online passive-aggressive algorithms. J. Mach. Learn. Rese. **7**(Mar), 551–585 (2006)

Proposed Model for Feature Extraction for Vehicle Detection

Padma Mishra[(⊠)] and Anup Girdhar[(⊠)]

TMV University, Pune, India
Padmamishra286@gmail.com, anupgirdhar@gmail.com

Abstract. A feature is a prominent interest point in an image that can be used for a different task processing of image besides computer vision based on processes for object recognition. The features could be extracted by mathematical models that detect deep variations in texture, detect edges, or color. The selected features must have global definition within the defined problem vehicle detection. The focus of this paper is on detection of vehicle, Extraction of Region of Interest for the feature which is represented globally their module might produce a model intended for encoding of images' features dependency technique can be applied. In this paper, We offer an extremely robust, capable, method aimed at creation of image feature vector for vehicle detection model system with both feature extraction also global feature representation method for both inter classes sameness and also the intra-class variation, thus to overcome the problem of multiplicity and ambiguity issues.

Keywords: Vehicle detection · Feature extraction · Region of Interest · Histogram of oriented gradients · Global features representation · Bag of features

1 Introduction

Through prompt increase of urbanization, traffic congestion, incident, also destruction posture is an excessive challenges aimed for traffic management systems. Visualization, by way of an evidence collection approach of realistic world environment, had involved ample consideration successful intelligent transportation scheme. In Computer vision methods exist primarily for used of gathering of traffic flow parameters and examine traffic behaviours used in traffic surveillance. Thus [1] the summary for background, ideas, simple techniques, most important issues, besides existing solicitations of parallel transportation man cooperative automated driving. The potential benefit of cooperative automated driving is that all vehicles jointly optimize their actions in order to improve traffic efficiency and safety. The transition from current-day driving to cooperative automated driving, has already started with Advanced Driver Assistance System for example Adaptive Cruise Control in addition lane observance assistance. Existing autonomous driving technologies. Thus the Video dispensation results be able to offer a several evidence resultant from real traffic world Trustworthy also robust detection of vehicle is a important element for surveillance of traffic. In that respect still several issues for vehicle detection in intelligent transportation systems [19].

© Springer Nature Singapore Pte Ltd. 2020
M. Singh et al. (Eds.): ICACDS 2020, CCIS 1244, pp. 258–271, 2020.
https://doi.org/10.1007/978-981-15-6634-9_24

Several vehicle occurrence also its poses make a problematic for training an unified model for detection. Multifaceted city environments, bad atmospheric condition, lighting alteration, besides the unfortunate/strong illumination circonstance aim take down the sensing demonstration [20]. Thus the particular traffic congestion, object are obstructed aside each another and so that separated transport aim easy hook on a respective car. Thus the constraint acquisition for recognition of object is similarly a serious concern. Thus the Recognition technique through complex constraints is commonly not hands-on. The improvements of machine learning methods will able to use to absorb the constraints. Now in [2], a weak supervised method aimed at target find existed offered. This technique does not requisite hand-operated assortment besides labeling of training examples. Algorithms boosting was stretched towards training of samples through probabilistic labels.

In this paper the discriminative size besides generalizability of the dictionary in adding processing quickness of Vehicle detection model system determined through on the dictionary size henceforth the performance of Bag of features vehicle detection model system besides hinge on dictionary scope. Other than one Distinct feature can be allocated to similar visual word owing towards reduced dictionary size; henceforth Results in decrease with discriminative abilities. A greater Dictionary misses the competence of generality Greater dictionaries lean towards to growth the processing overhead besides enhance additional consequence towards noise. We considered numerous dictionary sizes through respect towards their effect over Vehicle detection model system recognition rate besides processing speed in this work. Thus paper is planned sections wise as ensue: Sect. 2 considered related to work. The complete design of system through feature extraction besides Vehicle Detection Model System datasets are considered in Sect. 3. The ratio also the action are considered for planned Vehicle Detection Model System is discussed in Sect. 4. Section 5 provides conclusion.

2 Related Work

The basic idea behind the histogram oriented gradients information is that localized object arrival besides shape inside an picture is able to considered through the organisation of concentration gradient magnitudes as well as directions. Thus the histogram of oriented gradients divisions of an picture into minor spatial associated regions, so-called cells, and the Sect. 1-D histogram of gradient directions of each picture element within each compartment is accumulated, according to the gradient magnitudes. These histograms are then concatenated to get the descriptors. The histograms of each cell, can be contrast normalized aimed for improved invariance to lighting, brightening, shadowing etc., through calculative a amount of the strength through a greater area of image, so-called blocks, and then the point via this value towards normalize the entire cells inside the paper besides slab [7].

Dalal and Triggs proposed Dlagnekov [5] has considered scale invariant feature transform the difficultly of automobile form besides framework recognition aimed at aim of inquiring surveillance video collection for a uncomplete license plate number combined with some visual description of a car. Our planned methods intent render precious situational info for law implementation units in a assortment of civil structure. Overlapping blocks in which each cell approval various element to the ultimate descriptor vector, each normalized with respect to a different block. The normalized descriptor blocks are name to as histogram of oriented gradients descriptors. For our car detection pipeline, we use the following histogram of oriented gradients parameters: The histogram of arranged inclinations descriptor has a couple of key favorable circumstances over different descriptors. Since it works on nearby cells, it is invariant to geometric and photometric changes, aside from object direction. Such changes would just show up in bigger spatial areas. Besides, as Dalal and Triggs found, coarse spatial inspecting, fine direction examining, and solid neighborhood photometric standardization allows the individual body development of people on foot to be overlooked inasmuch as they keep up a generally upstanding position. The histogram of oriented gradients word is thus especially suitable for detection of humanoid in images. Petrovic and Cootes [4] has measured Raw pixels, outcome of Sobel edge, orientations of edges, Harris corner answer, normalized gradients, square mapped gradients using global representation. Author establish a comparatively simple set of extraction of attribute from units of motor vehicle forward-facing images be able to used for great performance verification besides recognition of vehicle category. Author defines the conceptualization besides subsequent scheme in flooded, besides the outcomes of tests examination a extensive variability of dissimilar characteristic. The concluding group is confident of identification rates of above 93% besides verification equal fault rates of less than 5.6% when verified on all over 1000 images comprising 77 dissimilar classes. The system is exposed to be robust aimed at a extensive range of weather in addition illumination conditions. Munroe and Madden [6] feature extraction canny edges using concatenation of edge image pixels. The outcomes of certain typical multi-class cataloging procedures are connected aimed at this problem. A unique class k-Nearest Neighbour cataloging procedure remains also implemented besides tested. Clady et al. [7] Oriented contour points from Sobel edges through modest link. This is also used by more complex algorithm included into OpenCV. Sobel is a well-known algorithm used for contour detection. Psyllose et al. [8] feature extraction Phase congruency (Make) + scale-invariant feature transform (Model) procedure remains matched by selected current procedure now systematical way. The experimental outcomes of pictures now Berkeley Segmentation Dataset besides at all sensed images demonstration that this procedure remains readily towards identify image features. Pearce and Pears [9] features extraction Canny edges, Harris corners, Square Mapped Gradients using global representation approach LNHS and concatenated-square mapped gradients, This method recursively segments the picture keen on quadrants, the element qualities now these quadrants are then added and privately standardized in a recursive, progressive style. Two diverse grouping approaches are explored; a k-closest neighbor classifier and a Naive Bayes classifier. Our framework can order vehicles with 96.0% precision, tried utilizing leave-one accessible cross-approval happening a steadfast dataset of 262 forward symbolisms of vehicles. Jang and Turk [10] are using Speed-Up Robust Features feature extraction approach and by means of Dictionary global representation methods. These procedures take been used hip different bag-of-words and structural matching approaches. This work demonstrates a recognition application, based upon the speed-up robust features feature descriptor algorithm, which fuses bag-of-words and structural verification techniques. The resulting system is applied to the domain of car recognition and achieves accurate (>90%) and real-time performance when searching databases holding thousands of pictures. Baran et al. [11] using scale invariant feature transform, speed-up robust features, edge histogram and by means of Dictionary created Sparse Vector of Existence Sums of global representation method, Author provides the full depiction of binary approaches aimed at MMR scenario. The advanced approaches have remained drilled besides through an experiment tested happening datasets ready aimed at training besides testing. Varjas and Tanacs [12] using square mapped gradients feature extraction approach and using Concatenated square mapped gradients global representation, advanced present methods founded on region of interests well-defined comparative towards the number plate. Square Mapped-Gradient highlights are removed from the district of premiums and acknowledgment is cultivated by arrangement using a learning set. The classifier is assessed utilizing ground truth information gave physically. By means of numerical reproductions we assessed the location resistance of the technique and proposed self-loader and completely oblivious strategy. Hsieh et al. [13] using accelerate hearty highlights, scale invariant component change, histogram of arranged inclinations utilizing Grid-based portrayal worldwide portrayal, Author has proposes another balanced accelerate vigorous highlights descriptor to advance the intensity of accelerate powerful highlights to recognize all conceivable even coordinating sets through a reflecting change.

Fraz et al. [14] using scale invariant feature transform and using Fisher Encoding based midlevel feature representation global representation approach, midlevel feature representation remains figured going on discriminate covers of the picture towards construct a cognition, the pictorial words of which remain used towards signify the figure inside that image. The writer devours planned image depiction technique has remained applied towards the application of cars make besides model recognition. Chen et al. [15] using Symmetric speed-up robust features and using Grid-based concatenated sparse representation global representation approach, Thus the paper offers a original symmetrical speed-up robust features descriptor towards to identify vehicles happening on roads besides put on the sparse representation aimed at application of automobile make-and-model recognition. Towards identify vehicles from roads, this paper suggests a symmetry transformation on speed-up robust features points toward identify entirely probable corresponding couples of symmetrical speed-up robust features points. Abdul Jabbar et al. [16] using speed-up burly characteristic and using Dictionary globular representation approach, Author has suggest besides estimate unexplored methods aimed at real-time programmed automobile make and model recognition created on a container of speeded-up vigorous features besides determination the appropriateness of these approaches aimed at object determination systems. These methods usage speed-up robust characteristic features of vehicles' front or else rear-facing imageries also recollect the leading representative characteristic now a dictionary.

Boonsim et al. [17] using histogram oriented ed gradients and using concatenated hog global representation approach, Author proposed a method with which identification of car make besides model on nightly through by means of accessible rear view features. Author takes offered one form classifier collaborative intended to classify a specific car model of notice since additional models.

3 Proposed Model

The Proposed model distributed hooked on two scheme training as well as testing/prediction scheme. The training scheme is utilized towards to train the model for recognition device via a subclass of the existing dataset, however prediction component identifies make besides ideal of vehicles an unseen before image. In this subdivision, we confer every component aimed at in cooperation with training also testing process. Detection of vehicle as well as Region of Interest Extraction also property extractions will be present similar intended for together training as well as testing work. Representation of features globally besides the components of classification are dissimilar in training besides testing process and it also produce a framework used for images encoding thus the features depending on the technique relevant. Likewise, class wise component besides creates a prototypical for example a outcome of training that exists then castoff through the testing to calculate the result intended for the new obtained images. Henceforth the pointer beginning training towards testing signify the usage of models in testing process produced for the duration of the training (Fig. 1).

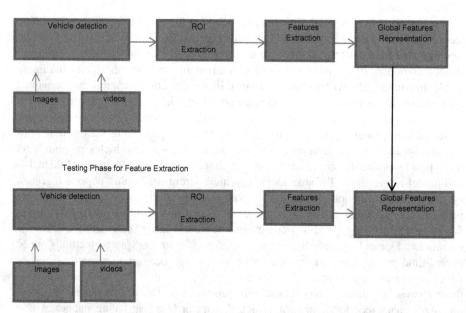

Fig. 1. Proposed model of Feature Extraction with Global Features Representation

3.1 Histogram of Oriented Gradient

Dalal and Triggs [1] introduced a feature descriptor based on directions of edges for robust human detection in 2005 named as Histograms of Oriented Gradients, besides consume stayed widely used subsequently before in numerous expansions on object recognition besides Vehicle Detection Model System. All object in addition figure currently an image remainders collected of a group of lines, thus we define everything in an picture through the dissemination of gradient directions. Basic theory behind the hog descriptor is to represent an image as the distribution of gradient direction. HOG splits the appearance hooked on minor associated. The Histograms of Oriented Gradients algorithm is able to be précised in the subsequent four stages:

- Gradient computation: The gradient provides the derivative of an image in both directions gradients. Thus the gradient magnitude and the gradient directions are calculated using as the first step of histogram of oriented gradients calculation. The x-gradient highlight the vertical edges (edge making a 90° angle with x-axis), x gradient highlight the horizontal edges (edge parallel to the x-axis) and all other edges are highlighted by x gradient as well as y gradient. The gradient magnitude generates edges and all the flat regions turn into the black background. Three gradient magnitudes besides gradient directions remain computed aimed at the color image the maximum gradient magnitude and corresponding direction are selected for further computation in histogram of oriented gradients. The gradient angles are between 0 and 180° if the absolute values of x and y are used, the absolute gradients are named as unsigned gradients. If the x gradient and y gradient are used without performing the absolute operation, the resulted gradient angles are between 0 and 360°, known as a signed gradient. Both the signed and unsigned gradients can be used to construct hog feature descriptor. If the bins in numbers are same for signed gradients and unsigned gradient; a large number of gradient directions are represented by single bin. If we increase the number of bins, which results in increased hogs feature descriptor size and increase the complexity of the training and testing process. The unsigned gradients outperform the signed gradient [1] in terms of classification accuracy. We have used unsigned gradient in our work.
- Orientation binning: The succeeding phase in the creation of hog descriptor is to construct a histogram for gradient directions. Each pixel in the window has two values; namely, a gradient magnitude and a gradient direction. The histogram contains 9 bins to accommodate gradient direction from 0 to 180° (bin 0, 20, 40... 160); the size of each bin is 20°. Different bin sizes are compared in [1] and it is determined that a bin size of 20° performs better as compared to other sizes. In the case of the signed gradient with the same bin size instead of an unsigned gradient, the histogram size is doubled, and therefore, the computation time increases at each phase. Every pixel in window adds weighted value to histogram based on orientation channel depending on its gradient magnitude. Histogram of oriented gradients does not add a pixel's gradient magnitude to one bin only; instead, histogram of oriented gradients distributes the gradient magnitude proportionally into two bins based on its gradient direction. For example, a pixel has the gradient magnitude of 5 and the gradient direction of 25°. The ratio of the distance of said pixel based on gradient direction from bin 20 and bin 40 is 0.25:0.75; hence the magnitude 1.25 is added to bin 40 and 3.75 to bin 20. For pixels having gradient direction of greater than 160°, the gradient magnitude is added to bin 160 and bin 0. The histogram representation lessens the impact of noise; individual pixel may have noise but its impact on the histogram is insignificant.

- Block Descriptor and Normalization: histogram of oriented gradients combines windows to make a bigger block. Each block consists of 4 windows (2×2). The blocks overlap with each other in both directions like sliding windows approach. Window 1, 2, 6, and 7 are combined in one block while window 2 and 7 are also part of next block. Every window is a part of four blocks except for the windows at corners (included in one block) and edges (included in two blocks). The histograms of windows are concatenated to form a bigger vector named as block descriptor. Since each block consists of four windows and each window's histogram has 9 bins (or represented by a 9×1 vector); hence the size of block descriptor is 36×1. Block descriptors are normalized individually to account for changes in illumination and contrast. Gradient magnitudes are sensitive to changes in lighting. In darker images, each pixel's intensity value is reduced, and therefore, the gradient magnitude affecting the histogram's magnitudes is reduced. We use normalization to make the descriptor independent of the lighting variations. Since the qualities of inclinations should privately standardized, we bunch the cells together into bigger, spatially associated squares.

HOG Feature Descriptor: Finally, all block descriptors are concatenated to form a single histogram of oriented gradients feature signifier for each picture. For example, we have an image of size 128×64. If we divide the image into windows of size 8×8 there will be total 16×8 windows and 15×7 blocks. Hence the total size of histogram of oriented gradients descriptor is a 3780-dimensional vector in this case ($15 \times 7 \times 36$). The histogram of oriented gradients blocks are computed similarly to scale invariant feature transform descriptors; however, Blocks of histogram of oriented gradients exist figured on solitary scale happening a dense grid fashion and short of orientation assignment, however scale invariant feature transform descriptors stay calculated aimed at scale-invariant key points besides remain rotated aimed at orientation alignment. The histogram of oriented gradients block descriptors exist used popular association towards encode spatial form information, while scale invariant feature transform descriptors are used singly. We consume used dissimilar window dimensions towards generate histogram of oriented gradients feature descriptor now our work. We have not combined more windows to make a larger related to block. As a result, we have created histogram of oriented gradients descriptor with the small property with respective to the standard histogram of oriented gradients descriptor besides enhanced the processing speed [18]. The typical histogram of oriented gradients generates a feature descriptor through a dimensions of 3780 elements aimed at a 128×64 image by way of discussed previous then the size of histogram of oriented gradients descriptor deprived of related to 1152 elements aimed at the similar picture besides identical assemble.

3.2 GIST Feature Descriptor

Hominids remain accomplished for categorizing an scene through a look deprived of allowing for the specifics current in image now a very rapid glimpse. For instance, subsequently observing tall buildings picture otherwise trees or else ocean, we can suddenly identify the scene deprived of thoughtful of the information or presence of additional objects. The GIST of a sight [2] discusses to the evidence matters collected now a glimpse (around 200 ms). The feature descriptor using GIST principal familiarize through Oliva and Torralba in [3], remains "a little dimensional portrayal of the scene, which doesn't require any type of division". The GIST descriptor was at first proposed for scene characterization. scale invariant component change and accelerate powerful highlights center around individual unmistakable focuses and the histogram of situated inclinations include descriptor is processed dependent on singular windows (fixes) and connected later, though GIST descriptor emphases happening the figure of a complete picture by way of a single object besides estimate the characteristic vector. The GIST signifier overlooks the existence of limited items besides their associations. Therefore GIST offers a all-inclusive depiction of a sight. The GIST feature signifier algorithm can be obtained by following three steps: The original image is initially transformed into m scales and n orientations; now we have m × n transformed images. Next, a set of Gabor filters is applied on all these m × n images. The Gabor filter examines the presence of potential specific frequency contents in any specific direction in an image. The dimensions of the output of the Gabor filter are same as the input image hence we generate m × n characteristic maps with identical dimension as of the image as input. For each feature maps are separate as grid producing multiple blocks. The values of the feature map in each block are averaged. Now, we have multiple averaged values representing each feature map. The division of characteristic maps with grid and averaging the values of each block in the grid reduces the dimensionality of the feature map. Lastly, the averaged values of all the blocks of all the feature maps are concatenated to construct the GIST descriptor or image feature vector. The GIST algorithm produces the feature descriptor with the same size for all the images regardless the image resolution. We have created GIST signifier through four measure besides eight position generating thirty-two converted images besides divided the feature maps into the 4 × 4 grid or 16 area, which generated a 512-dimensional characteristic vector (16 be an average of values × 32 feature maps).

3.3 Global Features Representation

The features provide the meaningful information about the image and are used to represent an image or object within an image. We define two types of feature representations, namely, local and global features. Local features are calculated for a part of the image on basis of prominent point(s) and local features represent only a part of a given image. The global features, on the other hand, are calculated for the entire image, and hence, represent the entire image. We have to combine the local features in order to represent the entire image, not just a patch. The feature representation which is globally is the method to association altogether the local features providing an appearance feature vector which signifies the whole image. For both local as well as global features are used to construct an image characteristic vector representing the entire image. Also, image feature vectors have the same size for a complete dataset that provides uniform representation in terms of a number of features/dimensions. The image feature courses exist later used to sequence the classifier and to predict the vehicle make and model. Although researchers have explored various feature extraction and global feature representation techniques to build discriminative and informative Vehicle Detection and Model representation, still very few approaches have been proven to show success across high degrees of multiplicity and ambiguity. Yet Vehicle Detection Model System techniques still undergo certain constraints like slow processing (non-real time processing of incoming images) and failures under conditions such as variations in lighting, weather, occlusion, etc. One of the goals of this research is to propose a highly robust, yet efficient, way for construction of image feature vector for Vehicle Detection Model System including both feature extraction and global feature representation technique, which handles both inter-class sameness also intra - class divergence, hence solving the multiplicity and ambiguity issues. Typically, the machine learning algorithms work with a fixed number of input features (dimensions) for training and testing datasets. and GIST image features for our Vehicle Detection Model System problem and constructed fixed length image feature vectors for every feature extraction technique. Our Vehicle Detection Model System model computes hog histograms for windows, and then, concatenates histograms to build histogram of oriented gradients feature descriptor vector. The simple concatenation provides the global features representation. We have divided the input images of the entire dataset into a same number of windows for histogram of oriented gradients descriptor to get fixed-size feature vector. The GIST procedure likewise concatenates the values to build the vector with feature and yields a feature vector of the same size for every input image regardless of the image resolution as discussed earlier. Hence, the global feature representation technique, concatenation, is a part of histogram of oriented gradients and GIST algorithm. We need to apply explicit global feature representation technique to build appearance feature vector that can represent the entire image included in the exercise dataset and analysis dataset homogeneously with the similar dimensionality.

4 Bag of Features

The concept of Bag of Words originally came from research on text analysis. Bag of words has been successfully used in document classification. The document/text is represented by a bag and important words are added into this bag without considering grammar specifics or the orders of words. The words collected in the bag can be considered as features. The features' frequency is used to train a classifier to predict the outcome of new cases. The Bag of words concepts can be applied to computer vision tasks in the same way. Since local image features can be considered as words abstracting the contents of an image, it is conceivable to build a dictionary of the local image features. A bag of visual words is a vector of frequencies of the visual (Fig. 2).

4.1 Algorithm

The pseudo code for the offline dictionary building step is given in
Pseudo-code representing the used dictionary generation.
Let I present the vehicle images included in training dataset with Nc number of classes.
Where Ii denotes the training images belonging to class i in training dataset.
I = {I1, I2,,,,, INc}

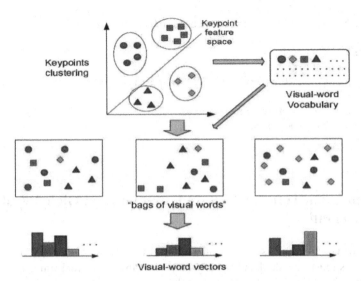

Fig. 2. Bag of features illustration

Image features are extracted from every jth image in Ii, and F represents all the features of the training dataset F = F_{ij}.

Algorithm 1: Building Dictionary

```
Step 1: Input: The set of images from all classes, I = {I1, I2,,,
INc},Dictionary Size SD
Step 2: Output: Dictionary D
Step 3: Initialize: F = {}
Step 4: Step 1: Collecting Local Features (SIFT OR SURF)
Step 5: for each class I £ [1; Nc]: do
Step 6: Initialize: Fi = {}
Step 7: for each image j £ [1; Ni]: do
Step 8: Fij ← Feature Extraction (Iij)
Step 9: Fi = Fi U Fij Fij is set of local features in image j
Step 10: end for
Step 11: F = F U Fi
Step 12: end for
Step 13: Step 2: Dictionary Building D ← Cluster (F, SD)
```

Algorithm 2: Histogram Creation for Global Feature Representation

```
Step 1: Input: An input Image I, Dictionary D, Dictionary Size SD
Step 2: Output: The computed BoW Histogram, H
Step 3: Initialize: H ← {0… 0}, where size of H = SD
Step 4: Step 1: Features Extraction (SIFT OR SURF)
Step 5: F ← Feature Extraction (I)
Step 6: Step 2: Histogram Creation
Step 7: for each feature f £ F: do
```

```
Step 8: k ← identify cluster with minimum distance with the feature
f
Step 9: H(k) ← H(k) + 1
Step 10: end for
Step 11: Step 3: Histogram Normalization
Step 12: m ← max(H)
Step 13: for each element in H: do
Step 14: H(b) ← H(b)/m
Step 15: end for
```

5 Computation Time for Feature Extraction and Global Feature Representation

The time for computation is significant factor aimed at every real-time application. Since features must be calculated for any new input images in real-time, we evaluate the detailed information on the computational time required for each step, starting with extracting the histogram of oriented gradients and GIST image features. In case of hog, we separate the picture into overlap-jointed blocks also the compute gradients and construct a histogram aimed at all chunk. All the blocks' histograms exist merely chain with each one of other for descriptor an image feature vector. Similarly, in the case of GIST, we convert the original image into a set of images with different scales and orientations and compute the Gabor Transform for each image. The transformed images are divided into grid of blocks and average is calculated for each block. Lastly, simple concatenation of averaged values of grids is castoff towards build image feature vector. The workings of histogram of oriented gradients besides GIST are discussed in the measured computational periods aimed at GIST also hog be present in Table 1. The block division configuration aimed at histogram of oriented gradients is given in second column of the table. The whole interval essential intended for computation for each one an design is render now sec for every hundred images. Such as we raise the numeral of area, procedure interval get increase. The computational time for GIST is about 2 times compared to the largest histogram of oriented gradients configuration and about 4 times compared to smallest histogram of oriented gradients configuration used. The exercise besides analysis datasets go through the similar procedure in event of histogram of oriented gradients and GIST features; henceforth the computational period essential remains the similar for exercise also analysis stages for together histogram of oriented gradients besides GIST.

Table 1. Computation time for Histogram Of Oriented Gradients and GIST (seconds/100 images)

Method	Configuration	Feature extraction
HOG	24 × 6	2.1
	30 × 6	2.74
	33 × 6	2.79
	24 × 9	2.85
	33 × 9	3.28
	45 × 9	3.91
	36 × 12	4.04
	45 × 12	4.56
GIST		8.91

5.1 Results Using Random Forest

The Random Forest algorithm works on base of growing a collection of decision trees named as the results using random forest. All decision trees are qualified through a unsystematic selected subcategory of training data besides a randomly particular attributes (features). During classification every decision tree in the random forest is traversed for every input feature vector and each tree predicts the class of the input vector based on its own configuration. Then the prediction results are combined, and an output class is selected using majority voting (Fig. 3 and Table 2).

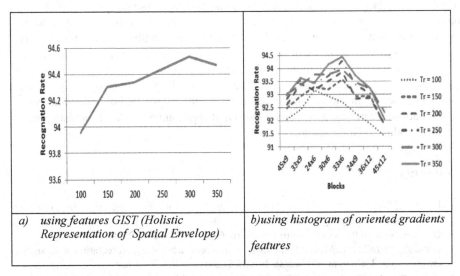

a) using features GIST (Holistic Representation of Spatial Envelope)	b) using histogram of oriented gradients features

Fig. 3. Recognition rate for–Vehicle Detection Model System using Random Forest

Table 2. Vehicle Detection Model System testing time (seconds/100 images)

	Configuration	Trees 100	Trees 150	Trees 200	Trees 250	Trees 300	Trees 350	Feature extraction
GIST		0.005	0.008	0.01	0.014	0.018	0.005	8.91
HOG	24 × 6	0.006	0.009	0.012	0.013	0.014	0.015	2.1
	30 × 6	0.008	0.009	0.011	0.014	0.016	0.017	2.74
	33 × 6	0.008	0.009	0.012	0.013	0.016	0.017	2.79
	24 × 9	0.007	0.01	0.011	0.016	0.018	0.019	2.85
	33 × 9	0.008	0.011	0.014	0.016	0.015	0.018	3.28
	45 × 9	0.007	0.012	0.011	0.013	0.016	0.018	3.91
	36 × 12	0.007	0.01	0.011	0.014	0.015	0.018	4.04
	45 × 12	0.007	0.009	0.011	0.014	0.016	0.018	4.56

6   Conclusion

The focus of this paper is objects as vehicle recognition for Extraction of Region of Interest also the feature extraction aimed at together exercise in addition analysis tasks. Global feature used for presenting dissimilarity in exercise procedure than in analysis procedure. Global feature representation segment also produce a framework intended for images encoding structures dependent on method used. The discriminative capacity and generalizability of the dictionary and processing speed of vehicle detection model system depend on the dictionary size; hence the performance of Bag of features vehicle detection model system also depends on the dictionary size and algorithm for building Dictionary proposed. More than one distinctive feature can be assigned to same visual word due to smaller dictionary size hence results in reduce discriminative capabilities. A larger dictionary loses the capability of generalization larger dictionaries tend to increase the processing overhead and add more penalties to noise. We analyze several dictionary sizes with respect to their effect over Vehicle Detection Model System recognition rate and processing speed in this work and also recognition rate used for dissimilar Random Forest configurations are presented via dissimilar styled appearances intended for histogram of oriented gradients, GIFT features.

References

1. Dalal, N., Triggs, B.: Histograms of oriented gradients for human detection. In: Computer Vision and Pattern Recognition, vol. 1. IEEE (2005)
2. Oliva, A.: Gist of the scene. Neurobiol. Atten. **696**(64), 251–258 (2005)
3. Oliva, A., Torralba, A.: Modeling the shape of the scene: a holistic representation of the spatial envelope. Int. J. Comput. Vis. **42**(3), 145–175 (2001)
4. Petrovic, V.S., Cootes, T.F.: Analysis of features for rigid structure vehicle type recognition. In: BMVC, vol. 2 (2004)

5. Dlagnekov, L., Belongie, S.J.: Recognizing Cars. Department of Computer Science and Engineering, University of California, San Diego (2005)
6. Munroe, D.T., Madden, M.G.: Multi-class and single-class classification approaches to vehicle model recognition from images. In: Proceedings of IEEE AICS (2005)
7. Clady, X., Negri, P., Milgram, M., Poulenard, R.: Multi-class vehicle type recognition system. In: Prevost, L., Marinai, S., Schwenker, F. (eds.) ANNPR 2008. LNCS (LNAI), vol. 5064, pp. 228–239. Springer, Heidelberg (2008). https://doi.org/10.1007/978-3-540-69939-2_22
8. Psyllos, A., Anagnostopoulos, C.-N., Kayafas, E.: Vehicle model recognition from frontal view image measurements. Comput. Stand. Interfaces 33(2), 142–151 (2011)
9. Pearce, G., Pears, N.: Automatic make and model recognition from frontal images of cars. In: 8th IEEE International Conference on Advanced Video and Signal-Based Surveillance (AVSS). IEEE (2011)
10. Jang, D.M., Turk, M.: Car-Rec: a real time car recognition system. In: IEEE Workshop on IEEE Applications of Computer Vision (WACV) (2011)
11. Baran, R., Glowacz, A., Matiolanski, A.: The efficient real- and non-real-time make and model recognition of cars. Multimedia Tools Appl. 74(12), 4269–4288 (2013). https://doi.org/10.1007/s11042-013-1545-2
12. Varjas, V., Tanács, A.: Car recognition from frontal images in mobile environment. In: 8th International Symposium on Image and Signal Processing and Analysis (ISPA). IEEE (2013)
13. Hsieh, J.-W., Chen, L.-C., Chen, D.-Y.: Symmetrical speed-up robust features and its applications to vehicle detection and vehicle make and model recognition. IEEE Trans. Intell. Transp. Syst. 15(1), 6–20 (2014)
14. Fraz, M., Edirisinghe, E.A., Saquib Sarfraz, M.: Mid-level representation based lexicon for vehicle make and model recognition. In: 22nd International Conference on Pattern Recognition (ICPR). IEEE (2014)
15. Chen, L.-C., et al.: Vehicle make and model recognition using sparse representation and symmetrical speed-up robust features. Pattern Recogn. 48(6), 1979–1998 (2015)
16. Siddiqui, A.J., Mammeri, A., Boukerche, A.: Real-time vehicle make and model recognition based on a bag of speed-up robust features features. IEEE Trans. Intell. Transp. Syst. 17(11), 3205–3219 (2016)
17. Boonsim, N., Prakoonwit, S.: Car make and model recognition under limited lighting conditions at night. Pattern Anal. Appl. 20(4), 1195–1207 (2016). https://doi.org/10.1007/s10044-016-0559-6
18. Nie, Z., Yu, Y., Jin, Q.: A vehicle logo recognition approach based on foreground-background pixel-pair feature. In: Pan, Z., Cheok, A.D., Müller, W., Zhang, M. (eds.) Transactions on Edutainment XIII. LNCS, vol. 10092, pp. 204–214. Springer, Heidelberg (2017). https://doi.org/10.1007/978-3-662-54395-5_18
19. Tang, Y., Zhang, C., Gu, R., Li, P., Yang, B.: Vehicle detection and recognition for intelligent traffic surveillance system. Multimedia Tools Appl. 76(4), 5817–5832 (2015). https://doi.org/10.1007/s11042-015-2520-x
20. Nur, S.A., et al.: Vehicle detection based on underneath vehicle shadow using edge features. In: 6th IEEE International Conference on Control System, Computing and Engineering. IEEE (2016)

Analysis of Feature Selection Methods for P2P Botnet Detection

Chirag Joshi$^{(\boxtimes)}$, Vishal Bharti, and Ranjeet Kumar Ranjan

DIT University, Dehradun, UK, India
{chirag.joshi,hod.cse,ranjeet.ranjan}@dituniversity.edu.in

Abstract. Botnets are one of the major threats today and one of the main reasons for this is its capability to hide in the network. It is not easy to detect Botmaster, the one who controlled botnets from a far end. There are different technologies and algorithms that are used for the detection of a botnet in a network. Some of the prominent techniques are based on machine learning algorithms. Machine learning have been proven in the past that they are the best in the business and also the leading techniques to detect botnet. In order to implement machine learning algorithms, the most important task is to analyze the dataset very well before using it. Feature selection techniques help in doing this. With the help of different feature selection techniques, we can find out the best criteria for the detection of a botnet. In a particular dataset, there are different numbers and types of features are present, and all these features don't contribute equally to the detection of the botnet. We need to find out the important features which will be more useful in building a botnet detection model. Many algorithms have been used in the past for botnet detection but most of them have used the different feature selection methods for different datasets. In this paper, we will be covering different feature selection methods and their analysis on the different botnets. Also, in the end, we will be comparing all these techniques and giving the best for a particular botnet.

Keywords: Feature selection · Machine learning · Botnet · SVM · K-means · Supervised learning

1 Introduction

A botnet can be explained as a collection of different types of threats that can be in combination or individually damaged by any computer system. They are managed by a botmaster who controls all the bots remotely through Command and Control (C & C). Mainly botnets are divided into three parts: - Centralized, P2P and Hybrid botnet.

Centralized botnet is the one that is controlled remotely by a botmaster. All the bot's come under this category have only one botmaster and every time instructions are coming from that one botmaster. At first, the botmaster infects a computer system and then instructs that computer system using Command and Control technique to infect the other systems to which it is connected. The computer systems which are connected to the first infected computer also get the instructions from the botmaster who is instructing the first infected computer. Figure 1 depicted the working of a centralized botnet.

© Springer Nature Singapore Pte Ltd. 2020
M. Singh et al. (Eds.): ICACDS 2020, CCIS 1244, pp. 272–282, 2020.
https://doi.org/10.1007/978-981-15-6634-9_25

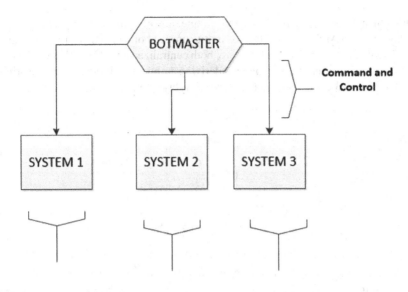

Fig. 1. Architecture of centralized botnet

P2P Botnet is the one which also works on the command and control technique but in this type of botnet every time a new botmaster born, when a botmaster starts working, it infected some computers and they all will get the command from the botmaster. The architecture of P2P botnet is described in Fig. 2.

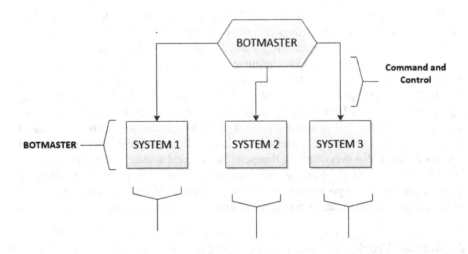

Fig. 2. Architecture of P2P botnet

Other than these two, there is one more architecture of botnet, which is also used a lot these days, Hybrid Architecture. Hybrid Architecture of botnet is a mixture of centralized and P2P botnet. In this type of architecture, both centralized and P2P architecture is used at any stage of infection. This complete underlying architecture is decided by the original botmaster. Figure 3 depicts the architecture of Hybrid Botnet.

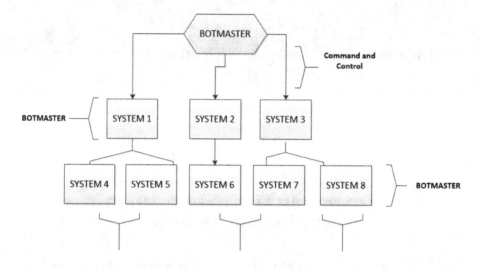

Fig. 3. Architecture of hybrid botnet

Data preprocessing is an essential task before applying any machine learning algorithm and in that feature, the selection is the most important task [1]. We always need to check the features of a dataset before using any machine learning algorithm because if we escape this step it will lead to the curse of dimensionality [2]. There are different types of feature selection algorithms depending on the use of classifiers. Some of them rank features or give priority to each of the features. Searching is also used for selecting important features from a large pool of features in a dataset. Ensemble techniques are also used for the feature selection which gives more accurate features in a dataset [3]. Also, we need to know that every dataset has a different type of data and different features therefore for every dataset there can be different feature selection algorithms [4]. The focal point of highlight determination is to choose a subset of factors from the information which can proficiently portray the information while decreasing impacts from clamor or unimportant factors and still give great forecast results.

2   Related Work

Different methods have been used for the elimination of features from a set of features. Lazzar et al. have presented Lot many methods are used for this like wrapper and filter method [5]. Yang et al. have proposed a new technique for the detection of the mobile botnet in which they have used multi-level feature extraction [6]. Alejandre et al. have presented a feature selection method based on Genetic Algorithm (GA) and C4.5 algorithm to select the best features from the feature set [7]. Some researchers also proposed a method to select features based on the traffic flow of the application layer(client) only [7]. Venkatesh et al. abuse two bundle size-based features, namely normal bytes per parcel and change of bytes per parcel, to identify HTTP botnets [8]. Some researchers also developed a new architecture Weasel which is completely encrypted. This architecture shows the use of flow duration, TCP flags and a number of packets for the detection of the botnet [9]. Mutual information-based information is also used for the selection of features [10]. Mitra et al. have shown an algorithm that uses the maximal information compression index between the value of feature pairs [11]. Fette et al. have presented a technique that uses URL and Javascript for feature selection [12]. Sequential techniques are also used for the feature selection. This algorithm uses the backtracking and optimal group of features by using a step-optimal technique [13]. Fuzzy rule-based techniques are also used for feature selection [14]. Some mainstream free criteria are separation measures, data measures, reliance measures, and consistency measures [15]. Late work right now centers around the steadiness records to be utilized for highlight choice, presenting measures dependent on Hamming separation [16]. Kalousis et al. present a broad similar assessment of highlight choice steadiness over various high-dimensional datasets [17]. Adriano et al. have presented tended to the serious issue where the uncooperative conduct in a group of halves and half specialists are distinguished and proposed the design of a decentralized screen to be inserted on the operators [18]. Sannasi Ganapathy et al. have proposed another component determination calculation called Intelligent Rule based Attribute Selection calculation and a novel order calculation named Intelligent Rule-based Enhanced Multiclass Support Vector Machine [19]. Gupta et al. have presented a Conditional random field-based feature selection method. In this method, each of the layers is responsible for a particular type of attack. Therefore, the probability of each feature will be calculated which will help in deciding the priority of a feature [20]. Md. Monirul et al. proposed a method that finds the best set of features based on sensitivity and selection algorithm [21]. A method named gradually feature removal has been used for the selection of best features in a KDD dataset which slowly removes less important features from a feature dataset [22]. Kwak et al. have proposed a method which uses Parzen window for best feature selection [23]. Archibald R et al. creators utilize the idea of RFE to infer a changed calculation for choosing highlights in hyper unearthly image data. In the SVM-RFE strategy [24]. Xu Z King et have shown the creators utilize the most extreme edge guideline (SVM) utilizing complex regularization issue advancement [25]. The authors also used cluster indicators for the selection of features. In these different clusters are made of features and cluster which has less indicator value has been removed. In the last, the cluster which remains will be used as the subset of features [26]. Multicriteria combination calculation is created which utilizes different component determination calculations to rank/score the highlights which are joined to acquire a powerful subset dependent on consolidating numerous classifiers to improve the precision [27]. The authors also propose separating the info highlights (in view of their component extraction methods) to get various classifiers and consolidate the forecasts to acquire an official conclusion [28]. Highlight choice systems show that more data isn't in every case great in AI applications. We can apply various calculations for the current information and with benchmark order execution esteems we can choose a last element determination calculation. For the current application, an element choice calculation can be chosen dependent on the accompanying contemplations: effortlessness, solidness, number of diminished highlights, characterization exactness, stockpiling and computational prerequisites. By and large applying highlight choice will consistently give advantages, for example, giving understanding into the information, better classifier model, improve speculation and distinguishing proof of unessential factors [4].

3   Dataset

Different datasets have been available which contain botnet and normal traffic. For our research, we have used CTU-13 Czech Technical University dataset. It contains a different type of botnet with normal traffic [29]. This dataset was captured in 2011. It consists of thirteen states of different botnets. The state has been captured in a PCAP file which is a network flow file. First, we need to convert this file into CSV file using WireShark or any other packet capture software, as shown in Fig. 4.

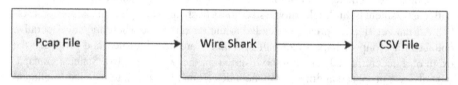

Fig. 4. Converting PCAP file into CSV file

Out of thirteen states, we have selected neris botnet state which contains botnet and normal traffic. A total of 11562 values have been used for the experiments. After converting the PCAP file into CSV file we have also done some data preprocessing and cleaning which gave us the data on which we can apply different algorithms. Datasets have various features namely Protocol, Source Address, Destination Address, Total packet, Total Byte, Source Byte.

All these features are important but we need to find out the best feature subset for the detection of botnet through this dataset.

4   Algorithms

Feature selection is always a vital task in machine learning classification algorithm. We will also use some feature selection algorithm before applying any machine learning algorithm. Below is the feature selection algorithm we will use in our analysis:

PCA: Principal Component Analysis (PCA) is a well-known dimensionality decrease procedure utilized in Machine Learning applications. PCA consolidates data from an enormous arrangement of factors into fewer factors by applying a type of change onto them. The change is applied so that straightly related factors get changed into uncorrelated factors.

Univariate Selection: Factual tests can be utilized to choose those highlights that have the most grounded association with the yield variable. The scikit-learn library gives the Select KBest class that can be utilized with a suite of various factual tests to choose a particular number of highlights. A wide range of measurable test examines be utilized with this determination technique.

Recursive Feature Selection: This algorithm recursively evacuates highlights, assembles a model utilizing the rest of the characteristics and ascertains model precision. RFE can work out the blend of credits that add to the expectation on the objective variable (or class).

Feature Importance: Sacked choice trees like Random Forest and Extra Trees can be utilized to evaluate the significance of highlights.

Correlation Matrix: Connection states how the highlights are identified with one another or the objective variable. Relationship can be certain (increment in one estimation of highlight builds the estimation of the objective variable) or negative (increment in one estimation of highlight diminishes the estimation of the objective variable) Heatmap makes it simple to distinguish which highlights are generally identified with the objective variable, we will plot a heatmap of associated highlights utilizing the seaborn library.

5 Methodology

In our methodology, we will first examine all the feature selection algorithm and find out the best subset of features. After that, we will apply all the classification algorithms without using the feature selection algorithm. In the next step, we will apply all the classification algorithm with the features which are extracted in the previous steps. We will record all the results in the tables. A block-diagram for the methodology used is shown in Fig. 5.

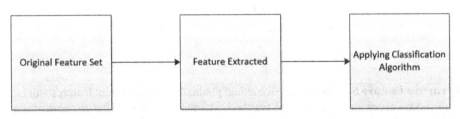

Fig. 5. Methodology

6 Experiment and Results

First, we apply feature selection algorithms and find out the best suitable feature subset out of the complete feature set. We applied each of the feature selection algorithms one by one and find out the result.

PCA: Let's start with Principle Component Analysis (PCA). As we know PCA gives us the ranking of the features according to the variance. Table 1 gives the result after applying PCA.

Table 1. Scores after applying PCA

S. No	Feature	Score (%)
1	Protocol	99.77
2	Source Address	20
3	Destination Address	9.76
4	Total Packet	2.12
5	Total Byte	0.62
6	Source Byte	0.025

Univariate Feature Selection: In this method we have selected three best features out of the 7 available features (Table 2).

Table 2. Scores after applying Univariate Feature Selection.

S. No	Features	Score
1	Protocol	3255.02
2	Source Address	6.48
3	Destination Address	1680.05
4	Total Packet	1.04
5	Total Byte	5.77
6	Source Byte	5.79

Recursive Feature Selection: In this method gradually less, important features will be removed on every step and we will get the best suitable features after the end of the algorithm. We have selected 4 features subset from the available features. The algorithm has given the results in the below format.

According to the above format we have made a table showing the scores of each feature (Table 3).

Selected Features: [True True True True False False]
Feature Ranking: [1 1 1 1 3 2]

Table 3. Ranking after applying Recursive Feature Selection

S. No	Features	Ranking
1	Protocol	1
2	Source Address	1
3	Destination Address	1
4	Total Packet	1
5	Total Byte	3
6	Source Byte	2

Feature Importance: In this method, a classifier like decision tree is used named "Extra Classifier." It estimates the important features from the feature set (Fig. 6).

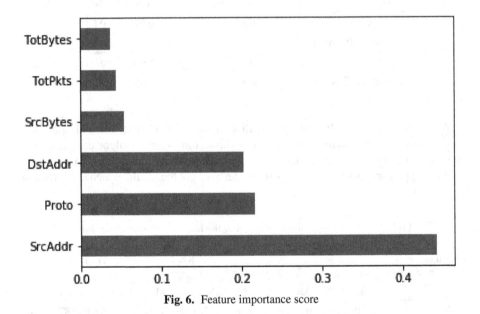

Fig. 6. Feature importance score

Correlation Heat Map: This algorithm gives the result on the basis of mutual information between features. This mutual information is termed as the correlation. The algorithm will plot a Heat Map which will show this property between features.

Table 4. Summarized Feature Subset after applying Feature Selection Algorithm

S. No	Algorithm	Feature subset
1	PCA	Protocol, Source Address
2	Univariate Feature Selection	Protocol, Destination Address and Source Byte
3	Recursive Feature Selection	Protocol, Source Address, Destination Address, and Total Packet
4	Feature Importance	Protocol, Source Address and Destination Address
5	Correlation Heat Map	Source Byte, Total Byte, Total Packet

Table 5. Shows the result of classification algorithm (without applying any of the feature selection algorithm)

S. No	Algorithm	Accuracy_Score	AUC	Accuracy rate
1	Support Vector Machine	0.85	0.91	0.82
2	Logistic Regression	0.77	0.82	0.63
3	K-Nearest Neighbor	0.97	0.98	0.99
4	Decision Tree	0.80	0.85	0.95

After this step, we will apply a classification algorithm. First, we will apply the classification algorithm without using any of the feature selection algorithms and then with the entire above-mentioned feature selection algorithm (Tables 4 and 5).

After applying classification algorithms, we have got the results which are shown in the Table 6.

Table 6. Shows the result after applying the feature selection algorithm

Classification Algorithm/Feature Selection	Support Vector Machine			Logistic Regression			KNN			Decision Tree		
	AS	AUC	AR	AS	AUC	AR	AS	AUC	AR	AS	AUC	AR
PCA	0.90	0.913	0.81	0.77	0.76	0.52	0.99	0.99	0.99	0.83	0.87	0.96
Univariate Feature Selection	0.78	0.82	0.72	0.77	0.81	0.62	0.96	0.97	0.99	0.85	0.88	0.89
Recursive Feature Selection	0.75	0.87	0.80	0.76	0.81	0.61	0.99	0.99	0.99	0.82	0.88	0.89
Feature Importance	0.78	0.82	0.79	0.76	0.80	0.60	0.99	0.99	0.99	0.84	0.89	0.89
Correlation Heat Map	0.76	0.71	0.73	0.77	0.68	0.36	0.98	0.98	0.99	0.77	0.50	0.99

7   Conclusions and Future Work

Among all the features, Protocol is the one which is predicting the accurate result. We can see this in the above tables that in Correlation Heat Map, when we remove protocol, the accuracy of all the classification algorithms has been reduced. K-Nearest Neighbor gives more accurate results than any other classification algorithm in all the cases. Logistic Regression is not performing up to the mark on this dataset. Its accuracy has improved when we use Univariate Feature Selection. Univariate Feature Selection has been performed well with all the classification algorithms. SourceByte Feature has been the one that is always reducing the accuracy of the algorithm. The accuracy of the K-Nearest Neighbor has been increased by the feature selection algorithms. Our results show that PCA and Univariate Feature Selection algorithms have performed better than any other algorithms. K-Nearest Neighbor has been the best algorithm amongst all the classification algorithms. In the future, we will combine a greater number of feature selection algorithms for improving the accuracy. Also, the different number of botnets can be combined to see which of the algorithms can performed better for this mix of botnet. Ensemble algorithms can also be used for the analysis of feature selection and detection of a botnet. More number of datasets can be used in the future for the analysis of various feature selection algorithms.

References

1. Saeys, Y., Abeel, T., Van de Peer, Y.: Robust feature selection using ensemble feature selection techniques. In: Daelemans, W., Goethals, B., Morik, K. (eds.) ECML PKDD 2008. LNCS (LNAI), vol. 5212, pp. 313–325. Springer, Heidelberg (2008). https://doi.org/10.1007/978-3-540-87481-2_21
2. Hughes, G.: On the mean accuracy of statistical pattern recognizer. IEEE Trans. Inf. Theory **14**(1), 55–63 (1968)
3. Guyon, I., Elisseeff, A.: An introduction to variable and feature selection. J. Mach. Learn. Res. **3**, 1157–1182 (2003)
4. Chandrashekar, G., Sahin, F.: A survey on feature selection methods. Comput. Electr. Eng. **40**(1), 16–28 (2014). https://doi.org/10.1016/j.compeleceng.2013.11.024
5. Lazar, C., Taminau, J., Meganck, S., Steenhoff, D., Coletta, A., Molter, C., et al.: A survey on filter techniques for feature selection in gene expression microarray analysis. IEEE/ACM Trans. Comput. Biol. Bioinform. **9**, 1106–1119 (2012)
6. Yang, M., Wen, Q.: A multi-level feature extraction technique to detect moble botnet. In: 2016 2nd IEEE International Conference on Computer and Communications (ICCC), 14 October 2016, pp. 2495–2498. IEEE (2016)
7. Alejandre, F.V., Cortés, N.C., Anaya, E.A.: Feature selection to detect botnets using machine learning algorithms. In: 2017 International Conference on Electronics, Communications and Computers (CONIELECOMP), 22 February 2017, pp. 1–7. IEEE (2017)
8. Kirubavathi Venkatesh, G., Anitha Nadarajan, R.: HTTP botnet detection using adaptive learning rate multilayer feed-forward neural network. In: Askoxylakis, I., Pöhls, H.C., Posegga, J. (eds.) WISTP 2012. LNCS, vol. 7322, pp. 38–48. Springer, Heidelberg (2012). https://doi.org/10.1007/978-3-642-30955-7_5
9. Garant, D., Lu, W.: Mining botnet behaviors on the large-scale web application community. In: 2013 27th International Conference on Advanced Information Networking and Applications Workshops, March 2013, pp. 185–190 (2013)

10. Hoque, N., Bhattacharyya, D., Kalita, J.: MIFS-ND: a mutual information-based feature selection method. Expert Syst. Appl. **41**(14), 6371–6385 (2014)
11. Hastie, T., Tibshirani, R., Friedman, J.: The Elements of Statistical Learning. SSS. Springer, New York (2009). https://doi.org/10.1007/978-0-387-84858-7
12. Fette, I., Sadeh, N., Tomasic, A.: Learning to detect phishing emails. Technical report, School of Computer Science, Carneige Melon University (2006)
13. Aha, D.W., Bankert, R.L.: A comparative evaluation of sequential feature selection algorithms, learning from data. In: Fisher, D., Lenz, H.J. (eds.) learning from data. Lecture Notes in Statistics, vol. 112, pp. 199–206. Springer, New York (1996). https://doi.org/10.1007/978-1-4612-2404-4_19
14. Zadeh, L.A.: Outline of a new approach to the analysis of complex systems and decision processes. IEEE Trans. Syst. Man Cybern. **SMC-3**, 28–44 (1973)
15. Liu, H., Motoda, H.: Feature Selection for Knowledge Discovery and Data Mining. Kluwer Academic, Boston (1998)
16. Dunne, K., Cunningham, P., Azuaje, F.: Solutions to instability problems with sequential wrapper-based approaches to feature selection. Technical report TCD-2002-28. Department of Computer Science, Trinity College, Dublin, Ireland (2002)
17. Kalousis, A., Prados, J., Hilario, M.: Stability of feature selection algorithms: a study on high-dimensional spaces. Knowl. Inf. Syst. **12**(1), 95–116 (2007). https://doi.org/10.1007/s10115-006-0040-8
18. Fagiolini, A., Valenti, G., Pallottino, L., Dini, G., Bicchi, A.: Decentralized intrusion detection for secure cooperative multi–agent systems. In: Proceedings of the 46th IEEE Conference on Decision and Control. IEEE, Amsterdam, pp. 1553–1558 (2007)
19. Ganapathy, S., Kulothungan, K., Muthurajkumar, S., Vijayalakshmi, M., Yogesh, P., Kannan, A.: Intelligent feature selection and classification techniques for intrusion detection in networks: a survey. EURASIP J. Wirel. Commun. Netw. **2013**(1), 1–16 (2013). https://doi.org/10.1186/1687-1499-2013-271
20. Gupta, K.K., Nath, B., Kotagiri, R.: Layered approach using conditional random fields for intrusion detection. IEEE Trans. Dependable Secure Comput. **7**(1), 35–49 (2010)
21. Kabir, M.M., Shahjahan, M., Murase, K.: A new hybrid ant colony optimization algorithm for feature selection. Expert Syst. Appl. **39**, 3747–3763 (2012)
22. Battiti, R.: Using mutual information for selecting features in supervised neural net learning. IEEE Trans. Neural Netw. **5**, 537–550 (1994)
23. Kwak, N., Choi, C.-H.: Input feature selection for classification problems. IEEE Trans. Neural Netw. **13**, 143–159 (2002)
24. Archibald, R., Fann, G.: Feature selection and classification of hyperspectral images with support vector machines. IEEE Geosci. Remote Sens. Lett. **4**, 674–677 (2007)
25. Xu, Z., King, I., Lyu, M.R.-T., Jin, R.: Discriminative semi-supervised feature selection via manifold regularization. IEEE Trans. Neural Netw. **21**, 1033–1047 (2010)
26. Zhao, Z., Liu, H.: Semi-supervised feature selection via spectral analysis. In: Proceedings of the 7th SIAM Data Mining Conference (SDM), pp. 641–646 (2007)
27. Mortazavi, A., Moattar, M.H.: Robust feature selection from microarray data based on cooperative game theory and qualitative mutual information. Adv. Bioinform. (2016). https://doi.org/10.1155/2016/1058305
28. Dietterich, T.: Machine learning research: four current directions. Artif. Intell. Mag. **18**, 97–136 (1997)
29. Garcia, S., et al.: An empirical comparison of botnet detection methods. Comput. Secur. J. **45**, 100–123 (2014)

ELM-MVD: An Extreme Learning Machine Trained Model for Malware Variants Detection

Pushkar Kishore[✉], Swadhin Kumar Barisal, Alle Giridhar Reddy, and Durga Prasad Mohapatra

Department of Computer Science, National Institute of Technology Rourkela, Rourkela, Odisha, India

{518CS1002,715CS2064,durga}@nitrkl.ac.in,
swadhinbarisal@gmail.com

Abstract. Malware variants are expanding at a fast pace and detecting them is a critical problem. According to surveys from McAfee, over 50% of the newly recognized malware are variants of earlier ones. Huge amount of miscellaneous malware variants compelled researchers to find a better model for detecting them. In this work, we propose an extreme learning machine trained model (ELM- MVD) for malware variants detection. We use the dataset comprising benign and malware executable names along with their features represented as a triplet of system calls. Along with that, we demonstrate that features in the form of a triplet vector are optimal while training a model. Feature reduction is done using an alternating direction method of multipliers (ADMM) technique. Finally, training is done on the ELM-MVD model and achieve 99.3% accuracy and 0.003 s detection speed.

Keywords: System calls · Malware variants detection · Extreme learning machine · Alternating direction method of multipliers

1 Introduction

Today, we deal with the strenuous security threat, i.e. malware. People stop using the traditional signature based malware detector, as they cannot detect evolved malware variants. As per the survey by Symantec [1], over 50% of the newer malware cases are the variants of the extant ones. In addition, the presence of malware leads to a lot of effort required in testing and software validation [11, 14]. The huge amount of miscellaneous malware variants compelled researchers to find a better model for detecting them. The current malware detection system needs a client to send the unidentified application sample to the cloud for verification. The detection system will disassemble or reverse engineer the sample to get special category of info known as system calls, opcode along with training a classifier. But the datasets that are available are obsolete now, so we need to have a new dataset that will properly represent the behavior of recent malware variants. There are numerous malware detection models based on Convolutional Neural Networks, Back-Propagation Neural Networks, Support Vector Machine, Logistic Regression, etc. but lags in providing exact and quick detection.

© Springer Nature Singapore Pte Ltd. 2020
M. Singh et al. (Eds.): ICACDS 2020, CCIS 1244, pp. 283–292, 2020.
https://doi.org/10.1007/978-981-15-6634-9_26

To address the issue, we develop a new dataset using a tool, NITRSCT (NITR system call tracer) [2], developed by us. The new dataset covers the behavior of the recent malware variants, represented in the form of a triplet of system calls. We also demonstrate that taking the value of **n = 3** provides a better malware detection model despite taking $n \in \{1, 2, 4\}$. Features reduction is done with the assistance of the alternating direction method of multipliers (ADMM) technique. Above indicated technique provides the weight of the features and by analyzing them; We freeze the number of features that can be used for training malware detectors.

The generalized radial basis function (GRBF) is specified in Eq. 1.

$$\Phi(x; c, r, \tau) = \exp(-||x - c||^{\tau}/r^{\tau}) \tag{1}$$

In Eq. 1, **x** is considered as a vector consisting of pattern's coordinates of the considered dataset, **c** is the location parameter for deciding kernel positions; **r** is defined as width, and τ is the real parameter. We train the ELM-MVD model using quick learning. First, the parameters that are present in the basis function are determined. The center's initialization is done by incoherently choosing patterns in the training dataset. Thus, the variables namely, **r** and τ are set accordingly whenever the center is initialized. After training is completed, we use it for predicting whether the sample is malicious or benign.

We organize the remaining part of this paper in this fashion: Sect. 2 shortly characterizes the related work. Section 3 introduces the mode adopted behind our proposed architecture. Section 4 highlights the experimental results. Section 5 emphasizes the comparison with related work. Section 6 emphasizes the threats to validity and Sect. 7 demonstrates the conclusions and future work.

2 Related Work

In this section, we take up some existing work related to our approach.

2.1 Malware Variants Detector

Researchers relied on machine learning to recognize malware variants. Cesare et al. [3] suggested that searching common and uncommon sets among control flow graphs would help in malware detection. Malware can be classified according to its family. So, Fan et al. [4] often constructed subgraphs of API calls of same class malware binary executable. However, the temporal flaws of reverse engineering make it tough to extract API calls every time. Zhang et al. [5] extracted opcode and API calls feature from the binary executable. They trained the Convolutional Neural Network (CNN) using opcode extracted from binary executable while API calls were used to train Back-Propagation Neural Network. The embedded features determined from the above process was used for training and testing a feature-hybrid based malware variant detector. Zhang et al. [6] built an opcode graph and topological features are extracted to detect malware attacking Android operating system. Niall et al. [7] used CNN for training the malware detector and classification was done on Dalvik opcode sequences. When Dalvik opcode sequences is longer, then the op-code embedding matrix used by them will not work. Stringhiniq

[8] proposed malware detection technique on file delivery networks. A semi-supervised Bayesian label distribution technique was used to distribute the file's character. But, delay in the propagation makes them prone to malware attacks as malware copies will propagate much faster than the file's reputation. Raff et al. [9] proposed byte (n = 2) Gram matrix for representing binary executable and used CNN for the detection. Kang et al. [10] had also used n-gram model for representing binary executable and used Support Vector Machine (SVM) for detecting malware. Zhang et al. [19] suggested a cost-conscious boosting technique for positive-unlabeled learning for malware detection. But, they have used opcode representation, which is only the static analysis of the binary executable.

2.2 System Call Based Detection

Maximum number of researchers are using system calls for malware detection. Rieck et al. [12] automatically identified those malware's classes, which are matched with the identical sequential system calls. Then, the undiscovered malware are assigned to the identified class. Xu et al. [13] represented system calls in the form of graphs and used graph kernels for computing similarities amidst binary executable. Then, these similarities value are fed into the SVM for classification. Kolbitsch et al. [15] built a graph depicting the data flow amidst system calls and checked analogy among graph for detecting malware. For system call technique to work, researchers captured system calls done by the binary executable while they are running in the sandbox. But, it costs them a lot of time in data preparation but malware's detection's accuracy is fine.

3 Methodology

In this section, we propose an extreme learning machine trained generalized radial basis function neural network model for malware variants discovery. We confer the architecture of our proposed malware detection model in Fig. 1. It consists of three steps: creation of dataset using NITRSCT, features reduction using an alternating direction method of multipliers and classification using the ELM-MVD model. By performing the above three steps, we can efficaciously detect malware variants. The description of the above three steps are discussed below:

3.1 Creation of Dataset Using NITRSCT

We have collected benign binary executable from 10 hosts in offices, computer laboratories, and isolated testbed for testing in real scenarios. The malware used for the experimental purpose is collected from VirusTotal[1] . Upon collecting system calls, we convert them into the features, which can be used in the dataset. Features are extracted from the list of system calls using n-gram technique. In n-gram model, we take 'n' consecutive system calls and consider it as a feature. We have considered three ML models and compared their true positive rate (TPR) by fixing false positive rate (FPR) at 10^{-5}.

---

[1] https://www.virustotal.com/.

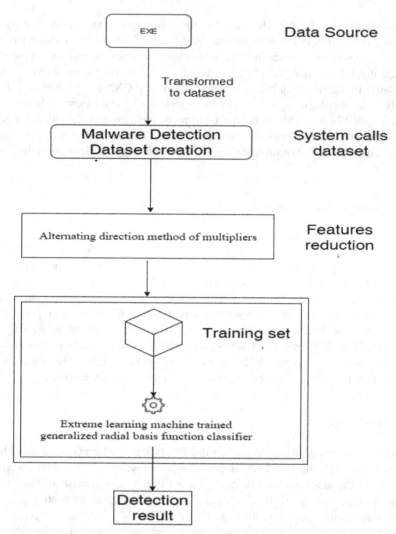

Fig. 1. Proposed architecture of our approach

But, we have varied the length of n, representing the length of the system calls in each feature. While selecting n = 1, TPR is 0.32, for n = 2, 0.85, for n = 3, it rises to 0.91 and it decreases to 0.81 whenever n = 4 is considered. From the above statistics, we can infer that n = 3 will be optimal for the length of system calls in each feature. After the features are generated, then the dataset[2] is finalized. Lastly, we take the logarithm of the features before training to reduce the differences between their values.

---

[2] https://github.com/pushkarkishore/NITRSCT/blob/master/data1.rar.

3.2 Features Reduction Using ADMM (Alternating Direction Method of Multipliers)

The ADMM framework [16, 17] is used for solving convex problems. This framework optimizes the optimization problem as given in Eq. 2.

$$\min : \psi(\beta) + \varphi(z) \text{ such that } C\beta + Dz = v \tag{2}$$

where, ψ and φ are the cost-defining terms in the objective function, while β and z are decision vectors.

First, we create an augmented Lagrangian using Eq. 3.

$$L(\beta, z, \gamma) = \psi(\beta) + \varphi(z) + \gamma^T(C\beta + Dz - v) + (0.5 * \rho)\|C\beta + Dz - v\|_2 \tag{3}$$

where, γ is the Lagrangian vector, and ρ is a positive penalty. The iterative steps for determining the value of $\{\beta, z, \gamma\}$ is executed using Eq. 4–6 respectively.

$$\beta^{k+1} = \arg \min L\left(\beta, z^k, \gamma^k\right) \tag{4}$$

$$z^{k+1} = \arg \min L\left(\beta^{k+1}, z, \gamma^k\right) \tag{5}$$

$$\gamma^{k+1} = \gamma^k + \rho(C\beta^{k+1} + Dz^{k+1} - v) \tag{6}$$

3.3 ELM-GRBF (Extreme Learning Machine–Generalized Radial Basis Function)

In ELM-GRBF [18], we randomly set the centers of the GRBFs by searching patterns in the training set. We fulfill two requirements here: (i) the smallest gap (d_N) between the distributions is mapped to the high value of the generalized radial basis function. (ii) Similarly, the farthest gap (d_F) between the distributions is mapped to the lower value. The d_N and d_F is evaluated using Eq. 7 and 8 respectively.

$$d_N = ((\delta)^2 * k)^{1/2} \tag{7}$$

where, δ is dimension's small residual distance and k is the count of inputs being considered.

$$d_F = \|C_i - C_j\| \tag{8}$$

where, i and j are the adjoining hidden node to each other. The shape parameter, τ is calculated using Eq. 9.

$$\tau = [\ln(\ln(\lambda)/\ln(1-\lambda))]/\ln(d_F/d_N) \tag{9}$$

where, λ is the user-defined parameter.

The width of the GRBF, r, can be evaluated using Eq. 10.

$$r = d_N/(-\ln(1-\lambda))^{1/\tau} = d_F/(-\ln(\lambda))^{1/\tau} \tag{10}$$

The hidden layer output matrix, H, of dimension (nxm) is estimated using Eq. 1. The output weight, β' is estimated with the help of Eq. 11.

$$\beta' = H^\dagger T \tag{11}$$

where, $H^\dagger$ is the pseudo-inverse of H and T is evaluated using Eq. 12.

$$T = H\beta \tag{12}$$

where, $\beta = (\beta^1, \beta^2, ..., \beta^j)_{mxj}$ and β^j is the weight vector of the connection between hidden node and the jth output node.

4   Experimental Results

We present the experiment to manifest the performance, efficiency and accuracy of the proposed model. First, we present the experimental setup, and then we discuss the performance of our model. We have also highlighted that our model performs much better than other state-of-the-art approaches.

4.1   Setup, Dataset and Hyper-parameters

We carry out the experiment on the same system. The version of the processor is Intel i5-3470 @ 3.20 GHz, the random access memory space is 16 GB, and the OS is Windows 10. We have implemented our model using Python programming language in which it does the matrix calculations using numpy package. The hyper-parameters, which are fixed by us, have a greater impact on the effectiveness of the model. We present the hyper-parameters used in our proposed model in Table 1.

Table 1. The hyper-parameters settings of our experiment

Detector	Hyper-parameter	Value
GenELMClassifier	hidden_layer	MLPRandomLayer (random_state = 0)
GenELMClassifier	binarizer	LabelBinarizer(−1, 1)
GenELMClassifier	regressor	None
Proposed model	hidden_layer	rbf_rhl (n_hidden = 100, random_state = 0, rbf_width = 0.01)

4.2   Performance Analysis of Malware Detection

The parameters, which we use for performance analysis of our proposed model, are classification accuracy, detection false positive rate, detection true negative rate, detection false negative rate, detection precision, detection recall, F1-score, training time cost and detection time cost. We compute the classification accuracy using Eq. 13. The recall of the model is the true-positive rate tested using Eq. 14, where true positive (TP) is the count of accurately classified binary malicious executables and false negative (FN) implies binary malicious executables misclassified as the benign one. TNR is the true negative rate tested using Eq. 15, where false positive (FP) is the number of binary benign executables uncategorized as binary malicious executables and true negative (TN) is the count of binary benign executables, which are accurately classified. FPR representing false positive rate, FNR representing false negative rate, Precision representing malware detector's precision and F1-score (evaluated using precision and recall) are evaluated using Eqs. 16–19, respectively.

$$accuracy = (TP + TN)/(TP + FN + TN + FP) \tag{13}$$

$$TPR(Recall) = TP/(TP + FN) \tag{14}$$

$$TNR = TN/(FP + TN) \tag{15}$$

$$FPR = FP/(FP + TN) \tag{16}$$

$$FNR = FN/(FN + TP) \tag{17}$$

$$Precision = TP/(FP + TP) \tag{18}$$

$$F1\text{-score} = (2 * precision * recall)/(precision + recall) \tag{19}$$

The performance assessment of our model is presented in Table 2.

Table 2. The values of performance parameters of our model

Sl. No.	Performance parameters	Value
1	Accuracy (%)	99.3
2	Recall (%)	99.3
3	TNR (%)	99.4
4	FPR (%)	0.6
5	FNR (%)	0.7
6	Precision (%)	99.3
7	F1-score (%)	99.3
8	Detection time (s)	0.003
9	Training time (s)	160

The algorithms proposed above will work in an effective manner as the real world malware variants detector are working. For the real-world similar dataset creation, we use malware samples from VirusTotal and benign ones from the real systems. Thus, the metrics calculated above will retain its value even in the real-world malware variants detector.

5 Comparison with Related Work

We have correlated the performance of our model with some contemporary state-of-the-art approaches and summarized them in Table 3.

Table 3. Comparison of performance of our approach with recent state-of-the-art approaches

Method	Accuracy (%)	Precision (%)	Recall (%)	1-FPR (%)	F1-score (%)	Detection time (s)	Training time (s)
Logistic [11]	77.5	98.6	55.8	99.2	69.6	0.007	14,633
Softmax [11]	78.9	83.1	72.5	85.3	77.4	0.006	14,105
CNN [9]	86.2	91.1	80.1	92.2	85.2	0.053	93,534
SVM [10]	85	82.8	88	81.7	85.2	0.006	609
Our approach	99.3	99.3	99.3	99.4	99.3	0.003	160

By correlating with the recent state-of-the-art approaches, we notice that our proposed model substantially enhances the classification accuracy, the detector's precision, the recall, the (1-FPR), the F1-score and the training time while retaining the detection accuracy. Accuracy is remarkably higher making it useful for industrial malware detection. Recall is 99.3%, which indicates that our proposed model correctly classifies 99.3% of the total relevant results. Considering the problem under consideration, we give higher emphasis to either precision or recall. Generally, we use a simplified metric, F1-score that is the harmonic mean of precision and recall. The value of F1-score from our approach is 99.3%. Specificity is another term for "1-FPR", which shows that binary benign executables being labeled benign is 99.4%. Its lower value will only block the binary benign executables, thus we consider it as an auxiliary parameter. Comparing with the above stated parameters, we observe that our proposed model is more suitable for malware detection.

6 Threats to Validity

For techniques to work with system calls, we need to capture the system calls called by the process during their execution on the sandbox. Therefore, collection of system calls is very tough and costlier process in terms of resources. Modern malware variants tend to hide their malicious behavior whenever they suspect that sandbox is present. Our model can eliminate this issue to some level. We analyze each act of malware inactiveness or sleepy behavior by making sandbox dynamically changing its time settings to deceive malware and stimulate its execution. The dataset, which is considered, has 0% incorrectly labeled data. However, we should analyze our model whenever the dataset will contain incorrectly labeled data.

7 Conclusions and Future Work

In this paper, we design an ELM trained model for malware variants detection (ELM-MVD). We have demonstrated that a vector of three consecutive system calls, when considered as a feature in dataset will be optimal for malware detectors. Feature reduction is done using alternating direction method of multipliers (ADMM). The experimental analyzed data obtained through rigorous testing convey that our malware detection model achieves 99.3% accuracy, while the other techniques achieve up to a maximum of 86.2%.

In the future, we can improve the cost of dataset creation and can detect newer sandbox-evading malware by enhancing the features of the sandboxes. We can consider a static analysis of the binary executable like API calls, opcode, etc. and design an ensemble detector to improve malware detection accuracy.

References

1. Symantec, Internet security threat report (2017)
2. Kishore, P., Barisal, S.K., Vaish, S.: NITRSCT: a software security tool for collection and analysis of kernel calls. In: IEEE Region 10 Conference (TENCON), pp. 510–515 (2019)
3. Cesare, S., Xiang, Y.: Malware variant detection using similarity search over sets of control flow graphs. In: IEEE 10th International Conference on Trust, Security and Privacy in Computing and Communications, pp. 181–189 (2011)
4. Fan, M., et al.: Android malware familial classification and representative sample selection via frequent subgraph analysis. IEEE Trans. Inf. Forensics Secur. **13**(8), 1890–1905 (2018)
5. Zhang, J., Qin, Z., Yin, H., Ou, L., Zhang, K.: A feature-hybrid malware variants detection using CNN based opcode embedding and BPNN based API embedding. Comput. Secur. **84**, 376–392 (2019)
6. Zhang, J., Qin, Z., Zhang, K., Yin, H., Zou, J.: Dalvik opcode graph based Android malware variants detection using global topology features. IEEE Access **6**, 51964–61974 (2018)
7. McLaughlin, N., et al.: Deep Android malware detection. In: Proceedings of the Seventh ACM on Conference on Data and Application Security and Privacy, pp. 301–308 (2017)
8. Stringhiniq, G., Shen, Y., Han, Y., Zhang, X.: Marmite: spreading malicious file reputation through download graphs. In: Proceedings of the 33rd Annual Computer Security Applications Conference (AC-SAC) (2017)
9. Raff, E., Barker, J., Sylvester, J., Brandon, R., Catanzaro, B., Nicholas C.: Malware detection by eating a whole EXE. Proceedings of arXiv:1710.09435 (2017)
10. Kang, B., Yerima, S.Y., McLaughlin, K., Sezer, S.: N-opcode analysis for Android malware classification and categorization. In: Proceedings of International Conference on Cyber Security and Protection of Digital Services (Cyber Security) (2016)
11. Barisal, S.K., Dutta, A., Godboley, S., Sahoo, B., Mohapatra, D.P.: MC/DC guided test sequence prioritization using firefly algorithm. Evol. Intell., 1–14 (2019)

12. Rieck, K., Trinius, P., Willems, C., Holz, T.: Automatic analysis of malware behavior using machine learning. J. Comput. Secur. **19**(4), 639–668 (2011)

13. Xu, L., Zhang, D., Alvarez, M.A., Morales, J.A., Ma, X., Cavazos, J.: Dynamic Android malware classification using graph-based representations. In: IEEE 3rd International Conference on Cyber Security and Cloud Computing (CSCloud), pp. 220–231 (2016)

14. Barisal, S.K., Behera, S.S., Godboley, S., Mohapatra, D.P.: Validating object-oriented software at design phase by achieving MC/DC. Int. J. Syst. Assur. Eng. Manag. **10**(4), 811–823 (2019). https://doi.org/10.1007/s13198-019-00815-8

15. Kolbitsch, C., Comparetti, P.M., Kruegel, C., Kirda, E., Zhou, X.Y., Wang, X.: Effective and efficient malware detection at the end host. In: USENIX Security Symposium, vol. 4, no. 1, pp. 351–366 (2009)

16. Boyd, S., Parikh, N., Chu, E., Peleato, B., Eckstein, J.: Distributed optimization and statistical learning via the alternating direction method of multipliers. Found. Trends® Mach. Learn. **3**(1), 1–122 (2011)

17. Afonso, M.V., Bioucas-Dias, J.M., Figueiredo, M.A.: An augmented Lagrangian approach to the constrained optimization formulation of imaging inverse problems. IEEE Trans. Image Process. **20**(3), 681–695 (2010)

18. Fernández-Navarro, F., Hervás-Martínez, C., Sanchez-Monedero, J., Gutiérrez, P.A.: MELM-GRBF: a modified version of the extreme learning machine for generalized radial basis function neural networks. Neurocomputing **74**(16), 2502–2510 (2011)

19. Zhang, J., Khan, M.F., Lin, X., Qin, Z.: An optimized positive-unlabeled learning method for detecting a large scale of malware variants. In: IEEE Conference on Dependable and Secure Computing (DSC), pp. 1–8 (2019)

Real-Time Biometric System for Security and Surveillance Using Face Recognition

Arvind Jaiswal[✉] and Sandhya Tarar[✉]

School of ICT, Gautam Buddha University, Greater Noida, India
arvindjaiswal116@gmail.com, tarar.sandhya@gmail.com

Abstract. This paper-based on real-time security and surveillance because today public security problems raised widespread concern. So due to the security issue, it can use the different types of biometrics for security and surveillance and hence facial recognition has important applications in the field of biometric and numerous systems related to security and surveillance. Many methods familiarized in this field, but some problems remain not recognized because that system is only used as a biometric in small scales, but in this paper, we focus on the live security of all public and private places and multiples uses. For this real-time security used the IP CCTV cameras for surveillance purposes for this need the image as a sample of a person for model training and basic details of the person who lives inside the city. Whenever need surveillance for any suspicious person or place inside the city. Then we can find out his/her present location, and all incident is monitor by IP CCTV Cameras. This all incident update in the centralized system so that we can find the live location of the particular person anytime and anywhere within the city. This security system has many applications in real life like crime control, terrorist alert, smart society, airport security, university campus and surveillance of public places. In the future, all the data collected from "The Unique Identification Authority of India" (UIDAI).

Keywords: Local binary pattern histogram (LBPH) · Haar-cascade detection · Computer vision · Image processing · Video surveillance · Smart security

1 Introduction

At present public security issues increased widespread concern so that due to security issues they can use the different types of biometric for security and the surveillance purpose because real-time security issue remains to increase due to the happening of various suspicious activity in the public and private places. And the security issue is a major problem in our society because lots of crimes happen every day like murder, kidnapping, chain snatching, rapes, and loitering, etc. And real-time security requirements play an important role in our society. Because day by day, a security issue may affect everybody lives, and many techniques are familiarized in these regards, and the face recognition plays an important role to solve this issue so that the facial recognition has important application in the field of biometric and numerous systems related to security

© Springer Nature Singapore Pte Ltd. 2020
M. Singh et al. (Eds.): ICACDS 2020, CCIS 1244, pp. 293–304, 2020.
https://doi.org/10.1007/978-981-15-6634-9_27

and surveillance here video surveillance plays a significant task in public security, and it is broadly used for supervising public incidents for wide-ranging an increasing number, of the surveillance IP CCTV (Closed-Circuit Television) cameras for being deployed in public areas example airports, railways stations, university campus, bank, shopping complex, and urban roads, etc. Mostly video surveillance system is used for a specific task like pedestrian recognition rather than entertainment purpose so the quality of IP Cameras is high resolution like 4K (3840*2160) are used in real scenarios. With the evolution in time, there is an extreme growth in the fields of technology which is helpful for smart society, and the smart people have the aim to minimize the human to machine interference, and this knowledge is very helpful for security requirements. In the future, it is not easy to collect the data from the individual persons, so that we can collect the data from "The Unique Identification Authority of India" (UIDAI).

2   Literature Review

For security purposes, we design the smart security system by using face recognition and the proposed model which provides security and surveillance with the help of facial recognition. Videos Surveillance has become a hot research field due to growth in a typical security issue. The identification and verification of a person's face is a useful method to mitigate security risk [1]. For this purpose, face recognition plays an important role and become an attractive field in computer-based application development in the past few decades. To identify a face, we extract the feature from the face. For this, there are three main tactics for feature extraction, appearance-based method, model-based method, and hybrid-based method as feature extraction are used [6]. In this paper, we can use the IP CCTV Cameras with high resolution to record all the incidents and moments within the city. But the same time problem is how can store the huge quantity of video data being generated each day for this some author use the VQA (Video quality assessment) for video compression [2]. It helps shrink video, size however it is at the same time harmful to the video quality. In this research paper, we focused on face recognition, and face recognition is growing as the most important research fields for the reason that of the comprehensive choice of application in the domains of commercial and law enforcement [11]. For this prospect first, we need to record all incidents in CCTV cameras. For face recognition first of we need to capture all the incidents or moments of the public place where we deployed the CCTV Cameras.

There are some steps for the face recognition process, and the first steps are movement recognition and position, and the rest of the steps depends especially on facial detection and recognition. In Videos, recognition, and localization of moving human face is completed through background subtraction models. More information can be brought into being in a study on background modeling [7]. In the background, subtraction has many restrictions, such as films with the terrible signal-to-noise ratio caused by low-resolution cameras, blur caused by jittering movement in the camera optic, surrounding environment noise, compression artifact [8]. Machine learning plays an important role in training the model and image processing help to extract information from digital images [9], to help multimodal mean live video streaming background modeling techniques. In facial recognition, most of the feed-forwards techniques for face recognition varies on skin color data [10]. Skin color pixels have value lies in between the ranges of 0.37 < r < 0.458 and 0.275 < g < 0.365. Before the face recognition, we can compare the face of the person to the existing sample collected inside the database. Human face recognition is an interesting area for the security perspective and challenging task in a real-time application for this researcher focus on a neural network to increase accuracy but in this research paper use the local binary pattern histogram (LBPH) for improving accuracy even using the traditional method. And face recognition is the latest field to solve the security issue. In this advance society popularity of security cameras in public areas is increasing and the reason behind this is fast and accurate [25] and no human involvement it detects automatically and it is used in a real-time scenario and at a time multiple faces detected. Here the face and facial features play important aspect for public security, which use in human-computer communication which is used to trace the face of the particular person and hence we can avoid the traditional security like fingerprints scanner, smart card, and password-based biometry which is easily hacked by professional hacker [14], but the face feature cannot hack easily. It is realistic and accurate for the market trends and hence the popularity of face recognition increases day by day. Since security is a significant issue for our society because today we are living in the digital world, where everything available digitally which is easily stealing by the person [23] so people authentication is necessary for a computer-based system. We can also monitor the visitor of an organization using the face feature, where we can verify the person before entering into the organization because these entering the building illegally like military headquarter, government building for a certain purpose [19], like stealing the organization assets or they are a spy of any country. For this purpose, we can use the video surveillance in the modern cities, because in this city there are lots of emergencies come any time so that to deal with these critical situations and hence to handle these emergencies we can monitor these cities by a camera with an intelligent system which monitor all the cities and give the emergency notification alarm in a critical situation [16], some common example like 26/11 Mumbai terror attack and 2015 in Paris where 129 people killed by a terrorist. Video surveillance also helps for the weapons detection and protect the civilians from the terrorist threats [24], which is an increase from last few decades which is a big issue for our cities and hence fast weapons detection technique is used for safety purpose with the help of ultrawideband radar. For this system compare the live face of the person to the sample present inside the database of the system, for

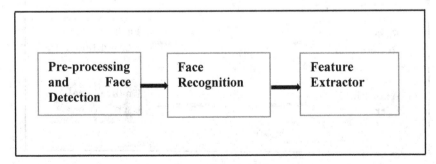

Fig. 1. The process of face recognition

this There are two types of comparison in facial recognition (1) Face Verification and (2) Face Identification (Fig. 1).

3 Methodology and Experiments

In this design, we can purpose a model that is used for security and surveillance where we can use Computer vision, Image processing, and machine learning. This all used to make an intelligent system that is helpful for the face recognition of a person whose data already present in the database.

Face detection: Face detection is used to deal with the image and it is used to training the database with the help of a haar cascade algorithm.
Face extractor: Face extractor is used to training the model, haar feature, and cascade classifier.

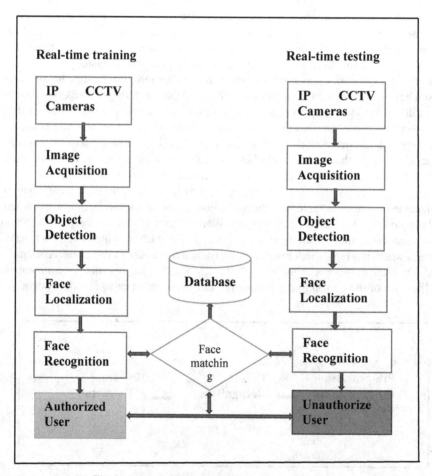

Fig. 2. Purposed model surveillance face recognition

Face recognition: Face recognition is the set of two tasks in real-time (1) Face identification and (2) Face verification (Fig. 2).

A. Computer vision: The computer vision aims to write down the computer programs that can interpret images and obtain information about the image. For the security perspective, we seek to automate the tasks that the human visual system can do. For training the model capture images for frontal face detection (Fig. 3).

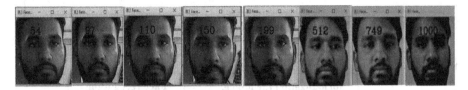

Fig. 3. Capture images for training

B. Image processing: After capturing the images with the help of computer vision our job is to extract information from digital images and videos, for this purpose, we can use image processing. Basically, in image processing, we can perform some operation on digital images to enhance and extract some information regarding that image, after that when a sample is collected. The collected sample BGR (Blue, Green, and Red), colored then all the sample dataset can be converted into grayscale. We require converting into grayscale is reducing complexity from 3D pixel value to (BGR) to a 1D value. The luminance of a pixel value of a grayscale image from 0 to 255. The conversion of a color image into a grayscale image is converting the BGR values (24 bit) into grayscale value (8 bit). After converting, grayscale we crop all the images for better recognition.

C. Machine learning: Machine learning methods based on an artificial neural network representing learning. Here machine learning is used to make the intelligent system. Because the IP CCTV cameras only record the video which is not used for real-time security, hence we cannot take the action at a time of crime scene and police officers follow the traditional method to see the recorded video, and after that, they take action. Which too lengthy process because of crimes happen every day all over the world. Millions of people travel every day from point A to point B by public transport. It is difficult to secure with the help of recorded videos, and hence the real-time surveillance is needed for smart society, and face recognition plays an important role in security and surveillance.

Training: Take samples of images for training images generated by the algorithm using "haarcascade_frontalface_defaults.xml" from Haar Cascades, so histograms are the same as a sample of images. Crops the images to 250*250 pixels values, and reduce the BGR images to Grayscales with 8bits per pixels. During testing, the algorithm again creates a histogram from the test image, and it compares that with the training histograms to try to get a match. All incidents recorded by the CCTV cameras regularly, but cameras cannot prevent the crime. But the security person monitors the CCTV cameras, and it is not possible to check all by the naked eye in real-time. In the future, we make a separate dangerous situation from the loaded video footage using behavior patterns and movement analysis. We design a system that senses the timely warning situation is escalated or out of control situation. In which we can recognize those situations in which people need help (Fig. 4).

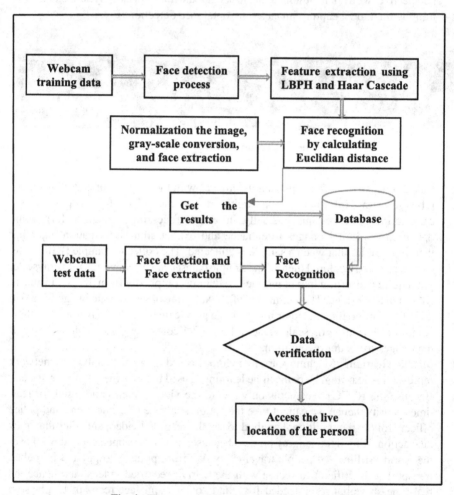

Fig. 4. The proposed live location of the persons

With the help of an intelligent. The surveillance system, we can stop the violence and attack, before doing this, but here we can train the data (collected sample) for facial recognition purposes. Whenever the person recognizes by the camera easily identified with the 90% accuracy in the future we can increase the accuracy from 90 to 100%. In this way, we can recognize the people with the help of facial recognition. After training the model we can start to predict the face of the model is confident more than 90% then recognize the face otherwise send the notification to the security police officer. After completing all the processes, now we can verify either it is the person whose data and basic details present in the database or not. If the data is not present in the database then we take the sample and basic details of the particular person. In this way, we can design a biometric for security and surveillance purpose. For this purpose, it is necessary to collect the basic details like name, age, sex, contact number, and samples of images of every person with different poses. When the all detail and sample is collected, of every person who lives in the city, after collecting the basic details we make the IDs of every person in the city. And these IDs helps us find the persons and also find the present place of every person who can live in the city and his/her moment inside the city, basically in this way surveillance all incidents of the particular person within the city. This idea also works for the country on a large scale to fulfill security requirements. To and find the present place of the people as well as all the moment in a real-time scenario. And this helps to make the smart society and smart public surveillance system. For this, purposes we can deploy all city with the IP CCTV cameras, with a high resolution like 4 K cameras that are making a start to be used in real-time. The resolution is high and the frame rate is 25 fps. The cameras have a high resolution because the movement of the person is fast and at the same time multiple numbers of faces, is recognized in one camera. But before this making the IDs of every individual, and we can deploy the cameras and every place inside the city. So that we can monitor all the public incident on the public places for public security. And we can surveillance the society from internal threats like kidnapping, loitering, rape, murder, and chain snatching, etc. This surveillance system is also deployed on the border area to control the cross -border terrorism attacks and refugees who cross the border. From one country to another country for employment and crime purpose.

We can also deploy this security system to airports for the people to visit in the country. Here we can launch the INDIA VISIT (India Visit and immigrant status indicator technology) the main aim of this how many foreigners travels gaining to India. Because a lot of foreigners visit India every day, and we cannot surveillance the people who come to India and leave India and sometimes the visa is expired, but they cannot leave the country and it is the crime. In this way we can notify foreigners before their visa, is expired we send the notification to leave the country in this way we can surveillance the refugee, as well as the people, visit for tourism purposes in the country. But the problem is that face recognition using the Biometric identification is the long-lasting process and the accuracy of the results. In this paper, we purposed the solution for an earlier face recognition process with accurate results. In this security, an enormous amount of surveillance videos are shots every day and most of them are inspected by people such as police officers or security personals then these people view surveillance videos. they are performing a certain task such as face recognition which performing these tasks they make the subjective evolution of the video quality, for example, they can easily identify the person in a video. But the problem is that how can store the huge amount of video data being generated daily it's a big problem for this some authors are using the compression techniques to compress the video to minimize the storage [2]. But it is not beneficial for accurate recognition because the accuracy is decreased When we compare the compressed or distorted face images with the uncompressed or undistorted than it is understandable that the compressed face will become more unnatural, as the compression level increase image quality is decreased, to solve this problem we can use the Hadoop, concept. Because here the storage (Big data) is the problem and the Hadoop are the solve these problems. The process in which we can find the live location of the person who moves anywhere in the city first IP CCTV cameras finds the location with the help of IP address of the cameras which locate in the particular areas and put the live location of the person in the centralized system which access by the security officer anywhere and anytime. After that when the next time we want to find the live location of the same person then we find in the next cameras which are situated nearly from the previous cameras, and these cameras recognize the person and again update his/her present location in the centralized system and this process going on and update all incident of the person with time. This helps us to make a smart surveillance system for our society as well as public places in the city.

4   Results and Discussions

For the face recognition purpose samples of images is needed for the further process so that we need the image of every person who lives inside the city this is done by webcam device and stored in the database, after that whenever any person comes in front of IP CCTV Cameras then device compares the person face with the existing face data already present inside the database and then saved in the database. The purposed system gets through the values of estimate and score comparison with a face classifier to test an image. If the approved person is obtained from the database, the present image is identical to the existing data. If a match is not obtained in the database, then the device is locked face of the person and send the notification to the security officer policeman. Besides, the data of an unauthorized person model will be added to the database. For this, a purpose we can use some tools to capture the images as well as background operation in the images collected in the database. Whenever the new face images taken by webcam devices are compared with image information in the database. The proposed effort is appropriate for offline and real-time environments. The proposed real-time dataset has an improved system. The results of our work are comparable with an approach by some authors, although deep learning-based convolutional neural networks (CNN) have shown better results those techniques are computational very inefficient. Finally, our objective is the work was low cost and better result. For this, we can use Python 3.7.1 in anaconda 2019, jupyter notebook, and some libraries like OpenCV 4.2.0.32 and Numpy 1.14.5 version (Figs. 5, 6 and Table 1).

Fig. 5. Face recognition

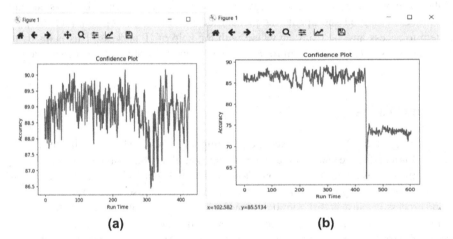

(a)   **(b)**

Fig. 6. (a) Accuracy of confidence plot. (b) Accuracy of not confidence plot

Table 1. Evaluation of face recognition

Accuracy	Percentage
Highest	93.67%
Average	89.27%

5 Application of Surveillance System Using Face Recognition

At present, the biometric-based security system has been significantly increased, especially in face recognition. Thus, face recognition application is a powerful way to perfectly and robustly give individual security such as with a smart society, smart home, smart card, smart campus, smart airports, law enforcement, public surveillance, or for entertainment.

Smart cards: Nowadays banking is the priority in daily life and a lot of people use the smart card for the money transaction. A lot of time we can forget ATM (automated teller machine) can we cannot proceed for a transaction for this we can use the facial recognition for the transaction of money when we haven't ATM or when the ATM is lost. The process in which we can find the live location of the person who moves anywhere in the city first IP CCTV cameras find a location with the help of the IP address of the cameras which locate in the particular areas and put in the centralized system which is accessible by the security police officer anywhere and anytime. After that when the next time we want to find the present location of the same person then another cameras which are situated near the previous cameras recognize the person and update the live location of the same person and update the location of the person in the centralized system and this process going on and update the present location of the person in the centralized system to access anyone from anywhere with the help of IP address. This helps us to make a smart surveillance system for our society as well as public places in a particular city.

6 Conclusions

The accuracy of the algorithm is exceptionally important, which showings that the LBPH classifier is an effective and exact face recognizer. The algorithm can provide access control throughout dissimilar lighting conditions (during day and night) because of the Haar Cascade frontal face profile. Here surrounding environment also affects the face recognition process because at different light conditions percentage of face recognition varies and result in change according to the different lighting condition. Although the shapes are just for frontal faces, it can detect correctly for slight head tilts (about 17–$22°$) and at 250 cm distance from the IP CCTV Cameras. The facial recognition method with haar cascade and LBPH (local Binary pattern histogram) method can enhance facial recognition with more than one face with accurateness up to 93.83%. The result varies between lighting conditions and the quality of the IP CCTV Cameras used. For better recognition with more confidence, the equipment or environment (IP CCTV Cameras and the room light/natural light) should be better quality. And the security and surveillance problems are resolved by this biometric system it provides real-time security. Facial images captured in a video streaming surveillance environment characteristically suffer from very poor quality. Besides because of the features of cameras, uncontrolled capturing surroundings may lead to ambient variations, such as lighting changes, face pose, light shadowing, and body or face motion blur. Consequently, the feature of facial captures using video surveillance cameras can influence the performance and efficiency of video-based facial recognition. We projected a video surveillance system that uses

LBPH characteristics and the Haar Casdae classifier. The successful implementation uses facial images and the proposed model was established under the exceptionally various condition and it performed efficiently and correctly.

Future work: In the future, the system may be extended on some bigger datasets using deep learning methods, such as convolution neural networks. (1) we can extend this biometric-based system for security and surveillance purposes for the whole country. (2) in the future, we can recognize the person with the help of Iris recognition, Voice recognition, and Retina scan. (3) In the future, we make a separate dangerous situation from the loaded video footage using behavior patterns and movement analysis. We design a system that senses the timely warning situation is escalated or out of control situation. In which we can recognize those situations in which people need help. With the help of an intelligent surveillance system, we can stop the violence and attack, before doing. (4) we can be deployed at the border for cross border terrorism and refugee (5) We can also deploy this security system to airports for the people to visit in the country. Here we can launch the INDIA VISIT (India Visit and immigrant status indicator technology) the main aim of this how many foreigners travels gaining to India.

References

1. Awais, M., et al.: Real-time surveillance through face recognition using HOG and feedforward neural networks. IEEE Access **7**, 121236–121244 (2019)
2. Heng, W., Jiang, T., Gao, W.: How to assess the quality of compressed surveillance videos using face recognition. IEEE Trans. Circuits Syst. Video Technol. **29**(8), 2229–2243 (2019)
3. Mokhayeri, F., Granger, E., Bilodeau, G.-A.: Domain-specific face synthesis for video face recognition from a single sample per person. IEEE Trans. Inf. Forensics Secur. **14**(3), 757–772 (2009)
4. Mantoro, T., et al.: Multi-faces recognition process using Haar cascades and eigenface methods. In: 2018 6th International Conference on Multimedia Computing and Systems (ICMCS), pp. 1–5 (2018)
5. Mai, G., et al.: On the reconstruction of face images from deep face templates. IEEE Trans. Pattern Anal. Mach. Intell. **41**(5), 1188–1202 (2019). https://doi.org/10.1109/TPAMI.2018.2827389
6. Roomi, M., Beham, M.P.: A review of face recognition methods. Int. J. Pattern Recogn. Artif. Intell. **27** (2013). https://doi.org/10.1142/s0218001413560053
7. Arroyo, R., Torres, J.J.Y., Bergasa, L.M., Daza, I.G., Almazán, J.: Expert video-surveillance system for real-time detection of suspicious behaviors in shopping malls. Expert Syst. Appl. **42**, 7991–8005 (2015)
8. Loke, S.C.: Astronomical image acquisition using an improved track and accumulate method. IEEE Access **5**, 9691–9698 (2017)
9. Kumar, M.P.R., Keerthi Sravan, R., Aishwarya, K.M.: Artificial neural networks for face recognition using PCA and BPNN. In: TENCON 2015 - 2015 IEEE Region 10 Conference, pp 1–6 (2015)
10. Wong, Y., Chen, S., Mau, S., Sanderson, C., Lovell, B.C.: Patch-based probabilistic image quality assessment for face selection and improved video-based face recognition. In: CVPR 2011 Workshops, Colorado Springs, CO, pp. 74–81 (2011)
11. Arya, S., Pratap, N., Bhatia, K.: Future of face recognition: a review. Procedia Comput. Sci. **58**, 578–585 (2015). https://doi.org/10.1016/j.procs.08.076,2015)

12. Kim, Y., Kim, H., Kim, S., Kim, H., Ko, S.: Illumination normalization using a convolutional neural network with application to face recognition. Electron. Lett. **53**(6), 399–401 (2017). https://doi.org/10.1049/el.2017.0023

13. Li, S.Z., Jain, A.K.: Handbook of Face Recognition. Springer, New York (2011). https://doi.org/10.1007/978-0-85729-932-1

14. Malik, D., Bansal, S.: Face recognition based on principal component analysis and linear discriminant analysis. Int. J. Res. Electron. Commun. Technol. (IJRECT 2016) **3**, 7–10 (2016)

15. Padilla, P., et al.: Electromagnetic near-field inhomogeneity reduction for image acquisition optimization in high-resolution multi-channel magnetic resonance imaging (MRI) systems. IEEE Access **5**, 5149–5157 (2017)

16. Shao, Z., Cai, J., Wang, Z.: Smart monitoring cameras driven intelligent processing to big surveillance video data. IEEE Trans. Big Data **4**, 105–116 (2018)

17. Abuzneid, M., Mahmood, A.: Enhanced human face recognition using LBPH descriptor, multi-KNN, and back-propagation neural network. IEEE Access, 1 (2018). https://doi.org/10.1109/ACCESS.2018.2825310

18. Meena, D., Sharan, R.: An approach to face detection and recognition. In: 2016 International Conference on Recent Advances and Innovations in Engineering (ICRAIE), pp. 1–6 (2016)

19. Satari, B.S., Rahman, N.A.A., Abidin, Z.M.Z.: Face recognition for security efficiency in managing and monitoring visitors of an organization. In: 2014 International Symposium on Biometrics and Security Technologies (ISBAST), pp. 95–101 (2014)

20. Jamil, N., Lqbal, S., Iqbal, N.: Face recognition using neural networks. In: 2001 Proceedings of the IEEE International Multi Topic Conference. IEEE INMIC 2001. Technology for the 21st Century, Lahore, Pakistan, pp. 277–281 (2001). https://doi.org/10.1109/INMIC.2001.995351

21. Khokher, R., Singh, R.C., Kumar, R.: Footprint recognition with principal component analysis and independent component analysis. In: Macromolecular Symposia, vol. 347 (2015). https://doi.org/10.1002/masy.201400045

22. Dalal, N., Triggs, B.: Histograms of oriented gradients for human detection. In: 2005 IEEE Computer Society Conference on Computer Vision and Pattern Recognition (CVPR 2005), San Diego, CA, USA, vol. 1, pp. 886–893 (2005). https://doi.org/10.1109/CVPR.2005.177

23. Meva, D., Kumbharana, C.K.: Identification of best suitable samples for training database for face recognition using principal component analysis with eigenface method. Int. J. Comput. Appl. **115**, 24–26 (2015)

24. Sakamoto, T., Sato, T., Aubry, P., Yarovoy, A.: Fast imaging method for security systems using ultrawideband radar. IEEE Trans. Aerosp. Electron. Syst. **52**, 658–670 (2016)

25. Xu, J., Denman, S., Sridharan, S., Fookes, C.: An efficient and robust system for multiperson event detection in real-world indoor surveillance scenes. IEEE Trans. Circuits Syst. Video Technol. **25**, 1063–1076 (2015)

26. Hou, B., Zheng, R., Yang, G.: Quick search algorithms based on ethnic facial image database, pp. 573–576 (2014). https://doi.org/10.1109/ICSESS.2014.6933633

An Effective Block-Chain Based Authentication Technique for Cloud Based IoT

S. Dilli Babu$^{(\boxtimes)}$ and Rajendra Pamula

Department of CSE, IIT ISM, Dhanbad, India
dillibooks@gmail.com

Abstract. Nowadays, cloud computing services and applications have created plenty of storage space to meet the demand and the utility for human life. However, these cloud services face serious security challenges against various attacks. Cloud computing services rely primarily on confidential data generated by devices connected to cloud account-specific personal information. These devices typically use security in an ID-based encryption scheme. The Identity-Encryption Technique (IDET) technique is an essential public cryptosystem that uses a user's identity information email or IP address. IDET uses signature authentication used in public-key encryption or public key infrastructure (PKI) instead of digital keys. The PKI uses a password for public key infrastructure (PKI) authentication. In contrast, the proposed model, an ID-based Block-chain Authority (IDBA) no need to maintain hash for authentication. Getting rid of Key management issues and avoiding a secure channel at the key exchange point is a functional problem.

Keywords: Block chain · ID based Encryption · Internet of Medical Things · Security · Privacy · and Public-Key Infrastructure

1 Introduction

Blockchain is a huge asset that demands cost-effective and sophisticated IT technology for better high level security. Cloud computing provides a stable, error-prone, and scalable environment for blockchain distribution management systems. The advantages of cloud storage include data cost, flexibility, access control, data reliability, scalability. Because of these features, each company moves its data to the cloud and uses the storage services provided by the cloud provider. The blockchain provides services as data security, data integrity, account or service traffic hijacking, insecure application programming interface (API), service denial (DOS). Malicious insiders, misuse of cloud services, shared technologies, and threats all affect adequate data security. Therefore, it is imperative to protect data from unauthorized access, alteration, or rejection of services. Proof of Ownership (PO) contains useful information such as data provider identification number (ID), data provider name, file upload date, and time. It used to identify the person with data in the uploaded block file. Cloud. It shows the source of the data file, and the person who downloads the data file can know the resource person who created it.

© Springer Nature Singapore Pte Ltd. 2020
M. Singh et al. (Eds.): ICACDS 2020, CCIS 1244, pp. 305–319, 2020.
https://doi.org/10.1007/978-981-15-6634-9_28

The various cryptographic schemes used for cryptography are ABE (Attribute-Based Encryption Standard) as Scheme 1, AES (Advanced Encryption Standard) as Scheme 2, ABE Scheme 3, and AES Hybrid. Moreover, a specific algorithm in the form of BCDS (blockchain-based cloud data security). It adds additional security to the data collected by Scheme 4 end users. The BCDS algorithm provides security through the cloud key and the session key through two different keys created by the cloud service provider (CSP). The BCDS algorithm consists of three different key lengths: 128 bits, 256 bits, and 512 bits. The length of the blockchain third party key (BTPK) and the session key are generated randomly in the cryptographic process. The BTPK is encrypted using the CSP session key, and the encrypted BTPK and session key sent to the data provided upon request. The data provider decrypts the session key with the BTPK and encrypts the data file using the decrypted BTPK. To decrypt the file, the data user must send a request to the CSP to receive the decryption key. The CSP sends the encrypted BTPK and the session key only if the data user is an authenticated and authorized data user. This technology reduces the number of unauthorized users in the blockchain cloud to reduces the maximum number of attacks in the cloud environment, such as a combination attack, brute force attack, and structured query language (SQL) injection attack. Cloud computing has many applications, allowing access to expensive applications and reducing the cost of installing and running computers and software. Users can place data anywhere. The Internet says that all users need to join the system. Cloud computing began as a tool for personal computing but is now widely used to access software online and online storage without worrying about infrastructure costs and processing power. Organizations can shut down and access their information technology (IT) infrastructure in the cloud. In addition to private companies moving to cloud computing, the government is moving its IT infrastructure to the cloud. The blockchain contains digital data from a variety of digital sources, including sensors, scanners, numerical modeling, video and mobile phones, digitizers, the Internet, email, and social networks. Cloud computing is an Internet-based innovation that simplifies various types of administration over the Internet, such as software, hardware, data storage, frameworks, and more. Cloud computing networks, virtualization, operating systems, and resources pose many security concerns. Scheduling, transaction management, load balancing, concurrent control, and memory management.

It allows for additional data storage and secure storage of reinforcement media. Cloud computing is an attractive hub for cybercrime. Cloud providers are required to ensure adequate security measures to prevent cybercrime. Many security issues arise when deploying these operators to the cloud. Cloud computing is a combination of blockchain. Blockchain gives users the ability to use Commodity Computing to process queries distributed across multiple datasets and to return the set promptly.

Due to the attractive features of security in blockchain cloud computing, blockchain becomes a useful and mainstream business model of cloud computing. In addition to the benefits at hand, the inclusion of previous components can lead to real cloud-specific security issues. Security is public security in the cloud, and users are delaying their transactions through the cloud. Security issues have hindered the improvement and widespread use of cloud computing. Understanding the security and security of opportunities in cloud computing, creating robust solutions is essential to its well-being. Cloud enables customers to switch from startups, reduce operating costs, and quickly access services and infrastructure resources when needed. In advertising and business, most industries use blockchain, but the basic features of security do not apply. If there is a security breach in the blockchain, it already has serious legal consequences and damages In many organizations, blockchain deployment is desirable and is useful for fraud detection. BigData-style analytics needs to address the challenge of detecting and preventing advanced threats and malicious intruders. Security challenges in a cloud computing environment categorized into network status, user authentication level, data level, and joint problems.

1.1 Cloud Computing Services

The cloud computing service feature provides an open standard for distributed service-oriented architecture. Cloud computing services use many XML-based technologies to convert diagnostic data into standard data and diagnostics databases, such as Simple Object Access Protocol (SOAP), Universal Description, Discovery, and Integration (UDDI). These basic standards used in cloud computing services. Figure 1 shows the process of cloud computing services.

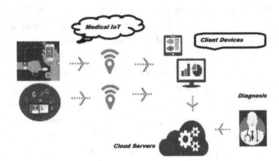

Fig. 1. Cloud service.

1.2 Introduction of the IDBA

Common classification challenges include the use of traditional security devices and various technologies. The data service for our outsourcing security provides access to our computing data. However, it requires verification to ensure that it is only accessible to authorized users. When users use the cloud environment, users rely on outsiders to make decisions about the data. Cloud suppliers must have adequate tools to prevent the use of customer data, which cannot be corrected. In all cases, any specific method may not completely prevent cloud suppliers from misrepresenting customer data. So a specific non-technical integration is required to achieve this. Customers need considerable confidence in their suppliers' specific skills and financial capabilities. The data encryption used to ensure access to the cloud and to ensure data security and seamless fighting. However, encoding increases integrity for traditional data usage services, for example, with simple text catchphrases on printed data or Planet data queries. Cloud computing is an industry where technology is proving to be a strategy for achieving cost and productivity transactions. It can make the wrong decision before it starts. Mature. Enterprise cloud security needs to carefully reviewed, security issues to reviewed, and plans made before implementing the technology. Researchers suggest establishing feasible plans for improved governance cooperation to address security issues and concerns successfully. The authors believe that switching to cloud computing is deliberate and gradual over time.

Structural diagrams of multi-level security systems using multi-level ABE and AES algorithms. The goal is to use cloud logic cryptographic techniques to protect sensitive data in the cloud and to develop a multi-level architecture that provides data protection in the cloud environment. Four levels of data security ensured by implementing various security services such as authentication, cryptography, and decryption. Only legitimate users are allowed access to cloud data. This security system has four steps to store data in a massive data cluster. Data providers and data users can enter their details on the registration form. Data providers and data users are allowed to login after registering on the system. Logging into the Cloud Service Provider (CSP) system is standardized. The CSP authorizes those credentials with those credentials and assigns encryption keys to authorized data providers. The second level of energy is defense. Only authorized persons are authorized to encrypt and upload data files. After authentication, the data file is encrypted using the hybrid ABE and AES algorithm and uploaded to the cloud.

Data cryptography is the third level of security. Only authorized persons are authorized to use the encrypted data stored. As the fourth step of security, authorized users request a decryption key and a one-time password (OTP) from the CSP when downloading data. The random number generated by the CSP is the mobile number of data sent or used to their email ID. After the OTP check, the appropriate algorithm can download and decrypt the data file using the standard key. After decryption, the user can view the data file which encryption of uploaded data files and encryption of downloaded data files. The data files are encrypted using hybridized attribute-based encryption (ABE) and advanced encryption standard (AES) algorithms. Upload data files from authorized data providers to cloud servers.

ABE is the philosophy of public-key cryptography, which enables users to encrypt and decrypt messages based on user attributes. ABE Project includes visualization, sender, and receiver and generates keys to encrypt or decrypt data from the sender and user. In this project, the Authority creates the keys to the properties; the Cryptographic Authority created these features of the Public Key (PK) and BTPK (MK). Hybrid AEB and AES Algorithm-Based Encryption (ABE) is a type of asymmetric cryptography that provides privacy to IT and security through the interplay of user features in the system. ABE systems typically include cloud service providers, data providers, and data users. CSP authenticates data providers, generates public and private keys of data users, and assigns keys to data providers and data users. The system policy created between the user attributes, which adds privacy to the system's authorized users. This project uses Ciphertext-Policy ABE (CPEB), where the data provider embeds the data access policy into the cipher and associates the data user attributes with its private key. The term cryptography refers to converting real data into the human-unreadable form, in decryption, is the conversion of encoded data into readable form. Only an authorized person can decode the source data by encrypting the data. So data privacy is achieved through cryptography.

Attribute-based cryptography usually has encryption features, including no complete data. The AES ABE with 128-bit keys used. Secret or secret key ciphers use the same key for encryption and decryption, so each sender and receiver must use the same secret key. After the ABE algorithm implemented, the AES executed. ABE uses a step to define identity derived from a set of attributes, for example, characters and messages encrypted using AES. Once the data file is encrypted, it uploaded to the cloud. Any request to read the data processed after the end-user decrypts it and reads the data from the requesting application. Works on a wide range of hardware and software platforms. It includes 8-bit and 64-bit platforms. It requires less memory to execute, making it more suitable for forest-space environments. This structure has the advantage of benefiting from teaching level equivalents.

IDBA is a new technology that makes it easy to design, implement, and manage health systems [2, 15]. IDBA provides user-health information and support functions to overcome network latency. IDBA can implement more logic and error-tolerant methods to meet customer needs (Fig. 2).

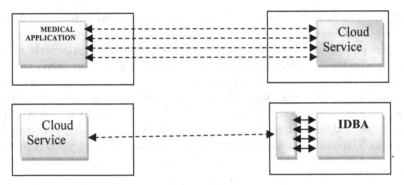

Fig. 2. IDBA and network load distribution

(B) Network latency exceeded

IDBA Network latency reduces robots (software automation). An example of a real-time system that responds to real-time situations. So latency is not acceptable. IDBA provides a solution; The IDBA sends all the information stored in it, and the IDBA acts as the destination controller so that all the information can be accessed directly from the IDBA (Fig. 3).

2 Preliminaries

2.1 Cloud Computing Services and Blockchain Security

Smart cities, smart grids, e-elites, and the industry rely through different devices across various modern-day population across roads, communities, and homes. The cloud data

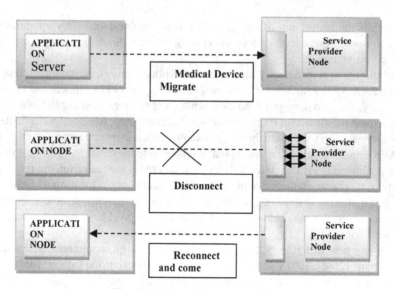

Fig. 3. IDBA and disconnected jobs

through based on their characteristics of user weather conditions, vehicle speed, traffic conditions, location, heart rate, blood pressure. The data being upload from respective owners periodically. For example, it is health care, doctors can actively monitor the patient's health and prescribe medication immediately to prevent delays and complications. To address this shortcoming, industry and education professionals focus on blockchain fundamentals and optimize blockchain-based models needed IoT based technologies. This network has several privilege modules to provide ABE. The ABE concept can be defined briefly, how to use blockchain. And then the structure of blockchain transactions. The feature-based cryptographic scheme supports privacy and access control through ABE single cryptography. ABE has four parties: Clusterhead, Blockchain Miners, Special Authorities, and Distribution Ledgers. Data from the CH sensors is collected and processed and encrypted before the transaction. CHd data is encrypted so that individual miners with the right features can view and verify transactions—for example, health care. So to decrypt and verify a minor transaction that has a "doctor" attribute or a "nurse" attribute. Additionally, once these transactions added to the blockchain, data will only be available to users with these "DOCTOR" or "NURSES" attributes. The app used to diagnose diabetes and cholesterol quickly. The data used for future research purposes, the application type allows the researcher to collect the right kind of data. The characteristics of CH cryptography determine the type of application. In this health care context, the CH selects the characteristics of the physician, nurse, hospital, location, and creates an entrance structure. In IDBA and cloud computing services systems, this algorithm not used with just one key and no IDBA public keys. It includes blockchains of various parties issued by the Certification Authority (CA).

2.2   ID-Based Broadcast Encryption

This section describes the Bi-Linear extension, two-line Diffie-Hellman option, ID-based Public Key infrastructure (PKI), and bone-Franklin encryption scheme. Security of the BDH-based ID-based Transmission Cryptographic Project.

Theorem: Let G1 and G2 are two periodic groups for some large-scale integrated queue. G1 Periodic Coordinator Group and G2 Periodic Coordinator Group. We believe that logarithmic problems are particularly severe for G1 and G2.

E: G1->G1, G2 is a pair that meets the following conditions:

(1)   Cond-1 :BiLinearity: E (AM, BN) = E (M, N) Ab, All M, N G1 and All A, B * N;
(2)   Cond-2: E (M, N) 1, M G1, N G1;
(3)   Cond-3: Calculation: The protocol for calculating E (M, N) for all M and N G1.

A serious Diffie-Hellman problem to determine M, aM, bM, cM G1, a, b, c Zn, c = ab Zn. The D-H property is satisfied with E (AM, BM) = E (M, M) Ab. The basis of the computational error is the protection of the algorithm, the Hellmann problem (CDHP). A B and Zq are randomized with G1 and selected in P1. The given calculation (P, AP, BP) is ABP G1. G1 is called the Group Diffie - Hellman (GDH) Group, and CDHP is called the Gap Diffie - Hellman Problem. DDHP. All A, B, C, Count E (P, P) and ABC (P, P).

2.3   ID-Based Public ID Infrastructure

Victims of Trusted Key Generation Centers (KGCs) have an ID-based public Key infrastructure. IDBA includes G1, G2, Public Key Generation PKG, Bi-Linear Coupling E: G1G1P2. E1: {0,1}*K1, E2: K2 {0,1}*. Setup: KGC randomly selects s * q number and sets Ppub = sP. The KGC system = G parameter G1, G2, q, P, P_Pub, H1, H2}, and publishes the master key.

Private Key Abstraction: The patient sends his identity to the Blockchain based Key generation Center (BKCG). BKGC calculates the patient's public key as BQID = E1 (ID) and returns it. This private key is SID = sQ_ID.

2.4   Bone-Franklin Technique

Bone-Franklin [1] technology allows private key SID holders to decrypt messages sent under private key QIDs. The original project was a kind of Identity-Based Encryption Technique (IDET) project in which the formation of BDH groups was challenging. Fujisaki has expanded its model-based strategy to protect against alternative cipher attacks by using Okamoto variants. Assume that the message is encrypted.

Setup: Calculate U = RP, where R = Q. Then calculate V = m H2 (e (Ppub, rQID))

Add Topper Cipher (U, V). Calculating with decryption VH2 (E (U, SID)) = V H2 (E (RP, SQID)) = V H2 (e (sP, rQId)) = VH2 (E (PPUB, RQID)) = M.

3   ID-Based Block-Chain Authority

The ID-based technique works here using the blockchain, and Cloud Service Provider (CSP). It selects two groups of system parameters, P, Q, G1, and G. The Ppub = Sp system calculates this public key and sends it to all registered users. WSP member. E1 selects two single-function CSP Calculate: {0, 1} * K1 and E2: K2 {0, 1} *.

3.1   IDBA and Cloud Computing Services

IDBA and Cloud Cloud Computing Services combine IDBA Patient Encryption Technology (IDET) and Cryptographic Analysis Data Encryption Algorithms. Where users launch IDBA for the cloud computing service system, every have identifier ID_I. The WSP can check if the IDBA is actually from a legitimate user.

(i)   Signature Plan

To sign the M {0, 1} * message, the patient UI randomization number r using BKGC using patient ID hash function E2: {0, 1}* Zq. The method works based on follows. Following the algorithm presented in the advanced section, the BKGC system installed by distributing each user's secret key through special security methods. The IDBA starts, the user creates a signature trio (ri, c, s) by signing the message with his secret signature key CID. RQI, C (H2 (Me, Re) + R), at least.

(ii)   Multi-party Calculation Plan

Selecta Pseudo random number k number

Assign j = R

Calculate j_i xi = ê (r QIDi, Ppub). Here is vector function.

Calculate the following polynomial function

fx (x) = (x − si1kxi1)/(xn − m) mod p = (x − x)) p

Where m = sic, (i -> 1 to n) (1 m l); m means the cloud computing service group H [m], which means all members are assigned to a cloud computing resource group.

We have the following equations:

(x − xi) = IC Mode p

So we get the following:

aok = (−xj)

a1k = (−xj). . ...

am − 2, am − 1, k = (−xj)

Amac = 1

{a0kP, a1kP, a2kP, ..., akP}

{P0k, P1k, P2k, ..., Pmk}

If (sID, R) = ê (sQID, rP) = if (rQId, Ppub == i)

Cke + xijCjk = Dk + Rk (aok, + a)

Cloud Computing Service Provider GK [K] should withdraw FK (X) from Cloud Computing Service Group. G [k1] Reopen fk1 (x) to add members to find the data set. As a result, cloud computing service providers can acquire new {iq and {ik1 and encrypt two cloud session keys. It reproduces the affiliate TK for members of the new cloud computing service group.

3.2 An Example of ID-Based Cryptocurrency

The normal oval curve represents the following:

y2 mode 37 $= $ x3 $+$ x $+$ 1 mode 37

In the group, G (0, 1) is selected. The smallest part of x (x, y) is EC. With this subtle change in mind, the new values of XR and YR are identical to the expression used to determine the slope. If y converts statistics too -y, it returns Pml. Now "#." Apply the first logarithmic idea to get the ASCII value.

The underlying problem in ID-based encryption schemes is lead management. The BKGC can create a user's private key, which can decrypt the cipher or create a signature for any message. Therefore, the model lacks user privacy and visibility. Patients and Private Key Generators (PKG) need a secure channel to issue private keys. Users combine some hidden parameters with actual diagnostic data. Finally, the patient calculates his private key by partially removing the private key with confidential information. That is, before submitting the request, some of the patient's hidden parameters are combined with actual clinical data and provided with covered clinical data. After this, the party reviews the request. During the evaluation, they make some diagnoses that are processed and refined. The Key release point is bind-blinding technology, which reduces the Key management problem and eliminates the need for a secure channel. It process by which users make requests through an insecure channel (this request may be one or more parameters of another party).

4 Implementation of ID-Based Signature Scheme

The trusted authority receives the User-Id (ID) in other authorities and verifies the patient's ID and partially provides the ID to the patient.

Signing: A party or organization uses its private key or signature message.

Checker: A party or organization that decides to accept or reject a public key or signed string.

4.1 Binding-Blind Technique (BBT)

Valid algorithms and valid (params, id, regid, em, id) perform the following functions:
    Calculate r $=$ r ((σID, p)). (PKID; -Rigid) c.
    If c $=$ h (m, r), accept the signature.

4.2 Cloud Computing Services for Security Integration

Zhang et al. The new IDBA proposes a security plan for cloud computing services. It is contrary to IDBA and provides an alternative approach to security policies without the use of Certification Authorities (CAs). It uses one key for the patient forever. This article uses a bone-Franklin ID-based public key project. It provides an important security framework for cloud computing services without key management. It provides secure channels for private key delivery.

4.2.1 Scheme Logic Setup

Cloud Computing Service Provider (CSP) parameters G1, G2, P, Q System, Select PPP = SP. For this strategy, the patient must register and subscribe to the WSP. Whenever he included in the group, the patient must provide his or her identity. Cloud computing service providers use the private key to authenticate the patient by sending sID = sQID. Connect with QID = H (ID) - A secure channel and a blind strategy to eliminate a major management problem. For N patients and L services, cloud computing service provider NLL Matrix S works as follows: If the patient is a member of the [cloud] computing service G [K], then SMK = 1 (1≤ kg, 1 IL m).

4.3 Verification Plan

Each patient has a unique ID. If we have a message, UI random number RR, ZQ, generator P AG1 and sign the patient message Q rQ ri, c (h 2 (m, re + r)) CD. Signature Triple (R, C, Mr). WSP Public Key Mats. Diagnostic data can be encrypted in two ways, depending on its size. On the other hand, for large diagnostic data sets, the diagnostic data is first encrypted using the session key. The session key is encrypted using the service encryption key. After receiving IDBA encrypted analysis data, it first decrypts the session key with its secret key. This analysis uses the session key to decrypt the data. When a member switches from G [k] to another group G [k1], the WSP must associate with another session key in the cloud computing service group. When a new member joins or an existing member exits or moves from one group to another, there may be three cases. When a member moves from one cloud to another, member G subscribes from one group to G [K], and G [K1] from another group is called K1 K1. Updating the CSP Group Registration Matrix. Cloud computing service providers have redesigned FK (X) functionality to remove G [K] members. Another polynomial function, FK1 (X) member, is added to the cryptographic analysis data set G [K1]. The FK (X) function is given by: Fk (x) mod p = m.

5 Performance and Compromise

Describes the results of comparing feature-based encryption (ABE) or advanced encryption (AES) and encryption times (milliseconds) using a hybrid ABE and AES algorithm for a certain number of files uploaded by cloud-based users. The results indicate that the encryption time required for ABE is comparable to the AES algorithm using the ABE algorithm or the AES encryption algorithm. NBE algorithm means that the AES algorithm used to re-encrypt encrypted and encrypted files. It takes more encryption than other algorithms. Depending on the number of files, encryption time can be increased or decreased. Most users use AEE with the ABE algorithm to provide better results for the number of files uploaded into the cloud to encrypt data files. 3.4 Configure results comparing decryption time to download several milliseconds (milliseconds) with ABE. Or AES, Hybrid ABE, and AES Hybrid with the Secure Hash Algorithm (SHA) algorithm for specific user file requests in the cloud. The graph shows that SHA requires a one-time password (OTP) with a hybrid ABE and AES decryption algorithm to decrypt the number of files compared to the ABE or AESL algorithm. More time is required, and this depends on the download that most users decide with the number of files from the cloud. The existing algorithm proposes to assign a unique key to each user, and the file owner requests the authentication of the required files using the proposed algorithm. Data decrypted using the AES algorithm because the decryption time using the hybrid ABE and the AES algorithm is longer than the other algorithms. The ABE algorithm again decrypts the data, and the time increases or decreases depending on the number of files. Decrypted files stored in the blockchain cloud. An authorized user provides the OTP data developed by the SHA algorithm for download. Therefore, for most users of the blockchain cloud, the number of downloaded files in a decrypted forest is many times greater than decryption performed by the ABE or AESL algorithm. Therefore, many users use FTP to decrypt data files using the AES algorithm, using the ABS algorithm to count the number of files downloaded in the cloud. It compares milliseconds to specific cryptographic time results with AES. It hybridizes AEB and AES algorithms with multiple files uploaded to the cloud. The proposed AES encryption algorithms require more time to encrypt the AES algorithm due to the added security of the encryption time required. Therefore, the number of files uploaded to the cloud has proven to be secure and secure. The BCDS algorithm, which uses cloud computing features, provides multiple security systems to Internet users through blockchain on-demand services. When a business or government agency or person using the cloud shares information, its security and privacy are suspected. Data security in the cloud concerns data user visibility. The company has its security policies that do not allow individual employees to access a certain amount of data. Information integrity is an essential part of information security systems. This service protects data from unauthorized editing or deletion. The goal of the job is to store data securely and securely and securely access the data. Four-tier security provides an effective security system for multiple users who can access data in the cloud. The data file is encrypted and uploaded to the cloud using a secure dynamic bit standard (BCDS) algorithm. DataProvider Allows users to upload secure data and access data from the Cloud Service Provider (CSP). Then this authorized person is called an authorized person who can access the data. When downloading a data file, the data user accesses the One-Time Password (OTP) to download and decrypt the data file using the download algorithm.

The resulting program implemented in the Intel Core 2 Duo 2.0 GHz processor with JDK 1.6.0 programming language. The Tool Module designed with Java Swing, which allows users to provide personal details such as usernames, email ids, and birth dates for verification. To register for cloud computing services, IDBA plans to study. The PKG (Private Key Generator) module designed to create public/private key pairs for registered users. Server module designed to serve patients. The server encrypts cloud computing services (usually HTML files) using the user's public key. It creates a private key (server) signature using ECDSA (Elliptic Curve Digital Signature Algorithm). Servers and users blindly associate one or more parameters (usually two large numbers) with a PKG (private key generator). Server.

Two experiments performed using RSA Custom Implementation. First, we tested the timing of generating keys for E2 and RSA at 192, 224, 256, 384 and 521 bits, and only for RSAs at 1024, 2048, 3072, 7680 and 15360. The second test we performed was to encrypt the 1K, 10K, 100K, and 1M size analysis data files for each algorithm. The test compared with the same key size, key size, and key power in the first experiment. Key Section and Cryptographic Analysis This section covers the results and analyzes of our experiments, including data encryption times. It is evident from Table 1 that the results of our experiment are slower than RSA. ECC vs. RSA during lead generation shown in Fig. 4. Notice that the quantity used here is logarithmic.

Table 1. Number of Session keys in RSA and ECC

No of cloud session keys in ECC	No of cloud session keys in RSA
14336	65536
20480	131072
24576	196608
28672	262144
32768	393216

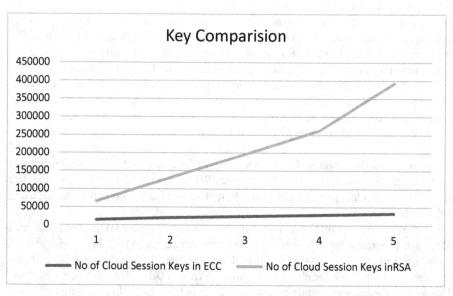

Fig. 4. Comparison of Number of Session keys in RSA and ECC

6 Conclusion

The cloud service and blockchain security provides a standard secure service. Useful measurement mathematical calculation tools used for sharing services and information. IDBA is part of the software and verification of user data. It provides visualization, authentication, and privacy along with user identification information that used as a public key for cryptography or signature verification. Secure channel-specific planning does not require private key delivery to eliminate critical management issues. It uses secure security logic and a binding-blinding method. We need to create MP typing services and software for IDBA to mimic the feasibility and operational model of reliable, secure, and secure technology. The user selects two unknown objects (usually two butt integrators) and calculates the standard parameters using its detection. It displays the parameters associated with the public key's private key (PKG). Private key behavior (PKG) is a semi-private key that communicates with the patient through a large channel. Finally, users combine this with their confidential information and create their private key. Therefore, we conclude that the use of the binding-binding method prevents key management and silences the existence of a secure channel for issuing private keys in ID-based cryptocurrencies. Other vital functions, such as alphanumeric characters, can be statistically mapped. In ID-based cryptography, oval particle dots are stable or compatible. Also, we try to use this appropriate measurement method to provide better security when displaying analytical data.)

References

1. Tama, B.A., Rhee, K.-H.: Tree-based classifier ensembles for early detection method of diabetes: an exploratory study. Artif. Intell. Rev. **51**(3), 355–370 (2017). https://doi.org/10.1007/s10462-017-9565-3
2. Alaba, F.A., et al.: Internet of Things security: a survey. J. Netw. Comput. Appl. **88**, 10–28 (2017)
3. Arias, O., et al.: Privacy and security in internet of things and wearable devices. IEEE Trans. Multi-Scale Comput. Syst. **1**(2), 99–109 (2015)
4. Arshad, H., Nikooghadam, M.: Three-factor anonymous authentication and key agreement scheme for telecare medicine information systems. J. Med. Syst. **38**(12), 136 (2014). https://doi.org/10.1007/s10916-014-0136-8
5. Atamli, A.W., Martin, A.: Threat-based security analysis for the internet of things. In: 2014 International Workshop on Secure Internet of Things (SIoT). IEEE (2014)
6. Atzori, L., Iera, A., Morabito, G.: The internet of things: a survey. Comput. Netw. **54**(15), 2787–2805 (2010)
7. Azaria, A., Ekblaw, A., Vieira, T., Lippman, A.: MedRec: using blockchain for medical data access and permission management. In: International Conference on Open and Big Data (OBD), pp. 25–30. IEEE (2016)
8. Bandyopadhyay, D., Sen, J.: Internet of things: applications and challenges in technology and standardization. Wirel. Pers. Commun. **58**(1), 49–69 (2011). https://doi.org/10.1007/s11277-011-0288-5
9. Botta, A., et al.: Integration of cloud computing and internet of things: a survey. Future Gener. Comput. Syst. **56**, 684–700 (2016)
10. Botta, A., et al.: On the integration of cloud computing and internet of things. In: 2014 International Conference on Future Internet of Things and Cloud (FiCloud). IEEE (2014)
11. Boussada, R., et al.: A secure and privacy-preserving solution for IoT over NDN applied to E-health. In: 2018 14th International Wireless Communications & Mobile Computing Conference (IWCMC). IEEE (2018)
12. Cachin, et al.: Blockchain, cryptography, and consensus. IBM Research, June 2017. https://www.itu.int/en/ITU-T/Workshops-andSeminars/201703/Documents/Christian%20Cachin%20bl blockchain-itu.pdf

13. Chen, H., et al.: SOUPA: standard ontology for ubiquitous and pervasive applications. In: 2004 The First Annual International Conference on Mobile and Ubiquitous Systems: Networking and Services, MOBIQUITOUS 2004. IEEE (2004)

14. Cognizant Technology Solutions (2017). https://www.cognizant.com/perspectives/how-blockchain-cantransform-life-insurance-processes

15. Dias, F.S., et al.: 22nd international symposium on intensive care and emergency medicine. Crit. Care **6**(1) (2002)

16. Diro, A.A., Chilamkurti, N.: Distributed attack detection scheme using deep learning approach for the Internet of Things. Future Gener. Comput. Syst. **82**, 761–768 (2018)

17. Diro, A.A., Chilamkurti, N., Kumar, N.: Lightweight cybersecurity schemes using elliptic curve cryptography in publish-subscribe fog computing. Mob. Netw. Appl. **22**(5), 848–858 (2017). https://doi.org/10.1007/s11036-017-0851-8

18. Diro, A.A., Chilamkurti, N., Veeraraghavan, P.: Elliptic curve based cybersecurity schemes for publish-subscribe internet of things. In: Lee, J.-H., Pack, S. (eds.) QShine 2016. LNICST, vol. 199, pp. 258–268. Springer, Cham (2017). https://doi.org/10.1007/978-3-319-60717-7_26

19. Fowler, S., Zeadally, S., Chilamkurti, N.: Impact of denial of service solutions on network quality of service. Secur. Commun. Netw. **4**(10), 1089–1103 (2011)

20. Gollakota, S., et al.: They can hear your heartbeats: non-invasive security for implantable medical devices. ACM SIGCOMM Comput. Commun. Rev. **41**(4) (2011)

21. Halperin, D., et al.: Security and privacy for implantable medical devices. IEEE Pervasive Comput. **7**(1), 30–39 (2008)

22. He, D., Kumar, N., Chen, J., Lee, C.-C., Chilamkurti, N., Yeo, S.-S.: Robust anonymous authentication protocol for health-care applications using wireless medical sensor networks. Multimedia Syst. **21**(1), 49–60 (2013). https://doi.org/10.1007/s00530-013-0346-9

23. Heer, T., et al.: Security challenges in the IP-based internet of things. Wirel. Pers. Commun. **61**(3), 527–542 (2011). https://doi.org/10.1007/s11277-011-0385-5

24. Hsu, C.-H., et al.: Efficient identity authentication and encryption technique for high throughput RFID system. Secur. Commun. Netw. **9**(15), 2581–2591 (2016)

25. http://bitfury.com/content/5-whitepapers-research/public-vs-private-pt1-1.pdf

26. http://dx.doi.org/10.1109/P2P.2013.6688704

27. Hussain, S., et al.: An efficient collision-resistant security mechanism for heterogeneous sensor networks. Internet Res. **19**(2), 227–245 (2009)

28. Sikorski, J.J., Haughton, J., Kraft, M.: Blockchain technology in the chemical industry: machine-to-machine electricity market. Appl. Energy **195**, 234–246 (2017)

29. Christidis, K., Devetsikiotis, M.: Blockchains and smart contracts for the internet of things. IEEE Access **4**, 2292–2303 (2016)

30. Linn, L.A., Koo, M.B.: Blockchain for health data and its potential use in health IT and healthcare-related research

31. Mettler, M.: Blockchain technology in healthcare: the revolution starts here. In: 2016 IEEE 18th International Conference on e-Health Networking, Applications, and Services (Healthcom), pp. 1–3. IEEE (2016)

32. Vasin, P.: Blackcoins proof-of-stake protocol v2 (2014)

33. Xia, Q., Sifah, E.B., Smahi, A., Amofa, S., Zhang, X.: BBDS: blockchain-based data sharing for electronic medical records in cloud environments. Information **8**(2), 44 (2017)

34. Tanenbaum, S., Van Steen, M.: Distributed Systems: Principles and Paradigms. Prentice-Hall, Upper Saddle River (2007)

35. Schwartz, N.Y., Britto, A.: The ripple protocol consensus algorithm, Ripple Labs Inc. White Paper, vol. 5 (2014)

36. Sun, et al.: Block chain-based sharing services: what blockchain technology can contribute to smart cities (2016)

37. Zhang, Yu., Wen, J.: The IoT electric business model: using blockchain technology for the internet of things. Peer-to-Peer Netw. Appl. **10**(4), 983–994 (2016). https://doi.org/10.1007/s12083-016-0456-1
38. Mukhopadhyay, U., Skjellum, A., Hambolu, O., Oakley, J., Yu, L., Brooks, R.: A brief survey of cryptocurrency systems. In: 2016 14th Annual Conference on Privacy, Security, and Trust (PST), pp. 745–752. IEEE (2016)
39. Zhang, Y., Wen, J.: An IoT electric business model, based on the protocol of bitcoin. In: 2015 18th International Conference on Intelligence in Next Generation Networks (ICIN), pp. 184–191. IEEE (2015)

Early Detection of Autism Spectrum Disorder in Children Using Supervised Machine Learning

Kaushik Vakadkar[1]([⊠]), Diya Purkayastha[1]([⊠]), and Deepa Krishnan[2]

[1] Computer Engineering, Mukesh Patel School of Technology Management and Engineering,
NMIMS University, Mumbai, India
kaushik.vakadkar@gmail.com, diyap@outlook.com
[2] Computer Engineering Department, Mukesh Patel School of Technology Management
and Engineering, NMIMS University, Mumbai, India
deepa.krishnan@nmims.edu

Abstract. Autism Spectrum Disorder (ASD) is a disorder which takes place in the developmental stages of an individual and affects the language learning, speech, cognitive, and social skills, and impacts around 1% of the population globally [14]. Even though some individuals are diagnosed with ASD, they can portray outstanding scholastic, non-academic, and artistic capabilities, which thus proves to be challenging to the scientists trying to provide answers to this. At present, standardized tests are the only methods which are used clinically, in order to diagnose ASD. This not only requires prolonged diagnostic time but also faces a steep increase in medical costs. In recent years, scientists have tried to investigate ASD by using advanced technologies like machine learning to improve the precision and time required for diagnosis, as well as the quality of the whole process. Models such as Support Vector Machines (SVM), Random Forest Classifier (RFC), Naïve Bayes (NB), Logistic Regression (LR) and KNN have been applied to our dataset and predictive models have been constructed based on the outcome. Our objective is to thus determine if the child is susceptible to neurological disorders such as ASD in its nascent stages, which would help streamline the diagnosis process.

Keywords: ASD · Autism · Machine Learning · Dataset · Preprocessing · Encoding · SVM · KNN · Random Forest · Logistic Regression · Confusion matrix · Precision · Recall

1 Introduction

Autism Spectrum Disorder occurs in the developmental stages of an individual and is a serious disorder which can impair the ability to interact or communicate with others. Usually, it impacts the nervous system, as a result of which the overall cognitive, social, emotional and physical health of the individual is affected [20]. There is a wide variance in the range as well as the severity of its symptoms. A few of the common symptoms the individual faces are difficulties in communication especially in social settings, obsessive interests and mannerisms which take a repetitive form. In order to identify ASD, an

© Springer Nature Singapore Pte Ltd. 2020
M. Singh et al. (Eds.): ICACDS 2020, CCIS 1244, pp. 320–329, 2020.
https://doi.org/10.1007/978-981-15-6634-9_29

extensive examination is required. This also includes an extensive evaluation and a variety of assessments by psychologists for children and various certified professionals.

A significant portion of the pediatric population suffers from ASD. In most cases, it can usually be identified in its preliminary stages, but the major bottleneck lies in the subjective and tedious nature of existing diagnosis procedures. As a result, there is a waiting time of at least 13 months from the initial suspicion to the actual diagnosis. The diagnosis takes many hours [9], and the continuously growing demand for appointments is much greater than the peak capacity of the country's pediatric clinics [5].

Owing to the gaps between initial concern and diagnosis, a lot of valuable time is lost as this disorder remains undetected. Machine Learning methods would not only help to assess the risk for ASD in a quick and accurate manner, but are also essential in order to streamline the whole diagnosis process and help families access the much-needed therapies faster.

We have structured our paper as follows: Sect. 1 includes the introduction to our project. Section 2 summarizes the literature survey performed. Sections 3 and 4 explain the working and methodology of the system we have proposed and its implementation. Section 5 portrays the inferences and results obtained. Lastly, Sect. 6 highlights our conclusions.

2 Review of Literature

Several studies have made use of machine learning in various ways to improve and speed up the diagnosis of ASD. In M Duda's [13] paper, forward feature selection coupled with under sampling was used to differentiate between autism and ADHD with the help of a Social Responsiveness Scale containing 65 items. Gopalkrishna Deshpande's [7] research involved using metrics based on brain activity to predict ASD. Soft computing techniques such as probabilistic reasoning, artificial neural networks (ANN) and classifier combination have also been used [15]. Many of the studies performed, have talked of automated ML models which only depend on characteristics as input features. A few studies relied on data from brain neuroimaging as well. In the ABIDE database, Milan N. Parikh [10] extracted 6 personal characteristics from 851 subjects and performed the implementation of a cross-validation strategy for the training and testing of the ML models. This was used to classify between patients with and without ASD respectively. In this study, have used five ML models to classify individual subjects as having ASD or No ASD, by making use of features, such as age, sex, ethnicity etc. and evaluated each of them to find the most optimal model.

The following table (Table 1) summarizes all the papers that were studied by identifying the key findings and limitations of each paper.

Table 1. Summary of literature review.

Paper	Key findings	Limitations
[13]	• Used forward feature selection and under sampling • Trained and tested six ML models on score sheets of 65 Social Responsiveness Scale from 2925 individuals having ASD or ADHD • Found that out of the 65 behaviours 5 were sufficient to distinguish ASD from ADHD with an accuracy of **96.4%**	• The dataset was compiled from primarily autism-based collections, as a result of which there was quite a significant imbalance, in favour of the ASD class
[7]	• Metrics based on brain activity used for prediction of ASD • Used **SVM** to obtain an accuracy of **95.9%** with 2 clusters and 19 features	• Constrained sample size/data set
[5]	• Uses **SVM** • Integrates ML algorithm inside ASD screening tool • Accuracy **97.6%**	• Imbalanced datasets • Small size of dataset, with 612 autism and 11 non autism cases
[1]	• Derived a novel algorithm which combines the structural and functional features • Chalked out various different representations of the functional connectivity of the brain • Results are an indication that combining multimodal features give the highest accuracy for distinguishing cases	• The ML models used show an increase in the accuracy of the prediction by a mere 4.2% for Autism, in comparison to the previous works carried out • Datasets suffer significantly from variations
[2]	• Makes use of **SVM**, **Naïve Bayes** and **Random Forest** classification methods • 95,577 records of children with 367 variables out of which 256 were found to be sufficient • Clearly delineates different attributes used • Created dataset having 4 classes (ASD: None, Mild, Moderate, Severe) • Highest accuracy of **87.1%** (2 class) and **54.1%** (4 class) achieved with J48 algorithm (decision tree)	• Does not predict the severity of ASD • Cursory set of attributes (conditions) used for identification of ASD which might not always necessarily translate to a case of ASD

(continued)

Table 1. (*continued*)

Paper	Key findings	Limitations
[18]	• Automated optimal feature selection using **Binary Firefly algorithm** (selected 10 out of 21 features as optimum) • No class imbalance problem (Among 292 instances in ASD children dataset, there are 151 instances with class 'yes' and 141 instances with class 'No') • Makes use of **NB, J48, SVM, KNN** models • Highest accuracy of **97.95%** achieved with **SVM**	• Missing instances in the ASD child dataset • Due to lesser number of instances in the dataset, there exists a chance of model overfitting on the dataset. • Certain disadvantages of swarm intelligence wrappers (Binary Firefly algorithm)
[21]	• Extracted **6 personal characteristics from 851 subjects in the ABIDE database** • Performed a cross-validation strategy which trains and tests ML models for classification • Classification performance was assessed by using parameters such as accuracy, AUC, • Sensitivity and specificity • 9 models were tested using 6 personal characteristics. The best performance was shown by the Neural Network Model with 0.646 as the mean AUC	• Data has been collected from 17 sites leading to heterogeneity. This might compromise the ML models • The small sex difference in ASD vs. controls observed is likely a function of the high incidence of ASD in males rather than a selection bias for this sub-study • Small size of dataset, i.e. 851 subjects

3 Working Model

Figure 1 demonstrates the general working and flow of our system. Various preprocessing techniques like noise removal, encoding and normalization can be applied, depending upon the dataset. As an optional step, feature engineering can also be performed, which involves choosing the most optimal features out of all the features present in the data set, to reduce data dimensionality in order to improve speed and efficiency. This can be either done manually or by using an automated algorithm, like the Binary Firefly Algorithm [19]. Once the data has been preprocessed, classification algorithms like Logistic Regression, Naïve Bayes, Support Vector Machine, K-Nearest Neighbors and Random Forest Classifiers are used to train the data set. The accuracy of each model is observed and compared. If the model performs well, then the training accuracy will be higher than the test accuracy. This model can then be deemed to be the best model and hence be used for further training and learning. Furthermore, several metrics like the F1 score and precision-recall values are also calculated for better evaluation of each algorithm. Finally, the result is obtained and deployed.

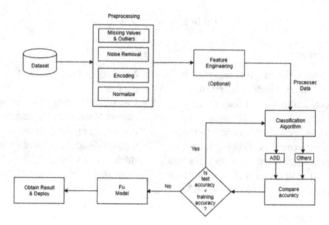

Fig. 1. Architecture of our proposed system

4 Methodology

4.1 Data Preprocessing

The dataset [3] that we have used has been developed by Dr. Fadi Thabtah [6] and it contains categorical, continuous and binary attributes. Originally, the dataset had 1054 instances along with 18 attributes (including class variable). Since the dataset contained a few non-contributing and categorical attributes, we had to preprocess the data. Preprocessing refers to the transformations applied to our data before feeding it to the algorithm. It is done to clean raw or noisy data and make it more suited for performing analysis. In order to deal with the categorical values, we are making use of label encoding. Label Encoding converts the labels into numeric form to make it machine-readable. Repeated labels are assigned the same value as assigned earlier. Five features having two classes (Ethnicity, Sex, Jaundice, Family_mem_with_ASD, and Class/ASD_Traits) have been selected to be binary label encoded.

4.2 Implementation

We split the dataset into two parts. The training set that consists of 80% of the data (843 samples) will be used to train the model. The remaining 20% of the data (211 samples) will be reserved for testing the accuracy and effectiveness of the model on data that the model has never seen before and will be referred as the testing data set.

After preprocessing the data, we tested five models, namely Logistic Regression, Naive Bayes, Support Vector Machine, K-Nearest Neighbors and Random Forest Classifier.

a. Logistic Regression Logistic Regression (LR)
Logistic Regression's primary aim is in finding the model with the best fit that describes the relationship between the binomial character of interest and a set of independent variables [12]. It makes use of a logistic function to find an optimal curve to fit the data points.

b. Naive Bayes (NB)
Based around conditional probability (Bayes theorem) and counting, the name "naive" comes from its assumption of conditional independence of all input features [11]. If this assumption is considered true, the rate at which a NB classifier will converge, will be much higher than a discriminative model like logistic regression. Therefore, the amount of training data required would be lesser. The main disadvantage of NB is that it only works well with limited number of features. Moreover, there is a high bias when there is a small amount of data.

c. Support Vector Machine (SVM)
Commonly used in classification problems, Support Vector Machine is based on the idea of finding the hyperplane that divides a given data set into two classes in the best possible way [17]. The distance from the hyperplane to the closest training data point is known as the margin. SVM aims to maximize the margin of the training data by finding the most optimal separating hyperplane [4].

d. K-Nearest Neighbors (KNN)
The KNN algorithm is based on mainly two ideas: the notion of a distance metric and that points that are close to one another are similar. Let x be the new data point that we wish to predict a label for. The KNN algorithm works by finding the k training data points closest to x using a Euclidean distance metric. KNN algorithm then performs majority voting to determine the label for the new data point x [8]. In our analysis, lower values of k (k = 1 to k = 10) gave us the highest accuracy.

e. Random Forest Classifier (RFC)
Random forest classifier is a flexible algorithm that can be used for classification, regression and other tasks as well [16]. It works by creating multiple decision trees on arbitrary data points. After getting the prediction from each tree, the best solution is selected by voting.

5 Results

1. **Dataset Analysis**
We plotted bar graphs in order to get a visual analysis of the dataset. In the first plot (Fig. 2), we can see that the number of toddlers who are ASD positive are those who do not have jaundice while birth. The count is almost 2–3 times more than that of jaundice born toddlers. Thus, we can infer that jaundice born children have a weak link with ASD. Also, from the bar plots obtained, we have observed that ASD is more prevalent in boys than in girls, by about 5 times.

For toddlers, most of the ASD positive cases are around 36 months of age. We can see that as the age increases, the number of ASD positive cases increase. From the graph it is evident that significant signs of autism occur at the age of 3 years (Fig. 3).

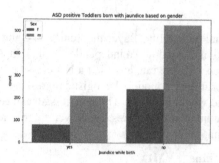

Fig. 2. ASD positive toddlers born with jaundice based on gender

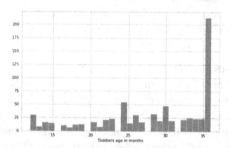

Fig. 3. Age distribution of ASD positive

2. Comparison of Models

The confusion matrix along with its parameters is shown below (Table 2).

Table 2. Confusion matrix for ASD prediction

Predicted	Individual has ASD	Individual does not have ASD
ASD is predicted	True positive	False positive
ASD is not predicted	False negative	True negative

Usually, in most predictive models, the data points lie in the following four categories:

i. True positive: The individual has ASD and we predicted correctly that the individual has ASD.
ii. True negative: The individual does not have ASD and we predicted correctly that the individual does not have ASD.
iii. False positive: The individual does not have ASD, but we predicted incorrectly that the individual has ASD. This is known as Type 1 error.
iv. False negative: The individual has ASD, but we predicted incorrectly that the individual does not have ASD. This is known as Type 2 error.

Shown below is a table (Table 3) comparing all the machine learning models we used-

Table 3. A comparison of the applied ML models

	LR	NB	SVM	KNN	RFC
Accuracy	97.15%	94.79%	93.84%	90.52%	81.52%
Confusion matrix	[57 5] [1 148]	[56 6] [5 144]	[52 10] [3 146]	[51 11] [9 140]	[45 17] [14 135]
F1 score	0.98	0.96	0.95	0.93	0.88

From the values obtained, we can thereby infer that Logistic Regression, giving the highest accuracy, is the best model for our current dataset. Logistic regression performs well when the training data size is small and it is binary in nature. The feature space is split linearly, and it works well even when only a few variables are correlated. However, Naïve Bayes assumes that all features are conditionally independent. Hence, if some of the features are interdependent, the prediction might be inaccurate.

In addition to accuracy, we have also found out the precision and recall values to provide a better insight. The F1 score has then been calculated by taking the weighted average of the precision and recall values. This score can vary between 0 and 1. The higher the F1 score, the better the model (a score of 1 is considered to be the best).

3. Precision and Recall Curves

Precision measures how accurate our positive predictions were i.e., out of all the points predicted to be positive how many of them were actually positive.

$$\text{Precision} = \text{True Positives}/(\text{True Positives} + \text{False Positives})$$

Recall measures what fraction of the positives our model identified, i.e., out of the points that are labelled positive, how many of them were correctly predicted as positive.

$$\text{Recall} = \text{True Positives}/(\text{True Positives} + \text{False Negatives})$$

Accuracy can be defined as the probability of the number of correct predictions made by the classifier. In other words, it is the fraction of correct predictions made out of the total number of predictions.

$$\text{Accuracy} = (\text{True Positives} + \text{True Negatives})/\text{Total}$$

Shown below are the precision and recall curves plotted for Logistic Regression (Fig. 4), Naïve Bayes (Fig. 5) and SVM (Fig. 6).

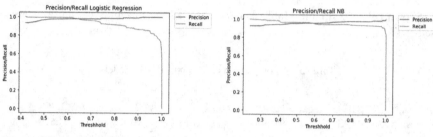

Fig. 4. Logistic Regression   Fig. 5. Naïve Bayes

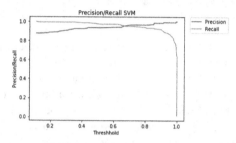

Fig. 6. Support Vector Machine

6 Conclusion

ASD behavioral assessment is a time-consuming phase that can be compounded by symptomatology overlaps. There is currently no diagnostic test that can quickly and accurately detect ASD, or an optimized and thorough screening tool that is explicitly developed to identify the risk between ASD and other similar disorders. We have designed an automated ASD prediction model with minimum behavior sets selected from the diagnosis datasets of each. Out of the five models that we applied to our dataset; Logistic Regression was observed to give the highest accuracy.

This research work has used a dataset which has less sample space. However, our research has provided useful insights in the development of an automated model that can assist the medical practitioners in detecting autism in children. In the future, we will be considering using a larger dataset to improve generalization. With this in consideration, our research has resulted in analyzing models that can accurately detect ASD in individuals with given attributes regarding the persons behavioral and medical information. These models can serve as benchmarks for any machine learning researcher/practitioner who is interested in exploring this dataset further or other data sets related to Autism Spectrum Disorder.

References

1. Sen, B., Borle, N.C., Greiner, R., Brown, M.R.: A general prediction model for the detection of ADHD and autism using structural and functional MRI. PloS one **13**, e0194856 (2018)

2. van den Bekerom, B.: Using machine learning for detection of autism spectrum disorder. In: 26th Twentieth Student Conference on IT, February 2017
3. Dataset: https://www.kaggle.com/fabdelja/autism-screening-for-toddlers
4. Support vector machines: the linearly separable case. https://nlp.stanford.edu/IR-book/html/htmledition/support-vector-machines-the-linearly-separable-case-1.html. Accessed 08 Oct 2019
5. Thabtah, F.: Machine Learning in Autistic Spectrum Disorder Behavioral Research: A Review and Ways Forward, pp. 1–20. Taylor & Francis (2017)
6. Ghiassian, S., Greiner, R., Jin, P., Brown, M.R.: Using functional or structural magnetic resonance images and personal characteristic data to identify ADHD and autism. PLoS one **11**, e0166934 (2016)
7. Deshpande, G., Libero, L.E., Sreenivasan, K.R., Deshpande, H.D., Kana, R.K.: Identification of neural connectivity signatures of autism using machine learning. Front. Hum. Neurosci. **7**, 670 (2013)
8. KNN Classification using Scikit-learn. https://www.datacamp.com/community/tutorials/k-nearest-neighbor-classification-scikit-learn. Accessed 08 Oct 2019
9. Kosmicki, J.A., Sochat, V., Duda, M., Wall, D.P.: Searching for a minimal set of behaviors for autism detection through feature selection-based machine learning. Transl Psychiatry **5**, e514 (2015)
10. Li, H., Parikh, N.A., He, L.: A novel transfer learning approach to enhance deep neural network classification of brain functional connectomes. Front. Neurosci. **12**, 491 (2018)
11. Naive Bayes for Machine Learning. https://machinelearningmastery.com/naive-bayes-for-machine-learning/. Accessed 08 Oct 2019
12. Logistic Regression. https://medium.com/datadriveninvestor/logistic-regression-18afd48779ce. Accessed 07 Oct 2019
13. Duda, M., Ma, R., Haber, N., Wall, D.P.: Use of machine learning for behavioral distinction of autism and ADHD. Transl. Psychiatry **6**, e732 (2016)
14. https://www.autism-society.org/what-is/facts-and-statistics/. Accessed 25 Dec 2019
15. Pratap, A., Kanimozhiselvi, C.: Soft computing models for the predictive grading of childhood autism—a comparative study. IJSCE **4**, 64–67 (2014)
16. Random Forests(r), Explained. https://www.kdnuggets.com/2017/10/random-forests-explained.html. Accessed 08 Oct 2019
17. Support Vector Machine—Introduction to Machine Learning Algorithms. https://towardsdatascience.com/support-vector-machine-introduction-to-machine-learning-algorithms-934a444fca47. Accessed 07 Oct 2019
18. Vaishali, R., Sasikala, R.: A machine learning based approach to classify autism with optimum behaviour sets. Int. J. Eng. Technol. **7**, 18 (2017)
19. http://www.downtoearth.org.in/news/health/amp/study-finds-genetic-variants-that-increase-adhd-risk-62281. Accessed 20 Dec 2019
20. https://www.helpguide.org/articles/autism-learning-disabilities/autismspectrumdisorders.htm. Accessed 20 Dec 2019
21. Parikh, M.N., Li, H., He, L.: Enhancing diagnosis of autism with optimized machine learning models and personal characteristic data. Front. Comput. Neurosci. **13**, 9 (2019)

Anatomical Analysis Between Two Languages Alphabets: Visually Typographic Test Transformation in Morphological Approaches

Mizanur Rahman[1(✉)], Md. Salah Uddin[1(✉)], Md. Samaun Hasan[1(✉)], Apurba Ghosh[1(✉)], Sadia Afrin Boby[2(✉)], Arif Ahmed[1(✉)], Shah Muhammad Sadiur Rahman[1(✉)], and Shaikh Muhammad Allayear[1(✉)]

[1] Department of Multimedia and Creative Technology, Daffodil International University, Dhaka 1207, Bangladesh
{mizan.mct,salah.mct,hasan.mct,apurba.mct,arif.mct, sadiur12-620}@diu.edu.bd, headmct@daffodilvarsity.edu.bd
[2] Department of Graphic Design, University of Dhaka, Dhaka, Bangladesh
afrin87.bob@mail.com

Abstract. We have many different types of languages and alphabets in the world, but the alphabet of each language has its own unique design. We tried to bridge between the two countries/languages by reflecting the differences in the alphabet in another language by highlighting its distinctive features. By using the two most commonly used languages like English & Arabic. Here the transformation process of flavors one into another language has been introduced by replication. The alphabet has the same characteristics in the alphabet of the two languages and reflected in other languages. In this study, we have completed the task of transforming the character flavor of English and Arabic as well as the Arabic language in English. The process is finding characteristic patterns and connecting to any other language alphabet to achieving new tastes.

Keywords: Language characteristic · Font anatomy · Typographic test exchange · Digital typography · Typeface design · Cognitive psychology · Typographic transformation · Visual perception of alphabet

1 Introduction

From prehistoric times, people became fascinated by the invention of the world, and have been continuously working on it. In the "Stone Age", they wrote on the rocks, and even in cave paintings, people painted inside the cave as a means of expressing their feelings. The language is called "Language is the dress of thought, being the case, Typography can be viewed as one of the swatches of fabric from which that dress is made" [1]. It is easily conceivable that the change in the way people use their daily clothing, it is not bizarre to rely on just one font design in digital typography. So if the new font style is as demanding as the time, then the pattern of the styles, if found in the mind is not sluggish.

© Springer Nature Singapore Pte Ltd. 2020
M. Singh et al. (Eds.): ICACDS 2020, CCIS 1244, pp. 330–339, 2020.
https://doi.org/10.1007/978-981-15-6634-9_30

In the current world, approximately 550000 + font styles have been created in the world [2, 9, 10, 21], in the same way; it is shown in many languages.

All people naturally welcome the change, in the midst of which a graphic designer is always on the lookout for change because his work is so needed that it can be done by uniqueness. We think a designer if following the formulated way to look out for new typographic creations, he will succeed to create typographic inventions. It is an unavoidable matter as a graphic designer when we face the subjective designing problem to solve the cultural and psychological approach. Language or typographic characteristics may be the best way to culturally and psychologically motivate the viewers. That way font style plays a powerful role in communication design. In this case, all graphic designers extremely concentrate on using typographic style. Beyond that, maximum designer creates lots of new typeface designs but he/she does not maintain the same theory.

Here is the main point of our study. In the world there are approximately 6906 languages, among them, 3500 languages are in written form [3, 17], each language has a different characteristic pattern. By finding this characteristic pattern and connecting it to any other language alphabet, it is possible to find the alphabetic taste of other languages in addition to the breakthrough in typography. In this study we explain how to work a typographic characteristic in other languages font. Although in this way it is possible to have characteristic transformation in all languages font. At this time we are trying to explain the transformation of two languages fonts' style.

2 Interpretation of Perceiving Ability to Recognize Cognitive Psychology

Although there are not plenty of ideas about the languages of the world, everyone has ideas about some languages or some writing methods for visual reasons; sometimes it is worth noting that some languages cannot read. But we can perceive only by seeing a few languages or alphabet which country or sub-race.

If we are more concerned with cognitive psychology, the human cerebrum follows the common need to dissect equivocal information and offer significance to visual upgrades, as arbitrary as they may show up. Where letterforms develop step by step and methodically from progressively combining segments, the human mind is animated and prodded into a procedure of speculating the visual data that is going to advance. Indeed, even the perusing procedure identified with static writings may include a procedure of speculating data. Subsequent to laying out a scope of hypotheses identified with the view of static writings [4]. Speculates that the saccadic eye development, which can prove related to the filtering of printed writings, might be a sign that includes perusing.

A British nervous system specialist looks at the view of equivocal pictures and claims the human mind can just process each conceivable translation in turn. This implies the human mind is compelled to process various potential understandings of vague symbolism in fast progression, which can be a dazzling encounter. Be that as it may, for what reason is it enrapturing? Whenever confronted with equivocal visual data, our brains are caught in our normal inclination of comprehending visual improvements [4].

2.1 Discussion as an Aspect of Visual Information

When designing communication materials, designers take into account not only the words used but also how the words are presented, the typeface of the words. What does it mean by what the word implies, as well as the color, the format, the visual expression of the typeface, etc.? All that matters, these factors depend on a lot of physical properties of typeface [18].

Advertisers are constantly looking for ways to increase the persuasive power of their advertisements. To that end advertisers are encouraged to match their advertising execution elements with their consumer motivations for processing brand information in advertisements. One less considered approach is to select the effective ingredients that will increase the motivation, scope and capacity of the customers.

The findings indicate that not only typography influences the consumer's ability to process information based on ad-based brand, but the effects of different typographic features are also highly interactive [5, 11–16].

3 Findings from the Relevant Research

The above discussion shows that the demand for typography is immense when it comes to the new feature. The effects of font manner are discussed and taken very seriously at the more relevant research. As a result of the design postulant changes in human perceptions, there are sufficient explanations in other research. One thing that is completely absent in their research is if the alphabet is manifest in the character of any other languages by the characteristics of the basic alphabet style or behavior of the different languages, the character of that language, it will increase the attractiveness to the other languages font design which is absolutely absent on this topic.

4 The Methodology of Our Proposed Font Designing Approach

The first thing is to choose two languages, then let's focus on the main task. We were able to complete the task based on the results obtained by analyzing the alphabets of the two languages in an orderly manner. The results obtained here are the structural analysis of the alphabet, the physical anatomy, visual look and characteristic of the alphabet. This study focuses on the subject of the two-language alphabet, which has been used as the main rule of the new font design. We have tried some experiments by some alphabet - without full fonts designing. By adapting this method, we think it is possible to transform the alphabet of one language into another (Fig. 1).

Step 1: Choose two languages
Step 2: Analyzing the two languages alphabets (physical anatomy, visual look and characteristic of the alphabet)
Step 3: Obtained results (common characteristic, Pattern, stroke gesture, Shape/Form, Anatomical Properties)
Step 4: Appling another language font

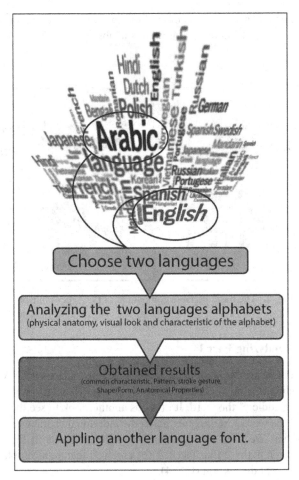

Fig. 1. Proposed methodology

4.1 Pre-talk About Two Languages

Arabic is composed from option to left – in opposition to English.

One segment we ought to become accustomed to in the Arabic content is that short vowels, for example, an, I or u (as antagonistic to the long vowels aa, uu and ii), are not yet demonstrated in the content. Arabic content by arrangement follows a joined on bent example. There is no comparability of the English literary substance we are presently perusing. Right now the letters have separate structures with spaces between them. There are no capital letters. In cursive content composing letters are consolidated by methods for a limit of turning into an individual from strokes (called ligatures). Thus, Arabic letters have somewhat selective structures, contingent upon whether they come toward the start, focus or surrender a word. A couple of letters don't join the accompanying letter, yet all Arabic letters are a piece of the past one. The cursive calligraphy style begins in boundless terms from the rearrangements of the signal of composing because of the way that-the tendency is an outcome of the speed and development of the motion. The calligraphic style is without trouble unmistakable in the Western world [6–8, 19, 20] (Fig. 2).

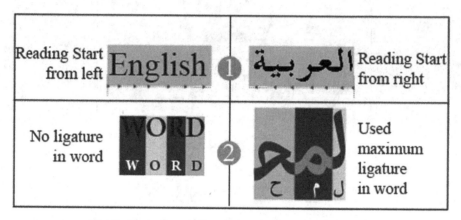

Fig. 2. The basic variable component of two types alphabet

4.2 Character Analyzing Part 1

There are four separate appearances in all Arabic characters in existence. When a letter is used at the beginning of a word, its appearance is changed from its previous appearance when used in the middle of the word. It gives us another look to see the same character used at the end of the word. On the other hand, the individual character is a very specific look from the other. We tried to visualize four different looks in Fig. 3. All Arabic alphabets can be seen in these four faces. The perceived feature available here that will be reflected in our next discussion (Fig. 4).

4.3 Character Analyzing Part 2

On the other hand in English alphabets are two separate views, one is capital letter another is small. In the case of script font ever we can see different looks which are seen in the Arabic cursive calligraphic font. Moreover, in English languages massive use of Capital & Small is mixed. In Fig. 5 we tried to visualize possible uses of English alphabets. It is pity difficult to connect two alphabets because there is much difference between one with another. Still, we have searched how to create consequences alphabetical tastes by morphological approach (Fig. 6).

4.4 Common Feature and Visual Perception

After analyzing the velocity & movement of the two characters, we have achieved some common features & perceptions, which are separately demonstrated in Fig. 7. The heir is the most relevant feature is half-circle, Circle, triangle, rectangle shapes used in two

Name	Initial	Medial	Final	Separate	Pronunciation
alif*	ا	ﻟ	ﻟ	ا	see opposite
baa'	ﺑ	ﺒ	ﺐ	ﺏ	b
taa'	ﺗ	ﺘ	ﺖ	ﺕ	t
thaa'	ﺛ	ﺜ	ﺚ	ﺙ	th
jiim	ﺟ	ﺠ	ﺞ	ﺝ	j
Haa'	ﺣ	ﺤ	ﺢ	ﺡ	H
khaa'	ﺧ	ﺨ	ﺦ	ﺥ	kh
daal*	ﺩ	ﺪ	ﺪ	ﺩ	d
dhaal*	ﺫ	ﺬ	ﺬ	ﺫ	dh
raa'*	ﺭ	ﺮ	ﺮ	ﺭ	r
zaay*	ﺯ	ﺰ	ﺰ	ﺯ	z
siin	ﺳ	ﺴ	ﺲ	ﺱ	s
shiin	ﺷ	ﺸ	ﺶ	ﺵ	sh
Saad	ﺻ	ﺼ	ﺺ	ﺹ	S
Daad	ﺿ	ﻀ	ﺾ	ﺽ	D
Taa'	ﻃ	ﻄ	ﻂ	ﻁ	T
DHaa'	ﻇ	ﻈ	ﻆ	ﻅ	DH
:ain	ﻋ	ﻌ	ﻊ	ﻉ	:
ghain	ﻏ	ﻐ	ﻎ	ﻍ	gh
faa'	ﻓ	ﻔ	ﻒ	ﻑ	f
qaaf	ﻗ	ﻘ	ﻖ	ﻕ	g
kaaf	ﻛ	ﻜ	ﻚ	ﻙ	k
laam	ﻟ	ﻠ	ﻞ	ﻝ	l
miim	ﻣ	ﻤ	ﻢ	ﻡ	m
nuun	ﻧ	ﻨ	ﻦ	ﻥ	n
haa'	ﻫ	ﻬ	ﻪ	ﻩ	h
waaw	ﻭ	ﻮ	ﻮ	ﻭ	w
yaa'	ﻳ	ﻴ	ﻲ	ﻱ	y
on alif	ﺃ	ﺄ	ﺄ	ﺃ	

Fig. 3. Four separated looks in the Arabic alphabets

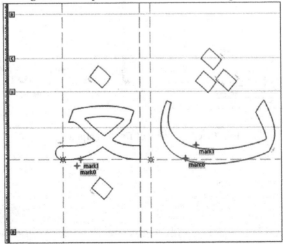

Fig. 4. Glyph view of Arabic alphabet.

Capital	Small	Capital	Small	Capital	Small	Capital	Small
A	a	H	h	O	o	V	v
B	b	I	i	P	p	W	w
C	c	J	j	Q	q	X	x
D	d	K	k	R	r	Y	y
E	e	L	l	S	s	Z	z
F	f	M	m	T	t		
G	g	N	n	U	u		

Fig. 5. Two separated looks in the English alphabets

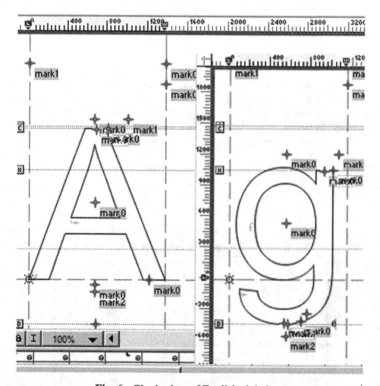

Fig. 6. Glyph view of English alphabet.

languages, Visual perception is the most universal in both alphabets. If we accumulate universal features from both characters and create new character design- then naturally the visual perception will be perceived as a targeted view of language test.

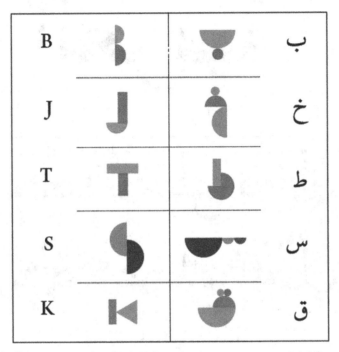

Fig. 7. Velocity and movement of the gesture in two types of alphabet.

5   Experimental Analysis by Two Languages

In this phase, we are trying to create visuals from the English word "Bangladesh" visually seeing Arabic letterform test. Since we want the English to be an Arabic test, then the Arabic font properties will need to be glued down. At this stage, the following Fig. 8 has shown some more mentioned features of the Arabic fonts. It is possible to change those features by reflection in the English language. It is noticeable in the picture is the main stem stroke and another anatomy such as a tooth, Tail, Stem, Occlusions, Dot with all are diacritics and Pseudo-word strictly maintained in this matter. Experimental result represents of English writing with Arabic alphabetical test. Our Subconscious mind at first shows Bangladesh word as an Arabic word, if we saw with little concentration to this word then visualize the writing is really English.

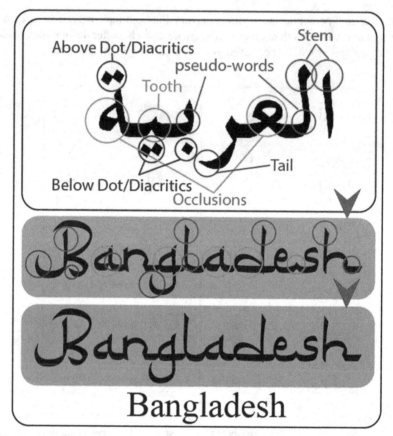

Fig. 8. Experimental representation of English writing with Arabic alphabetical Test.

6   Conclusion and Future Work

Finally, in this research we have proved how to make a cultural exchange by introducing visual collage between two languages. In this study, we created only two language test exchanges. It is possible to transform by the morphological approach in every language and our intention in this research is achieving qualitative font design in the alphabet of every language and reflection of taste variety. Finally, we have succeeded in this phenomenon. In the next journey, our uninterrupted attempt will be continued to the invention of a new methodology for achieving the goal. In the next steps, we will be trying to include language development theory which will create a new dimension for this research. Moreover, we have a plan to work with two different languages in the required approaches.

References

1. Johnson, S.: Abraham Cowley In The Lives of the English Poets. Everyman's Library #770, vol. 1, pp. 39–40. J. M. Dent & Sons, Ltd., London (1779–1781)

2. Extensis: How Many Fonts are There in the World? 10 June 2019. https://www.extensis.com/blog/how-many-fonts-are-there-in-the-world

3. Anderson, S.R.: How many languages are there in the world? (2010). https://www.linguisticsociety.org/content/how-many-languages-are-there-world

4. Hillner, M.: Time consciousness in relation to emergent typography. In: Fukuda, S. (ed.) AHFE 2019. AISC, vol. 952, pp. 71–79. Springer, Cham (2020). https://doi.org/10.1007/978-3-030-20441-9_8

5. McCarthy, M.S., Mothersbaugh, D.L.: Effects of typographic factors in advertising based persuasion: a general model and initial empirical tests. Psychol. Mark. **19**, 663–691 (2002)

6. The Guardian: Arabic alphabet table, 7 February 2010. https://www.theguardian.com/travel/2010/feb/07/learn-arabic-alphabet-table

7. Jamil, M.: The Beginner's Guide to Arabic (2010)

8. Zubair, K.M.A.: The rise and decline of Arabu Tamil language for tamil muslims. IIUC STUDIES **10**, 263–282 (2014). ISSN 1813–7733

9. Ross, F., Shaw, G.: Non-Latin Scripts: From Metal to Digital Type. St. Bride Foundation, London (2012)

10. Dogusoy, B., Cicek, F., Cagiltay, K.: How serif and sans serif typefaces influence reading on screen: an eye tracking study. In: Marcus, A. (ed.) DUXU 2016. LNCS, vol. 9747, pp. 578–586. Springer, Cham (2016). https://doi.org/10.1007/978-3-319-40355-7_55

11. Abidin, H.A.Z., Maaruf, S.Z.: Interactive visual art education pedagogical module: typography in visual communication. In: Luaran, J.E., Sardi, J., Aziz, A., Alias, N.A. (eds.) Envisioning the Future of Online Learning, pp. 171–182. Springer, Singapore (2016). https://doi.org/10.1007/978-981-10-0954-9_15

12. Meyrick, T., Taffe, S.: Authenticating typography in cultural festival brand marks. In: International Association of Societies of Design Research Conference 2019 (2019)

13. Liu, Y.-F.: I or We: the persuasive effects of typeface shapes: an abstract. In: Rossi, P., Krey, N. (eds.) AMSWMC 2018. DMSPAMS, pp. 193–194. Springer, Cham (2019). https://doi.org/10.1007/978-3-030-02568-7_53

14. Liu, Y.-F., Ling, I.-L., Jou, J.Y.H.: Cross-language comparison of the persuasive effects of typeface shapes: a conceptual framework. In: Petruzzellis, L., Winer, Russell S.S. (eds.) Rediscovering the Essentiality of Marketing. DMSPAMS, pp. 9–10. Springer, Cham (2016). https://doi.org/10.1007/978-3-319-29877-1_2

15. Brumberger, E.R.: The rhetoric of typography: the persona of typeface and text. Techn. Commun. **50**, 206–223 (2003)

16. Bi, R.: Effects of typographic variables on attitude measures in reading bilingual brands, Iowa State University (2014)

17. Infoplease: Most Widely Spoken Languages in the World, 28 February 2017. https://www.infoplease.com/arts-entertainment/writing-and-language/most-widely-spoken-languages-world

18. Ambrose, G., Salter, B.: The dichotomic tension of experimental typography. Paper presented at Typoday India, Mumbai, India (2019, in press)

19. https://i.pinimg.com/originals/6e/17/04/6e17042d99c018e5960206b69debf796.jpg. Accessed 12 Dec 2019

20. Abulhab, S.D.: Anatomy of an Arabetic type design. Visible Lang. **42**(2), 181–193 (2008)

21. Brandão, J.A., Almeida, C.M.: The italic style: understanding the shape through history. In: Rebelo, F., Soares, Marcelo M. (eds.) AHFE 2019. AISC, vol. 955, pp. 597–608. Springer, Cham (2020). https://doi.org/10.1007/978-3-030-20227-9_56

Auto Segmentation of Lung in Non-small Cell Lung Cancer Using Deep Convolution Neural Network

Ravindra Patil[1,2](✉), Leonard Wee[1], and Andre Dekker[1]

[1] Department of Radiation Oncology (MAASTRO), GROW – School for Oncology and Developmental Biology, Maastricht University Medical Centre (MUMC), Maastricht, The Netherlands
[2] Philips Research Bangalore, Bengaluru, India
patil.ravindra@philips.com

Abstract. Segmentation of Lung is the vital first step in radiologic diagnosis of lung cancer. In this work, we present a deep learning based automated technique that overcomes various shortcomings of traditional lung segmentation and explores the role of adding "explainability" to deep learning models so that the trust can be built on these models. Our approach shows better generalization across different scanner settings, vendors and the slice thickness. In addition, there is no initialization of the seed point making it complete automated without manual intervention. The dice score of 0.98 is achieved for lung segmentation on an independent data set of non-small cell lung cancer.

Keywords: Lung segmentation · NSCLC · Deep learning

1 Introduction

Lung cancer is the leading cause for cancer related deaths and is accompanied by a dismal prognosis with a 5-year survival rate at only 18% [1]. Out of all Lung cancer, Non-Small Cell Lung Cancer (NSCLC) accounts for 85% of the cases. Treatment monitoring and analysis [2] using computed tomography (CT) images is an important strategy for early lung cancer diagnosis and survival time improvement. In these approaches, accurate Lung anatomy and pathology region segmentation is necessary as it directly related to the treatment plan. After decades of development in imaging techniques, volumes of high-resolution images with low distortions are now more easily available. Despite development of approaches for lung segmentation in recent years [3–6], achieving accurate segmentation performance continues to require attention because of specific challenges. One such example is tumors have an intensity similar to that of lung wall; thus, they are difficult to distinguish using intensity values alone and also the structure of the lung changes based on the disease pathology such as consolidation, masses, pneumothorax or effusions.

There is considerable progress in developing Lung segmentation algorithms that have an ability to perform accurate delineation under different disease conditions [7].

© Springer Nature Singapore Pte Ltd. 2020
M. Singh et al. (Eds.): ICACDS 2020, CCIS 1244, pp. 340–351, 2020.
https://doi.org/10.1007/978-981-15-6634-9_31

In general, all the lung segmentation algorithms can be classified into following five sub categories (1) Intensity based (2) Shape based or Model based (3) Neighboring anatomy guided (4) Region based (5) Artificial Intelligence based [7]. The intensity based approaches are fast, intuitive and computationally efficient however these techniques fail during the pathological condition where there are attenuation variation. The shape based or the model based approaches provide a very good accuracy due to template mapping, however these algorithms are computationally inefficient and it's difficult to create representative training features. Neighboring anatomy based approach exploits the information of the spatial context of the neighboring organ of the lung such as rib cage, heart, spine for extracting the contours of the lung region, this approach is computationally expensive but provide good results when the intensity variation is mild to moderate. However, in case of the extreme diseased condition such as opacification of entire hemithorax this approach fails. Region growing approaches such as watershed transform, graph cuts and random walks are efficient but they tend to over segment. In recent times AI based (Machine Leaning and Deep Learning) approaches have become popular and in particular Deep Learning (DL) approaches due to the better accuracy that these algorithms achieve in ill defined pathologic conditions [8].

Recently, Convolutional Neural Networks (CNN) have been seen as a powerful tool for learning features from network layers [9]. CNN's act as a tool for learning discriminative features, which are useful in different image processing and computer vision tasks. CNN's need relatively less pre-processing compared to other known algorithms, which means the network learns the filters that in traditional algorithms had to be hand engineered [9]. The independence from prior knowledge and effort required in feature selection design is a major advantage. Moreover, CNN's include multi-layer processing, which ensures that the model learns the features at the granular level. In recent time, there has been progressive usage of CNN's for various medical segmentation tasks. Dou et al. [10] uses a 3D CNN, which focuses on the task of automatic nodule detection. DIAG Convnet by Setio et al. [11] provides automatic pulmonary nodule detection in CT images. The review article by Zhou et al. [12] lists various deep learning based medical segmentation approaches developed for different modalities and anatomies. Although, CNN's are explored for medical image segmentation, we did not come across a study with large-scale validation of the Lung segmentation approach with deep learning in diseased conditions.

The objectives of the current work are manifold. The first being to test the efficacy of the deep learning models on a large scale diseased Lung dataset to automatically segment the Lung region. Second, to compare DL approach with the traditional segmentation methods and comment on the complexity and efficiency. Third being, adding "explainability" to deep learning models so that the trust can be built on these models.

2 Methodology

In the traditional approach of Lung segmentation, the canonical steps are employed as shown in Fig. 1. However, there might be slight variation in each of the block based on the algorithm that is considered for implementation. Each of these blocks needs tuning based on the image type and the acquisition parameters of the scanner. Marker generation is used to define the region that is present inside and outside the Lung region; this is performed manually by marking the region using a seed point. Further, the image is preprocessed by applying the filters as well as HU thresholding to eliminate unwanted region. This is followed by running a segmenter algorithm (such as watershed, active contour etc.) and performing the morphological and post processing operations to correct the contours. These traditional approaches needs manual and empirical tuning and quite difficult to generalize on the large and varied dataset with different acquisitions and threshold values.

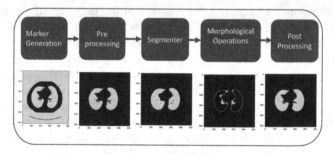

Fig. 1. Traditional lung segmentation approach

The approach adapted in this work is described in the section below, which is represented by the block diagram schematic as shown in Fig. 2. There are two phases in creation of the deep learning model, which aids in segmentation of the Lung anatomy. First being the training phase, where in the data is fed after preprocessing to the model, where the model is trained with the annotated ground truth (region of the Lung) as the reference. Further, the learnt model layers are analyzed using visualization to ascertain, what region of the image, model looked into to arrive at the delineation of the Lung region. In the scoring phase, the trained model is used to delineate the region of the lung on the unseen/live Lung CT scans.

2.1 Data

The data set used for model training which was obtained from The Cancer Imaging Archive (TCIA) repository of NSCLC patients [13]. This dataset contained pretreatment CT scans where in the lung regions were manually delineated by the radiation oncologist on the 3D volume. This data set will be referred as Lung 1. In total 422 subjects were used for training, which maps to ~42,019 Lung CT slices on which the algorithm was trained. Out of 422 subjects, randomly 300 subjects were used for model training and remaining 122 for model validation. The training data had following sub categories of NSCLC: Adenocarcinoma, Large cell carcinoma and Squamous cell carcinoma. The CT scanners used for imaging the subjects were from different vendors (Siemens, CNS Inc, Philips and GE) and the slice thickness varied from 0.65 to 5 mm with 512 * 512 resolution. Further, for testing the model, independent data set of Lung 2 was used which is marked as NSCLC Radiogenomics dataset in TCIA repository with 211 subject's [14].

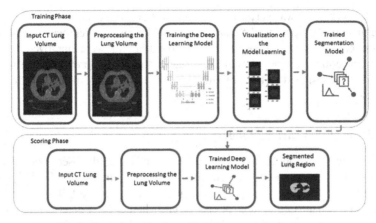

Fig. 2. Deep Learning based lung segmentation approach

The testing data had following different sub categories of NSCLC: Adenocarcinoma, Squamous cell carcinoma and NOS. The demographics of each of the subgroups of training and testing are mentioned in the Fig. 3.

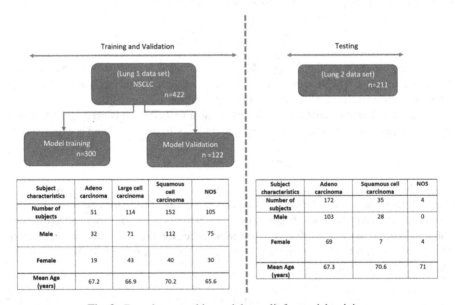

Training and Validation | Testing

(Lung 1 data set) NSCLC n=422

(Lung 2 data set) n=211

Model training n=300 | Model Validation n =122

Subject characteristics	Adeno carcinoma	Large cell carcinoma	Squamous cell carcinoma	NOS
Number of subjects	51	114	152	105
Male	32	71	112	75
Female	19	43	40	30
Mean Age (years)	67.2	66.9	70.2	65.6

Subject characteristics	Adeno carcinoma	Squamous cell carcinoma	NOS
Number of subjects	172	35	4
Male	103	28	0
Female	69	7	4
Mean Age (years)	67.3	70.6	71

Fig. 3. Data demographics and data split for model training

2.2 Pre Processing

The input CT volumes are processed to set non-anatomical regions such as air (with HU value below −1000) to 0, so that number of the pixel computations are reduced.

The data augmentation was performed in terms of translation and rotation (0 to 30°) to make the model robust against the data variations. No specific preprocessing steps such as de-noising, artifact corrections, system-based calibration were employed, which are typical in conventional approaches of Lung segmentation. The preprocessing approach was specifically designed to have minimum steps, to test the efficacy of convolution neural networks.

2.3 Training Deep Learning Model

The target of the current approach is, given a CT slice of a lung (diseased or non-diseased) the region of whole Lung need to be segmented. We employ modified U-Net [15] inspired CNN architecture to arrive at segmented region of the lung. We follow a pixel based classification mechanism that aims at classifying whether each pixel belongs to Lung region. The model architecture that is built was inspired from U-Net Convolution Neural Network (CNN), which consists of 18 convolutional layers, 4 central pooling layers with the convolutional kernel size of 3X3 in each convolutional layer. The schematic representation of the model architecture is as shown in the Fig. 4.

The convolutional layers perform convolutional operation on all input feature maps to obtain output features defined by the Rectified Linear Units (ReLU) activation function [16]. The feature map layer combination is defined by the Eq. (1)

$$f^j = ReLU\left(\sum_{i=1}^{m} C^{ij} * f^i + b^j\right) \tag{1}$$

Where f^i and f^j are the ith input feature map and j^{th} output feature map, respectively. We define C^{ij} as the convolutional kernel between f^i and f^j (* denotes the 2-D convolutional operation), b^j is the bias of the jth output feature map.

After each convolutional layer, a rectified linear unit (ReLU) is used as a non-linear activation function, this is added to bring non linearity to the model and is expressed as:

$$ReLU(z) = \max(0, z) \tag{2}$$

Further to the last convolutional layer, a fully connected layer is applied where each output unit connects to all inputs. This layer can capture correlations between different features produced by the convolutional layer. For achieving non-linearity and a two-class output classifier, the sigmoid function was used. Since the sigmoid function ranges from zero to one, it can be directly related to class probabilities making it ideal activation function for classification task.

$$sigmoid(z) = \frac{1}{1 + e^{-z}} \tag{3}$$

The goal of network training is to maximize the probability of the correct class. This is achieved by minimizing the dice coefficient loss function. The loss function is minimized during the model's training process. The weight updation was performed using Adaptive Moment Estimation (ADAM) algorithm [17]. Instead of adapting the parameter learning rates based on the average first moment as in RMSProp, ADAM makes use of the average of second moments of the gradients. Specifically, the algorithm calculates an exponential moving average of the gradient and the squared gradient, and the parameters β_1, β_2 control the decay rates of these moving averages. The initial value of the moving averages and β_1, β_2 values close to 1.0 (recommended) results in a bias of moment estimates towards zero. This bias is overcome by calculating the biased estimates and then calculating bias-corrected estimates. ADAM is an extension to stochastic gradient descent and converges faster than other stochastic optimization methods [18]. It also rectifies problem such as vanishing learning rate, slow convergence that other optimization problems face which leads to fluctuating loss function. The weights are updated based on the below equation

$$w^{(t+1)} \leftarrow w^{(t)} - \eta \frac{\widehat{m_w}}{\sqrt{\widehat{v_w}} + \varepsilon} \tag{4}$$

Where ε is a small number used to prevent division by zero.
And

$$\widehat{m_w} = \frac{m_w^{(t+1)}}{1 - \beta_1^t} \tag{5}$$

$$\widehat{v_w} = \frac{v_w^{(t+1)}}{1 - \beta_2^t} \tag{6}$$

Where m_w and v_w are estimates of the first moment and second moment of the gradients respectively. β_1 and β_2 are the forgetting factors for gradients and second moments of gradients, respectively.

The hyperparameters used for the model training are Optimizer = ADAM, Learning Rate = 1.0e−6, Metric = Dice score, Number of Epochs = 50, Batch Size = 2, Weight Initialization Method = Xavier initialization. The dice similarity coefficient (DSC) is used as the primary evaluation criteria for assessing the automatic segmentation accuracy; also, this is used as a loss function for the backward propagation in the proposed model. The DSC expressed as in Eq. (7) provides amount of overlap between two segmentation results [19], wherein G_t is the groundtruth segmentation and *Auto* is the automated segmentation performed by the trained model.

$$DSC = \frac{2 * V(G_t \cap Auto)}{V(G_t) + V(Auto)} \tag{7}$$

In the implementation, RT Structure of delineated lung region on each slice by the radiologist is considered as the ground truth and the region predicted by the model is overlaid on the RT Structure mask to arrive at the DSC.

2.4 Model Visualization

Interpreting the deep learning model and understanding model's rationale behind the decision process to arrive at the prediction is challenging. In order to ensure that the model is indeed looking into the relevant regions in the input image to arrive at the decision, a visualization engine was added on top of the built model. The approach used to build the model visualization was Gradient-weighted Class Activation Mapping (Grad-CAM) [20]. The Grad-CAM works on the class discriminative localization approach wherein it uses the gradients of target of the final convolutional layer to produce a coarse localization map, highlighting the regions considered in the image for predicting the concept. This is not only useful to know about the regions responsible for prediction but also aids in debugging the decision process in the networks. In essence, Grad-CAM takes into account the penultimate layer (layer before softmax) to interpret the decision of the CNN and identifies the respective filter activation for every spatial location (i, j) in the given image, further this is converted into the heat map based on the weights indicating the prominent regions. The mathematical aspect is depicted in the Eq. (8, 9).

$$W_k^c = \frac{1}{z} \sum_i \sum_j \frac{\partial Y^c}{\partial A_{ij}^k} \tag{8}$$

$$S^c = \frac{1}{z} \sum_i \sum_j \sum_k w_k^c A_{ij}^k \tag{9}$$

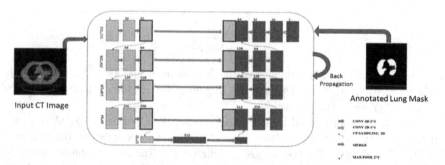

Fig. 4. Model architecture

The spatial score of a specific class S^c is the global average pooling over spatial location (i, j) for the gradient of respective class output Y^c with respect to the feature map A_{ij}^k. The spatial score is obtained by multiply the resulting value with the feature map along with its channel axis k. The $\sum$ describes the pooling and average operation and z is constant. The output of the sample GradCAM results can be seen in Fig. 5. It can be observed from the figure that location of the lungs are highlighted with red colour indicating the maximum activation of the filters in that region, mapping to understanding that the model has learnt the region of image that needs to be segmented.

3   Results and Discussion

The training data set containing 300 subjects of NSCLC, were fed to the U-Net inspired model for training and 122 subjects were used as the validation set. The training accuracy of 0.99 dice score was obtained after 50 epochs of training. The quantitative analysis resulted in average dice score of 0.98 on the independent test data of Lung 2 dataset,

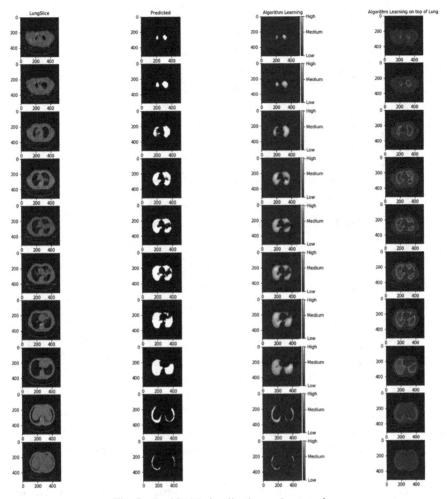

Fig. 5. GradCAM visualization on Lung region.

comprising of 211 subjects. The Bland-Altman plot showing the variation of difference between the ground truth dice score and predicted dice score vs mean of ground truth and predicted dice score of the test data is plotted in the Fig. 6. It can be observed that the differences are within mean ± 1.96 SD indicating that the segmentation performed by the model resembles the ground truth with high accuracy. Also, the box plot in the Fig. 7 on the test data shows the similar results indicating the average predicted dice score in the range of 1 to 0.98 without outliers being seen, thus suggesting the robustness of the model. The model was built using Keras (2.1.1) and Tensor flow (1.2.1) as backend in python 3.0. The machine configuration used for training the model had Nvidia GPU (Titan X 1080Ti) with the mentioned model hyper parameters. The scoring time on the same GPU configuration took one millisecond per lung slice.

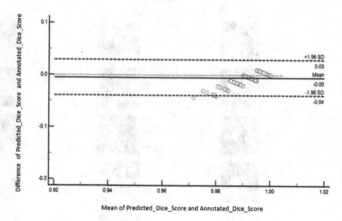

Fig. 6. Bland-Altman plot of dice score of lung anatomy

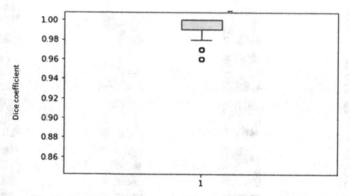

Fig. 7. Box Plot of dice score of lung anatomy on the test set

The sample results of segmented output can be observed in the Fig. 8. The first column is input CT slice, second column depicts annotation of the lung marked by the radiation oncologist, third column in the figure maps to the segmented output by our model, fourth being lung contour extracted from the predicted segmentation and fifth column is the over laid contour on the original Lung slice. It can be observed that the lung segmentation model is robust to the different anatomical structure of the Lung ranging from symmetric lung (rows d–f) to partial lung region being visible (row i–j), in the bottom two rows of the figure.

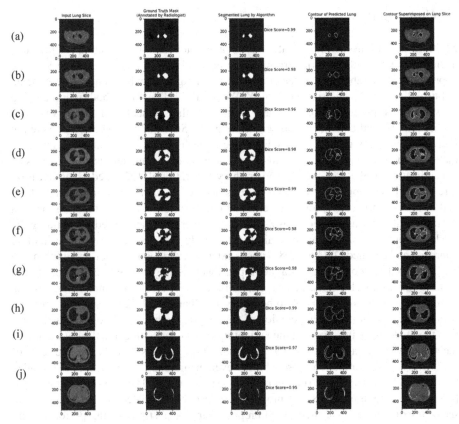

Fig. 8. Segmented Lung region and the associated ground truth masks

Table 1. Comparison with other lung segmentation approaches.

Title	Approach	Number of dataset(n)	Average dice score	Test dataset
Y. Wei G et al. [21]	Bresenham algorithm	97 subjects	0.95	Externally validated on 25 Subjects
Dai et al. [22]	Gaussian Mixture based model	Not mentioned	0.98	No external validation
Noor et al. [5]	Thresholding & morphological	96 subjects	0.98	No external validation
Zhang et al. [23]	Active contour	60 subjects	0.97	Externally validated on 60 subjects
Our approach	Deep Learning	422 Subjects	0.98	Externally validated on 211 subjects

4 Conclusion

Multiple algorithms in the literature have attempted the segmentation of the lung anatomy, the algorithms varies from simple thresholding to that of machine learning approaches. The Table 1 provides comparison of lung segmentation approaches being published recently. The prior work by Y. Wei et al. [21] performs both preprocessing as well as post processing of the scans to arrive at the segmentation and its validated on limited external data of 25 subjects leading to DSC of 0.95. However, as there are preprocessing steps used in this study, where in each of these steps needs to optimize based on the scan type, slice thickness and acquisition parameters leading to loss of generalization in the approach. Another study by Dai et al. [22] used Gaussian mixture model based approach; wherein there is need of manual seed point, initialization and custom preprocessing for noise elimination. The study does not mention the data characteristics, diseased conditions as well as number of subjects used for building the model, leading to questioning the claims of the study. Noor et al. [5] used thresholding approach and claimed DSC of 0.98 without external validation data set. In addition, the study focused only on a single scanner type as well as single slice thickness of 10 mm making the approach difficult to be generalized across all the scanner types, disease conditions and the variability. Another study by Zhang et al. [23] showed robustness of active contour model for delineation of the lung region with pathological conditions, also the model took into account the variability of scanner setting and the data. Further, the model was validated on the external data of 60 subjects resulting in the dice score of 0.97, with a robust validation approach. However, in this study there is a need to initialize the threshold value and perform preprocessing steps such as Gaussian smoothing. Compared to existing approaches, our model is exhaustively validated on the diseased data set with external data and there is no preprocessing or post processing performed, no domain-based adaptation is performed to arrive at the delineation. Our approach shows better generalization across different scanner settings, vendors and the slice thickness. In addition, there is no initialization of the seed point making it complete automated without manual intervention.

References

1. Siegel, R., Miller, K., Jemal, A.: Cancer statistics, 2016. CA-Cancer J. Clin. **66**, 7–30 (2016)
2. Aerts, H., et al.: Decoding tumour phenotype by noninvasive imaging using a quantitative radiomics approach. Nat. Commun. **5**, 4006 (2014). https://doi.org/10.1038/ncomms5006
3. Wei, Y., Shen, G., Li, J.J.: A fully automatic method for lung parenchyma segmentation and repairing. J. Digit. Imaging **26**(3), 483–495 (2013)
4. Dai, S., Lu, K., Dong, J., Zhang, Y., Chen, Y.: A novel approach of lung segmentation on chest CT images using graph cuts. Neurocomputing **168**, 799–807 (2015)
5. Noor, N.M., et al.: Automatic lung segmentation using control feedback system: morphology and texture paradigm. J. Med. Syst. **39**(3), 22 (2015)
6. Pulagam, A.R., Kande, G.B., Ede, V.K.R., Inampudi, R.B.: Automated lung segmentation from HRCT scans with diffuse parenchymal lung diseases. J. Digit. Imaging **29**(4), 507–519 (2016). https://doi.org/10.1007/s10278-016-9875-z

7. Awais, M. et al.: Segmentation and image analysis of abnormal lungs at CT: current approaches, challenges, and future trends. Radiographics (2015). https://doi.org/10.1148/rg. 2015140232

8. Chae, S.H., Moon, H.M., Chung, Y., Shin, J., Pan, S.B.: Automatic lung segmentation for large-scale medical image management. Multimed. Tools Appl. **75**(23), 15347–15363 (2016)

9. Yamashita, R., Nishio, M., Do, R.K.G., et al.: Convolutional neural networks: an overview and application in radiology. Insights Imaging **9**, 611–629 (2018). https://doi.org/10.1007/s13244-018-0639-

10. Dou, Q., Chen, H., Yu, L., Qin, J., Heng, P.A.: Multi-level contextual 3D CNNs for false positive reduction in pulmonary nodule detection (2016). IEEE Trans. Biomed. Eng. https://doi.org/10.1109/tbme.2016.2613502

11. Setio, A.A.A.: Pulmonary nodule detection in CT images: false positive reduction using multi-view convolutional networks. Med. Imaging IEEE Trans. https://doi.org/10.1109/TMI.2016.2536809

12. Zhou, T., et. al.: A review: deep learning for medical image segmentation using multi-modality fusion. Array **3**, 10004 (2019). https://doi.org/10.1016/j.array.2019.100004

13. Aerts, H.J., et al.: Data from NSCLC-radiomics [data set]. Cancer Imaging Arch (2019). https://doi.org/10.7937/K9/TCIA.2015.PF0M9REI

14. Shaimaa, B., et al.: Data for NSCLC radiogenomics collection. Cancer Imaging Arch. (2017). http://doi.org/10.7937/K9/TCIA.2017.7hs46erv

15. Ronneberger, O., Fischer, P., Brox, T.: U-Net: convolutional networks for biomedical image segmentation. In: Navab, N., Hornegger, J., Wells, William M., Frangi, Alejandro F. (eds.) MICCAI 2015. LNCS, vol. 9351, pp. 234–241. Springer, Cham (2015). https://doi.org/10.1007/978-3-319-24574-4_28

16. Xu, B., Wang, N., Chen, T., Li, M.: Empirical evaluation of rectified activations in convolutional network. arXiv preprint arXiv:1505.00853 (2015)

17. Goodfellow, I., Bengio, Y., Courville, A., Bengio, Y.: Deep Learning, vol. 1. MIT press, Cambridge (2016)

18. Ruder, S.: An overview of gradient descent optimization algorithms. arXiv preprint arXiv: 1609.04747 (2016)

19. Havaei, M., Guizard, N., Chapados, N., Bengio, Y.: HeMIS: hetero-modal image segmentation. In: Ourselin, S., Joskowicz, L., Sabuncu, Mert R., Unal, G., Wells, W. (eds.) MICCAI 2016. LNCS, vol. 9901, pp. 469–477. Springer, Cham (2016). https://doi.org/10.1007/978-3-319-46723-8_54

20. Selvaraju, R.R., Cogswell, M., Das, A., Vedantam, R., Parikh, D., Batra, D.: Grad-CAM: Visual Explanations from Deep Networks via Gradient-based Localization. arXiv:1610.02391 (2016)

21. Wei, Y., Shen, G., Juan-juan, L.: A fully automatic method for lung parenchyma segmentation and repairing. J. Digit. Imaging **26**, 483–495 (2013). https://doi.org/10.1007/s10278-012-9528-9

22. Dai, S., Lu, K., Dong, J., Zhang, Y., Chen, Y.: A novel approach of lung segmentation on chest CT images using graph cuts. Neurocomput. **168**(C), 799–807 (2015). https://doi.org/10.1016/j.neucom.2015.05.044

23. Zhang, W., Wang, X., Zhang, P., Chen, J.: Global optimal hybrid geometric active contour for automated lung segmentation on CT images. Comput. Biol. Med. **91**, 168–180 (2017). https://doi.org/10.1016/j.compbiomed.2017.10.005

Multiwavelet Based Unmanned Aerial Vehicle Thermal Image Fusion for Surveillance and Target Location

B. Bharathidasan[1](✉) and G. Thirugnanam[2](✉)

[1] Department of Electronics and Instrumentation Engineering, Annamalai University,
Chidambaram, India
ggtt_me@yahoo.com
[2] Department of Electronics and Communication Engineering,
Government College of Technology, Coimbatore, India
thirugnanam_me@yahoo.com

Abstract. A novel image fusion method in multiwavelet domain is proposed in this paper. The special frequency band and property of image in multiwavelet domain are employed for the image fusion algorithm. Due to the widespread use of digital media applications, multimedia refuge and the fusing has grown incredible important. Here in this research work, a low resolution multispectral and high resolution RGB image fused here is he new method to fuse that is proposed, to find out the armed person behind deep forest with surrounding trees. The picture is acquired from a wing which is new unmanned aerial vehicle (UAV) at 90 to 100 m distance in dark light surroundings. The combined effect of the texture resolution by a heavy decree RGB image and the thermal image taken by Dual Sensor Night Vision Goggle (DSNVG), to retrieve a fused IR RGB-thermal good image of the armed person. Inside this research work, The DSNVG is in construction to offer fusion of thermal imagery, to afford the profit of larger positional alertness due to developed risk discovery underneath nearly all battlefield outsides, like-minded with established bludgeon structure ranges, prolonged performance potential from high-light circumstances to sum dusk and through battlefield obstacles, increasing ability for municipal work. Here, Multiwavelet transform is being compared with wavelet packet for aerial vehicle fusion. In this work concludes that multiwavelet performs better than wavelet packet.

Keywords: Image fusion · Multiwavelet transform · Wavelet packet transform

1 Introduction

This fusing of Image technique by joining the considerable statistics beginning a establish of films of the identical spectacle within a restricted picture and the resultant diverse image is superfluous informative and complete than the host images. Input images could be the multisensor, occurence and more focus point [1]. The fused image is imaginary to place all important real terms from the original images. The artifacts introduced by

M. Singh et al. (Eds.): ICACDS 2020, CCIS 1244, pp. 352–361, 2020.
https://doi.org/10.1007/978-981-15-6634-9_32

the image fusion that supply to an erroneous psychiatry. Picture listing is the necessary fundamental procedure of the fusion technique. It also the procedure of transforming an assortment of sets of statistics into one classify coordination. Image fusion place submission in the pasture of steering management, goal discovery, and gratitude, medical judgment, satellite imagery for long distance sensing, military and inhabitant scrutiny [2]. These methods are confidential into picture element, feather, and decision makings.

The DSNVG is in construction to offer fusion of thermal imagery, to afford the profit of larger positional alertness due to developed risk discovery underneath nearly all battlefield outsides, like-minded with established bludgeon structure ranges, prolonged performance potential from high-light circumstances to sum dusk and through battlefield obstacles, increasing ability for municipal work [3].

The sieve deposit applied to put into practice wavelet transform and satisfies the properties of orthogonality, balance, little support and upper estimate sort concurrently. The DSNVG is in construction to offer fusion of thermal imagery, to afford the profit of larger positional alertness due to developed risk discovery underneath nearly all battlefield outsides, like-minded with established bludgeon structure ranges, prolonged performance potential from high-light circumstances to sum dusk and through battlefield obstacles, increasing ability for municipal work.

The DSNVG is in construction to offer fusion of thermal imagery, to afford the profit of larger positional alertness due to developed risk discovery underneath nearly all battlefield outsides, like-minded with established bludgeon structure ranges, prolonged performance potential from high-light circumstances to sum dusk and through battlefield obstacles, increasing ability for municipal work. Temperature pictures might be a precious gain along the illustration pictures. Temperature picture don't relays over the shadow, the productivity is the outcrop of temperature acquiring of the radiation of warmth of the substance [4, 5]. With the progress of new imaging sensing exist the necessitate of a carrying great weight amalgamation of all in a job imaging source. Fusing of images of illustration and temperature gathering results sums a innovative breadth in producing the final searching further unfailing. The DSNVG is in construction to offer fusion of thermal imagery, to afford the profit of larger positional alertness due to developed risk discovery underneath nearly all battlefield outsides, like-minded with established bludgeon structure ranges, prolonged performance potential from high-light circumstances to sum dusk and through battlefield obstacles, increasing ability.

The DSNVG is in construction to offer fusion of thermal imagery, to afford the profit of larger positional alertness due to developed risk discovery underneath nearly all battlefield outsides, like-minded with established bludgeon structure ranges, prolonged performance potential from high-light circumstances to sum dusk and through battlefield obstacles.

2   Wavelet Packet Transform

In WPT, the sub-bands LL1, LH1, HL1 and HH1 are decomposed further. The purpose is to provide time frequency plane to be partitioned added accurately [6]. A two level WPT produces 16 coefficients from LLA2 to HHD2 as shown in Fig. 2. LLA2 is the approximation coefficient and remaining 15 sub-bands are detail coefficients. These coefficients give supplementary resolution in time and increase the robustness and imperceptibility of the fusion method [7] (Fig. 1).

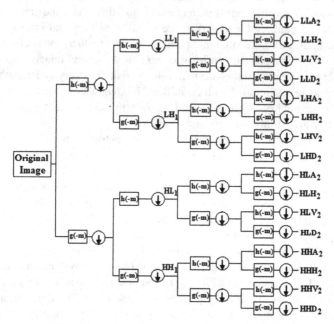

Fig. 1. Two level 2D WPT analysis filter banks

3 Proposed System

In this paper it is proposed to implement Multiwavelet and then applying new Image fusion technique to it [8].

3.1 Multiwavelet Transform

In WPT, with the help of the lower side of the tree structures otherwise the higher side of the branch the fundamental two-channel paper bank is simulated as shown in Fig. 2. This produces uninformed tree makeup with each tree analogous to a WPT basis. These bases are premeditated by in-between the repeated federation in steps of anecdotal sizes [9, 10]. Thin packed basics are consequently chiefly glowing bespoke to sour transmission part that have unlike deeds in diverse frequency steps. However, Packets are found to be less robust against image processing attacks and their shift variance property causes inaccurate extraction also. Hence, in this paper Multiwavelet Transform based image fusion is proposed and implemented [11].

Like wavelets, Multiwavelets rely on Multi Resolution analyses (MRA) like that of wavelets. Only to the large industry function $\varphi(t)$ and only wavelet celebration $\psi(t)$

present in wavelets using MRA, where as multiwavelets have large items have small industry celebrations underneath a single vetrified can be specified as,

$$\varphi(t) = [\varphi_1(t), \ \varphi_2(t), \ \ldots, \ \varphi_N(t)]^T \tag{1}$$

and large scale industry celebrations can be identified as,

$$\psi(t) = [\psi_1(t), \ \psi_2(t), \ \ldots, \ \psi_N(t)]^T \tag{2}$$

matrix dissolution First equation and waving second equation

$$\varphi(t) = \sum_K H(k)\varphi(2t - k) \tag{3}$$

$$\psi(t) = \sum_K H(k)\varphi(2t - k) \tag{4}$$

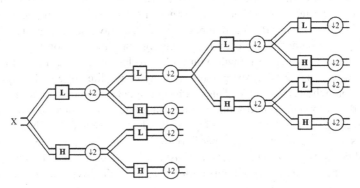

Fig. 2. Decomposition tree for two level using multiwavelet

Many of the literature specified that multiwavelets N is 2 i.e., scaling and wavelet functions are two. The sub-bands higher passing and lower passing are no. of matrices in its place of scaling findings. First figures reveals that a paper savings that decreasing these pictures to a single level. The input is to be a multiple filter that has. Taps of size N × N filter deposits matrices,

The DSNVG is in construction to offer fusion of thermal imagery, to afford the profit of larger positional alertness due to developed risk discovery underneath nearly all battlefield outsides, like-minded with established bludgeon structure ranges, prolonged performance potential from high-light circumstances to sum dusk and through battlefield obstacles, increasing ability for municipal work (Fig. 3).

LL_1LL_2	LL_1HL_2	HL_1LL_2	HL_1HL_2	H_1L_1	H_1L_2
LL_1LH_2	LL_1HH_2	HL_1LH_2	HL_1HH_2		
LH_1LL_2	LH_1LH_2	HH_1LL_2	HH_1HL_2	H_2L_1	H_2L_2
LH_1LH_2	LH_1HH_2	HH_1LH_2	HH_1HH_2		
L_1H_1		L_1H_2		H_1H_1	H_1H_2
L_2H_1		L_2H_2		H_2H_1	H_2H_2

Fig. 3. Two sided multiwavelet analysis

4 Image Fusion

Here there are one or more input images that are combined to form only one image that consists of all the features of that input images [12]. In this research paper, fusion is done to join the images of RGB and temperature based gray scale image. The resulting mixed image accommodates uniformly the finding out armed person behind group of trees. An pioneering mixing growth with multiwavelet move toward shut to is done. N of multiwavelets is 2 as it is suggested in many of the literatures. The input has to be two vectors when N is double. This can be possible by moreover dividing ones and threes model individually across findings or doing one watercourse of input into two watercourse or filter previously the timing small industry host to detect the steady guess that results second screams of span partially of the input. The protuberance image of a widened multiwavelet produces imaging fusing scheme is available in second figure. RGB is deliberate as Image 1, and temperature image is applied as second picture (Fig. 4).

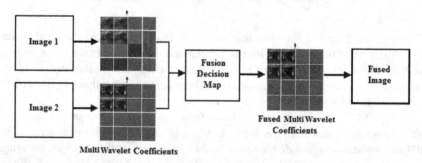

Fig. 4. Procedure of multiwavelet fusing images

In WPT, the sub-bands LL1, LH1, HL1 and HH1 are decomposed further. The purpose is to provide time frequency plane to be partitioned added accurately. A two level WPT produces 16 coefficients from LLA2 to HHD2 as shown in Fig. 2. LLA2 is the approximation coefficient and remaining 15 sub-bands are detail coefficients.

These coefficients give supplementary resolution in time and increase the robustness and imperceptibility of the fusion method [13].

5 Fusion Rule

The DSNVG is in construction to offer fusion of thermal imagery, to afford the profit of larger positional alertness due to developed risk discovery underneath nearly all battlefield outsides, like-minded with established bludgeon structure ranges, prolonged performance potential from high-light circumstances to sum dusk and through battlefield obstacles, increasing ability for municipal work:

$$W_j^a = \alpha * W_{Aj}^a + \beta * W_{Bj}^a \tag{5}$$

W – sub-bands of the image, A and B locate for IR and Thermal images are in straight, j is the height of wavelet and the principles of $\alpha + \beta = 1$.

Manner in brains that larger frequency sub-bands located of a particulars, the reseacher developed greater repeat sub-bands are improved constituent to place sideways extra excellent values [14, 15].

$$W_j^d = \begin{cases} W_{Aj}^d, & \left| W_{Aj}^d \right| \geq \left| W_{Bj}^d \right| \\ W_{Bj}^d, & \left| W_{Aj}^d \right| \leq \left| W_{Bj}^d \right| \end{cases} \tag{6}$$

6 Simulation Results

Image 1 is the infra red image and image 2 is the thermal image as shown in Figs. 5 and 6 of size 256×256. Y, U, V components are obtained from input image 1 and for fusion only Y component is taken and the figures as shown in Figs. 7, 8 and 9. With the use of Multiwavelet transform, Image 1 is sub-divided into two levels and it is shown in Fig. 10 and in Fig. 11. Similarly, Multiwavelet transform decomposes image 2 and it is displayed in the Figs. 12 and in Fig. 13. Fusion rule which is proposed here in this work, is applied on seven subtitutes of first and second pictures. The mean is calculated between low-frequency components in two images in lower oreder sub-bands. Fused multiwavelet sub-band image is retrieved after the process of fusing, and immediately reverse MWT is shown in Fig. 14 as it is called as Y component of the compound image as indulged in Fig. 15. After that Y component mixed picture element is converted to an RGB mixed image as directed in Fig. 16. PSNR and Normalized Correlation are compared and tabulated in Table 1 intended for wavelet packet and multiwavelet transform.

Fig. 5. Input image 1 (Infra Red image)

Fig. 6. Input image 2 (Thermal image)

Fig. 7. Y component of input image1

Fig. 8. U component of input image1

Fig. 9. V component of input image1

Fig. 10. One level Multiwavelet decomposition

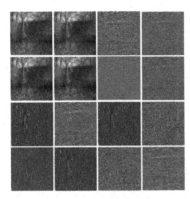

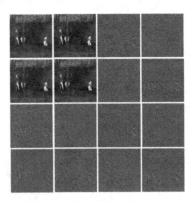

Fig. 11. Two level Multiwavelet decomposition of image1

Fig. 12. One level Multiwavelet decomposition of image2

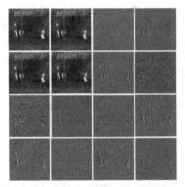

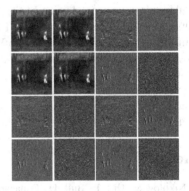

Fig. 13. Two level Multiwavelet decomposition of image2

Fig. 14. Fused coefficients

Fig. 15. Output Y component of image fusion

Fig. 16. RGB fused image

Table 1. Performance comparison of wavelet packet and multiwavelet

Input images	PSNR (dB)		Similarity measure		MSE	
	WPT	Multiwavelet	WPT	Multiwavelet	WPT	Multiwavelet
Proposed UAV Images	45.9472	47.1903	0.9478	0.9686	0.9112	0.8821
INSAT	45.9327	47.2843	0.9459	0.9612	0.9198	0.8892
LANDSAT	43.4233	46.3925	0.9362	0.9589	0.9723	0.8997
PAN-MS	44.7331	46.9276	0.9312	0.9525	0.9610	0.8933

7   Conclusion and Future Work

Multiwavelet transform based thermal image fusion is attempted in this paper work. Simulation results arrived using MATLAB reveals that the supremacy of the proposed Multiwavelet based image fusion to WPT. The results from Table 1 show that the imperceptibility is 47 db when compared to WPT and it shows the better results. As per the similarity measure is concern 96% is the proposed method values when compared to previous methods. This fusion technique found to be successful for the fusion in the survival of intrusion. In this work, Multiwavelet transform based thermal image fusion is attempted. In the future work, it is proposed to attempt Framelet transform, a new novel technique which is to be applied for the image fusion for UAV images.

References

1. Nikolov, S., Hill, P., Bull, D., Canagarajah, N.: Wavelets for image fusion. In: Petrosian, A.A., Meyer, F.G. (eds.) Wavelets in Signal and Image Analysis. CIVI, vol. 19, pp. 213–241. Springer, Dordrecht (2001). https://doi.org/10.1007/978-94-015-9715-9_8
2. Vekkot, S., Shukla, P.: A novel architecture for wavelet based image fusion. World Acad. Sci. Eng. Technol. **57**, 372–377 (2009)
3. Li, H., Manjunath, B.S., Mitra, S.K.: Multisensor image fusion using the wavelet transform. Graph Models Image Process. **57**(3), 235–245 (1995)
4. Pu, T., Ni, G.: Contrast based image fusion using the discrete wavelet transform. Opt. Eng. **39**(8), 2075–2082 (2000)
5. Xiong, Z., Ramchandran, K., Orchad, M.T.: Wavelet packet image coding using space-frequency quantization. IEEE Trans. Image Process. **7**, 160–174 (1998)
6. Bedi, S.S., Agarwal, J., Agarwal, P.: Image fusion techniques and quality assessment parameters for clinical diagnosis: a review. Int. J. Adv. Comput. Commun. Eng. **2**, 1153–1157 (2013)
7. Qiguang, M., Baoshul, W.: A novel image fusion method using contourlet transform. In: Proceedings, 2006 International Conference on Communications, Circuits and Systems, vol. 1, pp. 548–552 (2006)
8. Chandana, M., Amutha, S., Kumar, N.: A hybrid multi-focus medical image fusion based on wavelet transform. Int. J. Res. Rev. Comput. Sci. **2**, 1187–1192 (2011)
9. Tu, T., Huang, P.S., Hung, C., Chang, C.: A fast intensity-hue-saturation fusion technique with spectral adjustment for IKONOS imagery. IEEE Trans. Geosci. Remote Sens. **1**(4), 309–312 (2004)

10. Saleta, M., Catala, J.L.: Fusion of multispectral and panchromatic images using improved IHS and PCA mergers based on wavelet decomposition. IEEE Trans. Geosci. Remote Sens. **42**(6), 1291–1299 (2004)
11. Geetha, G., Mohammad, S.R., Murthy, Y.S.S.R.: Multifocus image fusion using multiresolution approach with bilateral gradient based sharpness criterion. Int. J. Comput. Sci. Technol. **10**, 103–115 (2012)
12. Bindu, C.H., Prasad, K.S.: Performance analysis of multi source fused medical images using multiresolution transforms. Int. J. Adv. Comput. Sci. Appl. **3**, 54–62 (2012)
13. Godse, D.A., Bormane, D.S.: Wavelet based image fusion using pixel based maximum selection rule. Int. J. Eng. Sci. Technol. **3**, 5572–5577 (2011)
14. Ping, Y.L., Sheng, L.B., Hua, Z.D.: Novel image fusion algorithm with novel performance evaluation method. Int. J. Syst. Eng. Electron **29**, 509–513 (2007)
15. Sahu, D.K., Parsai, M.P.: Different image fusion techniques – a critical review. Int. J. Mod. Eng. **2**, 4298–4301 (2012)

Investigating Movement Detection in Unedited Camera Footage

Samuel Sciberras and Joseph G. Vella[(✉)]

Faculty of ICT, Department of Computer Information Systems,
University of Malta, Msida, Malta
{samuel.sciberras.15,joseph.g.vella}@um.edu.mt

Abstract. Digital evidence from CCTVs is an aid in crime scene investigations and there is a demand for more automation. This paper describes a system that detects motion induced events within a video clip based on user-defined criteria, such as filtering by colour and size of the moving object and then extracts features and regions where events have been detected. Post processing includes finding association rules between objects that appear simultaneously in a clip based on their colour. All processing techniques follow best practices.

The available Wallflower dataset is used for evaluation, and confusion matrices are computed by comparing the results achieved by this system against the ground truth values for each image sequence. Ranges of effective pre-processing parameter values were set for erosion, dilation and background subtractor threshold and the system was tested across a wide array of parameter values. For each combination, measures are extracted and the F1 Score is calculated.

The lowest and highest F1 Score obtained across all image sequences were of 67% and 95% respectively. It is noted that the image quality of clips and background affect the F1 scores.

Keywords: Computer vision · Movement detection · Digital forensics

1 Introduction and Background

Analysis of video footage from Crime Scene Investigation (CSI) is an important part of digital forensics and one which is constantly improving, as investigations rely on such media to support evidence [1]. However, the investigators still have to manually review the video footage to detect events of interest. This review is very time consuming, prone to human error, subjective, and inefficient.

Video footage is also vital in the CSI as it can provide indications towards identification of potential suspects and witnesses, vehicles used by identified individuals and any associated detections (e.g. a person following another). Video footage and the detections are very effective evidence in a court trial. There are numerous CSI that have been resolved by video footage from one or many video sources; one example is untangling the events in a violent street brawl [2].

This paper describes the design, methodology adopted for evaluating and technologies used to implement the system are described. Finally, the results and observations from the evaluation process are reported.

© Springer Nature Singapore Pte Ltd. 2020
M. Singh et al. (Eds.): ICACDS 2020, CCIS 1244, pp. 362–371, 2020.
https://doi.org/10.1007/978-981-15-6634-9_33

2 Background

2.1 Digital Forensics

Digital Forensics is the means of gathering and interpreting information from digital media or digital devices as evidence, that can be used in building a crime case [3]. Digital Image and Video Forensics is a branch of Digital Forensics that deals with discovering and analysing facts about an image or a video. Examples of forensic studies include detecting alterations, discovering or recovering hidden data, and tracing back the origins of the media [4].

2.2 Computer Vision and Digital Image Processing

As video content grows, the field of computer vision is much more in demand as there is an ever-growing demand for video analysis techniques and image understanding [5]. Computer vision techniques can be applied to several different areas, such as motion, and object and pedestrian detection.

Digital Image Processing is the act of manipulating and changing images that are stored digitally on some storage media, using a digital computer. Such manipulations range from a simple image rotation to more complicated operations such as colourising a greyscale image.

2.3 Motion Detection

Motion detection in a video clip is the process of identifying motion of objects that are present in two or more sequential images [6]. There exist many approaches for detecting motion, and several algorithms have been proposed. These are still challenged by content of a video clip and therefore effects the quality of the detected results. Such challenges include illumination changes, dynamic backgrounds, occlusion, clutter, camouflage and inclement weather [7].

A popular technique used for detecting motion is Background Subtraction. It has been used successfully for indoor and outdoor applications [8]. The technique works by selecting an image from an video sequence, called a background model, and then subtracting it from the current frame in the sequence. This will generate a binary image, where the white pixels represent the differences between the current frame and the background model.

2.4 Data Mining

Data mining is the combined process of analysing gathered data and transforming that data into meaningful, reduced, and useful information.

Association rule mining was introduced in 1993 and is still one of the most widely used pattern-discovery methods [9]. It can be described as discovering associations between two or more objects from within a given database populated with transactions. In order to accept an association rule, it has to satisfy the rules of confidence and support. Confidence is a percentage of transactions containing Y and X with regard to the overall number of transactions containing Y. Support on the other hand is a percentage of transactions containing both X and Y together from the overall number of transactions [10].

2.5 Similar Systems

There are existing and similar systems that incorporate motion detection such as those found in real-time camera systems. Some of these camera systems use Passive Infrared (PIR), which detect motion based on the heat emitted by objects in contrast to the heat of the background. Other systems use Computer Vision to detect motion, which offer more advanced features [11].

3 Aims, Objectives and Requirements

The aim here is to develop a computerised system that automates the process of detecting motion event details in video evidence collected from stationary cameras. Furthermore, the system has to offer functionalities such as filtering the results by colour and size of targets, analysing features within the detected areas, and mining association rules between moving objects that appear within a video based on their colour. Due to the diversity of scenes, environments, and quality of captured footage, there cannot be a definite pre-processing and post-processing configuration that fits all videos. For this reason, the end user is able to change the default configurations through parameters settings.

The system is required to provide a time and resource efficient method for automating the process of detecting motion events in the specified video file. To measure the system's capabilities an evaluation is conducted using an oracle type of comparison against the results returned by our system. To run the evaluation a publicly available data set is used; the Wallflower data set [12] contains different types of image sequences.

To address the turnaround time for processing a video sequence and extracting the events therein, automation and in some cases lightly supervised automation is being entrenched in the system. To address the quality and consistency of motion event detection, the system depends less on human intervention and adopts best practices in computer vision and motion detection during processing; human intervention is known to decrease the quality of work with load due to subjectiveness, tiredness, distractions and other effects that deter the focus of the investigator. By relieving the investigator's time from mundane task of reviewing all the video sequence the system allows an investigator to conduct multiple runs each specifying different search criteria whose settings are tuned through his expertise.

The proposed system is an interactive system that analyses digital video extracted from stationary cameras, such as CCTV surveillance systems, for evidence through automatic detection of events based on visual movement. The user is given options to aid and direct the motion detection process, such as pre-processing options on the binary images generated, change the image resolution for analysis, and even the parameters to adjust the background subtraction. Once motion is detected, the system then allows the user to choose their next action from the following; parametric motion detection, detected object feature analysis, and association rule mining based on objects detected.

If the user opts for supervised automation, then parametric motion detection is available to filter the results by selecting one or more colours, and minimum and maximum sizes of the objects to be reported. The user is allowed to input the minimum number of pixels that match the parameters, to provide further filtering. If any of the set parameters are satisfied, then the system outputs the results to the user, marking the region where the parameter selected is satisfied using a bounding rectangle.

For detected object feature analysis, the user selects a frame where motion was detected, and subsequently select the area highlighting the moving object for further analysis. The results are shown in a list, displaying the detected features and a percentage representing the degree of confidence in feature detection.

If the user chooses to perform association rule mining, they are able to modify the support and confidence levels with which the algorithm mines association rules. Once an association rule instance is found it is presented in textual format displaying the degree of confidence with which this implication holds.

For each process run, set parameters and results obtained are stored in a database to enable explanation of the conditions that lead to these detections.

4 Design and Implementation

The system is designed to have four major modules to meet the requirements.

4.1 Motion Detection

The motion detection process is made up of three phases. The first phase is collecting the input video properties; this meta data is used to drive processing the images contained in the input video file. The video file is then split into individual frames, representing each image in the video file. Each image is then resized to a resolution specified by default or set by the user. Even though downsizing images causes loss in quality, it is a trade-off for boosting the time performance of the detection process.

The motion detection process, the second phase of motion detection, is implemented using a combination of best practices reported in the literature. In order to detect changes between two frames, background subtraction is used to generate a binary image highlighting differences between two frames being the reference frame and the input frame, by colouring different pixels white. Along with background subtraction, a background modelling technique is used to build and update a reference frame, which adapts to the changing background throughout the video timeline to represents the ideal background from which to subtract each consequent frame from. The library OpenCV offers various ways to apply background subtraction. The option chosen here was the Background Subtractor MOG2 object [13] which uses an improved Gaussian Mixture-based background modelling technique [14]. MOG2 accepts the following parameters: the number of frames to consider for background modelling, a threshold value to check how fitting a pixel is to the background using the squared Mahalanobis distance function, and whether to mark shadows in the binary images. The default values for each of the parameters are 500, 32, and 'False' respectively.

Once the binary images are generated, the morphological operators, such as erosion and dilation [15], are applied to the image as specified by the user. After these operations on the binary images, the system sifts the binary images for areas of interest marked by white pixels. Another process is involved to coalesce objects of interest that are effectively one; the Connected Components algorithm [16] is used to identify separate objects found within the frame (called Labels). When a Label satisfies the motion detection parameter and is deemed to be a moving object, an object structure representing the motion event is created for later access. After all the video timeline is analysed for events, time intervals in seconds and the frames where motion is detected are presented to the user. All the configurations and results are also stored in the database for the run.

The third phase of the motion detection process is that of gathering and reporting the results. The user is able to view the results in either of two perspectives. First, by choosing to save the images where motion is detected to storage. This allows for further review of each single frame. Second, by choosing time intervals, in a range format and frames, from the original video for each event.

4.2 Parametric Motion Detection

The second module enables parametric motion detection based on user set filters; these include selecting one or more colours, and a minimum and maximum size of the target object. Once the parameters are set by a user and the motion detection process is run, results are presented with a list of images that meet the set user criteria with the area satisfying user constraints marked by a bounding rectangle.

The implementation of the parametric motion detection module is composed of three functions. Two of the functions are dedicated to the types of filtering available; namely colour and size. The third function supports colour filtering which compares the values of a pixel against a classification of colours, to determine whether a pixel belongs to a certain colour group using the HSV (hue, saturation, value) colour space. Specifically, for each marked object the image is iterated, and each of its pixel's is checked for satisfaction of any of the parameters set by the user. Similarly, size filtering takes place by comparing the height and width of the Label to those set by the user.

4.3 Object Feature Analysis

The third module is feature analysis of any detected object. This is implemented with the use of an API provided by Google (Google Cloud Platform Service [17]). The user is allowed to crop a region that has been detected in a frame which is then sent using the API and the system awaits the identification of features.

The API available from the Google, the Vision API, is provided with an image, and it returns a dictionary providing information about the features detected, such as the description, and the confidence score.

4.4   Association Rule Mining

The final module provides association rule mining. In order to find association rules, the FP Growth algorithm [18] was adopted. This finds associations between objects based on their colour, and if they appear simultaneously in the video satisfying a certain degree of confidence, then they are flagged in an association rule. The confidence level can be user defined.

In order to implement the association rule mining module, the PyFPgrowth library is used. This library offers the functionality to perform the FP Growth algorithm on a set of transactions. These transactions are set based on the colours of the objects that appear simultaneously throughout parts of the video.

4.5   User Interface

The User Interface design for this system is influenced by the available system's features and expected usage so as to accommodate the various processes available, depending on and related to the flow of processing.

The user interface is implemented using the PyQt5 and PyQt5-tools libraries. PyQt5 is a library that allows you to create a user interface using objects such as sliders, text fields and buttons.

In order to keep the user interface responsive while the system is processing a request, multi-threading had to be implemented. The user interface runs on the main thread, whilst the additional triggered processes are executed on daemon threads. Threading was done using the in-built Python library.

4.6   Database

All of the above modules post results to a database, where the end user can then review the results without having to perform the processes again. The database was also used for debugging and evaluating the system. Evaluation was done by developing a separate script to automate inputting ranges of different pre-processing operation values, rather than having to do each process with each different variable through the User Interface. An important role of the database is to provide the basis for a custody system and also to exactly explain results (i.e. from which input file and which settings has the process executed with).

For this purpose, the library PyGreSQL was imported into the solution. A dedicated script was implemented to integrate the system with the database. There are seven functions used to store results into the database and for evaluating results.

Implementation of the system was done using Python 3.6.4 programming language. Libraries and APIs used for computer vision and matrices are OpenCV, NumPy and Google Cloud Platform (Vision API). PyGreSQL and PyFPgrowth were used for database connectivity and association rule mining respectively.

5   Evaluation, Results and Testing

The evaluation was conducted on the motion detection process, as it is the basis of the whole system. The data set used for evaluation is the publicly available and very realistic Wallflower data set [12]. An oracle table for each of the used image sequences was encoded by manually enumerating the motion events.

In order to make sure that the system is functioning as intended, a Unit Testing method is employed. When conducting Unit Testing one can test specific pieces of code, such as a function, or a set of functions [21]. For this purpose, the standard python library 'unittest' is used. All the main functionalities are tested using assertions, excluding the database functions and user interface.

5.1 Evaluating Motion Detection

The method of evaluation is based on the use of confusion matrices and calculating the accuracy, precision and recall on the four classifications: false positive, false negative, true positive and true negative. Once recall and precision are calculated, we are then able to compute the F1 Score [19], which is a harmonic average of precision and recall.

The settings used for evaluating the motion detection process were as follows:

- 500 frames background subtraction modelling;
- Shadows set to 'False';
- 15 pixels minimum area of object to detect;
- Frames resolution is unchanged;

Due to allowing the end user to set the system's settings to fit their requirements, it would be impossible to evaluate the full variety of all the possible configurations for pre-processing and detection process settings available. After carefully testing and analysis the effects of erosion, dilation and background subtractor threshold, it evidently showed that extreme iterations of dilation and erosion operations as well as low or high threshold values makes the motion detection pointless. For these reasons, the evaluating process for the system is conducted on combinations of three separate ranges for erosion, dilation and background subtractor. The ranges chosen are 0 to 9 for erosion and dilation, and 25 to 34 for background subtractor threshold.

From the results obtained for the Time of Day image sequence, it can be observed that although the background is constantly changing, due to the fact that it is a very slow and gradual change it does not impact the performance of the system as much as the Light Switch and Foreground Aperture image sequences do. In the latter image sequences, a sudden change occurs where the background model needs to quickly adapt to a completely new reference frame. This results in a number of false positives, hence having a significant negative effect on the Precision measure, and subsequently the F1 score. In the Moved Object image sequence, we do not experience any change in the background model, hence testing the motion detection process only. The results obtained for each image sequence are depicted in Fig. 1.

From the results obtained it can be concluded that every video file will have its own best configuration that will output the best results, meaning that there is no common combination of pre-processing settings that will output the best result for all video input to the system. It was also noted that the system had considerable amounts of false positives after motion is detected. This is due to the background adapting to the changes occurring, being marked as motion.

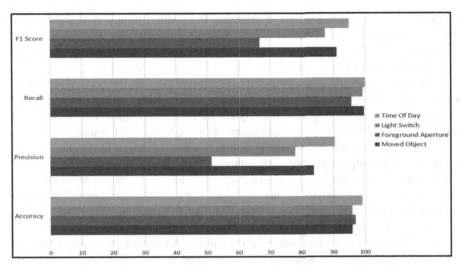

Fig. 1. Bar chart showing the best measured results for each image sequence

5.2 Evaluating Association Rule Mining

In order to evaluate the association rule mining module, a staged video was created. The video shows rectangles moving in and out of the scene, each with different colours, however with a bias that green and red are to appear together more frequently, but not always. The results achieved were correct, as the objects of colours green and red both imply one another, meaning that there is an association between them.

5.3 Google Cloud Vision Evaluation

H. Hosseini, B. Xiao and R. Poovendran conducted an experiment in order to evaluate the robustness of the Google Cloud Vision API [20]. The authors first tested the API using the original images and recording the results returned. Then, they modified the images by adding impulse noise until the results returned by the API fail to recognize the contents of the image input. The authors conclude by stating that the Google Cloud Vision API can be deceived when using images with noise added to them. However, when the modified image is restored, the API outputs very similar results to those returned when using the original image.

6 Conclusions and Future Work

The aim was to develop a system that would automate the process of manually reviewing digital video evidence, while still allowing the end user to control the retrieval by modifying the available pre-processing and post-processing parameters.

The evaluation process showed that the least and most F1 Score percentage obtained were those of 67% and 95% respectively. The average F1 Score of all results is 58%, which does not reflect the system's success rate. This is due to the fact that the configured parameters play a significant part in achieving the optimal results, however one can also affect the results negatively when system is misconfigured. For this reason, a brief analysis of the video is required to identify the parameter configuration settings in order to achieve better results.

The OpenCV library was very useful in development of the final system. It is relatively easy to learn and use due to the documentation provided by the developers and community. PostgreSQL too was also very effective when used to process the data in a quick and efficient manner, and an indispensable value for chain of custody.

Several features have been implemented in this motion detection system, however due to the area of digital forensics and computer vision being so widespread, there are many other possible improvements that can be done to make this system better and even more effective. Examples of such improvements are: adapt the square pixel kernel used for morphological operations erosion and dilation according to the resolution of the frame; implement object tracking; optimization of process by using main memory caching; add more background subtraction algorithms for the user to choose from; and add more event types to association rule mining rather than just the colours of objects.

References

1. Xu, J., Su, Y., You, X.: Detection of video transcoding for digital forensics. In: International Conference on Audio, Language and Image Processing, Shangai (2012)
2. Video Forensics - Case study from our archive, Manchester Video Limited, 22 08 2014. https://www.manchestervideo.com/2014/08/22/video-forensics-case-study-from-our-archive/. Accessed 27 Mar 2020
3. Kessler, G.: Advancing the Science of Digital Forensics. Computer 45(12), 25–27 (2012)
4. Rocha, A., Scheirer, W., Boult, T., Goldenstein, S.: Vision of the unseen: current trends and challenges in digital image and video forensics. ACM Comp. Surv. 43/4, 1–42 (2011)
5. Liu, Q., Li, X., Elgammal, A., Hua, X., Xu, D., Tao, D.: Introduction to computer vision and image understanding the special issue on video analysis. Comput. Vis. Image Underst. 113(3), 317–318 (2009)
6. Singla, N.: Motion detection based on frame difference method. Int. J. Inf. Comput. Technol. 4(15), 1559–1565 (2014)
7. Shaikh, S.H., Saeed, K., Chaki, N.: Moving object detection approaches, challenges and object tracking. Moving Object Detection Using Background Subtraction. SCS, pp. 15–23. Springer, Cham (2014). https://doi.org/10.1007/978-3-319-07386-6_3
8. Wu, J., Trivedi, M.: Performance characterization for Gaussian mixture model based motion detection algorithms. In: IEEE International Conference on Image Processing (2005)
9. Hipp, J., Guntzer, U., Nakhaeizadeh, G.: Algorithms for association rule mining - a general survey and comparison. ACM SIGKDD Explor. Newslett. 2(1), 58–64 (2000)
10. Fournier-Viger, P., Lin, J.C.-W., Kiran, R.U., Koh, Y.S., Thomas, R.: A survey of sequential pattern mining. Data Sci. Pattern Recogn. 1(1), 54–77 (2017)
11. Ansaldo, M.: How your home security camera detects motion. TechHive. https://www.techhive.com/article/3263662/home-tech/how-your-home-security-camera-detects-motion.html. Accessed 3 Apr 2018
12. Wallflower: Principles and Practice of Background Maintenance, Microsoft. https://www.microsoft.com/en-us/research/publication/wallflower-principles-and-practice-of-background-maintenance/

13. OpenCV Background Subractor MOG2. Open Source Computer Vision. https://docs.opencv.org/master/d7/d7b/classcv_1_1BackgroundSubtractorMOG2.html. Accessed 28 Mar 2020
14. Zivkovic, Z.: Improved adaptive Gaussian mixture model for background subtraction. In: Pattern Recognition, Cambridge, UK (2004)
15. Morphological Transformations, Open Source Computer Vision. https://docs.opencv.org/trunk/d9/d61/tutorial_py_morphological_ops.html. Accessed 25 Mar 2020
16. Structural Analysis and Shape Descriptors, Open Source Computer Vision. https://docs.opencv.org/master/d3/dc0/group__imgproc__shape.html. Accessed 28 Mar 2020
17. Google Cloud, Google. https://cloud.google.com/
18. Borgelt, C.: An implementation of the FP-growth algorithm. In: Proceedings of the 1st International Workshop on Open Source Data Mining: Frequent Pattern Mining Implementations, pp. 1–5 (2005)
19. Shung, K.P.: Accuracy, Precision, Recall or F1? Towards Data Science, 15 03 2018. https://towardsdatascience.com/accuracy-precision-recall-or-f1-331fb37c5cb9. Accessed 28 Mar 2020
20. Hosseini, H., Xiao, B., Poovendran, R.: Google's cloud vision API is not robust to noise. In: Machine Learning and Applications (ICMLA), Cancun, Mexico (2017)
21. Runeson, P.: A survey of unit testing practices. IEEE Softw. **23**(4), 22–29 (2006)

Time Series Forecasting Using Machine Learning

Ruchi Verma[✉], Joshita Sharma, and Shagun Jindal

Department of Compter Science and Engineering, Jaypee University of Information Technology, Waknaghat Solan, HP 173234, India
ruchi.verma@juit.ac.in

Abstract. Forecasting is an essential part of any business as extensive amount of data is available, one needs to combine statistical model with machine learning to improve accuracy, throughput and overall performance. In this paper a time series forecasting approach is used with machine learning techniques to forecast the store item demands. $SARIMA(0,1,1)X(0,1,0)_{12}$ model is used with parameters (0,1,0,12) referring to seasonalcomponents of series combined with ARIMA (0,1,1) for trend components. We trained our model taking past 4 year values of store items and predicted sales for next year.

Keywords: Sales prediction · Machine learning · Time series · SARIMA model · Grid search · Akaike information criterion

1 Introduction

Forecasting is a data science task that is central to many activities within an organization. Faced by challenges like sudden rise in demands, high expectations, unknown demand and high costs, every organization needs plans ahead of time to efficiently allocate the scarce resources to maximize the benefits and meet the demands in the peak times. It helps in goal setting as well as measuring performance over time for further improvements. Since 1960s various statistical models have been presented for this purpose like linear regression, ARIMA.

The increasing availability of huge amounts of data and firms striving to excel in business has made forecasting a difficult and complex task. Nowadays, with great advancement in AI and machine learning, forecasting can be done with machine learning which reduces the cost of labour and automation leading to less time consumption.

In this paper, effort is made to predict one year sales of store item from given 5 years of data taken from website Kaggle [1]. Various trends, seasonal variation are considered for monthly, yearly plots of data items. Machine learning techniques are used to fit Autoregressive Integrated Moving Average (ARIMA) and Seasonal Autoregressive Integrated Moving Average (SARIMA) models.

© Springer Nature Singapore Pte Ltd. 2020
M. Singh et al. (Eds.): ICACDS 2020, CCIS 1244, pp. 372–381, 2020.
https://doi.org/10.1007/978-981-15-6634-9_34

2   Literature Survey

In this section, we present a brief overview of contributions which focus on time series analysis and forecasting. Ariyo et al. [2] have used uses autoregressive integrated moving average (ARIMA) models to successfully predict stock prices over time. ARIMA (1,0,1) was selected as the best model for Zenith bank stock index. He concluded how ARIMA model has a strong potential for short-term prediction and can compete favourably with existing techniques for stock price prediction. Nochai and Nochai [3] forecasted three types of prices of palm oil i.e. farm price, wholesale price and pure oil price for 2000–2004 in Thailand. Main focus was on fitting ARIMA model by minimizing mean absolute percentage error. Hong [4] used machine learning hybrid of RNNs and SVMs, namely RSVR, to forecast rainfall depth values. It is observed that this results in great forecasting performance which concludes RSVRCPSO model provides a promising alternative for forecasting rainfall values. Pavlyshenko [5] explained machine learning models for sales predictive analytics. TheThe effect of machine-learning generalization consists in the fact of capturing the patterns in the whole set of data which is effective for small number of predictions. Etuk and Mohammed [6] used a similar model but with different approach. SARIMA(0,0,0)x(0,0,1)$_{12}$ was used to predict monthly rainfall in Gadaref, Sudan. Data was already stationary and other parameters were estimated by ACF/PACF plots. Luo et al. [7] used SARIMA model to predict Cucumber price. SARIMA(1,0,1)x(1,1,1)$_{12}$ was fitted with fitting error 17%. It was deduced that SARIMA model is feasible for short-term prediction of vegetable prices. Jarrett and Kyper [8] used the methods of ARIMA-Intervention analysis toanalyse and draw conclusion concerning how the stock market price index in China behave over time. Various parameters of ARIMA were found by ACF/PACF plots after data was made stationary. Further, AIC value indicated the usefulness of the model fitted. Guha and Bandyopadhyay [9] gave an inside view of application of ARIMA time series model to forecast future values of gold. He pointed how ARIMA is best for short run as model needs to be retrain to capture the new trends in the data over years. He used six different models to predict values and fitted ARIMA(1,1,1) which was the chosen according to lowest value of root mean square error value.

3   Approach

Time series is ordered sequence of useful data such as temperature, stock price and item demands over regular spaced interval of time. ARIMA models are a type of models that have the ability to predict stationary as well as non-stationary data that is their mean, variance and autocorrelation change over time. This gives it an upper hand as most practical data we need to work with is non-stationary. In SARIMA (Seasonal ARIMA) another parameter for seasonal trend is added. It compromises of making data stationary, model identification, parameter estimation followed by model's use for future prediction. Following are the steps needed.

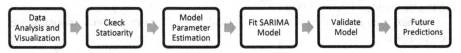

3.1 Data Analysis and Visualization

Time Series Analysis is the art of extracting meaningful statistics from the time series data. It is helpful in determining the structure and factors behind the given time series data. Therefore making a better decision in choosing model for prediction. Following is plot between sales of item in store with time (Fig. 1).

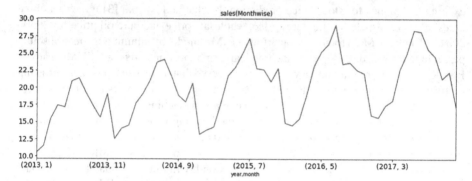

Fig. 1. Plot of store item demands w.r.t. month

From the above observations we can infer that the sales increase with year, which means there is a trend. Also, the sales are high in the months of June and July, which indicates seasonality. By looking at the graphs we can say that all stores show some seasonality and trend. But it is hard to say precisely just by looking. To get a better understanding we do ETS decomposition. Time series decomposition is a mathematical procedure which transforms a time series into multiple different time series. It originally means extracting trend and seasonality components.

The original time series is usually split into three components:

- Seasonal: These are patterns that repeat within fixed periods of time. For example, a website might receive more visits during weekend than weekdays.
- Trend: The underlying trend of the metrices. A website increasing in popularity must show an upwards trend.
- Random: These are the residuals of the original time series after seasonal and trend is removed.

We can assume additive model or multiplicative model to perform decomposition. In additive model, trend is linear and seasonality is constant over time whereas it is increasing or decreasing at non-linear rate in multiplicative model. From previous section, we deduce to use multiplicative model in ETS decomposition (Fig. 2).

We confirm from the ETS decomposition that there is an upward trend and sales shows seasonality of yearly cycle.

The next step is we need to check for stationarity. If the data is not stationary, we need to make it.

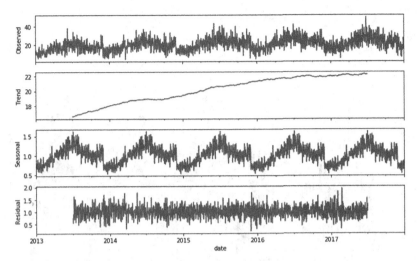

Fig. 2. ETS decomposition of sales into three components

3.2   Check for Stationarity

Stationary data has mean, standard deviation constant over time and autocorrelation independent of time. In general, it is preferred to have data stationary for forecasting as ARIMA predicts according to previous values i.e. its own lags and lagged forecast errors. There are many statistical test to check whether data is stationary or not. The Augmented Dickey-Fuller test is a type of statistical test called a unit root test. The intuition behind a unit root test is that it determines how strongly a time series is defined by a trend. The null hypothesis of the test is that the time series can be represented by a unit root, that it is not stationary. The alternate hypothesis is that the time series is stationary [10].

Null Hypothesis:If failed to be rejected, it suggests the time series has a unit root, meaning it is non-stationary. It has some time dependent structure.

Alternate Hypothesis: The null hypothesis is rejected; it suggests the time series does not have a unit root, meaning it is stationary. It does not have time-dependent structure.

We interpret this result using the p-value from the test as follows:-

p-value > 0.05: Fail to reject the null hypothesis, the data has a unit root and is non-stationary.

p-value ≤ 0.05: Reject the null hypothesis, the data does not have a unit root and is stationary.

We use Augmented Dickey Fuller test to get the summary statistics such as rolling mean as follows (Fig. 3).

Test Statistic = −2.987278, p-value = 0.036100, Lags used = 20.00000.

Since p-value is very close to the threshold and rolling mean and standard deviation is not constant over time, we can use methods to make it stationary. One way to make the data stationary is use differencing where we subtract value at an instant by previous value at a particular instant. In our case, we take first difference and find the values and plot them (Fig. 4).

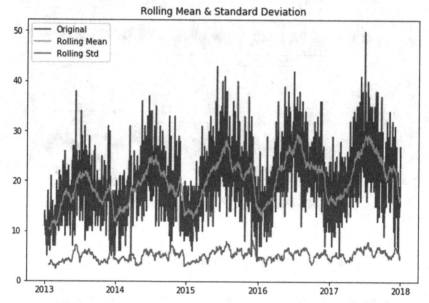

Fig. 3. Rolling mean and standard deviation of sales data for 5 years

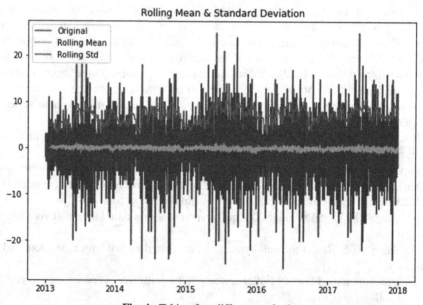

Fig. 4. Taking first difference of sales

Test Statistic = $-1.520810e+01$, p-value = $5.705031e-28$, Lags Used = $2.000000e+01$

This gives us our desired result and gives us constant mean and variance. Now we can move further with the prediction part. Also we can infer that the minimum differencing required to make data stationary (d) is 1.

3.3 Model's Parameter Estimation

In ARIMA model, we have three parameters

- p is the parameter associated with auto-regressive (AR) aspect of model.
- d is the parameter associated with the integrated part of model.
- q is the parameter associated with the moving average part of model (MA).

As observed previously our model has a seasonal component as well, in this case we have three more parameters P, Q and D and m which are similar to p, q, and d but corresponds with the seasonal component of model.m is the parameter in SARIMA that denotes the no. of steps for a single seasonal period which is found out to be 12 i.e. yearly seasonal cycle. To estimate parameters we can either do ACF/PACF plots or do a grid search putting an algorithm. We proceed with grid search. The pyramid-arima library for Python allows us to perform grid search over multiple values of parameters in a range to extract the best fitting model. Akaike information criterion (AIC) is an estimator of how well the model fits for given dataset in comparison to each other (Table 1).

Table 1. Grid search results

ARIMA (p,d,q)	SARIMA (p,d,q,m)	AIC	Fit Time (in sec)
(0,1,0)	(1,1,1,12)	171.706	0.643
(0,1,0)	(0,1,0,12)	167.845	0.015
(1,1,0)	(1,1,0,12)	159.195	0.137
(0,1,1)	(0,1,1,12)	150.423	0.536
(0,1,1)	(1,1,1,12)	152.756	0.732
(0,1,1)	(0,1,0,12)	148.756	0.196
(1,1,1)	(0,1,0,12)	149.579	0.488
(0,1,2)	(0,1,0,12)	149.827	0.307
(1,1,2)	(0,1,0,12)	151.018	0.504
(0,1,1)	(1,1,0,12)	150.482	0.645

Lower the AIC value, better thee model will fit. We choose the parameters with the lowest AIC value. This helps us get the closest predictions becauseas lowest value of AIC means model is better relative to other. We can clearly see from the table that lowest value AIC and BIC is found model with $p = 0$, $d = 1$, $q = 1$ and $P = 0$, $D = 1$, $Q = 0$, $m = 12$ (Table 2).

Table 2. Summary of fitted model

SARIMA model	Log likelihood	AIC	BIC	HQIC
$(0,1,1)X$ $(0,1,0)_{12}$	−71.378	148.756	154.306	150.845

3.4 Validating Model

Grid search results not only in estimation of parameters but also returns a model object that we can fit to train the data. We train with data by giving monthly data of past 4 years (from 2013 to 2016) and test data for 1 year (2017). Now we make predictions for 207 to validate our model (Fig. 5, Tables 3 and 4).

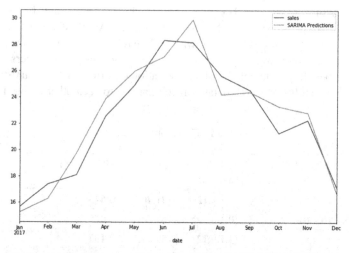

Fig. 5. Monthly actual sales vs. predicted values using SARIMA$(0,1,1)X(0,1,0)_{12}$ model

We check error using RMSE (root mean square error) which is 1.228 which is a measure of how far the actual values are from predicted values. We can deduce that out of 12 actual values only 1 value is different gving us about 92% acurracy. Also from the graph we can see that our predicated values are close to actual sales.

Table 3. Predcited values of test data

Date	Predicted sales
2017-01-01	15.241402
2017-01-02	16.274773
2017-01-03	19.757531
2017-01-04	23.890865
2017-01-05	25.983338
2017-01-06	27.090865
2017-01-07	29.886563
2017-01-08	24.209144
2017-01-09	24.390865
2017-01-10	23.305918
2017-01-11	22.824198
2017-01-12	16.757531

Table 4. Actual values of test data

Date	Actual sales
2017-01-01	15.645161
2017-01-02	17.392857
2017-01-03	18.096774
2017-01-04	22.566667
2017-01-05	24.935484
2017-01-06	28.333333
2017-01-07	28.161290
2017-01-08	25.612903
2017-01-09	24.533333
2017-01-10	21.290323
2017-01-11	22.266667
2017-01-12	17.193548

4 Result

The final step after validating model is to apply the model to predict sales for the year 2018 (Fig. 6).

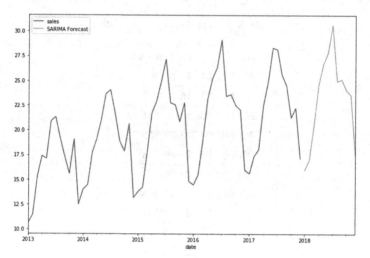

Fig. 6. Final monthly prediction of store item

5   Conclusion

Time Series Forecasting is similar to any other predictions but with an extra complexity time. This is why it is often neglected as time is difficult to handle. In this work, SARIMA $(0,1,1)(0,1,0)_{12}$ was found out to be the best candidate model for predicting one year of store items demand using last 5 years of store item salestaken from Kaggle. This demonstrates the potential of SARIMA models to predict sales and demands which can give businesses a good start and reduce the uncertainty of demands so as to make profits. Moreover, with machine learning techniques, future forecasts can be easily predicted with accuracy. We fitted the model by directly training the model on past dataset with machine learning.

References

1. Store Item Demand Forecasting Challenge, kaggle.com, June 2018. https://www.kaggle.com/c/demand-forecasting-kernels-only/overview
2. Adebiyi, A., Adewumi, A., Ayo, C.: Stock price prediction using the ARIMA model. In: Presented at 16th International Conference on Computer Modelling and Simulation (2014)
3. Nochai, R., Nochai, T.: Arima model for forecasting oil palm price. Presented at 2nd IMT_GT Regioonal Conference on Mathematics, Statistics and Applications University Sains Malaysia, Penang, 13–15 June 2006
4. Hong, W.C.: Rainfall forecasting by technological machine learning models. Appl. Math. Comput. **200**(1), 41–57 (2008)
5. Pavlyshenko, B.M.: Using stacking approaches for machine learning models. In: Proceedings of the 2018 IEEE Second International Conference on Data Stream Mining & Processing (DSMP), Lviv, Ukraine, 21–25 August 2018
6. Etuk, E.H., Mohamed, T.M.: A seasonal arima model for forecasting monthly rainfall in gezira scheme, sudan. J. Adv. Stud. Agric. Biol. Environ. Sci. **2**(7), 320–327 (2014)

7. Luo, C.S., Zhou, L.Y., Wei, Q.F.: Application of SARIMA model in cucumber price forecast. Appl. Mech. Mater. **373–375**, 1686–1690 (2013)
8. Jarrett, J.E., Kyper, E.: ARIMA modeling with intervention to forecast and analyse Chinese stock prices. Int. J. Eng. Bus. Manag. **3**(3), 53–58 (2011)
9. Guha, B., Bandyopadhyay, G.: Gold price forecasting using ARIMA model. J. Adv. Manage. Sci. **4**(2), 117–121 (2016)
10. Li, S.: An End-to-End Project on Time Series Analysis and Forecasting with Python, towardsdatascience.com, July 2018. https://towardsdatascience.com/an-end-to-end-project-on-time-series-analysis-and-forecasting-with-python-4835e6bf050b?gi=b708004bef61

Improving Packet Queues Using Selective Epidemic Routing Protocol in Opportunistic Networks (SERPO)

Tanvi Gautam[1]([⊠]) and Amita Dev[2]

[1] University School of Information and Communication Technology,
Computer Science and Engineering, Guru Gobind Singh Indraprastha University,
Dwarka 110078, New Delhi, India
tanvigautam99@gmail.com
[2] Indira Gandhi Delhi Technical University for Women, Kashmere Gate,
New Delhi 110006, India
amita_dev@hotmail.com

Abstract. Opportunistic networks are an extension of Adhoc networks and sub-class of MANET wherein network possess intermittent connectivity and store – carry – forward mechanism. During routing, if a packet transmission has been initiated by a node through Epidemic to all its neighboring nodes by producing a multiple copy resulting in the network resource consumption early. In Epidemic routing mechanism a multiple copy of a message packet has been created in the direction of inflate the message packet delivery ratio contrary it also enhances additional buffer capacity of a network. By proposing a SERPO approach the packet transmission happens only on a selected route which find out with the help of Dijkstra's algorithm and for minimizing the packet congestion using Weighted Fair Queuing. In this paper we propose an outline to improve the delivery ratio, transmission delay and access delay of a packet in a network using SERPO. We broadly simulate the proposed scheme in ONE simulator and compared it with Epidemic variants using the network performance metrics message transmission (packet delivery), message relaying time (transferring time) and access delay ratio respectively.

Keywords: Adhoc network · Buffer management scheme · Epidemic routing protocol · MANET · Opportunistic Networks

1 Introduction

Wireless message transmission model broadly classified into two classes: infrastructure and infrastructure less network. Infrastructure network belongs to network that possess characteristics as: presence of the central administrator and fixed/stationary nodes whereas in infrastructure less network absence of central administrator and dynamic/mobile nodes [1]. Infrastructure less networks are of various kinds such as; Adhoc Network [2–4], Opportunistic Networks (OppNets) [5], Wireless Sensor Network

© Springer Nature Singapore Pte Ltd. 2020
M. Singh et al. (Eds.): ICACDS 2020, CCIS 1244, pp. 382–394, 2020.
https://doi.org/10.1007/978-981-15-6634-9_35

(WSN) [6], Vehicular Adhoc Network (VAN) [7] and so on. Opportunistic Networks has been well-thought-out of a subclass of DTN and an extension of MANET. The main characteristics of an OppNets are as follow:

i.   Subclass of DTN/Adhoc and an extension of MANET. From DTN it inherited the feature of intermittent connectivity (i.e. no pre-defined routing path between source and destination node) whereas MANET inherited feature is dynamic topology of the network.
ii. Transitional nodes (i.e. intermediate) used store – carry – forward fashion.
iii. Network link recital is extremely flexible or risky.
iv. Dynamic network topology.
v.   Tolerant of extended interruptions and high fault frequency.

The network link performance in an intermittent environment is extremely flexible or risky which results in the worse performance of TCP/IP and Internet protocols hence OppNets arise as ancient routing algorithms fail in such environment. The transitional nodes trail store – carry – forward (i.e. transitional nodes store the message packet till it comes into communication range with other nodes) for message packet transmission in an intermittent region. Epidemic create a multiple copy of a message packet forwarding through transitional node to destined node through flooding mechanism [8]. Such methodology helps in improving the message delivery ratio but it also enhances more network resource consumption such as power, battery, energy, buffer and so on. But if in a case new message packet has been arrived in a network it causes old message packet drop from a buffer resulting in dropping message delivery ratio and irrelevant network resource consumption [9]. The main encounter of Epidemic towards utilizing the less network resources and to improves the packet delivery ratio (PDR) by utilizing the minimum network bandwidth. The most promising OppNet routing protocols are Epidemic variants [8, 10–12] and we propose the Epidemic variant for improvising the network performance metrics message transmission (packet delivery) ratio, message relaying (packet transferring) time and message access delay ratio which helps in optimizing the less network resource consumption resources. In our proposal we propose a strategy to improve the packet delivery ratio (PDR), packet transmission time and packet access delay by enhancing a buffer strategy scheme on the basis of encountering a nodes distance using Dijkstra's algorithm[13] along with the improving buffer management scheme using the Weighted Fair Queuing mechanism instead of FIFO for message packet queuing [14]. As a result, it also improves the network resource consumption, packet delivery, transmission time and access delay ratio respectively because as soon the transmission starts, node start receiving the message copies moment buffer space is not sufficient to hold the packet node will not accept it and primarily it is called as Selective Epidemic RP (SERPO). Remainder segment of paper is systemized as per follows. Unit II, boons the contextual and assessment of interrelated work. Unit III give the detailed explanation of proposed approach SERPO. Unit IV is fermenting an enactment of planned method SERPO using ONE simulator along with the observation and results. Section 5 we accomplish and stretch the forthcoming course of our proposed work.

2 Contextual and Associated Work

Herein we extant an outline of numerous work that has been proposed by researchers for Epidemic protocol in OppNets namely First Protocol [15], Epidemic [8], Spray and Wait [16], Energy efficient N Epidemic [10], Prioritized Epidemic Routing [11], Improving energy consumption of Epidemic Routing [12].

2.1 First Contact [15]

The promising one is to let the source node carrying packet all the way to its destined node by moving utterly in an environment. If uncertainly in matter of path unavailability, then certainly the packet waits till the path becomes available for packet transmission and as soon the path availability is there then finally packet has been dropped (first). The pro of this scheme is it remove the local copy of packet whereas cons are it has poor packet delivery ratio and large network resource consumption.

2.2 Epidemic [8]

Flooding based scheme in which a multiple copy of packet has been generated and transmitted to all transitional nodes. Each node holds the packet till it comes into communication range with the transitional node. Each node holds the packet that has been originated and packets received by the other nodes. Those entries that has been locally set by a host prepared in a form called as bit vector "Summary Vector" (SV). The instant both transitional nodes originate into communiqué series through each further, node with smaller identifier inductees an "anti – entropy" assembly. During such session node exchanges the summary vector table with each other. Working of Epidemic shown below in Fig. 1.

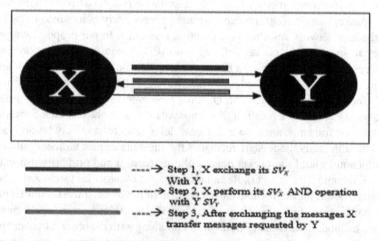

Step 1, X exchange its SV_x With Y.

Step 2, X perform its SV_x AND operation with Y SV_y

Step 3, After exchanging the messages X transfer messages requested by Y

Fig. 1. Working of epidemic routing protocol

The pros of this scheme are that packet delivery ratio is high within minimum amount of time. The cons of this scheme are that it incurs maximum buffer space requirement and resource consumption.

2.3 Spray and Wait [16]

For controlling the flooding level in Epidemic routing [8] an interesting technique [16] has been proposed. Within two phases packet has been initiated: Spray and Wait. In Spray phase L duplicate packets broaden through the foundation hop and respite further transitional nodes receives the duplicate packet from the source node in L interval distinct delays. During Wait phase if other transitional nodes didn't have this duplicate packet, then such node accomplishes the direct transmission. The pros of this scheme are a smaller number of packet transmissions. The cons are large network resource consumption.

2.4 Energy Efficient N – Epidemic Routing Protocol for Delay Networks [10]

Energy efficient n – Epidemic routing protocol for delay networks [10] mechanism packet transmission only done when it has n – neighboring nodes. On the basis of valid n – value obtained the method of packet forwarding only that helps in maintaining the energy consumption of node, in former mentioned case packet forwarding occurs for reducing the network energy consumption. By choosing appropriate n value it also reduces the number of packet transmissions. The pros are it improves the packet transmission time but if n value is small then large number of transmission else smaller or no transmission at all which impact the packet delivery ratio.

2.5 Prioritized Epidemic Routing for Opportunistic Networks [11]

In [11] PREP is being an extended variant of Epidemic [8]. To enact a fractional message bundles followed by DTNRG speech [17] intended for communication and obliteration. The dependency of precedence purpose is on four inputs: latest taken rate from basis to end point, latest taken cost from source, packet finish and era period respectively. Average availability act as a novel metric which is generally use for inter – node cost computation and priority assigned to a node. To each available link, average availability value is epidemically flooded, in case of packet replication packet dropping occurs by a node. Pros is it incurs less resource consumption and con is poor packet delivery ratio.

2.6 Improving Energy Consumption of Epidemic Routing in Delay Tolerant Networks [12]

The rationale behind this [12] approach is that network performance metric i.e. message transmission ratio is relatable with long life and in enhancing the network life time which in turns increases the long life of network. If a node utilizes less network resources (energy) only then network lives longer. The approach is packet transmission done by a likelihood node which possess high energy and free buffer because such node lives for longer time in a network and packet stays in a buffer for longer time. If a transmitting node energy and buffer space is less in comparison with the receiving one then such node will not receive the packet even if it has copy of it. The pro of this scheme is less network resource consumption but it significantly affects the packet delivery ratio.

3 Proposed Approach

Priorly getting into detail regarding the proposed approach, we first summarizing the observations that has been concluded in Epidemic routing [8] which impose as a reason for proposing a new approach. The observations drawn as follows:

i.   IMEP layer is responsible for informing to all transitional nodes about routing path availability.
ii. CDF (i.e. Cumulative Distributive Function) [18] is used for measuring the amount of resource consumption i.e. the number of nodes traversed by a source node for transferring the message to the final destination.
iii. It can be observed that to limit buffer space requirement we need to minimize the latency ratio and number of traversed nodes i.e. packet reach to the destined node by traversing minutest figure of transitional nodes.

Therefore, it can be concluded that to limit the resource consumption it is required to limit the buffer space requirement. Primarily argument is that the performance of Epidemic routing [8] is outstanding in comparison with Delay Tolerant Network [3] and MANET [4] and for ensuring 100% packet delivery ratio. But it also consumes lot of network resource consumption while a multiple copy of message has been generated. Then, question arise is that how we simply fix the shortcomings in Epidemic [8] where its performance weakens in terms of network resource consumption (such as bandwidth/storage considerations).

   The key idea behind SERPO is that it follows a reverse approach proposed approach is that packet lives longer in a network if a network possess sufficient buffer storage then only it resolves and upsurge the packet transfer ratio. If a packet does less buffer utilization in a network then it increases the life of node, network, packet and minimizes less network resource consumption respectively. Choosing the optimal buffer space during packet transmission is a key of increasing a life of network. In our scheme SERPO, it is basically divided into three phases: (1) Apprehension of network configuration that qualifies in assessing the link availability from source node to destined and cost computation helps in determining cost from source to destination node followed by OR operation on the basis of node cost value (2) Packet queuing, Precedence and Dissemination in this scheme packet queuing is done using Weighted Fair queuing scheme afterwards node precedence selection is done on the basis of which dissemination of packet completed. The parameters used to show in Table 1 with notations

3.1 Apprehension of Network Configuration

In this phase, a node determines the link availability among two nodes which is essential in maintaining the bidirectional link (i.e. transmitting packet in both direction among

Table 1. Characteristics of packet queue mechanism [14] categories

Packet queuing mechanism name	Pros	Cons
Priority queuing [14]	Higher priority packet always processed first	No service for lower priority traffic
Weighted Round Robin [14]	Fair buffer allocation	Knowledge of packet size is required
Weighted Fair Queuing [14]	Good packet delivery ratio No starvation	Low priority traffic incurs a minimum amount of delay
Custom Queuing [14]	Each queue processed in order no priority has been taken into consideration	Maximum queue size is 16
Class Based WFQ [14]	Fair buffer allocation with FIFO basis	No priority queue for real time traffic

two nodes) which can be done using link discovery algorithm [7]. With each link a novel metric is attached called as link metric. Link metric can be used in determining the routing path availability among two nodes (i.e. whether the link will be available in future or not). Link metric can be calculated by Eq. (1):

$$LA = \frac{L_u}{L_T} \tag{1}$$

LA: Link Availability Metric
L_u: Time for which link was active in past
L_T: Total time for which link was active

Once the link is active after determining the value of LA, node sends a note to the transitional node called "ACT" and store it in a hash index table using Epidemic [8]. In minimizing network overhead only, a sync each node is permitted in each time of link availability and also helps in giving the knowledge about network topology. In cost determining phase for determining lowest cost route using Dijkstra's algorithm [13] metric called as Route Cost (RC). After determining both value link availability and route cost, we get the node cost value (NV) which can be determined below using Eq. (2).

$$NC = LA + RC \tag{2}$$

The moment we determined NC, then we choose minimum value of node with the help of which we determine the shortest available routing path.

3.2 Packet Queuing, Precedence and Dissemination

The second phase has been conducted into three subphases: packet queuing, precedence and dissemination. Packet queuing mechanism in Epidemic [8] is done on FIFO [14] basis. There are various queuing mechanisms: Priority queuing (PQ), Weighted Round Robin (WRR), Weighted Fair Queuing (WFQ), Custom Queuing and Class Based WFQ respectively [14]. In our proposed approach the chosen one queuing mechanism is WFQ because no starvation, less resource consumption and priority has been assigned to a packet on the chosen metric (*i.e. Node Cost*). The characteristics of queuing categories has been defined below in Table 1.

During Epidemic [8] execution in ONE [19] it has been observed that on the buffer size = 25 M the bottleneck has been observed in community model. Bottleneck can be defined as "congestion in the network that impeded traffic flow." Bottleneck in a network measured with the help of a metric called "Buffer occupancy rate (BOR)" using Eq. (3). Bottleneck probability (BPR) can be defined below using Eq. (4). LM: Lively messages can be defined as the number of messages which has been initially transmitted. RM: Relay messages can be well-defined by means of numeral messages transmitted.

$$BOR = \frac{Number\ of\ LM}{Total\ number\ of\ messages} \tag{3}$$

$$BPR = \frac{Number\ of\ AM}{Total\ number\ of\ messages} \times 100 \tag{4}$$

For exploring the characteristics of Epidemic [8] using FIFO [14] queuing mechanism under a different number of scenarios. We first explore the robustness of Epidemic by varying the buffer size (5 M, 10 M, 15 M, 20 M, 25 M) with 60, 80, 100 and 120 nodes respectively. In Fig. 2 we study a bottleneck probability of FIFO packet queuing in Epidemic [8] for observing the number of packets delivered and the instance on 20 M, delivered packets start drop which represents the possibility of arising a congestion in a network resulting in bottleneck in OppNets. In, Table 2 observations has been drawn represent the lively messages (LM), relay messages (RM), aborted messages (AM), Buffer occupancy rate (BOR) and Bottleneck probability ratio (BPR) respectively.

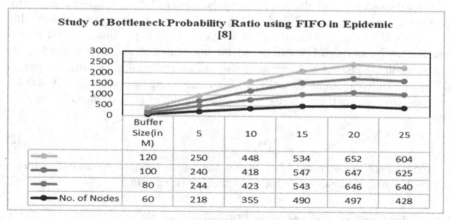

Fig. 2. Study of bottleneck using FIFO in Epidemic [8]

Table 2. Observed values using FIFO in Epidemic

S. No.	No. of nodes	LM	RM	AM	BOR (%)	BPR (%)
1	60	8119	6681	1438	26.8	48.63
2	80	1548	1261	287	10.5	19.6
3	100	23587	19535	4048	2.611	2.76
4	120	33326	27719	5607	2.1	0.82

From Fig. 2 it has been studied that on 20 M buffer size, bottleneck has been increased in the network as the packet dropping is being started. The observations that has been concluded from Fig. 2. can be tabularized below in Table 2 for buffer size = 5 M.

Due to the observations that has been drawn in Table 2 after studying FIFO queuing in Epidemic and on studying the characteristics from Table 2, we use WFQ [14] in our proposed approach which employs a fair allocation of buffer space to packet by giving an assumption to the packet length (in our approach packet size = 20). The working scheme of WFQ [14] shown diagrammatically below in Fig. 3. It supports variable length packet due to which large buffer space has been not allocated to the large packet size and for smaller packet not small buffer space, the optimal buffer size has been chosen. After assigning the buffer size packet assemble at input port (ingress), then each packet queue follows share output port bandwidth and packet transmission-initiated bit by bit. On the basis of bit by bit packet transmission at output port (egress) packet reassemble. After packet gets reassemble at the egress port, classifier helps in the queueing of the packet on the basis of smallest packet finish time. The packet with smallest finish time has high priority and packet with highest finish time has less priority, priority of packet has been assigned on the basis of Table 3. Priority Values.

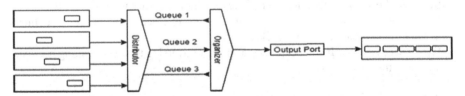

Fig. 3. Weighted Fair Queuing (WFQ) [14] packet queuing mechanism

Packet datagram of message packet shown below in Fig. 4 which basically comprises of three fields: Source/Destination address, Node number, Priority Value.

Source/Destination Address	Packet No.	Class of System (Priority Value)

Fig. 4. Packet datagram of message packet

Table 3. Priority value on the basis of Class of System

Type of a class	Priority Value
Emergency situations such as military, disaster and road accidents	1
Less number of packets in network	0

The field description has been defined below:

a. Source/Destination address: Denotes the node source/destination address.
b. Packet No.: Denotes the Node number of a packet.
c. Class of System (Priority Value): Priority of a packet has been decided on the basis of packet type which is defined below in Table 3.

4 Simulation Results

We present the simulation of proposed approach SERPO in ONE simulator [19], a comparative analysis with Epidemic [8], Energy efficient N Epidemic (N Epidemic) [10], and Prioritized Epidemic Routing for Opportunistic Routing (P Epidemic) [11] respectively. The constraints castoff intended for imitation by considering most of the ONE [19] parameters in default mode. The generation of nodes were random in a square area A and node mobility has been chosen a well-known Shortest Path Based Movement mobility model with the packet payload size (1000 bytes) and Time to live is 20 min. Along with the comparative analysis of SERPO we studied its performance with help of three parameters: packet delivery ratio, packet transmission time and packet access delay respectively.

a. Transfer ratio/Message Transmission Ratio (PDR-Packet Delivery Ratio): Ratio of transferred packets to total number of message (packet) distribution.
b. Packet transmission time: It is a synonym of packet overhead ratio and can be defined as total transmission time of a packet delivery.
c. Packet access delay: It can be defined as packet traversing time of traveling between source to the destination.

Below, Fig. 5 presents transfer ratio/message transmission ratio (PDR-Packet Delivery Ratio) analysis with the existing routing protocols to show that the performance of proposed approach SERPO is best. By applying the WFQ packet queuing mechanism [14] observation drawn which indicate the message delivery ratio of SERPO seems to be improved by 0.021%, 0.09%, 0.48% on n = 20 (where n: buffer size [M]) with comparison of Epidemic, N Epidemic and P Epidemic respectively. On n = 40, again the PDR is improving by 0.61%, 0.001% and 0.54% with comparison of Epidemic, N Epidemic and P Epidemic respectively. But at n = 60, PDR is dropping in Epidemic, N Epidemic and P Epidemic respectively as soon as n = 80 then it seems to be improving again by 0.156% and 0.947% with Epidemic and P Epidemic respectively not in N Epidemic. Again, on n = 100, PDR seems to be increase by 21%, 41.5% and 11.28% respectively with comparison of Epidemic, N Epidemic and P Epidemic respectively. This shows that there is no much of buffer consumption and resource usage in network. Therefore, it can be concluded that our proposed approach SERPO reduces the network consumption and improves the packet delivery rate using WFQ [14]. With the proposed approach SERPO it has been observed that bottleneck issue is being improved.

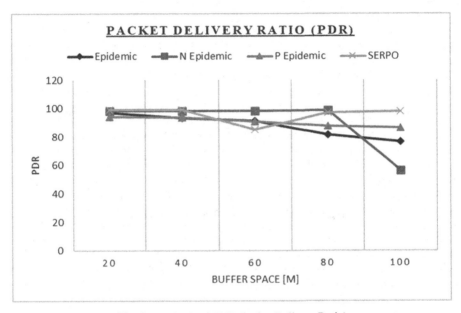

Fig. 5. Analysis of PDR (Packet Delivery Ratio)

As of Fig. 6 observation can be concluded that packet overhead ratio of SERPO is best on n = 20, 40, 60, 80, 100 with comparison of Epidemic and N Epidemic but on performance analysis with P Epidemic it has been seen that on n = 20 performance of SERPO is poor but it seems to be improving upto n = 40–100 by 0.136%, 0.077%, 0.349%, 0.479% respectively. It shows that network overhead ratio is less which in turn reduces network resource consumption and overall improves the occurrence of bottleneck probability. Below in Fig. 7 presents the packet access delay performance metric analysis of SERPO that on n = 20,40,60,80,100 its performance is best with comparison of Epidemic, N Epidemic and P Epidemic respectively. But in SERPO access ratio on n = 40 it increases by 0.1% within comparison of n = 20 and it seems to be again decreased by 0.1% on n = 60. Again, it increases on n = 80 by 1.01% and again decreases by 0.12% on n = 100. SERPO improves the packet access delay ratio due to fair allocation of buffer size to the packet queues after observing the FIFO [14] bottleneck analysis. Again, it increases on n = 80 by 1.01% and again decreases by 0.12% on n = 100. SERPO improves the packet access delay ratio due to fair allocation of buffer size to the packet queues after observing the FIFO [14] bottleneck analysis.

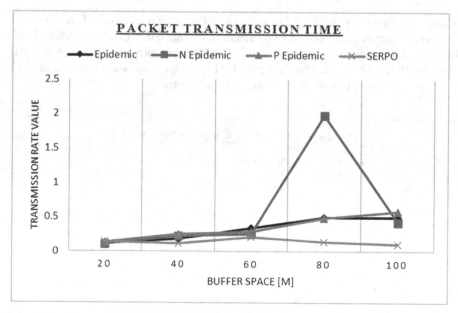

Fig. 6. Analysis of Packet Transmission Delay

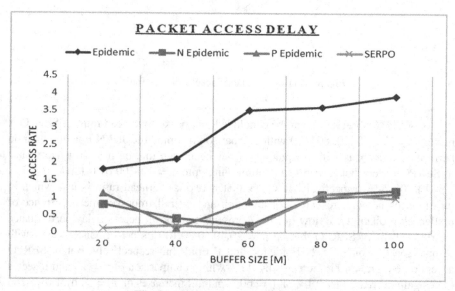

Fig. 7. Analysis of packet access delay

5 Conclusion and Future Work

It can be concluded that performing the simulation in ONE [19] simulator the average of packet delivery ratio is 4.785 i.e. which means that SERPO increase the delivery performance by 479% average. The analysis of performance metric i.e. Packet Transmission Delay is average in performance within comparison of all existing routing protocols. It has been shown that on buffer size $= 20\,M$, congestion has been controlled by making the packet queue size $= 20\,M$ following by Weighted Fair Queuing mechanism i.e. it reduced the overall latency in a network. The simulation shows that SERPO performance is better and increases the delivery ratio by 479% respectively. We can give the conclusion that performance of SERPO is better than existing routing protocols. The advancement can be done in proposing a new buffer management approach by reducing the network overhead ratio and resource consumption. Additional examination of planned method SERPO desirable for diverse simulation settings like varying buffer dimensions, packet dimensions and under various mobility models.

References

1. Rinne, J., Keskinen, J., Berger, P., Lupo, D., Valkama, M.: Wireless energy harvesting and communications: limits and reliability. In: 2017 IEEE Wireless Communications and Networking Conference Workshops (WCNCW) (2017). https://ieeexplore.ieee.org/document/7919070

2. Malarkodi, B., Gopal, P., Venkataramani, B.: Performance evaluation of adhoc networks with different multicast routing protocols and mobility models. In: 2009 International Conference on Advances in Recent Technologies in Communication and Computing (2009). https://ieeexplore.ieee.org/document/5328071

3. Ukey, N., Kulkarni, L.: Energy based recent trends in delay tolerant networks. In: Mishra, D., Nayak, M., Joshi, A. (eds.) Information and Communication Technology for Sustainable Development. LNNS, vol. 9, pp. 141–150. Springer, Singapore (2018). https://doi.org/10.1007/978-981-10-3932-4_15

4. Maratha, B.P., Sheltami, T.R., Salah, K.: Performance study of MANET routing protocols in VANE. Arab. J. Sci. Eng. 42, 3115–3126 (2017). https://doi.org/10.1007/s13369-016-2377-y

5. Huang, C.M., Lan, K.C., Tsai, C.Z.: A survey of opportunistic networks. In: Proceedings of the 22nd International Conference on Advanced Information Networking and Applications (WAINA), 25–28 March. Biopolis, Okinawa, Japan (2008)

6. Kim, B., Park, H., Kim, K., Godfrey, D., Kim, K.: A Survey on real-time communications in wireless sensor networks. Wireless Commun. Mobile Comput. 2017, 1–14 (2017). https://www.hindawi.com/journals/wcmc/2017/1864847/

7. Al-Sultan, S., Al-Doori, M., Al-Bayatti, A., Zedan, H.: A comprehensive survey on vehicular Ad Hoc network. J. Netw. Comput. Appl. 37, 380–392 (2014). https://dl.acm.org/citation.cfm?id=2565407

8. Vahdat, A., Becker, D.: Epidemic routing for partially connected ad hoc networks. Technical Report CS – 2000 – 06, Department of Computer Science. Duke University, Durham, NC (2000)

9. Spyropoulos, T., Psounis, K., Raghavendra, C.S.: Single – copy routing in intermittently connected mobile networks. In: Proceedings of the Sensor and Ad Hoc Communications and Networks (SECON), pp. 235–244, October 2004

10. Lu, X., Hui, P.: An energy-efficient n-epidemic routing protocol for delay tolerant networks. In: Networking, Architecture and Storage (NAS), 2010 IEEE Fifth International Conference (IWCMC), 2013 9th International, pp. 731–735, July 2013
11. Yamazaki, T., Yamamoto, R., Miyoshi, T., Asaka, T., Tanaka, Y.: PRIOR: prioritized forwarding for opportunistic routing. IEICE Trans. Commun. **100**(1), 28–41 (2017)
12. Bista, B.: Improving energy consumption of epidemic routing in delay tolerant networks. In: 2016 10th International Conference on Innovative Mobile and Internet Services in Ubiquitous Computing (IMIS) (2016). https://ieeexplore.ieee.org/document/7794476
13. Fuhao, Z., Jiping, L.: An algorithm of shortest path based on Dijkstra for huge data. In: 2009 Sixth International Conference on Fuzzy Systems and Knowledge Discovery (2009). https://ieeexplore.ieee.org/document/5359145
14. Singh, A.K, Meenu: A survey on congestion control mechanisms in packet switch networks. In: 2015 International Conference on Advances in Computer Engineering and Applications (2015). https://ieeexplore.ieee.org/document/7164833
15. Jain, S., Fall, K., Patra, R.: Routing in a delay tolerant network. In: Proceedings of the ACM SIGCOMM, pp. 145–158 (2004)
16. Spyropoulos, T., Psounis, L., Raghavendra, C.S.: Spray-and-wait: an efficient routing scheme for intermittently connected networks. In: Internet Request for Comments, RFC Editor, RFC 6693, August 2012. http://www.rfc-editor.org/rfc/rfc6693.txt, http://www.rfc-editor.org/rfc/rfc6693.txt
17. Cerf, V., et al.: Delay-tolerant network architecture, April 2007. Internet RFC 4838
18. Tanyer, S.: The cumulative distribution function for a finite data set. In: 2012 20th Signal Processing and Communications Applications Conference (SIU) (2012). https://ieeexplore.ieee.org/document/6204462?section=abstract
19. Keranen, A., Ott, J., Karkkainen, T.: The ONE simulator for DTN protocol evaluation. In: Proceedings of the 2nd International Conference on Simulation Tools and Techniques, ser (Simutools 2009), pp. 55:1–55:10. ICST, Brussels, Belgium, Belgium: ICST (Institute for Computer Sciences, Social – Informatics and Telecommunications Engineering) 2009

Heart Disease Prediction System Using Classification Algorithms

Sarthak Vinayaka and P. K. Gupta[✉]

Jaypee University of Information Technology, Waknaghat 173 234, India
juitsarthak@gmail.com, pkgupta@ieee.org

Abstract. Heart disease and stroke have had an impact on 28:1% of total deaths in India in 2016 as compared to 15:2% in 1990. With the rising use of learning algorithms, In this edition paper, we have developed a system for predicting heart disease that can predict heart disease by using a modified random forest algorithm. The proposed algorithm is trained with a dataset consisting of 303 instances which help to predict the occurrence of heart disease with an accuracy of 86:84% and can be implemented in the medical field to improve the overall diagnosis about heart disease.

Keywords: Random forest · Machine learning · Heart disease · Accuracy

1   Introduction

In the past, heart disease was more common in the age between 50–70 years but now it is rapidly spreading among the youth and getting more common countrywide. With the evolution of technology, the youths are becoming more dependent on the excessive use of technology that impacts an adverse effects on their health. It is because their physical activities level has reduced and also their eating habits have changed a quiet lot and getting more prone to use of fast foods and packed foods which in turn increases the chances of heart disease. According to research conducted by WHO, it is estimated that due to heart disease 12 million death occurred worldwide. The death caused by heart disease is sudden or without any warning or symptoms. With a heart prediction The pathologist information will also help to assign weight to the most impactful attribute. More useful the attribute for deciding the heart disease will get more weight further, it will also help the doctors to provide more accurate decisions by using their knowledge and using the proposed model for heart disease prediction. Several studies in the past have achieved some accuracy by applying Artificial Neural Network (ANN) and other Machine Learning based algorithms over the existing dataset, and concluded that due to lack of data relevant to the accuracy obtained is not maximized more [5,6,14]. The purpose of this paper is to increase the accuracy by using a different approach in the Random Forest algorithm

© Springer Nature Singapore Pte Ltd. 2020
M. Singh et al. (Eds.): ICACDS 2020, CCIS 1244, pp. 395–404, 2020.
https://doi.org/10.1007/978-981-15-6634-9_36

that will fit best in the provided dataset and also considering the non-linear dependency of an attribute in the dataset. Proposed the methodology will help to get better accuracy by adjusting the number of trees and the depth of the trees in the random forest algorithm.

2   Data Source

In this paper, the data used is provided by the machine learning repository of UCI [17]. In the dataset, the total number of instances is 303 out of which healthy instances were 164 and the remaining instances belong to heart disease. Also, there exist 14 clinical features for each example.

2.1   Feature Illustration

Table 1 represents 14 clinical features used for the proposed system. In the used dataset these 14 features are categorized into 8 symbolic and 8 numeric features.

Table 1. Various clinical features and their description [17]

Clinical features	Description
Age	Instance age in years
Sex	Instance gender
Cp	Chest pain type
Trestbps (mm Hg)	Resting blood pressure
Chol (mg/dl)	Serum cholesterol
Fbs	Fasting blood sugar
Restecg	Resting electrocardiographic results
Thalach	Maximum heart rate achieved
Exang	Exercise induced angina
Oldpeak	ST depression induced by exercise relative to rest
Slope	The slope of the peak exercise ST segment
Ca	Number of major vessels (0–3) colored by flourosopy
Thal	3 = normal; 6 = fixed defect; 7 = reversible defect
Target	Diagnosis of heart disease or not

3   Related Work

In this section, we have discussed the various classification algorithms like Naive Bayes, KNN, decision tree, random-forest etc. as used and implemented by the other authors in their work.

3.1 Naïve Bayes

In [3], Dulhare has used the Naive Bayes algorithm for heart disease prediction and achieved accuracy 87:91% with the use of a proposed modified algorithm known as Naive Bayes + PSO. In [1], Anbarasi et al. have also used the Naive Bayes classification algorithm that performs consistently before and after the reduction of features for the proposed model and gained an accuracy of 96:5%. In [10], Medhekar et al. have used Naive Bayes classification algorithm based on Bayesian Theorem and calculated the accuracy of 89:58% in their results. In [2], Cherian et al. have proposed a Heart disease predictive system and concluded that Jelinek-mercer smoothing technique is the more effective than Naive Bayes for predicting patients with heart disease. they have achieved 78% accuracy with naive Bayes whereas with Laplace Smoothing it comes around 86%. In [18] Vembandasamy et al. have applied the Naive Bayes algorithm and concluded that this algorithm provides 86:4198% of accuracy with minimum prediction time.

3.2 K-Nearest Neighbours

In [4], Ketut et al. have used the KNN algorithm and concluded that the most important variables came from the data used they are: Exang, CP, and Sex, and concluded with some variations as the most important. They have achieved an accuracy of 81:85% with 8 parameters, and 80:61% with 13 parameters. In [9], Marimuthu et al. have done the analysis for prediction of heart disease and it was proposed to use the k value determined by the square root of the recognition number. They have achieved an accuracy of 83:60% with the proposed algorithm. In [7], Gagandeep et al. have used the K nearest neighbor algorithm and presented that by increasing the number of n- neighbors it provides a very small difference in the average and achieved accuracy was 86%.

3.3 Decision Tree

In [8], Kirmani and Ansarullah have used decision tree-based classification algorithm and achieved 79:1% of accuracy in their results with the use of equal width discretization Information Gain Decision Tree. In [13], Dangare and Apte have used the J48 Decision Tree algorithm. This algorithm uses a pragmatic way to build a decision tree. The idea is to create a tree that provides flexibility, accuracy. In [11], Pandey et al. have also used the decision tree algorithm and obtained the accuracy of Pruned J48 Decision Tree which is better than the simple approach. The results show that fasting blood sugar is the most important attribute it provides a better distinction that conflicts with other qualities but it does not provide better accuracy.

3.4 Random Forest

In [12], Patil and Kinariwala have used the Random forest algorithm and used a weighted voting method which provides accuracy of 74:19% with classical RF, 79:42% with modified RF, and 83:6% with weighted RF. In [15], Singh et al. have used the Random forest algorithm and kept the minimum splits to 10 and the accuracy achieved is 85:81%. In [16], Lavanya et al. have used the random forest algorithm and obtained the accuracy of 53:7736% in predicting the class label of unknown records.

4   Our Proposed Methodology

Random Forest is another learning algorithm that also applies to the nonlinear tendency of the data set as well as provides a better outcome compared to the decision tree algorithm. Random Forest is made up of large quantities of trees along with deliberately random inputs. Proper adjustments should be needed to get better results in the random forest so that by changing parameters such as randomness, number of trees, and the maximum depth, the accuracy could be increased (Fig. 1).

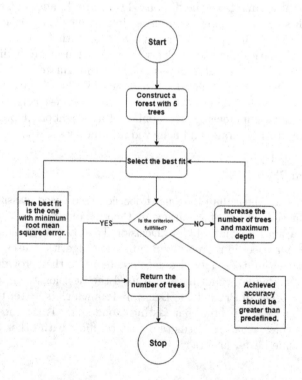

Fig. 1. Our proposed algorithm flow chart.

In the proposed methodology, we have considered the various cases as mentioned follows by changing the amount of trees, and the height of trees to observe the validity of the proposed method.

- Case 1: Amount of trees = 5, Height of tree = 5
- Case 2: Amount of trees = 5, Height of tree = 25
- Case 3: Amount of trees = 10, Height of tree = 60
- Case 4: Amount of trees = 50, Height of tree = 80
- Case 5: Amount of trees = 100, Height of tree = 90
- Case 6: Amount of trees = 110, Height of tree = 110
- Case 7: Amount of trees = 200, Height of tree = 130
- Case 8: Amount of trees = 300, Height of tree = 130

The accuracy obtained for Case 1 to Case 6 varies from 82:89, 82:29, 81:57, 84:21, 86:84, and 86:84 respectively. The accuracy achieved is 86:84% which is highest. Whereas, for Case 7 and Case 8 accuracy is 84:21, and 85:52 respectively which is lower in comparison to Case 6, and also it takes a double amount of time as expressed in Table 2. From Table 2, it is also lucid that accuracy does not follow any trend. Here, we can say that the nature of dataset decides what should be the number of trees, depth of trees and randomness that will bestfit and give

Table 2. Various scenarios for the proposed methodology

Number of trees	Depth	Accuracy (%)	Time (Seconds)
5	5	82.89	0.053
5	25	82.29	0.06
10	60	81.57	0.08
50	80	84.21	0.32
100	90	86.84	0.61
110	110	86.84	0.66
200	130	84.21	1.25
300	130	85.52	1.59

the highest accuracy these parameters may change for the different datasets. We can find the perfect combination of hit and trial method. From Table II, it is also clear that accuracy does not follow any trend. Here, we can say that these factors such as randomness, number of nodes, number trees, and depths depending on the nature of the data and is different from one data to another. The perfect combination of these parameters can only be obtained through the sheer brute force.

5   Performance Analysis

In this section, we have discussed the following parameters of the performance analysis of the proposed algorithm.

- TP (true positive): is a test result that observes the state when the state is present.
- TN (true negative): is a test result that does not observe the state when the state is absent.
- FP (false positive): is a test result that observes the state when the state is absent.
- FN (false negative): is a test result that does not observe the state when the state is present.

The above-mentioned parameters TP, TN, FP, and FN are used to calculate the Accuracy, Precisionscore, F1score, and Recallscore of the proposed algorithm. Results obtained for performance analysis are shown in Fig. 3, Fig. 4, Fig. 5 and Fig. 6.

$$AccuracyScore = \frac{TP + TN}{TP + TN + FP + FN} \tag{1}$$

where TP is True Positives, TN is True Negatives, FP is False Positives, and FN is False Negatives.

$$PrecesionScore = \frac{TP}{TP + FP} \tag{2}$$

$$F_1 Score = 2 * \frac{PrecisionScore * RecalScorel}{PrecisionScore + RecallScorel} \tag{3}$$

$$RecallScore = \frac{TP}{TP + FN} \tag{4}$$

6   Experiments and Results

This section represents the various results obtained after the implementation of the proposed approach on the dataset. During implementation, We have applied the proposed classification algorithm using python and its libraries like sci-kit learn, pandas, NumPy, and matplotlib. The obtained results for the proposed results are shown in Fig. 2, Fig. 3, Fig. 4 and Fig. 5 where we proposed algorithm has achieved an accuracy of 86:84%, F1score = 0:88, Recallscore = 0:93, and Precisionscore = 0:85. Here, Fig. 6 also represents the comparison of the proposed approach with the previously published techniques and represents the highest achieved an accuracy of 86:84%.

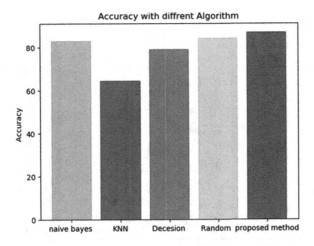

Fig. 2. Comparing accuracy of proposed approach with different algorithms.

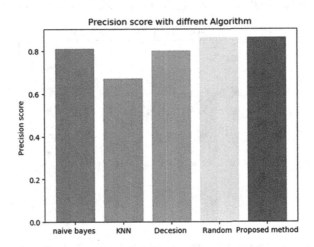

Fig. 3. Comparing precision score of proposed approach with different algorithms.

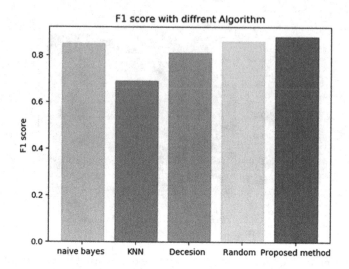

Fig. 4. Comparing F1 score of proposed approach with different algorithms.

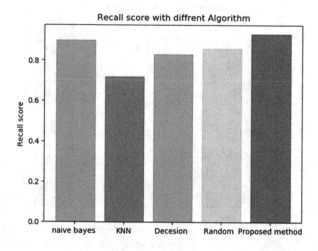

Fig. 5. Comparing Recall score of proposed approach with different algorithms.

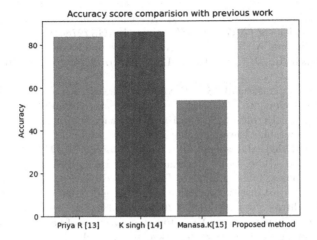

Fig. 6. Comparing accuracy of proposed approach with previously proposed techniques.

7   Conclusion and Future Work

In this work, we have applied various machine learning algorithms onto the dataset and measured the accuracy while predicting heart disease. We have achieved the highest accuracy of 86.84% with the proposed modified random forest algorithm. The proposed algorithm equally respond better in real-time and its accuracy can be increased by collecting more data and by implementing other deep learning based techniques and convolutional neural network. This machine-based prediction based methodology will help to reduce human errors while detecting heart disease.

References

1. Anbarasi, M., Anupriya, E., Iyengar, N.: Enhanced prediction of heart disease with feature subset selection using genetic algorithm. Int. J. Eng. Sci. Technol. **2**(10), 5370–5376 (2010)
2. Cherian, V., Bindu, M.S.: Heart disease prediction using Naïve Bayes algorithm and laplace smoothing technique. Int. J. Comput. Sci. Trends Technol. **5**(2), 68–73 (2017)
3. Dulhare, U.N.: Prediction system for heart disease using Naive Bayes and particle swarm optimization. Biomed. Res. (India) (2018). https://doi.org/10.4066/biomedicalresearch.29-18-620
4. Enriko, I.K.A., Suryanegara, M., Gunawan, D.: Heart disease prediction system using k-Nearest neighbor algorithm with simplified patient's health parameters. J. Telecommun. Electron. Comput. Eng. **8**(12), 59–65 (2016)
5. Gupta, P., Maharaj, B.T., Malekian, R.: A novel and secure iot based cloud centric architecture to perform predictive analysis of users activities in sustainable health centres. Multimed. Tools Appl. **76**(18), 18489–18512 (2017)

6. Gupta, P., Tyagi, V., Singh, S.: Predictive Computing and Information Security. Springer, Heidelberg (2017). https://doi.org/10.1007/978-981-10-5107-4

7. Kaur, G., Sharma, A., Sharma, A.: Heart disease prediction using KNN classification approach. Int. J. Comput. Sci. Eng. (2019). https://doi.org/10.26438/ijcse/v7i5.416420

8. Kirmani, M.M., Ansarullah, S.I.: Prediction of heart disease using decision tree a data mining technique. IJCSN Int. J. Comput. Sci. Netw. 5(6), 885–892 (2016)

9. Marimuthu, M., Deivarani, S., Gayathri, R.: Analysis of heart disease prediction using various machine learning techniques. In: Advances in Computerized Analysis in Clinical and Medical Imaging (2019). https://doi.org/10.1201/9780429446030-13

10. Medhekar, D.S., Bote, M.P., Deshmukh, S.D.: Heart disease prediction system using Naive Bayes. Int. J. Enhanced Res. Sci. Technol. Eng. 2(3), 1–5 (2013)

11. Pandey, A.: A heart disease prediction model using decision tree. IOSR J. Comput. Eng. (2013). https://doi.org/10.9790/0661-1268386

12. Patil, P.P.R., Kinariwala, P.S.A.: Automated diagnosis of heart disease using data mining techniques. Int. J. Adv. Res. Ideas Innov. Technol., 560–567 (2017)

13. Dangare, C.S., Apte, S.S.: Improved study of heart disease prediction system using data mining classification techniques. Int. J. Comput. Appl. 47(10), 44–48 (2012). https://doi.org/10.5120/7228-0076

14. Singh, G.A.P., Gupta, P.K.: Performance analysis of various machine learning-based approaches for detection and classification of lung cancer in humans. Neural Comput. Appl. 31(10), 6863–6877 (2018). https://doi.org/10.1007/s00521-018-3518-x

15. Singh, Y.K., Sinha, N., Singh, S.K.: Heart disease prediction system using random forest. In: Singh, M., Gupta, P., Tyagi, V., Sharma, A., Ören, T., Grosky, W. (eds.) Advances in Computing and Data Sciences, pp. 613–623. Springer Singapore, Singapore (2017). https://doi.org/10.1007/978-981-10-5427-3_63

16. Lavanya, T., Satyanarayana, N., Manasa, K.: Heart disease prediction using random forest algorithm, August 2018. https://doi.org/10.5281/ZENODO.1400571, https://zenodo.org/record/1400571

17. UCI: UCI Machine Learning Repository: Heart Disease Data Set (2019)

18. Vembandasamy, K., Sasipriya, R., Deepa, E.: Heart diseases detection using Naive Bayes algorithm. Int. J. Innov. Sci. Eng. Technol. 2(9), 441–444 (2015)

Data Sciences

Graph Database and Relational Database Performance Comparison on a Transportation Network

Jinhua Chen, Qingyu Song, Can Zhao, and Zhiheng Li[✉]

Department of Automation, Tsinghua University, Beijing 100000, China
{cjh18,sqy18,zhaoc17}@mails.tsinghua.edu.cn,
zhhli@mail.tsinghua.edu.cn

Abstract. Facing the problem of structuring irregular data in the big data era, graph databases are a powerful solution to handle link relationship without costly operations and enjoy great flexibility as data model changes. Though it's well-known that graph databases have superior performance in a certain area than relational databases, little effort has been put into investigating the detail of these advantages. In this paper, we report a systematic performance study of graph databases and relational databases on a transportation network. We design a database benchmark considering traversal and searching performance to evaluate system performance in different data organizations, initial states, and running modes. Our results show that graph databases outperform relational databases system in three main graph algorithms testing. Furthermore, we discuss the reasonable practice in applications based on graph databases from our experiment results.

Keywords: Graph databases · Relational databases. Neo4j · Benchmark · Graph algorithm

1 Introduction

We are living in a digital world where data is everywhere. With the rapid development of information technology and internet industry, mountains of data are produced every day. In the Big Data era, the way how data is stored in an efficient and appropriate approach is one of the key fundamental technical considerations. Acting as the central data storage and manipulating platform, database is the core of these technical considerations.

Traditionally, databases have been designed with the relational model since Edgar F. Codd's milestone paper on the introduction of relational model [1]. As a result, relational database systems are designed and served in various industry information systems. However, this kind of traditional database practice is not a one-fits-all solution. It has huge limitations and disadvantages when it comes to cases where relationships are queried and processed frequently. The reason behind this is, data in the relational model is normalized to strictly support the ACID transactions for preserving data consistency [2]. When complex relationships are involved in queries, many costly join operations will

© Springer Nature Singapore Pte Ltd. 2020
M. Singh et al. (Eds.): ICACDS 2020, CCIS 1244, pp. 407–418, 2020.
https://doi.org/10.1007/978-981-15-6634-9_37

be made, which causes heavy overhead. What's worse is, big data will get bigger, and relevant relationships between these connected entities will grow exponentially, so is the overhead.

Compared with relational databases, graph databases are based on the graph model, fitting the object-oriented applications more naturally. Graph databases depend less on schema regulations, and they are easy to handle link relationship without costly join operations. In addition, the performance advantage of graph databases over relational databases is stable, which means graph database performance keeps constant as data size grows while relational database will meet exponential overhead growth. In addition, graph database enjoys great flexibility as data model changes. It's convenient to add changes to the existing database structures without heavy modifications on a current model [12].

As stated above, graph database has many advantages over relational database in certain areas. However, many of these are stated in a more theoretical manner, there is a lack of practical assessment on the arguments. Besides general performance comparison between graph database and relational database, it's more necessary to focus on some more specific scenes, where graph database is perfectly suitable in those fields. In this paper, we focus on the transportation field, where performance comparison experiments of three variants of the shortest path algorithm are conducted on a transportation network. We choose Neo4j as the representative of the graph database systems and PostgreSQL as the representative of the relational database systems. We provide a high-level overview of Graph database knowledge in *Sect. 2*, followed by a description of database benchmark consideration. Related research work is presented in *Sect. 3*. In *Sect. 4* we illustrate the research methodology, including the descriptions of dataset, experiment design and experiment setup. Experiments results are then shown in *Sect. 5.1*. More detailed discussions of the results are presented in *Sect. 5.2*. Finally, we draw some conclusions in *Sect. 6*.

2 Graph Database

2.1 Graph Database System

According to [3], graph databases are those in which data structures for the schema and instances are modeled as graphs or generalizations of them, and data manipulation is expressed by graph-oriented operations and type constructors. Graph database is based on graph model, which generally defined as a collection of nodes and edges, denoted as $G = (V, E)$, where V is a set of vertices and E is a set of edges. Edge e_i in E connects two vertices in a directed or undirected way, it can be expressed as a triple (s, e, v_i), where s, e $\in V$ and v_i is a real value as edge value. Conceptually the property graph model is the simplest data model of expressing and describing connected information [11].

Property is an important data unit in graph database. In graph database, property information is supported on node entity and edge relationship by one or more key-value pairs. It extends a single edge value on general graph model. This kind of data model is property graphs. It is the most popular data model in the industry, and is becoming increasingly prevalent in academia [20]. Property field on nodes or edges enables another layer of abstraction to the structure to provide convenience for possible common queries and many graph algorithms like shortest path algorithm.

The world we live in is highly interconnected, and graph structure is everywhere. It has been deployed on many leading companies to meet their business requirements. Customized recommendation system [5], artificial intelligence system [6] and fraud detection system [7] are build based on the graph database.

2.2 Graph Query Language

Most relational databases use a dialect of SQL as their query language while the graph database world has a few query languages to choose from. Cypher, SPARQL, GraphQL, and Gremlin are some popular graph database query languages.

Cypher is a declarative query language for property graphs, created for the Neo4j graph database system. As it's declarative, the query optimizer automatically chooses the strategy that is predicted to be the most efficient. It has now been implemented commercially in other products such as SAP HANA Graph, Redis Graph, Agens Graph, and Memgraph. The language therefore is used in hundreds of production applications across many industry vertical domains [19]. SPARQL is a SQL-like declarative query language created by W3C to query RDF (Resource Description Framework) graphs. GraphQL is a query language created by Facebook for APIs that is not specific to graph databases. Gremlin is a query language for Apache TinkerPop. It's a graph traversal DSL (Domain Specific Language) that can be either declarative or imperative.

2.3 Database Benchmark

A database benchmark is a standard set of executable instructions that are used to measure and compare the relative and quantitative performance of two or more database management systems through the execution of controlled experiments [8]. Standard database benchmarks include TP1 benchmark, TPC-A, and TPC-B benchmarks to estimate the performance of online transaction processing. A series of new benchmark standards, TPC-C, TPC-D, TPC-H, TPC-R, and TPC-W are designed to meet the needs of emerging industry or academic requirements [9]. While these are well-designed and effective database benchmarks in different domains, they are designed and developed for relational databases. Paper [14] analyzes the important aspects for designing a graph database benchmark roundly. This paper presents a benchmark designed to measure the performance of the graph database system and the relational database system based on the principles discussed in [14]. Details of this benchmark setting are presented in *Sect. 4.2.*

3 Related Work

To our knowledge only a handful of researches on graph databases performance has been conducted. The Neo4j organization has made an official performance comparison testing for several graph algorithms covered by its graph algorithms library [4]. Paper [13] provides a comparative analysis of the Neo4j graph database with the relational database MySQL. It measures the retrieval times of queries by neo4j and MySQL on graphs of hundreds of nodes. Paper [10] makes a graph community discovery algorithms comparison in Neo4j with a Regularization-based Evaluation Metric. Those community discovery algorithms are implemented in Java over Neo4j. Paper [12] makes the shortest path algorithm performance comparison in graph and relational database. In the work of [4], the algorithms performance comparison is made on a single Neo4j platform. The algorithms are not implemented in a relational database system to make a cross-platform comparison, so is the work in [10]. Paper [12] conducts a cross-platform algorithm performance comparison between a graph database system and a relational database system. However, it only implements one shortest path algorithm on both platforms due to platforms constraint (it's the only algorithm implemented on both platforms) at that time (2014). With the development in these years, we are now able to implement more algorithms in these platforms to make further research.

4   Research Methodology

4.1   Dataset

In order to comprehend the effectiveness of graph databases under transportation domain, we use an urban traffic road network. Euclidean relationship among nodes and edges is not concerned. A detailed description is as follows.

The Shenzhen urban traffic network dataset with 36,968 nodes and 86,230 roads is provided by the OpenStreetMap project [16]. The edges represent real roads in urban traffic network. The nodes are start and end of edges that represented real intersections,

Table 1. Shenzhen urban traffic network dataset description

Data fields	Description	Data type
ID	Edge identifier	Integer
D	Road direction: 'B': two-way 'FT': one-way, source to target 'TF': one-way, target to source	String
SN	Start node identifier	Integer
EN	End node identifier	Integer
C	'FT' direction road length	Double
RC	'TF' direction road length (if 'B' direction, same to Cost)	Double
X1	Start node X-axis	Double
Y1	Start node Y-axis	Double
X2	End node X-axis	Double
Y2	End node Y-axis	Double

crossroads, and destinations of roads, etc. In this study, detail information of real roads such as road name, deflection angle, and road kind are removed. Topological properties include nodes and edges that corresponding to real road network, edges' weight that takes real road length as value and direction of edges expressed by start and target nodes are preserved. Traffic network's Euclidean feature is reflected by coordinates of start and target nodes, coordinate reference system of which is WGS 84. Data fields' detail description is shown in Table 1 below. Table 2 gives a sample data.

Table 2. Sample data

ID	D	SN	EN	C
14	B	25956	25956	800.0838
RC	**X1**	**Y1**	**X2**	**Y2**
800.0838	244719.8	2486072	245413.4	2486398

4.2 Experiment Design

4.2.1 Principle

Since relational databases are significantly different from graph databases in their data structure, a new database benchmark is necessary for testing the applications based on graph structure. According to the principles of designing a graph database benchmark discussed in [14, 15], traversal performance, the time needed to find a set of edges that meet a condition traversal, graph analysis, communities, and connected components finding are some useful performance considerations. In addition, different graph queries are not homogeneous, we should try to build balanced operations. The shortest path algorithm is an ideal algorithm for benchmark experiments. To make balanced testing, different versions of shortest path algorithms that traverse the graph in different ways will be included in the benchmark. In our work, we set several diverse graph algorithms to build our benchmark.

4.2.2 Platforms

We choose Neo4j as the graph database experiment platform, it is one of the most popular property graph databases that stores graphs natively on disk and provides a framework for traversing graphs and executing graph operations [19]. Neo4j is open-source, lightning-fast and easy to use with its powerful and productive graph query language and high-level programming language APIs. It's has proved to be one of the most efficient graph database systems [14]. Another strength of Neo4j is its native graph storage and graph processing framework [4], which is faster than a non-native framework database.

4.2.3 Algorithms

Algorithms with different time complexity and space complexity are adopted to comprehensively compare Neo4j with PostgreSQL. Both the Neo4j and PostgreSQL databases implement some graph algorithms, such as the Dijkstra shortest path algorithm. Based on the intersection of the native algorithm libraries of Neo4j and PostgreSQL's official documents, algorithms of different complexity are proposed which are A* algorithm [17] and Dijkstra algorithm [3]. The best time complexity of Dijkstra algorithm is $O(nlogn)$. The best time complexity of the A* algorithm [2] is $O(mlogn)$, where m represents number of neighbors. In transportation domain, m can take the value 4 as the reason of the four routes in a crossroad which are going straight, turning left, turning right and turning around.

Besides the normal Dijkstra shortest path algorithm, the K-shortest paths algorithm is another useful shortest path algorithm. It computes single-source paths for a graph with non-negative relationship weights. In our experiment, the implementation of K-shortest paths algorithm uses Dijkstra algorithm to find the shortest path and then proceeds to find K-1 deviations of the shortest paths.

4.2.4 Execution

The whole process of the algorithm execution can be divided into the following three steps. (1) Load the graph into the database systems. (2) Run the algorithm on the platform. (3) Consume results. We launch the discussion of algorithm execution by introducing the concept of projection.

A projection is a (sub) graph of interest of a property graph. The graph algorithms library in Neo4j provides two kinds of projection: (1) label-based projection and (2) Cypher-based projection. Label-based projection extracts all nodes with a given label while Cypher-based projection extracts a subgraph based on two Cypher queries. Neo4j has the technique to make efficient loading (quickly load the relevant subgraphs from Neo4j into the dedicated data structures) by using low-level APIs of the graph database to avoid memory churn and preferably only use primitive numeric types [18].

Time-consuming of loading a topology structure into a target system is an important indicator for system performance. Though it's possible to make use of the efficient loading technique to load a graph, we choose to set the loading stage as an independent initial step to avoid extra overhead. We do not set parameters for graph projection, so that the algorithms run on the complete graph dataset.

In terms of the result consumptions, the graph algorithms library in Neo4j offers three kinds of result consumptions: (1) write back results, (2) tabular aggregated results, and (3) tabular streamed results. The non-streaming procedures write back results to the property graph in the database as node properties or relationships. The tabular aggregated results procedures report the various statistics for the computed metrics and for operations. The tabular streamed results procedures return tabular results as a tuples stream for further processing [18]. For a more efficient execution, we use the non-streaming procedures to run our algorithms, all procedures are set to write back their results into the graph dataset.

In our experiment, we compare performance of these algorithms in Neo4j and Post-greSQL platform in different aspects, including different graph size setting and different initial state setting.

4.3 Experiment Setup

The experiment is carried out under a single PC with CPU of Inter Core i7-6700T @2.80 GHz, DDR4 8 GB RAM, 203.8 GB disk. The operating system is Ubuntu 16.04.3 LTS 64-bit. Database versions are Neo4j 3.5.3 under Java 1.8 JVM. PostgreSQL 9.5.17 with PostGIS 2.2.1 and PgRouting 2.1.0 extension. We have used mapping rules from ontology files to regulate the Neo4j database, which can greatly reduce the required storage space. PostgreSQL database management system uses the spatial PostGIS extension for standardized datatype geometry and PgRouting extension for corresponding algorithms.

5 Research Results

5.1 Experiment Results

5.1.1 Load Speed

We organize a small size graph structure (3300 edges and corresponding nodes) and a big size graph structure (100000 edges and corresponding nodes), they are organized in a topology-similar manner. In fact, various loading methods can be used to import data into these systems, we organize the traffic network data as CSV format and use two systems' CSV import module to load these graphs into Neo4j and PostgreSQL database management systems respectively. The loading speed by the size of graph is shown in Fig. 1.

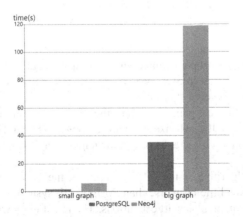

Fig. 1. Loading speed by the size of graph

5.1.2 Algorithm Comparison

We set the initial state of the system in different situations to simulate cool start and heat start. More specifically, we restart the whole system to make a cool start environment, and proceed algorithm after former running calculation to make a heat start environment. Graph size is also a chief consideration in experimenting, we set different start nodes and end nodes to characterize different graph size. In this follow experiment instance, a running instance with a start node at 8401 and an end node at 65555 represents an algorithm execution on a big graph size structure, the other one simulates an execution on a smaller graph. As demonstrated in Fig. 2, the shortest path algorithm is executed 6 times consecutively in each type of graphs. Neo4j outperforms PostgreSQL system at the same graph by nearly 30%. With the iterations grow, both systems get better performance with slight fluctuations. PostgreSQL system performance is more stable after the first two executions.

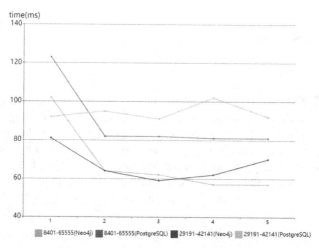

Fig. 2. Shortest path algorithm overall result

Focusing on the cool start performance and subsequent heat start performance, result in Fig. 3 shows that both PostgreSQL system and Neo4j system have better average heat start performance than the cool start one.

Comparing with shortest path algorithm, the K-shortest path algorithm execution shows less stability in small graph size structure as shown in Fig. 4. However, executions of K-shortest algorithm on big graphs have shown more stability after several iterations as shown in Fig. 5.

The last tested algorithm is the A* algorithm, at the initial cool start execution, Neo4j performance is a little bit worse than PostgreSQL system, and then shows great performance in subsequent heat start executions. The result is shown in Fig. 6.

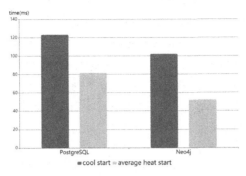

Fig. 3. Shortest path algorithm in different initial state

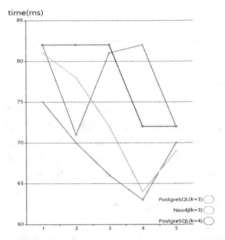

Fig. 4. K-shortest path algorithm (27393-32721)

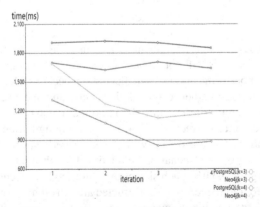

Fig. 5. K-shortest path algorithm (12484-52588)

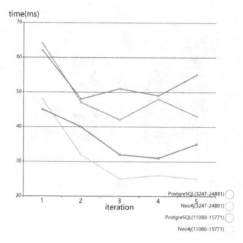

Fig. 6. A* algorithm overall result

5.2 Result Discussion

There are various ways to import data into a Neo4j system, including raw Cypher language create operation, CSV import module, batch-inserter, batch-importer, and Neo4j-importer. For small size (less than ten thousand nodes) graph, raw create operation is appropriate. Batch inserter is only available in Java language, it must stop Neo4j running instance to perform insertion, which is the same as batch-importer and Neo4j-importer. These three tools enjoy fast insertion speed at ten thousand nodes per second. In our loading experiment, target graph node size is less than ten thousand, thus its best practice is using the CSV import module to load data.

As stated before, three graph algorithms used in this paper are chosen to conduct our experiments not only because they fit our database benchmark well, they have their implementations on both systems as well. To ensure the consistency of the algorithm, the implementations of these algorithms in different platforms should be seriously reviewed. Since both database systems have the source code openly available, it could be easily concluded that both have similar implementations. Since slight difference is inevitable, we consider these differences will not violate our experiment principles.

Neo4j gains better performance at the smaller dataset because it looks only at records that are directly connected to other records. It does not scan the entire graph to find the nodes that meet the search criteria while the relational databases search all the data to meet the search criteria. Moreover, Neo4j's default configuration is optimized for smaller localized traversal queries, so it will get better performance in smaller graph size structure. In the big dataset scenario, the relational graph database needs to search all nodes in a larger graph so it's much slower than the Neo4j's executions.

Indexes needed for rapid queries are constructed after the initial cool start execution, so the heat start executions always outperform the cool start one with big advantage. Some fluctuations after several iterations of execution may have something to do with the specific topology structure of certain area. In most cases, algorithm execution will reach a stable execution time after several iterations.

6 Conclusion

In this paper, graph database and relational database system are compared roundly based on our carefully designed and looked through database benchmark. Three shortest path algorithms are chosen to perform the comparison experiments. It can be concluded from the results of experiments that Neo4j outperforms PostgreSQL system by average nearly 30% in our transportation network model since it does not scan the entire graph to find target record. However, by using the same import method, Neo4j needs more time to load the road network in both the small-size data and big-size data. The heat start execution is always recommended to pursue better performance. Specific graph structure may influence the performance of graph algorithms, especially in small-size graph.

We can draw a conclusion from our research that the transportation network is appropriate and recommended to store and handled in graph database systems like Neo4j. The transportation field is an ideal application area of graph databases. To gain deeper insights in this fitness, we notice that the transportation network is a graph structure naturally and it needs great flexibility to add a road sometimes. These are two core aspects of graph databases advantages. The graph structure ensures a graph modeling in graph databases. The need for schema modification is also well-handled in graph databases. In fact, it's a reasonable inference that graph databases shine in the fields where data interconnectivity or topology is important, such as social network and recommender system. Further study on graph databases' performance in these fields can be explored as future work.

Acknowledgments. This work was supported in part by the National Natural Science Foundation of China (61790565).

This work was supported in part by the National Key R&D Program of China (No. 2018YFB0105100).

This work was supported in part by the Science and Technology Innovation Committee of Shenzhen (JCYJ20170412172030008).

References

1. Codd, E.F.: A relational model of data for large shared data banks. Commun. ACM **13**(6), 377–387 (1970)
2. Jaiswal, G.: Comparative analysis of relational and graph databases. IOSR J. Eng. **3**(8), 25–27 (2013)
3. Angles, R., Gutierrez, C.: Survey of graph database models. ACM Comput. Surv. **40**(1), 1–39 (2008)
4. Neo4j Corporation: Efficient-graph-algorithms-neo4j.Neo4j Blog, July 2019. https://neo4j.com/blog/efficient-graph-algorithms-neo4j/
5. Neo4j Corporation: Walmart Optimizes Customer Experience with Real-time Recommendations. Neo4j Blog, July 2019. https://neo4j.com/case-studies/walmart/?ref=blog
6. Neo4j Corporation: Neo4j Powers Intelligent Commerce for eBay App on Google Assistant. Neo4j Blog, July 2019. https://neo4j.com/case-studies/ebay-shopbot/?ref=blog
7. Neo4j Corporation: Real-time Graph Analysis of Financial Data Creates Potential for Millions in Fraud Detection Savings. Neo4j Blog, July 2019. https://neo4j.com/case-studies/fortune-500-financial-services/?ref=blog

8. Gray, J.N.: The Benchmark Handbook for Database and Transaction Processing Systems. Morgan Kaufmann, Burlington (1993)
9. Seng, J.-L., Yao, S.B., Hevner, A.R.: Requirements-driven database systems benchmark method. Decis. Support Syst. **38**(4), 629–648 (2005)
10. Kanavos, A., Drakopoulos, G., Tsakalidis, A.: Graph community discovery algorithms in Neo4j with a regularization-based evaluation metric. In: Proceedings of the 13th International Conference on Web Information Systems and Technologies (WEBIST 2017), pp. 403–410 (2017)
11. Rodriguez, M.A., Neubauer, P.: Constructions from dots and lines. Bull. Am. Soc. Inf. Sci. Technol. **36**(6), 35–41 (2010)
12. Miller, J.J.: Graph database applications and concepts with Neo4j. In: SAIS 2013 Proceedings (2013)
13. Almabdy, S.: Comparative analysis of relational and graph databases for social networks. In: 2018 1st International Conference on Computer Applications & Information Security (ICCAIS), pp. 1–4 (2018)
14. Dominguez-Sal, D., Martinez-Bazan, N., Muntes-Mulero, V., Baleta, P., Larriba-Pey, J.L.: A discussion on the design of graph database benchmarks. In: TPCTC 2010, pp. 25–40 (2010)
15. Dominguez-Sal, D., Urbón-Bayes, P., Giménez-Vañó, A., Gómez-Villamor, S., Martinez-Bazán, N., Larriba-Pey, J.: Survey of graph database performance on the HPC scalable graph analysis benchmark. In: Web-Age Information Management, pp. 37–48 (2010)
16. Haklay, M., Weber, P.: Openstreetmap: User-generated street maps. IEEE Pervasive Comput. **7**(4), 1–39 (2008)
17. Dijkstra, E.W.: A note on two problems in connection with graphs. Numer. Math. **1**(1), 269–271 (1959)
18. Allen, D., et al.: Understanding trolls with efficient analytics of large graphs in Neo4j. In: BTW 2019, pp. 399–396 (2019)
19. Francis, N., et al.: Cypher: an evolving query language for property graphs. In: SIGMOD 2018 Proceedings of the 2018 International Conference on Management of Data, pp. 1433–1445 (2018)
20. Libkin, L., Martens, W., Vrgoč, D.: Querying graphs with data. J. ACM **63**(2), 1–53 (2016)

Optimizing Creative Allocations in Digital Marketing

Shubham Gupta$^{(\boxtimes)}$, Anshuman Gupta, Parth Savjani, and Rahul Kumar

Research and Development, MiQ Digital India, Bangalore 560001, KA, India
{shubhamgupta,anshuman,parth,rahul}@miqdigital.com
http://www.wearemiq.com

Abstract. Establishing the best strategy to optimize and test digital advertising campaigns is essential to the success of every marketing campaign. One common "test-and-learn" approach is creative optimization through which advertisers can generate the highest possible ROI on their advertising spends. Due to the uncertainty in determining the most effective creative a priori to a campaign, companies experiment with various strategies. Marketing firms try to distribute their creatives to both explore (sample more information) and exploit (the current data). The aim is to dynamically explore which creative is best suited to a specific audience by running multiple parallel experiments and exploit it in the post-experimentation phase. This explore/exploit trade-off is best explained by the Multi-Armed Bandits (MAB), the fundamental pillar in this discourse. MAB relies on Reinforcement Learning to converge on a solution with the least opportunity costs. Over time, we have tested key model parameters which can help in delivering campaign goals efficiently with improved uplift. We propose a customized MAB solution that has the potential to offer at least 50% uplift in a marketing KPI relative to traditional MAB policies through dynamic creative optimization.

Keywords: Multi-armed bandit · Online advertising · A/B testing · Reinforcement learning · Artificial intelligence

1 Introduction

In the era of digitization [1], organizations are moving more towards digital spend to market their products compared to offline strategies. This shift has unintentionally caused a hard-hitting competition in the market across the globe. Marketing teams gather resources and try their best to optimize budget allocation for different marketing campaigns. Marketing teams are trying every possible strategy to tackle the uncertainty but because of huge competition and huge user base to target, it's practically not possible to consider every possibility. One such uncertain area in digital marketing is Creative Optimization [2]. There are several approaches to this problem like A/B testing, Split testing, Multi-variant testing [3], which identifies the best variant or creative among different variants by allocating equal budgets on live traffic, but with high opportunity costs.

© Springer Nature Singapore Pte Ltd. 2020
M. Singh et al. (Eds.): ICACDS 2020, CCIS 1244, pp. 419–429, 2020.
https://doi.org/10.1007/978-981-15-6634-9_38

While an A/B test allows marketing firms to learn which major formatting of a site or piece of content is most engaging, multi-variant tests allow them to zone in on which tiny details are most engaging by showing audiences variations that only have subtle differences. To tackle high opportunity costs in a more "intelligent" way, a reinforcement learning [4] technique called Multi-Arm bandits [5] has gained more popularity in recent times because it produces faster results by using machine learning algorithms to dynamically allocate traffic to variations that are performing well while allocating less traffic to underperforming variations. It tries to solve the "explore-exploit" problem by minimizing the regret function [5]. Multi-arm Bandits (MAB) policies like epsilon-greedy, Thompson sampling, UCB [5] try to find the best performing arm and allocate most of the resources to that arm to get high Return on Investment (ROI).

Yet there is a problem with this approach. Using the traditional MAB approach we are not considering the next best-performing arms and don't allocate budget to them. In this paper, we will be talking about the importance of allocating budget on these arms and what level of impact they can have on ROI. Using the simulated data, which is a representation of data we observe in digital campaigns, we tried to optimize traditional MAB regret function [6] and proposed our custom MAB approach. With the results obtained from the experiments conducted on different marketing campaigns, a minimum of 50% lift was seen over time compared to a traditional MAB policy.

2   Problem Formulation

2.1   Traditional Multi-arm Bandits

In the traditional Multi-Arm Bandit problem [5] there are A arms among which we have to identify the arm which gives the best reward. Let c_i be an arm in the experiment such that $c_i \in \{1, 2, ...A\}$ and $R_{c_i,t}$ be the reward of ith arm at time t and $R_{c_i,t} \in [0, 1]$. Each arm has an unknown distribution of reward and has an unknown expected reward μ_{c_i}. Over multiple trials at different time steps, i.e., t = 1,2,... we can infer $R_{c_i,t}$ to be a random variable for arm i at time t. Assuming $R_{c_i,t}$ and $R_{c_i,s}$ to be identically distributed and $R_{c_i,t}$ and $R_{c_j,t}$ to be independent, we can refer μ_{c_i} to be the expected reward of arm i. If arm k were to be the best arm and total T trials occurred in the experiment, the total expected reward from the experiment ideally would have been:

$$R_{ideal} = E\left[\sum_T R_{c_k}\right] = \mu_k T \tag{1}$$

But, in real scenarios, we don't know which is the best arm to pull. So we choose different arm at different point of time in the experiment based on our *observations* on previous actions so as to get, what we say as a *policy* [4]. Thus. a *policy* can be defined as sequence of arms pulled at different time, i.e., $p = \{c_{i,1}, c_{i,2}....c_{i,T}\} \; \forall i \in \{1, 2, ...A\}$. Thus, total expected reward generated from our policy p can be

$$R_p = E\left[\sum_T R_{c_i,t}\right] \tag{2}$$

Now when we have both cumulative rewards after T trials in the experiment we would like to measure the success of the experiment by analyzing the gap between R_{ideal} (the reward generated if we would have known what was the ideal arm before starting the experiment) and R_p (reward generated over different trials as we learnt about different arms) i.e., regret. Then τ, the regret [6], can be formulated as,

$$\begin{aligned}
\tau &= R_{ideal} - R_p \\
&= E\left[\sum_T R_{c_k}\right] - E\left[\sum_T R_{c_i,t}\right] \\
&= \mu_k T - E\left[\sum_T R_{c_i,t}\right] \\
&= E\left[\sum_T \max_i (R_{c_i,t}) - \sum_T^p R_{c_i,t}\right]
\end{aligned} \tag{3}$$

The above regret holds true when there is only one single best arm throughout the experiment, i.e., we are assuming that the rewards of each arm follow an unknown random distribution. These are what we call as **Stochastic Multi-arm Bandits** [7]. In this paper, we are following the principles of **Adverserial Multi-arm Bandit** [8]. The important part of the adversarial setting is not that we are making a deterministic assumption, but rather that we are not making any assumptions about the distribution of the rewards. It's possible that the adversary is behaving stochastically. Accordingly, the above regret function can be changed to

$$\tau = E\left[\sum_T \max_{i,t} (R_{c_i,t}) - \sum_T^p R_{c_i,t}\right] \tag{4}$$

An adversarial bandit problem [8] is a pair (A, x, y), where A represents the number of actions, and x is an infinite sequence of payoff/reward vectors $x = x(1), x(2), \ldots$, where $x(t) = (R_{c_1}(t), \ldots, R_{c_A}(t))$ is a vector of length A and $R_{c_i,t} \in [0, 1]$ is the reward of action i on step t and y is an infinite sequence of activation vectors $y = y(1), y(2), \ldots$, where $y(t) = (\phi_{c_1}(t), \ldots, \phi_{c_A}(t))$ is a vector of length A and a step function which activates when an arm is selected based on a policy p . In best case scenario if we know the best arm at any time step t which is giving maximum reward $\phi_{c_i,t}^{ideal}$ can be,

$$\phi_{c_i,t}^{ideal} = \begin{cases} 1 & c_i = \max_i (x(t)) \\ 0 & c_i \neq \max_i (x(t)) \end{cases}$$

Thus, if we take the dot product of the two vectors we reformulate (4),

$$\tau = E \left[\sum_T^{ideal} x(t)^T \cdot y(t) - \sum_T^p x(t)^T \cdot y(t) \right] \tag{5}$$

2.2 Proposed Approach

The goal of any Multi-arm Bandit Problem [5] is to minimize this regret (4), to achieve maximum cumulative reward for a given scenario. But with the digital marketing scenario, we can't follow the traditional policies like UCB, exp3 [5]. Traditional policies like exp3, in each trial t, take only the arm which gives maximum reward at that time. Thereby inherently ignoring the next best set of arms which could have possibly increased the cumulative reward or decrease the regret. Because of the disruptive marketing data and a small budget, there is a need to correctly identify the best set of arms. There is always a particular order or rank of each arm at a particular time t which is a much better approximation of a bandit than just considering one best arm. There will never be a discrete arm selected and thus, $\phi_{c_i,t} \in [0,1]$. In the best-case scenario, if we assume $N_{c_i}(T)$ to be the proportion of times each arm i was pulled out of total pulls for all arms till time T, (5) can be rewritten as:

$$\tau = R_{ideal} - R_p$$
$$= E \left[\sum_T^{ideal} x(t)^T \cdot y(t) - \sum_T^p x(t)^T \cdot y(t) \right]$$
$$= E \left[\sum_i R(c_i, t) \right] * N_{c_i}(T) - E \left[\sum_i R(c_p, t) \right] * f_{c_i} \tag{6}$$
$$= \sum_i \mu_i * N_{c_i}(T) - E \left[\sum_i R(c_p, t) \right] * f_{c_i}$$

where , μ_i is the expected reward for i^{th} arm and f_{c_i} is the proportion of times arm i was pulled in the policy p out of total pulls for all arms.

The above regret (6) is a variant of the traditional Multi-Arm Bandit. Instead of selecting a discrete arm, we now have the flexibility of selecting an arm partially. Hence, in our approach arm selection is not binary. We can assign weight to each arm $\ni \sum_i f_{c_i} = 1$. Therefore, using Eq. (6), we try to distribute the weights over time such that over a longer period we can get top-performing arms with higher weights. Hence, the problem of identifying a single best arm at each time t becomes a problem of allocation of f_{c_i} to each arm at time t.

Our task now is to approximate $f_{c_i} \cong N_{c_i}(t)$ till any given point of time. Given the above regret function (6), we can model f_{c_i} with a Dirichlet Distribution. We can deduce that f_{c_i} at ant point of time t is the mode of Dirichlet distribution [9] for that arm and we can say that the sum of weights of all arms in the distribution is 1, i.e., $\sum_i f_{c_i} = 1$. The system is bounded,i.e., any change in the parameters of Dirichlet distribution of that creative will also have an impact on the weight of that creative. Hence,

$$f_{c_i} = Dirichlet(\alpha_{c_i}) = \Psi(\alpha_{c_i})$$

3   Learning the Dirichlet Parameters

To estimate Dirichlet parameters we followed some assumptions and experimented. The objective of the experiment was to define the best possible approach which would lead to faster convergence to the parameters of the distribution.

Let us assume, $\alpha_{c_i,t}$ to be the Dirichlet parameter for the arm i at time t and is dependent on historical rewards, i.e.,

$$\alpha_{c_i,t} = f(R_{c_i,t}, R_{c_i,t-1}, R_{c_i,t-2}, R_{c_i,t-3}, ...)$$

and all arms follow below mentioned assumptions :

- For an arm i,if, $R_{c_i,t} < R_{c_i,t+1} < R_{c_i,t+2}$ is true, then $\alpha_{c_i,t} < \alpha_{c_i,t+1} < \alpha_{c_i,t+2}$ should also be true
- If $R_{c_1,t}, R_{c_2,t}, R_{c_3,t}, ...$follow a non-decreasing order, then $\alpha_{c_1,t}, \alpha_{c_2,t}, \alpha_{c_3,t}, ...$ should also follow this order
- If there is a crossover in rewards between two arm, then corresponding alphas should also exhibit this change.

3.1   Experiment

In the experiment we devise a framework to measure the relative reward of different creatives. This way we can observe their ranking and learn the Dirichlet parameters. Let us define a benchmark ϕ at time t, to be the median reward of all the creatives at time t, i.e.,

$$\phi_t = median(R_{c_1,t}, R_{c_2,t}, R_{c_3,t}, ..., R_{c_4,t}) \tag{7}$$

The percentage change in reward, $\delta_{c_i,t}$, for each creative with respect to the median reward ϕ_t at time t can be defined as:

$$\delta_{c_i,t} = (R_{c_i,t} - \phi_t)/\phi_t * 100 \tag{8}$$

$\delta_{c_i,t}$ is the incremental value which will help the algorithm to identify the best set of performing arms. Different approaches like comparing the percentage change in rewards of creatives at consecutive times or difference in ranks of rewards of creatives at different points of time can also be used in determining $\delta_{c_i,t}$ and better handling of uncertain fluctuations of rewards of each creative.

Given below is the custom MAB algorithm for digital marketing scenario which uses the existing skeleton of traditional MAB Eq. (5) so as to finally convert it to Eq. (6) above.

Algorithm

1. Given $\gamma \in [0,1]$, initialize the Dirichlet parameters $\alpha_{c_i} = 1$ for i = 1, ..., A.
2. In each round t:
    (a) Set $f_{c_i}(t) = (1 - \gamma) * \Psi[\alpha_{c_i}](t) + \frac{\gamma}{A}$ for each arm i
    (b) Distribute the budget B to all arms according to their distribution of $f_{c_i}(t)$.
    (c) Observe reward vector $x(t)$.
    (d) Calculate $\delta_{c_i}(t)$ for each arm i
    (e) Set $\alpha_{c_i}(t+1) = \alpha_{c_i}(t) + \delta_{c_i}(t)$ for each arm i

---

Here, γ is the explore-exploit parameter [10] which is used quite commonly in MAB framework. The second step in each round t, takes care of the exploitation and exploration part of the MAB problem by using experiment parameter γ, which can be tweaked as experiment progresses. Higher value of γ indicates that the experiment is more towards exploration phase while lower γ guides the experiment towards exploitation. $\gamma = 0.5$ supports balanced exploration and exploitation.

For the initial round, we can distribute equal budget to each arm and observe the rewards. For each arm i we can calculate $\delta_{c_i,t}$ incremental value and update its $\alpha_{c_i,t+1}$, the Dirichlet parameter accordingly. This process can go on indefinitely or till t rounds, depending on the experiment run time or objective of the experiment.

4   Data

The data used in the experiment is a simulated representation of digital marketing data. In this data, there are 12 creatives or advertisements named Ad1, Ad2 and so on. Each advertisement has a different phrase and image yet pertain to a single marketing theme such that the audience seeing these images remains similar. Over time audience's response to each of these advertisements was recorded and was used to perform online training to exp3 MAB policy and our custom policy. Refer to Fig. 1 to see the sample data which tells about the total impressions, clicks, and conversion for each advertisement shown on a given day. CTR refers to the click-through rate and CVR refers to the conversion rate for a given advertisement. We will be using **metric**, which is the sum of CTR and CVR, as our reward and will be treating it as the key performance indicator(KPI) to optimize on.

Throughout the course of the experiment there is randomness in the rewards of creatives which can be seen in Fig. 2 and Fig. 3.

Figure 2 tells about the distribution of rewards for each creatives throughout the course of the experiment. As we can see the mean reward for each creative is almost same, the need to seek the correct ranking of different creatives, as explained in Sect. 2.2, seems necessary in our case.

date	Advertisement	impressions	clicks	conversions	CTR	CVR	metric
2019-05-13	Ad1	6662.0	0.0	8.0	0.000000	0.001201	0.001201
2019-05-14	Ad1	6958.0	0.0	2.0	0.000000	0.000287	0.000287
2019-05-13	Ad10	6709.0	0.0	6.0	0.000000	0.000894	0.000894
2019-05-14	Ad10	6821.0	1.0	3.0	0.000147	0.000440	0.000586
2019-05-13	Ad11	6646.0	0.0	1.0	0.000000	0.000150	0.000150
2019-05-14	Ad11	6873.0	1.0	2.0	0.000145	0.000291	0.000436
2019-05-13	Ad12	6710.0	0.0	0.0	0.000000	0.000000	0.000000
2019-05-14	Ad12	6765.0	1.0	2.0	0.000148	0.000296	0.000443
2019-05-13	Ad2	6708.0	1.0	0.0	0.000149	0.000000	0.000149
2019-05-14	Ad2	6834.0	0.0	4.0	0.000000	0.000585	0.000585
2019-05-13	Ad3	6748.0	0.0	0.0	0.000000	0.000000	0.000000

Fig. 1. Sample data

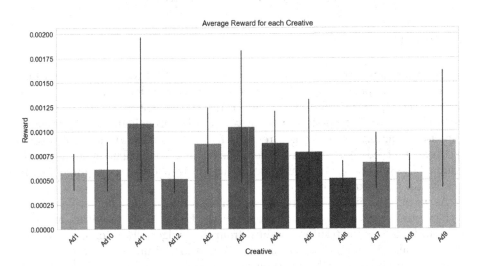

Fig. 2. Bar Plots of average reward for each creative

Figure 3 further supports the randomness in rewards of different creatives when plotted daily. The audience had different creative each day which they liked most.

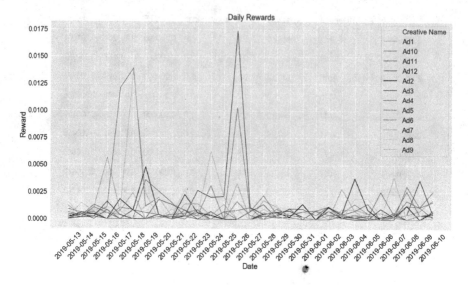

Fig. 3. Daily rewards for each creative

5   Results

In the experiment, we are trying to compare the EXP3 MAB policy and our optimized MAB policy. Equal budget is spent each day for almost one month on both the algorithms and at the end of the experiment we expect each policy to give the highest possible cumulative reward. Throughout the experiment, responses from the audience, i.e., rewards for each creative at the end of each day were observed and fed back to the modeling algorithms. The modeling algorithms then update their recommendations and provide updated weights for each creative for the next day of serving.

As we progress in the experiment we can see in Fig. 4, the cumulative reward of the Optimized MAB algorithm takes over the EXP3 MAB algorithm in a few days only. This means returns obtained from the budget allocation done by the optimized algorithm are greater than exp3, and in few more days they exceed so much that the number becomes quite impactful.

At the end of the experiment, we saw an uplift of more than **100%** using our approach compared to the traditional Multi-arm bandit approach. The cumulative reward of our approach was **0.049** compared to a single winning MAB approach with EXP3, i.e., **0.028**. Concerning time to converge, our approach took over exp3 early on and provided considerable lift over time. The results obtained are only valid when arms have similar expected rewards. In the case of different expected rewards, exp3 will always perform better.

When we look at the final weights of creatives, $\alpha_{c_i,t}$ at the end of the experiment Fig. 5, we can see that the weights from optimized algorithm we proposed, correlate more with the average reward of each creative Fig. 2 compared to

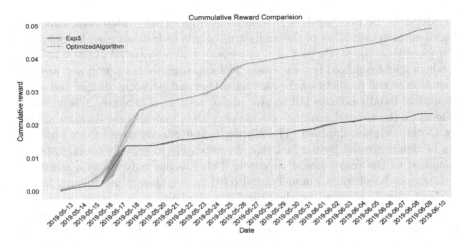

Fig. 4. Performance comparison

average selection proportion obtained from EXP3 algorithm, which further validates our approach for this kind of data.

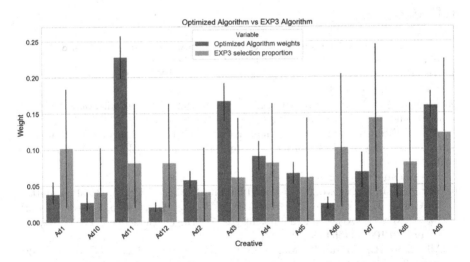

Fig. 5. Weight of each advertisement at the end of the experiment

6   Conclusion and Future Work

Digital marketing firms are confronted with various marketing challenges to optimize their marketing campaigns [12,13]. To help them in their endeavor and to reduce the opportunity costs, our approach can be utilized. An overview of traditional Multi-arm bandits with its mathematical formulation is presented and the proposed method and how it is different from existing Multi-arm bandits is described in detail. Most of the multi-arm bandit techniques used for optimization focuses on the single best arm to identify and allocate all of the budget on it, which might not hold in some cases. The approach we have proposed provides a solution to the problem where each advertisement performs similar and where traditional bandit policies fail to give higher returns. Therefore, it is necessary to identify the natural ranking of arms over time and allocate budgets according to the weights obtained. The efficiency and effectiveness of our approach can be demonstrated by the fact that we were able to provide considerable lift compared to the conventional approach. This has not only led to a significant performance boost for our model but also led to a significant incremental impact on client-specific strategy for their marketing budget.

However, assuming a single context [11] (meaning, the online behavior of users) for different groups of users in a marketing strategy is not ideal. The research done above was solely to optimize the campaign budget distribution and reduce the opportunity costs. What warrants additional research is how can we split our audience of different contexts and run our approach of multi-arm bandits on each context, i.e., multi-learners so that we can get tailored performance for each group of users in a marketing campaign.

References

1. Kiani, G.R.: Marketing opportunities in the digital world. Internet Res. **8**(2), 185–194 (1998). https://doi.org/10.1108/10662249810211656
2. Kulkarni, A.P., Gavlovski, A.S., Zhang, Z., Zeng, G.X. : U.S. Patent Application No. 15/199,386 (2018)
3. Gofman, A.: Consumer driven multivariate landing page optimization: overview, issues and outlook. The IPSI BgD Trans. Internet Res. **3**(2), 7–9 (2007)
4. Sutton, R.S. : Introduction: the challenge of reinforcement learning. In: Reinforcement Learning, pp. 1–3. Springer, Boston (1992). https://doi.org/10.1007/978-1-4615-3618-5_1
5. Bergemann, D., Valimaki, J.: Bandit problems (2006)
6. Vermorel, J., Mohri, M.: Multi-armed bandit algorithms and empirical evaluation. In: Gama, J., Camacho, R., Brazdil, P.B., Jorge, A.M., Torgo, L. (eds.) ECML 2005. LNCS (LNAI), vol. 3720, pp. 437–448. Springer, Heidelberg (2005). https://doi.org/10.1007/11564096_42
7. Allesiardo, R., Féraud, R., Maillard, O.-A.: The non-stationary stochastic multi-armed bandit problem. Int. J. Data Sci. Anal. **3**(4), 267–283 (2017). https://doi.org/10.1007/s41060-017-0050-5
8. Auer, P., Cesa-Bianchi, N., Freund, Y., Schapire, R.E.: The nonstochastic multi-armed bandit problem. SIAM J. Comput. **32**(1), 48–77 (2002). https://doi.org/10.1137/S0097539701398375
9. Kotz, S., Balakrishnan, N., Johnson, N.L. : Continuous multivariate distributions, Volume 1: models and applications, vol. 1. John Wiley (2004). (Chapter 49: Dirichlet and Inverted Dirichlet Distributions)

10. Reverdy, P., Wilson, R.C., Holmes, P., Leonard, N.E. : Towards optimization of a human-inspired heuristic for solving explore-exploit problems. In: 2012 IEEE 51st IEEE Conference on Decision and Control (CDC), pp. 2820–2825. IEEE, December 2012

11. Kolluri, V., DoRosario, A.: Context-and behavior-based targeting system. U.S. Patent Application No. 10/990,754 (2005)

12. Thomas, I.: Using multi-armed bandit experimentation to optimise multichannel digital marketing campaigns. Appl. Mark. Anal. **3**(2), 146–156 (2017)

13. Martín, M., Jiménez-Martín, A., Mateos, A.: The multi-armed bandit problem under delayed rewards conditions in digital campaign management. In: 2019 6th International Conference on Control, Decision and Information Technologies (CoDIT) (2019)

Big Data Analytics for Customer Relationship Management: A Systematic Review and Research Agenda

Sarika Sharma[✉]

Symbiosis Institute of Computer Studies and Research, Symbiosis International (Deemed University), Atur Centre, Model Colony, Pune 411016, India
sarika4@gmail.com

Abstract. In today's dynamic business scenario, customers have the power to rule the market on their terms and conditions. Customer Relationship Management (CRM) plays an imperative role by covering all methods and measures to have a better customer understanding, and to make the most of this knowledge in applications like production and marketing. With the emergence of big data, it brings a whole new inclusion of CRM strategies which can support customization of sales, personalization of services, and customer interactions. The paper aims to study the extant state of big data analytics for customer relationship management through the method of systematic literature review. Thematic analysis from the relevant studies is done and a framework is proposed as an outcome of the study. This framework can be used to analyze the present state of research in area of big data analytics and CRM and also future directions for the further research are provided in the paper.

Keywords: Customer relationship management · CRM · Big data · Systematic review · Framework

1 Introduction

Big data is an area of recent research which academics and practitioners are taking into account, the means through which they can explore to develop various strategies to bring in competitive advantages to the enterprises. The field of big data is growing with a fast pace and is gaining importance over the recent years. This also led to the rapid growth in literature on data analytics which is now attracting a stable stream of research in the area and journal publications. Customer relationship management is an added advantage to the companies, who want to understand their customers better by exploring the customer insights by generating useful information from data, which is an essential part of CRM process, and can be achieved by data mining. Data mining tools and techniques are implemented to find the hidden customer information from customer data by generating patterns and leading to an effective CRM.

The adoption of big data is done in an enormous way, mainly in the services sectors where managing customer relationship is a key to the survival of business in market.

© Springer Nature Singapore Pte Ltd. 2020
M. Singh et al. (Eds.): ICACDS 2020, CCIS 1244, pp. 430–438, 2020.
https://doi.org/10.1007/978-981-15-6634-9_39

Therefore it is critical to analyze and examine the application of big data in context of CRM strategies of an organization. A frontline mechanism in organization, which deals with customer interactions, requires broad support. Mostly it needs accurate data analytics to make sure that potential customers are being engaged in meaningful transactions. Big data analytics is therefore becoming a hot topic of interest in information systems, management, computer science, and social sciences. Because of widespread adoption of social media, social networks, and smart mobile devices, this trend is mainly attributed to the integrated information systems.

Management of a superior customer relationship management system in any organization mainly comprises of the advanced concepts, tools and techniques, and strategies related to the customer relationship management. CRM system is having tools with latest technology which can provide organization with a mechanism to understand their existing customers as well as the potential customers. Its practices aims to deliver a particular set of activities that might encourage them to make decisions and transactions based on customer data [1]. The linkage between big data analytics and CRM process has explored and been hypothesized in research literature. The CRM process aims to provide the insights about customer by stressing on the relevance of focusing on a single view of the customer.

Big data is considered as a vital and potential enabling factor for various business process, strategies, and innovation. It is now treated as a promising form of value creation for any business through the mechanisms specific to the business. As a matter of fact, the current increased data availability in terms of volumes, variety, and velocity has triggered the innovations at a great level. The data characteristics are typically associated with the concept of big data and the quality of data does affect its various applications. Maintaining big data is an added advantage to the firms as it allows them to develop advanced segmentation such as micro-segmentations. These are helpful in the real time applications, which can be further exploited to refine the various CRM activities of any organizations. Competitive performance gains can be achieved by making use of CRM technologies when integrated to the businesses [2].

Customer relationship management is mainly being contextualized within technology solutions. This is often described as analogous to a data driven and information-enabled form of the concept of relationship marketing. The CRM can be distinguished as: Operational CRM, involved in the current customers' interactions; Analytical CRM, enables the process of decision-making, by customer data transformation and analysis; and Collaborative CRM, which by leveraging inter-departmental teamwork and communication within a firm leads to an improved customer experience.

The coordination of customer data is being done using a multi-channel perspective and the CRM process. This process has three primary dimensions namely, relationship initiation, the maintenance, and the termination. The various applications of big data are developed in context of CRM. Some are already developed, and can be applied in most of the situations. Big data as well as big data analytics are transforming businesses which are customer centric and customer-facing. These businesses are generating and increasingly collecting large volumes of customer data. The data in terms of their interactions or shopping behavior can be useful for enabling the real-time decision making process. Business organizations are compiling the customer data spread among the various heterogeneous sources of data, which is often external and not structured.

2 Literature Review and Research Gap

The emerging field of big data analytics and CRM is being explored by the researchers, but still it is not reported in its extant form. This paper aims to conduct a systematic literature review and present a framework combining big data analytics and CRM technologies together. Some authors [3] aimed at compiling and presenting some of the different analytics tools, techniques, and methods which can lead to the advancements of CRM processes [4]. Suggest that big data-enabled CRM initiatives could require several changes in the pertinent critical success factors. They proposed to adopt an explorative approach, which can ease the hype created around the field of big data. Authors [2] in their paper have provided a systematic literature review that can lead to the demonstration of the mechanisms used for big data analytics for customer relationship management. Some Authors have also explored the other dimensions of CRM, which are not focusing on the technological approaches only. They have identified CRM as a business strategy supported by an information system, data. Noting these literature gaps there is a need of study which can report the existing state of big data analytics in CRM. The proposed study aims to fill this gap and seek to present the relevant literature in form of a framework.

3 Methodology

The systematic review is carried with these steps as followed (see Fig. 1):

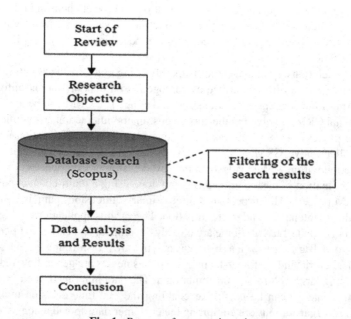

Fig. 1. Process of systematic review

Step 1: Research objectives or question
To propose a conceptual framework for big data analytics and customer relationship management.

Step 2: Search Strategy
The database selected for this is Scopus, which covers wide range of research articles on these topics. The present research connects the area of big data analytics and customer relationship management, and Scopus suits the multi disciplinary nature of this study. The database is searched using following search string:

"Big data" AND CRM

The search for articles was carried on 12 February 2020. The search was done for conference paper, article, and book chapter, which resulted in 112 numbers of articles.

Step 3: Assessment of quality of studies
These 112 articles were accessed for the quality and relevance by the author. Some articles which only talked about big data or CRM were discarded. Some articles talked only about data mining but not about CRM were also excluded. Articles covering both big data and CRM were only selected for further review. As a result 29 articles were found to be within the scope of present study and were included for next step. It can be observed that the research on big data analytics and CRM is a relatively new field and has gained pace in last five years as number of articles published are 22.

Step 4: Data Analysis and summarization
The research articles are compiled and presented in Table 1.

Step 5: Interpretation and findings
The research studies are first scanned for knowing the data and creating the initial impression. Then the codes and sub themes are generated, and based on the sub-themes the related ones are combined together are arranged in a thematic framework. The framework is based on the inductive content analysis and the research can be organized in following themes. It is also presented as a graph representing the number of papers (Fig. 2). Also the studies are arranged and presented as per the framework in Table 2.

I Processes
I(A) For Big data Analytics
I(B) For CRM
II Architecture
III Data Quality

From Table 2 and Fig. 2 it can be analyzed that there are almost same number of studies for processes related to the big data analytics and the customer relationship management. Both the imperative areas are addressed by the researchers and contribution of research papers are also there. Regarding the architecture also sufficient number of studies are dedicated. But it can be seen that data quality, although very crucial for the big data and CRM is not that well acknowledged as there are a few studies in this area.

Table 1. Compilation of research studies

S. no.	Author(s) and year of publication	Main findings
1	[6] Abu Ghazaleh and Zabadi (2020)	Explored the role of internet of things for big data and its impact on CRM in the modern customer services. They developed the analytic hierarchy framework for planning and establishment to determine the factors affecting the implementation of internet of things for big data
2	[7] Grambau et al. (2019)	Reference Architecture combining processes and heterogeneous data sources. They proposed a framework for social media data which can be implemented for collecting, processing, and analyzing this data. They included the machine learning approach for the same
3	[8] Serrano et al. (2019)	Their study addressed the issue of customer relationship management in hospitality industry. Mainly talked about identification and duplication problem. They proposed a generic framework to improve the data quality for customer relationship management
5	[9] Rawat et al. (2019)	Social media is being used as a competitive edge tool by the companies by effective use of CRM, digital marketing, and search engine optimization. The authors attempted to explore the use and optimization of social media using big data to achieve corporate goals
6	[10] Arco et al. (2019)	Suggested a big data and artificial intelligence framework for customer interactions
7	[11] Chiranjeevi et al. (2019)	They developed a system to evaluate the customer satisfaction through bot interactions. That system involves usage of bots to handle multiple customer interactions therefore enhancing customer relationship management
8	[12] Manigandan et al. (2019)	They suggested how business intelligence can be acquired and implemented for developing the competitive edge in business. Developed techniques and some applications for data management and analysis
9	[13] Perera et al. (2018)	Challenges of handling big data according to velocity, volume and variety. Focused on big data analytics by using big data analytical framework and data mining techniques
10	[14] Elyusufi et al. (2018)	Used the Complex Events Process (CEP) architecture that perfectly meets this need. They propsed to intercept the customer behavior and interactions and analyze them in the real time
11	[15] Ballestero et al. (2018)	Developed customer relationship management system by using advanced big data analytics for robust client profiling
12	[16] Zerbino et al. (2018)	Explored the role of big data and its impact on customer relationship management and also suggested some critical success factors
13	[17] Shrivastava et al. (2018)	Addressed the requirement of system and data sets for developing strategy. Available architectures on big data and CRM also studied
14	[18] Shrivastava et al. (2018)	Capturing customers by knowing them well in place of first predicting churn and then taking action, procedures
15	[19] Francisco et al. (2017)	A comparison and analysis among total data quality management (TDQM) and total information quality management (TIQM) while putting focus on the overall data quality problems faced in the context of a CRM
16	[20] Gallego and De-Pablos-Heredero (2017)	Impact over the power of data applied to selling strategies.
17	[21] Gončarovs and Grabis (2017)	Addressed the people involved by exploring the collaboration between business representatives and data scientists in the refinement of the prediction models iteratively
18	[22] Orenga-Roglá and Chalmeta (2016)	Developed and suggested a methodology to implement social CRM business processes through the use of the social CRM computer system
19	[23] Gómez-Mateu et al. (2016)	Examined the impact, current challenges, and opportunities of big data in biomedical research
20	[24] Ennaji et al. (2016)	Designed a social intelligence framework which can be used for extracting, as well as consolidating the reviews which are expressed through social media
21	[25] Ennaji et al. (2016)	Designed a multi-agent framework which can analyze and extract views and opinions from the social media

(continued)

Table 1. (*continued*)

S. no.	Author(s) and year of publication	Main findings
22	[26] Tiefenbacher and Olbrich (2016)	Suggested that capabilities for leveraging big data should be developed which can lead to the creation of valuable relationships with customers
23	[27] Xu and Chu (2016)	Suggested a multi-agent system from a data mining and explored their applications from a customer relationship management perspective
24	[28] LiK et al. (2015)	Talked about the architecture and suggested that enterprises have a choice to make regarding big data applications and choosing the type of stack
25	[29] Daif et al. (2015)	Existing architectures, and subsequently present an architecture with features
26	[30] Jung and Jung (2015)	The person cloud computing system connected with the smart phone to the method that a customer can confirm the real-time analysis data which is the advantage of the big data analysis
27	[31] Xu and Chu (2015)	Introduction of multi-agent systems from data mining perspective and the value of data mining for the customer relationship management
28	[32] Zhang et al. (2014)	Suggested mass security for customer data, while focusing on design and analysis of customer churn warning model by making use of various data mining technologies
29	[5] Deutsch (2012)	Data quality problems in the customer relationship management system. Worked upon the tradeoff between the volume of data and the quality of data maintained along with the scaling issues

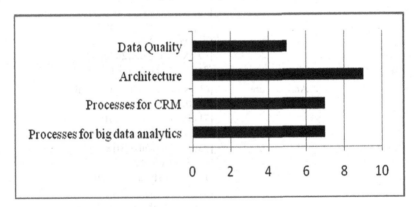

Fig. 2. Number of studies representing framework

4   Conclusions and Future Scope

The present research paper contributes to the field of big data analytics and CRM by: a) Compiling the research studies in the field; b) Arranging the studies in themes and proposing a framework; c) Providing insights on future research areas. The future studies can be taken up in these upcoming areas and more studies can explore its further dimensions. The areas identified are in processes of implementing big data for CRM. Architecture for big data can be another emerging area of research. Quality of data plays vital role and can affect the outcomes and implementation of data mining algorithms, hence to explore data quality enhancing mechanisms can be an area of research. Despite a rigorous and comprehensive effort towards the chosen topic, researchers in no way claim it to be a holistic one.

Table 2. Studies arranged in the proposed framework

Framework		Author (s)
I	Processes	
	I(A) Big Data	[6] Abu Ghazaleh and Zabadi (2020)
		[12] Manigandan et al. (2019)
		[16] Zerbino et al. (2018)
		[21] Gončarovs and Grabis (2017)
		[23] GÃ³mez-Mateu et al. (2016)
		[26] Tiefenbacher and Olbrich (2016)
		[30] Jung and Jung (2015)
	I(B) CRM	[9] Rawat et al. (2019)
		[10] Arco et al. (2019)
		[15] Ballestero et al. (2018)
		[18] Shrivastava et al. (2018)
		[22] Orenga-Roglá and Chalmeta (2016)
		[27] Xu and Chu (2016)
		[31] Xu and Chu (2015)
II	Architecture	[6] Abu Ghazaleh and Zabadi (2020)
		[7] Grambau et al. (2019)
		[11] Chiranjeevi et al. (2019)
		[13] Perera et al. (2018)
		[14] Elyusufi et al. (2018)
		[17] Shrivastava et al. (2018)
		[24] Ennaji et al. (2016)
		[30] Jung and Jung (2015)
		[32] Zhang et al. (2014)
III	Data Quality	[8] Serrano et al. (2019)
		[19] Francisco et al. (2017)
		[20] Gallego and De-Pablos-Heredero (2017)
		[32] Zhang et al. (2014)
		[5] Deutsch (2012)

References

1. Nisar, G., Prabhakar, T.: Trains and twitter: firm-generated content, customer relationship management and message framing. Transp. Res. Part A Pol. Pract. **113**(c), 318–334 (2018)
2. Mikalef, P., Pappas, Ilias O., Krogstie, J., Giannakos, M.: Big data analytics capabilities: a systematic literature review and research agenda. Inf. Syst. e-Bus. Manage. **16**(3), 547–578 (2017). https://doi.org/10.1007/s10257-017-0362-y

3. Gončarovs, P.: Data analytics in crm processes: a literature review. Inf. Technol. Manage. Sci. **20**(1), 103–108 (2017)
4. Ee-Zerbino, P., Aloini, D., Dulmin, R., Mininno, V.: Big data-enabled customer relationship management: a holistic approach. Inf. Process. Manage. **54**, 818–846 (2018)
5. Deutsch, T.: Big data: data quality's best friend? IBM Data Manage. Mag. **7**(A32) (2012)
6. Abu, G.M., Zabadi, A.M.: Promoting a revamped CRM through Internet of Things and big data: an AHP-based evaluation. Int. J. Organ. Anal. **28**(1), 66–91 (2020)
7. Grambau, J., Hitzges, A., Otto, B.: Reference architecture framework for enhanced social media data analytics for predictive maintenance models. In: Proceedings - 2019 IEEE International Conference on Engineering, Technology and Innovation, ICE/ITMC 2019, pp. 392–400 (2019)
8. Serrano, L., Ballestero, P., Romero, S., Ruiz, C., Ãlvarez, J.L.: Entropic statistical description of big data quality in hotel customer relationship management. Entropy **21**(4), 419 (2019)
9. Rawat, S., Jindal, S., Moorti, R.S., Mangal, Y., Saxena, N.: Change in IT world with the evolution of social media using Big Data. Proceedings of the 2018 International Conference on Communication, Computing and Internet of Things, IC3IoT 2018, pp. 408–412 (2019)
10. Arco, M.D., Presti, L.L., Marino, V., Resciniti, R.: Embracing AI and big data in customer journey mapping: From literature review to a theoretical framework. Innov. Mark. **15**(4), 102–115 (2019)
11. Chiranjeevi, H.S., Shenoy, M.K., Diwakaruni, S.S.: Evaluating the satisfaction index using automated interaction service and customer knowledgebase: a big data approach to CRM. Int. J. Electron. Custom. Relationsh. Manage. **12**(1), 21–39 (2019)
12. Manigandan, E., Shanthi, V., Kasthuri, M.: Parallel clustering for data mining in CRM. Adv. Intell. Syst. Comput. **808**, 117–127 (2019)
13. Perera, W.K.R., Dilini, K.A., Kulawansa, T.: A review of big data analytics for customer relationship management. In: 2018 3rd International Conference on Information Technology Research, ICITR, pp. 1–6 (2018)
14. Elyusufi, Z., Elyusufi, Y., Aitkbir, M.: Customer profiling using CEP architecture in a big data context. In: ACM International Conference Proceeding Series, vol. 64, pp. 1–6 (2018)
15. Ballestero, P., Serrano, L., Ruiz, C., Romero, S., Ivarez, J.L.: Using big data from customer relationship management information systems to determine the client profile in the hotel sector. Tourism Manage. **68**(1), 187–197 (2018)
16. Zerbino, P., Aloini, D., Dulmin, R., Mininno, V.: Big data-enabled customer relationship management: a holistic approach. Inf. Process. Manage. **54**(3), 818–846 (2018)
17. Shrivastava, P., Sahoo, L., Pandey, M.: Recognition of telecom customer's behavior as data product in CRM big data environment. Smart Innov. Syst. Technol. **79**(1), 165–173 (2018)
18. Shrivastava, P., Sahoo, L., Pandey, M., Agrawal, S.: Development of policy designing technique by analyzing customer behavior through big data analytics. Adv. Intell. Syst. Comput. **710**(1), 573–581 (2018)
19. Francisco, M.M.C., Alves-Souza, S.N., Campos, E.G.L., De Souza, L.S.: Total data quality management and total information quality management applied to customer relationship management. In: ACM International Conference Proceeding Series, pp. 40–45 (2017)
20. Gallego, C., De-Pablos-Heredero, C.: Customer relationship management (CRM) and big data: a conceptual approach and their impact over the power of data applied to selling strategies. Dyna (Spain) **92**(3), 274–279 (2017)
21. Gončarovs, P., Grabis, J.: Using data analytics for continuous improvement of CRM processes: case of financial institution. In: Kirikova, M., Nørvåg, K., Papadopoulos, George A., Gamper, J., Wrembel, R., Darmont, J., Rizzi, S. (eds.) ADBIS 2017. CCIS, vol. 767, pp. 313–323. Springer, Cham (2017). https://doi.org/10.1007/978-3-319-67162-8_31

22. Orenga-Roglá, S., Chalmeta, R.: Social customer relationship management: taking advantage of Web 2.0 and big data technologies. SpringerPlus **5**(1), 1–17 (2016). https://doi.org/10.1186/s40064-016-3128-y
23. GÃ3mez-Mateu, M., et al.: Big data in biomedical research. In: Perspectives from the biostatnet-CRM Workshop. Boletin de Estadistica e Investigacion Operativa **32**(3), pp. 257–277 (2016)
24. Ennaji, F.Z., El Fazziki, A., El Alaouiel Abdallaoui, H., Sadiq, A., Sadgal, M., Benslimane, D.: Multi-agent framework for social CRM: extracting and analyzing opinions. J. Eng. Sci. Technol. **12**(8), 2154–2174 (2017)
25. Ennaji, F.Z., El Fazziki, A., Sadgal, M., Benslimane, D.: Social intelligence framework: Extracting and analyzing opinions for social CRM. In: Proceedings of IEEE/ACS International Conference on Computer Systems and Applications, AICCSA, pp. 1–7 (2016)
26. Tiefenbacher, K., Olbrich, S.: Capabilities and impediments to leverage customer value from data - A case study from the automotive industry. In: International Conference on Information Systems 2018, ICIS 2018 (2018)
27. Xu, L., Chu, H.-C.: Big data analytics toward intelligent mobile service provisions of customer relationship management in e-commerce. J. Comput. (Taiwan) **26**(4), 63–72 (2016)
28. Li, K., Deolalikar, V., Pradhan, N.: Big data gathering and mining pipelines for CRM using open-source. In: Proceedings - 2015 IEEE International Conference on Big Data, IEEE Big Data, pp. 2936–2938 (2015)
29. Daif, A., Eljamiy, F., Azzouazi, M., Marzak, A.: Review current CRM architectures and introducing new adapted architecture to big data. In: 2015 International Conference on Computer and Computational Sciences, ICCCS 2015 (2015)
30. Jung, L.-S., Jung, D.-H.: A study on application plans of big data to improve customer satisfaction in auto maintenance industry. Inf. (Jpn) **18**(6), 2679–2684 (2015)
31. Xu, L., Chu, H.-C.: The cooperation mechanism of multi-agent systems with respect to big data from customer relationship management aspect. In: Nguyen, N.T., Trawiński, B., Kosala, R. (eds.) ACIIDS 2015. LNCS (LNAI), vol. 9011, pp. 562–572. Springer, Cham (2015). https://doi.org/10.1007/978-3-319-15702-3_54
32. Zhang, Y., Wang, Y., He, C., Yang, T.: Modeling and application research on customer churn warning system based in big data era. Int. J. Multimed. Ubiquit. Eng. **9**(9), 281–298 (2014)

Agricultural Field Analysis Using Satellite Surface Reflectance Data and Machine Learning Technique

Medha Wyawahare, Pranesh Kulkarni[(✉)], Aditya Kulkarni, Ankit Lad, Jayant Majji, and Aayush Mehta

Department of Electronics and Telecommunication, Vishwakarma Institute of Technology, Pune, India
{medha.wyawahare,aditya.kulkarni18,ankit.lad18,pavan.majji18, aayush.mehta18}@vit.edu, kulkarnipranesh1767@gmail.com

Abstract. The environmental, social, and economic problems confronting agriculture today are symptoms of agricultural industrialization. In this study, the agricultural field is analyzed using satellite surface reflectance data. This technology facilitates monitoring of crop vegetation by spectral analysis of satellite images of different sites and crops which can track positive and negative dynamics of crop development. Using this analysis, the field can be categorized into different categories rating its potency to grow crops, which helps the user to get detailed information about the current condition of the field. For the analysis, we have used Landsat 8 data. We have used the Google Earth Engine to import the data from the ground station. The indices we have used for this study are Normalized Difference Vegetation Index (NDVI), Modified Soil Adjusted Vegetative Index (MSAVI) and Normalized Difference Water Index (NDWI) and average rainfall data. For clustering the data, we have implemented k-means clustering algorithm. We have collected data from over 6 years and by taking mean values we classified the agricultural fields into different categories according to their quality.

Keywords: Surface reflectance · Machine learning · Spectral indices · Earth engine · Satellite field monitoring · Vegetation indices

1 Introduction

Satellites that capture pictures of the earth's surface also are equipped with sensors named Active and Passive. They are utilized to acquire surface reflectance data which can be beneficial for various applications. In the recent government agricultural surveys that are challenging, expensive and tedious to study and analyze each and every agricultural field. Also, it is beneficial for insurance and private companies which need to visit a certain place for client claim verification which is hassle full job. In these scenarios satellite data portray an important role by simplifying the job that can otherwise be stressful as we only need to do is enter the geographical coordinates of a specific agricultural field as an input. The study of spectral reflectance responses can be used to analyze the crop

© Springer Nature Singapore Pte Ltd. 2020
M. Singh et al. (Eds.): ICACDS 2020, CCIS 1244, pp. 439–448, 2020.
https://doi.org/10.1007/978-981-15-6634-9_40

health, moisture content, field vegetation, spectral indices using the Landsat 8 sensors. They can help calculate the Normalized difference vegetation index (NDVI) and the Modified Soil-Adjusted Vegetation Index (MSAVI2) which is the simplified version of the Modified Soil-Adjusted Vegetation Index (MSAVI) algorithm is actually invented to deal with the soil brightness problem. NDVI is the spectral index which is estimated by using a red and near-infrared band and the MSAVI index is also used for crop health monitoring. The last index Normalized Difference Water Index (NDWI) which can measure the change in water or moisture content in leaves by using Near Infrared (NIR) and Short Wave Infrared (SWIR) bands. These indices can create a rough outline of an agricultural field moreover by combining these indices with the rainfall data collected through the Climate Hazard Group Infrared Precipitation with Station Data (CHIRPS) one can predict the overall quality of an agricultural field and can get information about the fertility of land by analyzing 6 years of data. We have used google earth engine to import the multispectral data from United States Geological Survey (USGS) datasets. We have designed one web application to acquire data for a specific interested land. By processing these data vegetation indices that are mentioned above are then calculated. In this work, we have developed a python application for processing this data. We have developed a machine learning classification model to classify the agricultural lands based on vegetation. We have used the K-means clustering algorithm to classify the data. So, after calculating the spectral indices the average values are calculated and are fed to a machine learning model that can predict the quality of the field.

2 Related Work

C. Yang and G. L. Anderson used Landsat Multispectral Scanner data so as to know how Landsat works and how the images are derived from the satellite [1]. Wiegand and Richardson compared eight vegetation indices and four individual bands calculated from Landsat Multi-Spectra Scanner (MSS) data. They concluded that all the vegetation indices and corresponding bands contributed to grain yields. In one technique, Chang and Liu observed direct and indirect changes in spectral responses caused by changes in water content of plants, pigments, change in photosynthesis activity and chlorophyll levels fluorescence indices and their internal association [2]. They also have discussed some common approaches and the new techniques in applying spectral reflectance and indices for analyzing water levels and physiological activities such as a change in chlorophyll in plants. In one of the studies, three different vegetation indices of RapidEye imagery are taken into consideration for the classification of crop type and the effect of each index on the accuracy of classification was observed [3]. The three indices NDVI, the Green Normalized Difference Vegetation Index (GNDVI) and the Normalized Difference Red Edge Index (NDRE) were studied and on that basis crops are classified. All these three indices incorporated the near-infrared band. Support vector machines were used for the classification of crop type. The first sampled fields were classified with all these three indices. Then by performing classification with each single index, the contribution of each index in classification was also observed. The author concluded that the highest classification accuracy was obtained by combining all the three indices. Results show that the NDRE index contributes more to classification. It was also concluded that Rapid-Eye imagery is highly preferred for agricultural as well as forestry uses since it has red and NIR bands. In one study machine learning techniques are implemented to remotely sensed imagery to train predictive models for prediction of vegetation health [4]. This tool was built to increase the capacity for vegetation health monitoring in data-scarce regions. The author had processed 11 years' imagery of the Moderate Resolution Imaging Spectroradiometer (MODIS) dataset for building the model. In this technique, the Enhanced Vegetation Index (EVI) was used for the prediction of crop health. In one of the techniques, Han et al. studied the relationship between drought and precipitation, temperature, vegetation with the help of a random forest machine learning algorithm [5]. In this study, the author tested standardized precipitation index (SPI), relative soil moisture (RSM), NDWI, normalized multi-band drought index (NMDI), the normalized difference drought index (NDDI) for verifying effectiveness in drought monitoring. The author designed a new index for drought monitoring by combining all these indices. The conclusion of the study was the combined index of drought monitoring was very effective in drought monitoring. Table 1 describes the different approaches for field vegetation analysis or crop monitoring. Cici Alexander estimated land surface temperatures (LST) from the reflectance of a thermal band of Landsat 8 in his study [6]. The author calculated spectral indices by using all the combinations of the first seven spectral bands. Their correlations with LST are analyzed. Correlation between land cover and LST is estimated [7]. In this study, the author concludes that NDVI has the strongest correlation (0.77 to 0.86) with LST. Keeping in mind the impacts on ecosystem functions Nobuyuki Kobayashi et al. (2020) developed an administration that helps the crop cover maps with accurate space-time information [8]. According to the author's work, the subject that has gained some weight in the paper is remote sensing, which is used to obtain crop information at local or global scales. The study of cyclic and seasonal natural phenomena or vegetation phenology can be extracted from the spectral indices, which designate the combinations of spectral measurements at different wavelengths. These spectral indices are obtained through the sentinel 2A image data which gives the reflectance data which in turn helps to get the current situation of the crop. The authors have also highlighted the importance of each index for identifying each crop. They have arrived at a judgment from their work that the use of spectral indices has made better the classification accuracy, they also called attention to an important point which is the use of integrated reflectance and spectral indices enhanced the effects related to large sets of correlated variables and decreased the accuracy. Maria Romero et al. (2018) have carried out experiments in a vineyard in the Shangri-La region, located in Yunnan province in China, these experiments were conducted to estimate midday stem water potentials of grapevines [9]. For evaluating the correlations between stem water potentials and the vegetation indices, statistical methods and machine learning methods are used. But according to the author, there weren't any strong correlations found by using simple machine learning techniques. However, by using Artificial Neural Networks the correlation found out was high as reported by the authors. The author concludes that by using this model one can analyze a plant by plant basis to identify sectors of stress within the vineyard. This model can be used for optimal irrigation management. In one of the studies, Raí A. Schwalbert et al. (2020) implemented different algorithms such as multivariate linear regression, random forest algorithm and neural networks for forecasting soybean yield using NDVI, EVI, land surface temperature and precipitation as independent variables [10]. The author was successful to conclude that it is beneficial to integrate statistical techniques, remote sensing data and weather records to field survey data for forecasting soybean yield.

Table 1. Researchers and their approaches

Author	Algorithms or spectral indices used
Üstüner et al.	NDVI, GNDVI, NDRE
Emily Burchfielda, John J. Nayb and Jonathan Gilliganc	EVI
Hongzhu et al.	SPI, RSM, NDWI, NMDI, NDDI
Cici Alexander	NDVI, SWIR
Raí A. Schwalbert et al.	NDVI, EVI

However, the classification of a given piece of land into different categories by combining appropriate indices by efficient way remains an unsolved problem. In this work, we have developed a novel method to predict the quality of land by analyzing 6 years' satellite data. In this study, we have combined NDVI, MSAVI, and NDWI indices and rainfall data.

3 Proposed Methodology

3.1 Vegetation Indices

Vegetation indices are obtained from satellite and drone data by analyzing multispectral imagery bands. Satellite has two types of sensors i.e. passive and active. Passive which measures sunlight reflected from earth's surface whereas active sensors emit radiations and measures reflected wave [11]. In this way surface reflectance, data can be calculated. Human eyes can visualize visible light but satellite sensors have the ability to sense infrared or invisible light. So it plays a big role in crop monitoring. There are many vegetation indices for crop health monitoring and drought monitoring. For our study we have chosen the following three indices:

A. NDVI
B. MSAVI
C. NDWI

NDVI was originally founded by NASA in 1970 [12]. In plants, chlorophyll absorbs visible light from the spectrum and leaves of plants reflect NIR light [13]. So areas with dense vegetation or plants with good photosynthesis activity will reflect less visible light and more NIR light. NDVI ranges from −1 to 1 [14]. Land covers having NDVI more than 0.5 are considered as healthy or dense vegetation areas. Fields with exposed soil have NDVI nearly in between 0.1 to 0.2. Unhealthy or dead plants may have values less than 0.3 or less than 0 [14]. NIR and RED bands are required for calculating the NDVI index. NDVI can be calculated by the Eq. (1). In this equation, NIR represents the near infrared band and RED represents red band.

$$NDVI = \frac{(NIR - RED)}{(NIR + RED)} \tag{1}$$

The main drawback of NDVI is the brightness of the soil [15]. So MSAVI is used to correct soil brightness due to which sensitivity to photosynthesis decreases [13]. It has the same application such as drought analysis, crop monitoring and yield prediction. MSAVI can be calculated by the Eq. (2). In this equation, NIR represents the near infrared band and RED represents the red band.

$$MSAVI2 = \frac{(2 * NIR + 1 - \sqrt{(2 * NIR + 1)^2 - 8 * (NIR - RED)})}{2} \tag{2}$$

NDWI is generally used for drought analysis, analyzing water level in water reservoirs and groundwater level monitoring. NDWI index can be also used for estimating water content in plant bodies [13]. NDWI index can be calculated by using the Eq. (3). Where SWIR represents the shortwave infrared band.

$$NDWI = \frac{(NIR - SWIR)}{(NIR + SWIR)} \tag{3}$$

3.2 Web Application

We have used USGS Landsat 8 surface reflectance tier 1 multispectral data for our study. We have used the Google Earth Engine platform to get access to these photographs and datasets. Google earth engine is a platform for academic, non-profit research organizations and Non - Governmental Organizations (NGOs) which provide satellite imagery for their analysis and studies [16]. More than 40 years of geospatial data is available on this platform. It provides Application Programming Interface (API) and other tools for developing application development and for research purposes. We can import datasets directly from their cloud storage. This platform allows us to deploy a web application for analysis purposes. We have designed one web application that takes geographical coordinates as an input and exports multispectral data of a specific piece of land in comma-separated values (CSV) format. For designing this web application, we have imported USGS Landsat 8 surface reflectance tier 1 dataset and CHIRPS Daily: climate hazards group infrared precipitation with station data for precipitation records. We can get the wavelength of each specific band from exported spreadsheet.

3.3 Landsat 8 Spectral Bands

Landsat 8 has two main sensors: Operational land imager (OLI) and Thermal infrared sensor (TIRS) [17]. The operational land imager can create nine spectral bands and TIRS can produce two thermal bands. Table 2 shows detailed information about Landsat 8 bands. In which bands B2, B3, B4 are visible bands that can be used to render the RGB image.

Table 2. Landsat 8 multispectral bands

Name of band	Wavelength	Description
B1	0.435–0.451 μm	Band 1 (ultra-blue) surface reflectance
B2	0.452–0.512 μm	Band 2 (blue) surface reflectance
B3	0.533–0.590 μm	Band 3 (green) surface reflectance
B4	0.636–0.673 μm	Band 4 (red) surface reflectance
B5	0.851–0.879 μm	Band 5 (near infrared) surface reflectance
B6	1.566–1.651 μm	Band 6 (shortwave infrared 1) surface reflectance

3.4 Python Application Development

For calculation of vegetation indices, we have designed python application which enables a user to import CSV file which consists of multispectral data and precipitation data. To read the data from datasheets, we have used a python library called panda. Then, we have implemented different algorithms for calculation of vegetation indices such as NDVI, MSAVI2, NDWI by taking specific bands into consideration. By processing 6 years' data we have calculated mean, maximum, percentage change values which can help out in analysis of any agricultural field. By processing precipitation data, we have calculated average rainfall at that location. After that, we have done a small survey and we randomly chose some agricultural sites from the state of Maharashtra for our study. We have sampled 140 fields in total throughout the course of the project. While choosing fields of interest we ensured that all types of fields should be included in the dataset. Then after collecting multispectral data of selected fields, we recorded mean values of all indices and average rainfall for each field by using the same python tool and prepared datasheet containing those values.

3.5 Classification Algorithm

In order to develop the classification model, a detailed study of range of vegetation indices for a particular type of vegetation was carried out. The mean values of vegetation indices and average rainfall are the input parameters of the machine learning model. As there is no previous data to compare with, the model uses unsupervised methods of learning to find common features in the fed data. Unsupervised learning is one of the machine learning technique, where you do not need to supervise the model [18]. It mainly deals with the unlabeled data. In our study, we have implemented K-means clustering algorithm for unsupervised learning. K-means clustering is an algorithm that finds patterns in the dataset and forms of clusters of data points that fit together [19]. In our model, all parameters are plotted against one another to see the relationship between each other. The relevant indices are then clustered using the Scikit library of Python and the analysis report can be observed. This study clusters the data points to divide them into 4 major clusters. Each land (data point) is a part of only one cluster. The clusters are formed on the basis of the variation of vegetation indices. After the clusters have been formed, each data point belongs to the specific class which afterward predicts the quality index of the field. Now, it is further classified in order to find the probability of every land lying in its cluster. For doing so, the algorithm of multiclass classification: Support Vector Machines is applied. This model will take the input as mean values of NDVI, MSAVI2 and NDWI and average precipitation and will predict the class of a corresponding field which we call as quality index. In this way, we have designed a python tool to analyze the current and past scenario of an agricultural field and to predict the overall quality of a field.

4   Result and Discussion

We have calculated vegetation indices MSAVI, NDVI and NDWI for different sampled fields using spectral data collected for the respective fields. Using our tool, we found out that a healthy field has a MSAVI index between 0.55–0.95. It is also observed that the average values of MSAVI which lie between 0.6 to 0.75 and the highest values lie between 0.85–0.95. NDWI is not directly affected by the vegetation a field but it shows a few important factors. In the case of healthy vegetation, areas observed values of the NDWI index are greater than 0 but not greater than 0.5. You can refer Fig. 1 for more details it represents the statistical values calculated for the area with healthy vegetation.

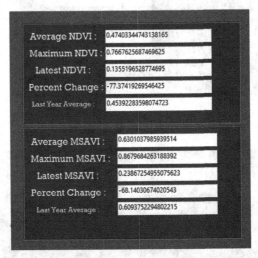

Fig. 1. Statistical data of Vegetation Indices values of healthy vegetation area calculated by the python tool developed for the study.

For barren and dead lands, we found that the MSAVI index has a mere range of 0.1 to 0.40 and the average value of MSAVI for such fields is a mere 0.25. Results show a similar trend in the case of the NDVI index. Results show that the mean value of the NDVI index is in-between 0 to 0.15 and the maximum NDVI index in the case of dead vegetation is around 0.1 to 0.25. All the values of the NDVI index for dead vegetation lies between 0 to 0.25. It is observed that the NDWI values of unhealthy vegetation areas are less than 0. Refer Fig. 2 for more detailed information, the figure represents data of area with unhealthy vegetation. Through the clustering algorithms, the data is classified into 4 clusters. Further, through the Support Vector Machines algorithm, the probability of each land in its cluster is calculated. According to the statistics shown in Fig. 3, the accuracy of the classification model is 80% and on further using the predict probability function, the confidence scores of predictions are calculated. From the predictions of the machine learning model it is observed that the predicted quality index lies in the range of 1 to 4. Range of quality index can be increased by increasing number of clusters, but it would decrease the accuracy of the model and hence it is not recommended.

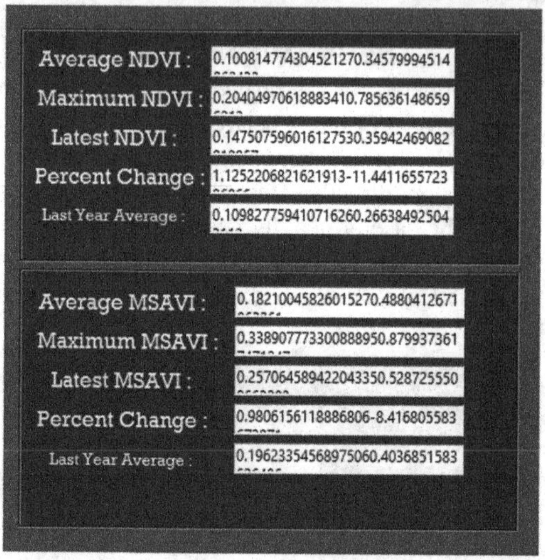

Fig. 2. Statistical data of Vegetation indices values of unhealthy vegetation area.

```
5
              precision    recall  f1-score   support

           0       0.73      0.94      0.82        17
           1       0.92      0.67      0.77        18

    accuracy                           0.80        35
   macro avg       0.83      0.80      0.80        35
weighted avg       0.83      0.80      0.80        35

The accuracy of the model is 80.0 %
```

Fig. 3. Statistics of the machine learning model.

According to the results obtained from study, it is observed that the MSAVI index is the most decisive index to the vegetation field. It is comparatively larger in the case of healthy vegetation areas. For dense vegetation field, it shows higher mean value. Whereas in case of unhealthy vegetation mean value of MSAVI is relatively lower. MSAVI isn't affected by the brightness of soil and hence it can be considered as a good tool for crop health monitoring index. NDVI values show some trends while classifying agricultural fields on the basis of vegetation. It is higher for healthy fields where as low in case of barren fields, but due to soil brightness error, it contributes less in the classification of fields. The Last parameter we have studied was NDWI which is actually a water index used for drought prediction. But we can use it as a classification factor as it shows a specific trend. It lies towards the negative side for dead vegetation and it is positive in case of healthy vegetation areas. We can classify the agricultural field by using mean values of indices as features. The Overall quality factor of the agricultural field is 1 for extremely healthy vegetation areas and it is 4 for completely dead vegetation areas. We can classify every and any given field in this range depending on all the different vegetative index. Through the analysis of parameters, it is noticed that the green or dead vegetation is simply a measure of the accuracy that the predicted result needs to be compared to. Buffer time of landsat8 is around 16 days i.e. the satellite images are updated once within 16 days, which is quite a long time. Multispectral data can be collected by satellite as well as drones but the accuracy of satellite data is less than that of drones but using drones can be a costly as well as a complicated affair. There is a huge importance of having a huge data set with a lot of data points because it helps increase the accuracy of the tools and makes it more efficient.

5 Conclusion

In this study, agricultural fields from different states in India were studied. The study concludes that the value of a quality factor is affected by both the current as well the past situation of a field, and hence they are the most decisive parameters to predict and conclude the health and quality of an agricultural field. The accuracy of the model calculated by the support vector machine algorithm was 80%. This can be increased by increasing the number of data points and it depends upon the distribution of data. One more inference was obtained that the MSAVI index is the most important factor affecting the most in the prediction of quality of the field but when it is combined with other vegetative indexes the accuracy and the efficiency increases. The output produced through our algorithm is a quality factor that indicates the nature of the field and all of this can be done without physically visiting the place by just a few clicks on our tool. This tool can be very useful for government organizations, Insurance companies and banks for client verification claims and loan grants.

6 Future Scope

We have developed this technique on the windows platform. We can deploy a web application for this purpose. One can also build this system on the android platform. If we use more vegetation indices we could get a more precise analysis of the quality of the land and field vegetation. The accuracy of the model can be increased by using neural network. This tool can be further used for price prediction of the agricultural land.

References

1. Yang, C., Anderson, G.L.: mapping grain sorghum yield variability using airborne digital videography. Precis. Agric. **2**, 7–23 (2000). https://doi.org/10.1023/A:1009928431735

2. Chang, L., et al.: A review of plant spectral reflectance response to water physiological changes. Chin. J. Plant Ecol. **40**(1), 80–91 (2016). https://doi.org/10.17521/cjpe.2015.0267

3. Üstuner, M., Sanli, F.B., Abdikan, S., Esetlili, M.T., Kurucu, Y.: Crop type classification using vegetation indices of rapideye imagery. Int. Arch. Photogramm. Remote Sens. Spatial Inf. Sci. **XL-7**, 195–198 (2014). https://doi.org/10.5194/isprsarchives-xl-7-195-2014

4. Burchfield, E., Nay, J.J., Gilligan, J.: Application of machine learning to the prediction of vegetation health. Int. Arch. Photogramm. Remote Sensing Spatial Inf. Sci. **XLI-B2** (2016). https://doi.org/10.5194/isprsarchives-xli-b2-465-2016

5. Han, H., Bai, J., Yan, J., Yang, H., Ma, G.: A combined drought monitoring index based on multi-sensor remote sensing data and machine learning. Geocarto Int. 1–16 (2019). https://doi.org/10.1080/10106049.2019.1633423

6. Alexander, C.: Normalized difference spectral indices and urban land cover as indicators of land surface temperature (LST). Int. J. Appl. Earth Observ. Geoinf. **86**, 102013 (2020). https://doi.org/10.1016/j.jag.2019.102013, ISSN 0303-2434

7. Veeraswamy, G., Nagaraju, A., Balaji, E., Sridhar, Y.: land use land cover studies of using remotesensing and gis a case study in gudur area nellore district, andhrapradesh. Int. J. Res. **4** (2017)

8. Kobayashi, N., Tani, H., Wang, X., Sonobe, R.: Crop classification using spectral indices derived from Sentinel-2A imagery. J. Inf. Telecommun. **4**(1), 67–90 (2020). https://doi.org/10.1080/24751839.2019.1694765

9. Romero, M., Luo, Y., Su, B., Fuentes, S.: Vineyard water status estimation using multispectral imagery from an UAV platform and machine learning algorithms for irrigation scheduling management, Comput. Electron. Agric. **147**, 109–117 (2018). https://doi.org/10.1016/j.compag.2018.02.013, ISSN 0168-1699

10. Schwalbert, R.A., Amado, T., Corassa, G., Pott, L.P., Prasad, P.V., Ciampitti, I.A.: Satellite-based soybean yield forecast: integrating machine learning and weather data for improving crop yield prediction in southern Brazil. Agric. Forest Meteorol. **284**, 107886 (2020). https://doi.org/10.1016/j.agrformet.2019.107886, ISSN 0168-1923

11. Fu, W., Ma, J., Chen, P., Chen, F.: Remote sensing satellites for digital earth. In: Guo, H., Goodchild, M.F., Annoni, A. (eds.) Manual Digit. Earth, pp. 55–123. Springer, Singapore (2020). https://doi.org/10.1007/978-981-32-9915-3_3

12. Anyamba, A., Tucker, C.: Historical perspectives on AVHRR NDVI and vegetation drought monitoring. Remote Sensing of Drought: Innovative Monitoring Approaches (2012). https://doi.org/10.1201/b11863

13. Brecht. Remote Sensing Indices (2018). https://medium.com/regen-network/remote-sensing-indices-389153e3d947

14. Jena, J., Misra, S., Tripathi, K.: Normalized Difference Vegetation Index (NDVI) and its role in Agriculture (2019)

15. Xue, J., Baofeng, S.: Significant remote sensing vegetation indices: a review of developments and applications. J. Sensors **2017**, 1–17 (2017). https://doi.org/10.1155/2017/1353691

16. Google earth engine website. https://earthengine.google.com/

17. Barsi, J.A., Schott, J.R., Hook, S.J., Raqueno, N.G., Markham, B.L., Radocinski, R.G.: Landsat-8 thermal infrared sensor (TIRS) vicarious radiometric calibration. Remote Sensing **6**, 11607–11626 (2014). https://doi.org/10.3390/rs61111607

18. Bao, W., Lianju, N., Yue, K.: Integration of unsupervised and supervised machine learning algorithms for credit risk assessment. Exp. Syst. Appl. **128**, 301–315 (2019). https://doi.org/10.1016/j.eswa.2019.02.033, ISSN 0957-4174

19. Li, Y., Wu, H.: A clustering method based on k-means algorithm. Phys. Proc. 1104–1109 (2012). https://doi.org/10.1016/j.phpro.2012.03.206

Sponsored Data Connectivity at the Network Edge

Ivaylo Atanasov, Evelina Pencheva[(⊠)], Ivaylo Asenov, and Ventsislav Trifonov

Faculty of Telecommunications, Technical University of Sofia, Kliment Ohridski blvd. 8,
1000 Sofia, Bulgaria
{iis,enp,vgt}@tu-sofia.bg, ivaylo.asenov@balkantel.net

Abstract. Sponsored data connectivity enables third parties to pay for specific
content traffic used by mobile subscribers. It requires network's intelligence for
specific traffic detection, usage monitoring, reporting and credit management. The
paper presents an approach to define Application Programming Interfaces (API)
for sponsored data connectivity deployed at the edge of the mobile network. API
enable setting and updating a chargeable party at session setup or during the session
The API design follows the Representational State Transfer architectural style. A
feasibility study is provided which illustrates the API practicability. The latency
introduced by the proposed API is evaluated by emulation.

Keywords: Multi-access Edge Computing · Network function
programmability · Representational State Transfer · Behavioral equivalence ·
Latency evaluation

1 Introduction

Fifth generation (5G) mobile networks offer considerable capabilities to provide rich
multimedia context to mobile devices. The forecast for mobile data traffic per month in
2025 is about 160 exabytes, which is more than four times the monthly usage at the end
of 2019 [1]. Sponsored data connectivity is a way for content or service providers to
advertise their products and to encourage subscribers to use more mobile data services.
Subscribers are offered sponsored access to various types of content, electronic books,
games, movies, trailers, and sports events. The sponsored data access is paid by third
party. In this way, by subsidizing the costs of the subscribers' broadband traffic, the
consumption of advertising and content on the mobile network is stimulated [2, 3].
Criteria for sponsored data connectivity may depend on access to web portal or mobile
application, subscribers' demographic profile, location etc.

The impact of sponsored data connectivity on network operators, content providers
and subscribers is analyzed in [4, 5]. Effective business models and winning strategies
are investigated in [6–8]. Sponsored data connectivity requires network's intelligence for
specific traffic detection, usage monitoring, reporting and credit management. The functionality for sponsored data connectivity in 5G system as a part of policy and charging
framework is defined in [9].

© Springer Nature Singapore Pte Ltd. 2020
M. Singh et al. (Eds.): ICACDS 2020, CCIS 1244, pp. 449–462, 2020.
https://doi.org/10.1007/978-981-15-6634-9_41

In this paper, we study capabilities for programmability of sponsored data connectivity at the network edge. 5G system enables exposure of policy and charging functionality. Exposure of core network functionality empowers programmability and it is achieved through Application Programming Interfaces (API) [10, 11]. 5G adopts distributed core closer to the network edge and thus reducing latency and backhaul traffic [12, 13]. Further improvement of network performance can be achieved by deployment of Multi-access Edge Computing (MEC). As a key technology for 5G, MEC enables provisioning of cloud intelligence close to where it is needed [14, 15]. The paper proposes an approach to design MEC-based API for sponsored data connectivity. The rest of the paper includes description of the proposed API functionality illustrated by typical use case scenarios, API data model including supported resources, methods and data types, as well as feasibility study illustrating the approach, and evaluation of API performance metrics.

2 Sponsor Connectivity Data API at the Network Edge

The mobile edge host, which is located at the network edge, provides the required Network Function Virtualization (NFV) infrastructure with cloud resources. The necessary functionality required to deploy mobile edge applications on the NFV infrastructure is provided by the mobile edge platform. It also hosts mobile edge services and enables applications to consume and provide services. The MEC host can be co-located with distributed 5G core and the core network functions can share the same NFV infrastructure with the mobile edge applications.

We propose a new mobile edge service named Sponsored Data Connectivity Service (SDCS) which enables mobile edge applications to change the chargeable party from the beginning of a session or during the session. This core functionality can be exposed by the core Network Exposure Function (NEF).

The following example illustrates the SDCS functionality at the network edge. A multimedia content provider wants to advertise the product (e.g. on-line movies or video games) for mall visitors at the point of cell. In order to encourage the visitors to consume the content, the content provider subsidize the mobile data traffic related to movie or video game trailers. The free multimedia content enables the subscribers to choose whether to buy the advertised content or not. The content provider pays for advertising data traffic or shares the advertising revenue with the telecom operator.

Figure 1 illustrates the information flow for the above scenario.

1. When setting up a session between Application Server (AS) via NEF, the mobile edge application requests from the SDCS to become a sponsor (chargeable party) for the session to be setup.
2. The SDCS in turn invokes the NnefChargeableParty_Create operation of NEF and requests to be notified about usage monitoring events.
3. The NEF authorizes the request and triggers packet data session modification.
4–5. After successful packet data session modification, the NEF assigns an identifier to the chargeable party session and responds to the SDCS, which in turn sends a response to the mobile edge application.

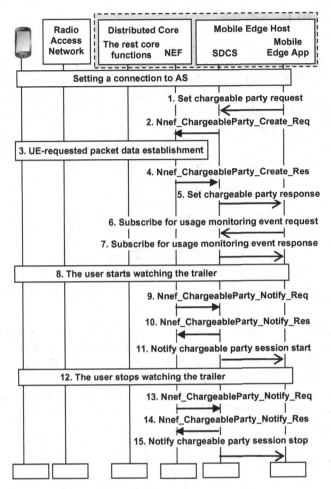

Fig. 1. Flow of application-initiated change of chargeable party at the data session start and subsequent notifications about usage of network resources

6–7.   The mobile edge application subscribes to receive notifications about usage monitoring.

8.   The subscriber starts watching the free movie trailer.

9–11.   The NEF notifies the SDCS about the event using the Nnef_ChargeablePartyNotify operation, and the SDCS in turn notifies the application.

12.   The user stops watching the free movie trailer.

13–15.   The NEF notifies the SDCS, which in turn reports the event to the mobile edge application. The network charges the sponsor for the data traffic (not shown).

3 Service API Description

This section provides a detailed service description by typical use cases illustrating the API functionality.

The proposed SDCS API follow the adopted REST (Representational State Transfer) architectural style, where resource is a key concept. Figure 2 illustrates the SDCS resource structure, where all resources follow the SDCS URI. The SDCS URI can be discovered using service directory.

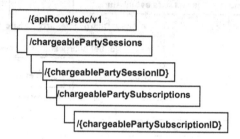

Fig. 2. Structure of resources of the Sponsored Data Connectivity Service

The chargeablePartySessions resource represents all AS sessions with option to change chargeable party. The resource supports HTTP methods GET and POST.

A mobile edge application can retrieve information about all AS sessions with changed chargeable party by sending a GET request to the chargeablePartySessions resource. The SDCS responds with 200 OK where, the message body contains a list of all AS sessions with changed chargeable party.

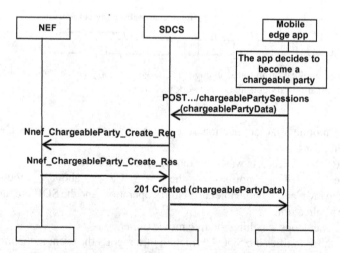

Fig. 3. Flow of initial chargeable party change at AS session setup

Figure 3 shows the flow of setting the chargeable party at AS session setup. When the mobile edge application wants to request to subsidize the data traffic from the beginning of an AS session or during the AS session, the application sends a POST request to the resource representing chargeable party sessions. The body of the request contains chargeablePartyData data type. The SDCS requests setting the chargeable party at the beginning of AS session and reporting of usage monitoring events. Upon receiving successful response from NEF, the SDCS creates a resource representing the AS session with chargeable party and responds to the mobile edge application with 201 OK message which includes the URI of the created resource.

Table 1 shows the attributes of the chargeablePartyData data type which maps on a JSON structure.

The mobile edge application may decide to update the sponsoring status of an existing AS session. In order to do this, the application sends a PUT request to the resource representing the respective chargeable party session. The request body indicates the sponsoring status change. The SDCS interacts with NEF to enforce the requested update and on successful response, the SDCS sends a 200 OK response to the mobile edge application.

The chargeablePartySessionID resource, representing an existing AS session with changed chargeable data, supports also GET method used to retrieve information about existing sponsored data and DELETE method used to remove an existing AS session with sponsored data.

The chargeablePartySubscriptions resource represents all subscriptions for reporting accumulated usage of sponsored data for existing AS session. The resource supports GET method, which retrieves all existing subscriptions for reporting accumulated usage of sponsored data and POST method, which creates a new subscription. Figure 4 shows the flow of creating a new subscription for reporting accumulated usage for an AS session with chargeable party set.

The subscriptionData data type defines the usage monitoring information and URI where the application wants to receive notifications. Optionally, the mobile edge application may include the subscription duration. On receiving the request, the SDCS creates a resource representing the respective subscription.

The chargeablePartySubscriptionID resource represents an existing subscription for reporting accumulated usage for an AS session with chargeable party set. It supports GET, PUT and DELETE methods.

In case of active subscription, the mobile edge application is notified about events related. To notify the application, the SDCS sends a POST request to the address provided in the subscription request. The request body contains accumulatedUsage data type, which is a JSON structure. Table 2 shows the attributes of the accumulatedUsage data type.

Table 1. Attributes of ChargeablePartyData data type

Attribute name	Type	Cardinality	Meaning
>timeStamp	TimeStamp	0..1	TimeStamp
>asID	String	1	Application Server Identifier
>userAddress	URI	1	The address of the user
>trafficFlowDescription	Structure	1..n	IP flow description
>>flowID	String	1	Indicates the IP flow
>>direction	Enumerated	1	Indicates the uplink or downlink direction of the flow; 0 = uplink, 1 = downlink, 2 = both
>>sourceAddress	String	1	The IP address of the source
>>destinationIPAddress	String	1	The IP address of the destination
>>sourcePort	Integer	1	The port of the source
>>destinationPort	Integer	1	The port of the destination
>>protocol	String	1	The protocol used
>sponsorID	String	1	The unique sponsor identifier
>sponsoringStatus	Boolean	1	Shows whether the sponsoring is enabled or disabled
>usageMonitoringInfo	Structure	1	Information about usage monitoring
>>unitVolume	Structure	0..1	Information about volume of sponsored data in bytes
>≫grantedUnitsDL	Integer	0..1	The number of granted bytes in downlink
>≫grantedUnitsUL	Integer	0..1	The number of granted bytes in uplink
>≫usedUnitsDL	Integer	1	The number of used bytes in downlink
>≫usedUnitsUL	Integer	1	The number of used bytes in uplink
>>time	Structure	0..1	Information about time period of sponsored data
>≫grantedTime	Integer	1	The time period granted in seconds
>≫usedTime	Integer	1	The time period used in seconds.
>appInsID	String	1	The unique application instance identifier
>requestID	String	1	The request identifier allocated by the application

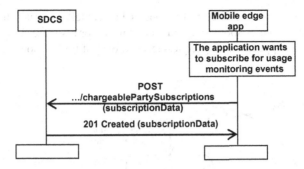

Fig. 4. Flow of subscription for notifications about usage monitoring

Table 2. Attributes of accumulatedData data type

Attribute name	Type	Cardinality	Meaning
>timeStamp	TimeStamp	0..1	TimeStamp
>reportedEvents	Structure	1..n	Indicates the reported events
>>flowID	String	1	The IP flow identifier
>>reportedEvent	String	1	Indicates the reported event
>>accumulatedUsage	Structure	0..1	Usage information corresponding to the event
>≫volumeDL	Integer	0..1	Downlink data bytes
>≫volumeUL	Integer	0..1	Uplink data bytes
>≫duration	Integer	0..1	The amount of time in seconds.
>startTime	TimeStamp	0..1	Indicates the start time of sponsored data
>stopTime	TimeStamp	0..1	Indicates the stop time of sponsored data

4 Feasibility Study

In order to illustrate the practicality of the Sponsored Data Connectivity API, models representing the sponsoring status are proposed. The models reflect the mobile edge platform and application views. Both models are formally described, and it is proved that the models behave the same way, i.e. they are synchronized.

Figure 5 shows a simplified model of the sponsoring status supported by a mobile edge application. In Null state, the sponsoring of the data traffic is not activated. On application trigger (e.g. at AS session setup) the mobile edge application requests a change of chargeable party. In SettingChargeableParty state, the application waits for response of previously sent request. In WaitForSponsoredTrafficStart state, the sponsoring status is enabled, the application subscribes for receiving notifications about sponsored data traffic and waits for traffic start. In WaitForSponsoredTrafficStop state, the application waits for sponsored data traffic stop. In UpdateChargeableParty state, the application has requested an update of sponsoring status and waits for response. In DeleteChargeableParty state, the application has requested a removal of traffic sponsoring and waits for response.

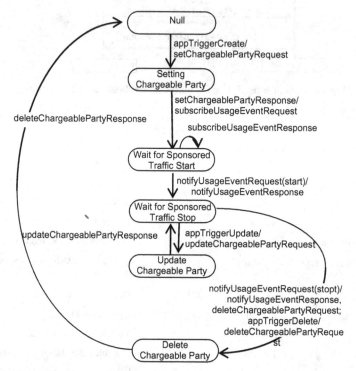

Fig. 5. Model of the sponsoring status supported by a mobile edge application

Figure 6 shows the simplified model of the sponsoring status supported by a mobile edge platform.

In Idle state, the sponsoring status is disabled. In CreateASsessionWithChargeableParty state, a chargeable party transaction is activated in the network and the mobile edge platforms waits for response. In WaitForASsessionTraffic, the sponsoring status is enabled, and the mobile edge platform waits for notification about sponsored traffic start. In ASsessionTrafficMeasuring, the network measures the accumulated traffic usage. In UpdateASsessionWithChargeableParty state, a procedure for sponsoring status update is activated in the network and the mobile edge platforms waits for response.

Each of the models is formally describes as a quadruple of set of states, set of actions, set of transitions and set of initial states. For the sake of simplicity, the names of states and actions are substituted by short notations given in brackets.

By T_A it is denoted a formal description of the model representing the sponsoring status supported by a mobile edge application. $T_A = (States_A, Actions_A, Transitions_A, s_A^0)$, where:

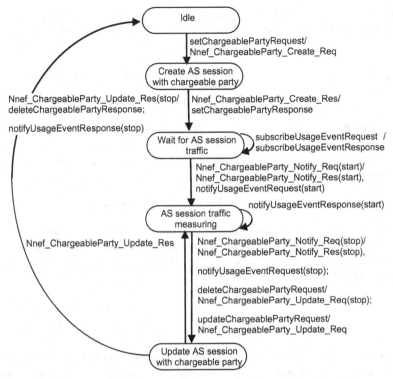

Fig. 6. Model of the sponsoring status supported by a mobile edge platform

$States_A$ = {Null [s_1^A], SettingChargeableParty [s_2^A], WaitForSponsoredTrafficStart [s_3^A], WaitForSponsoredTrafficStop [s_4^A], UpdateChargeableParty [s_5^A], DeleteChargeableParty [s_6^A]};

$Actions_A$ = {appTriggerCreate [t_1^A], setChargeablePartyResponse [t_2^A], subscribeUsageEventResponse [t_3^A], notifyUsageEventRequest(start) [t_4^A], appTriggerUpdate [t_5^A], notifyUsageEventRequest(stop) [t_7^A], appTriggerDelete [t_7^A], updateChargeablePartyResponse [t_8^A], deleteChargeablePartyResponse [t_9^A]};

$Transitions_A$ = {(s_1^A, t_1^A, s_2^A), (s_2^A, t_2^A, s_3^A), (s_3^A, t_3^A, s_3^A), (s_3^A, t_4^A, s_4^A), (s_4^A, t_5^A, s_5^A), (s_4^A, t_6^A, s_6^A), (s_4^A, t_7^A, s_6^A), (s_5^A, t_8^A, s_4^A), (s_6^A, t_9^A, s_1^A)};

$s_A^0 = s_1^A$.

Short names of states and actions are given in brackets.

By T_M it is denoted an LTS representing the multiparty multimedia session state model supported by the network $T_M = (States_M, Actions_M, Transisions_M, s_M^0)$, where:

$States_M$ = {Idle [s_1^M], CreateASsessionWithChargeableParty [s_2^M], WaitForASsession-Trafic [s_3^M], ASsessionTrafficmeasuring [s_4^M], UpdateASsessionWithChargeableParty [s_5^M]};

$Actions_M$ = {setChargeablePartyRequest [t_1^M], Nnef_ChargeableParty_Create_Res [t_2^M], subscribeUsageEventRequest [t_3^M], Nnef_ChargeableParty_Notify_Req(start) [t_4^M], notifyUsageEventResponse(start) [t_5^M],

Nnef_ChargeableParty_Notify_Req (stop) [t_6^M], deleteChargeablePartyRequest [t_7^M], updateChargeablePartyRequest   [t_8^M],   Nnef_ChargeableParty_Update_Res [t_9^M], Nnef_ChargeableParty_Update_Res(stop) [t_{10}^M], notifyUsageEventResponse(stop) [t_{11}^M]};

$Transitions_M = \{(s_1^M, t_1^M, s_2^M), (s_2^M, t_2^M, s_3^M), (s_3^M, t_3^M, s_3^M), (s_3^M, t_4^M, s_4^M), (s_4^M, t_5^M, s_4^M), (s_4^M, t_6^M, s_5^M), (s_4^M, t_7^M, s_5^M), (s_4^M, t_8^M, s_5^M), (s_5^M, t_9^M, s_4^M), (s_5^M, t_{10}^M, s_1^M), (s_5^M, t_{11}^M, s_1^M)\};$

$s_M^0 = s_1^M.$

Formal model verification is conducted using the concept of weak bisimulation. The aim is to prove that both models expose behavioural equivalence, i.e. the application model on sponsoring status simulates the mobile edge platform model on sponsoring status and vice versa [16].

Proposition: T_A and T_M have a weak bi-simulation relationship.

Proof: To prove the existence of bi-simulation relationship it requires identification of pairs of states in both models such that the respective transitions match each other. By $R_{A\&M} = \{(s_1^A, s_1^M), (s_2^A, s_2^M), (s_3^A, s_3^M), (s_4^A, s_4^M)\}$. Then the following functional matching between the transitions of T_A and T_M exists:

1. The mobile edge application requests to set a chargeable party and the mobile edge platform triggers Nnef_ChargeableParty_Create procedure: for $(s_1^A, t_1^A, s_2^A) \; \exists \; (s_1^M, t_1^M, s_2^M)$.
2. The mobile edge application is notified when the mobile edge platform receives a report for successful enabling of sponsoring status, and the application subscribes for events related to accumulated usage: for $(s_2^A, t_2^A, s_3^A), (s_3^A, t_3^A, s_3^A) \; \exists \; (s_2^M, t_2^M, s_3^M), (s_3^M, t_3^M, s_3^M)$.
3. The mobile edge application is notified when the mobile edge platform receives a report for sponsoring traffic start: for $(s_3^A, t_4^A, s_4^A) \; \exists \; (s_3^M, t_4^M, s_4^M), (s_4^M, t_5^M, s_4^M)$.
4. The mobile edge application initiates an update of sponsoring status and the mobile edge platform triggers Nnef_ChargeableParty_Update procedure: for $(s_4^A, t_5^A, s_5^A), (s_5^A, t_8^A, s_4^A) \; \exists \; (s_4^M, t_8^M, s_5^M), (s_5^M, t_9^M, s_4^M)$.
5. The mobile edge application initiates a removal of sponsoring and the mobile edge platform triggers Nnef_ChargeableParty_Update procedure: for $(s_4^A, t_7^A, s_6^A), (s_6^A, t_9^A, s_1^A) \; \exists \; (s_4^M, t_7^M, s_5^M), (s_5^M, t_{10}^M, s_1^M)$.
6. The mobile edge application is notified when the mobile edge platform receives a report for sponsoring traffic stop: for $(s_4^A, t_6^A, s_6^A), (s_6^A, t_9^A, s_1^A) \; \exists \; (s_4^M, t_6^M, s_5^M), (s_5^M, t_{11}^M, s_1^M)$.

Therefore, T_A and T_M have a weak bi-simulation relationship.   ∎

5 Experimental API Emulation

One of the Key Performance Indicators of MEC is latency which is measured per service [17]. The service processing time influences the service latency in terms of time required by the mobile edge platform and application to process the requests and responses.

An example HTTP request to set a chargeable party for AS session looks like the following:

```
POST /appRootExam/sdc/v1/chargeablePartySessions HTTP/1.1
Host: example.com
Accept: application/json
Content-type: application/json
Content-length: 551


{"timeStamp":"2020-01-19   03:14:07'UTC","asID":"BBCCDD12","userAddress":"/subscribers/331278","trafficFlo
wDescription":{"flowID":"ABCD67","direction":2,"sourceAddress":"168.212.226.204","destinationAddress":
"168.212.223.150","sourcePort":80,"destinationPort":80,"protocol":"AppleLowLatencyHLS"},"sponsorID":"XXYY
ZZ","sponsoringStatus":"enable","usageMonitoringInfo":{"unitVolume":{"grantedUnitsDL":100000,"grantedUnits
UL":1000,"usedUnitsDL":0,"usedUnitsUL":0},"time":{"grantedDuration":180,"usedDuration":0},"appInsID":"AABB
CCDD","requestID":"F15BC"}
```

The corresponding HTTP response looks as:

```
HTTP/1.1 201 Created
Location: example.com/appRootExam/sdc/v1/chargeablePartySessions/2233ABF15
Content-type: application/json
Content-length: 575


{"timeStamp":"2020-01-19   03:14:07'UTC","asID":"BBCCDD12","userAddress":"/subscribers/331278","trafficFlo
wDescription":{"flowID":"ABCD67","direction":2,"sourceAddress":"168.212.226.204","destinationAddress":
"168.212.223.150","sourcePort":80,"destinationPort":80,"protocol":"AppleLowLatencyHLS"},"sponsorID":"XXYY
ZZ","sponsoringStatus":"enable","usageMonitoringInfo":{"unitVolume":{"grantedUnitsDL":100000,"grantedUnits
UL":1000,"usedUnitsDL":0,"usedUnitsUL":0},"time":{"grantedDuration":180,"usedDuration":0},"appInsID":"AABB
CCDD","requestID":"F15BC", "sessionID":"FFEELLMM"}
```

The experimental setup includes client and server parties for the above request and response. The client is implemented in Java and runs on Intel® Core™ i7-3770 CPU @ 3,4 GHz, 8 cores, 8 GB RAM, Ubuntu. The server runs on Intel® Core™ i7-9750H CPU @ 2.60 GHz, 6 cores, 16 GB RAM, Ubuntu OS. The interface between the client (e.g. mobile edge applications) and the server (mobile edge platform) is based on REST. The server side is implemented using Eclipse Vert.x and Redis. Vert.x is an open source event-driven application framework [18]. Redis is an open source in memory data structure storage [19]. The server party consists of 1 to 8 of verticle copies and a Redis storage. Each verticle copy supports a REST-based interface and a Redis client. The Redis clients write in the Redis storage. The Redis storage is configured to work in a single node as master and no clustering is provided at all. The dedicated memory for Redis is 1 GB.

The experiment is conducted for 100 000 operations (POST requests and relevant 201 OK responses). Figure 7 shows the latency measured in *nanoseconds* for 99% of the operations. There is a small number (about 1%) of operations for which the latency is quite high. The emulation is done for 1, 4 and 8 verticle copies. The operations latency shown in Fig. 7 ranges from say 130 ns to more than 5 ms for the last percent.

Figure 8 shows the number of operations vs latency. The experiment is conducted for 1, 4 and 8 verticle copies. It is obvious that the average latency is decreased with the increase of verticle copies.

The results of emulation highlight the benefits of deploying sponsored data connectivity control at the network edge.

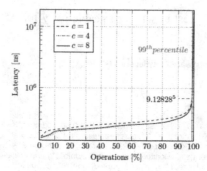

Fig. 7. Latency as a function of operation numbers (c stands for verticle copies)

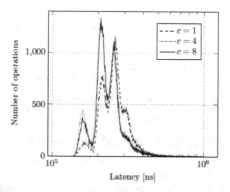

Fig. 8. Number of operations versus latency measured for 1, 4 and 8 verticle copies

6   Conclusion

Sponsored Data Connectivity enables sponsoring specific content data traffic for mobile subscribers without impacting their data plan allowance. It is beneficial for all the parties involved, telecom operators, content providers and subscribers.

In this paper, an approach to programmability of sponsored data connectivity at the network edge is proposed. MEC is a promising technology for introducing the required programmability in the vicinity of end user, where the content is used. MEC provides the required secured and manageable environment for running mobile edge applications. Using the advantages of NFV, RESTful API which enable setting and updating a chargeable party at the Application Server session setup or during the session, are defined. The API are described by resources and supported methods, information exchanged and data types, which demonstrate the API functionality. As a feasibility study, models representing the sponsoring status supported by a mobile edge application and the network are proposed, and formally verified. As an experimental implementation, the API latency is evaluated by emulation. The results show the low-level latency introduce by the proposed API.

Acknowledgments. This work was supported by grant of the project KP-06-H37/33 2019, funded by Bulgarian National Science, Ministry of Education and Science.

References

1. Ericsson, Ericsson Mobility Report (2019). https://www.ericsson.com/4acd7e/assets/local/mobility-report/documents/2019/emr-november-2019.pdf
2. Bangera, P., Hasan, S., Gorinsky, S.: An advertising revenue model for access ISPs. In: IEEE Symposium on Computers and Communications (ISCC), Heraklio, pp. 582–589 (2017)
3. Wang, W., Xiong, Z., Niyato, D., Wang, P.: A hierarchical game with strategy evolution for mobile sponsored content/service markets. IEEE Global Communications Conference GLOBECOM, pp. 1–6 (2017)
4. Zhang, L., Wu, W., Wang, D.: TDS: time-dependent sponsored data plan for wireless data traffic. In: IEEE International Conference on Computer Communications (INFOCOM), pp. 1–6 (2016)
5. Xion, Z., Feng, S., Niyato, D., Wang, P., Leshem, A., Han, Z.: Joint sponsored and edge cashing content service market: a game-theoretical approach. IEEE Trans. Wireless Commun. **18**(2), 1–16 (2019)
6. Maillé, P., Tuffin, B.: analysis of sponsored data in the case of competing wireless service providers. In: 9th International Conference on Network Games, Control and Optimization NETGCOOP, New York City, United States, pp. 1–16 (2018)
7. Maillé, P., Tuffin, B.: Wireless service providers pricing game in presence of possible sponsored data. In: 15th International Conference on Network and Service Management (CNSM), Halifax, Canada, pp. 1–14 (2019)
8. Vyavahare, P., Manjunath, D., Nair, J.: Sponsored data with ISP competition. Cornell University Library, pp. 1–2 (2019). arXiv:1906.00581
9. GPP Technical Specification Group Services and System Aspects; Policy and charging control framework for the 5G System (5GS); Stage 2, Release 16, v16.3.0 (2019)
10. Llorens-Carrodeguas, A., Cervello-Pastor, C., Leyva-Pupo, I., Lopez-Soler, J.M., Navarro-Ortiz, J., Exposito-Arenas, J.A.: An architecture for the 5G control plane based on SDN and data distribution service. In: Fifth International Conference on Software Defined Systems (SDS), Barcelona, pp. 105–111 (2018)
11. Barakabitze, A.A., Ahmad, A., Mijumbi, R., Hines, A.: 5G network slicing using SDN and NFV: a survey of taxonomy, architectures and future challenges. Comput. Netw. **167**(106984), 1–40 (2020)
12. Nguyen, V., Grinnemo, K., Taheri, J., Brunstrom, A.: On load balancing for a virtual and distributed MME in the 5G core. In: IEEE 29th Annual International Symposium on Personal, Indoor and Mobile Radio Communications (PIMRC), Bologna, pp. 1–7 (2018)
13. Raveendran, N., Zha, Y., Zhang, Y., Liu, X., Han, Z.: Virtual core network resource allocation in 5G systems using three-sided matching. IEEE In: International Conference on Communications (ICC), Shanghai, China, pp. 1–6 (2019)
14. Nakazato, J., Tao, Y., Tran, G.K., Sakaguchi, K.: Revenue model with multi-access edge computing for cellular network architecture. In: Eleventh International Conference on Ubiquitous and Future Networks (ICUFN), Zagreb, Croatia, pp. 21–26 (2019)
15. Pencheva, E., Atanasov, I., Velkova, D., Trifonov, V.: Application Level User Traffic Control at the Mobile Network Edge, Open Innovation Association 24th FRUCT (Finnish-Russian University Cooperation in Telecommunications) Moscow, Russia, pp. 312–327 (2019)

16. Liu, X., Yu, T., Zhang, W.: Logics for bisimulation and divergence. In: Baier, C., Dal Lago, U. (eds.) FoSSaCS 2018. LNCS, vol. 10803, pp. 221–237. Springer, Cham (2018). https://doi.org/10.1007/978-3-319-89366-2_12
17. ETSI GS MEC-IEG 006 Mobile Edge Computing; Market Acceleration; MEC Metrics Best Practice and Guidelines, v1.1.1 (2017)
18. Eclipse Vert.x. https://vertx.io/
19. Redislab 5.0.7. https://redis.io/

Dynamic Bidding with Contextual Bid Decision Trees in Digital Advertisement

Manish Pathak$^{(\boxtimes)}$ and Ujwala Musku

Research and Development, MiQ Digital India, Bangalore 560001, KA, India
{manishpathak,ujwala}@miqdigital.com
http://www.wearemiq.com

Abstract. Real-time bidding (RTB) has been one of the most prominent technological advances in online display advertising. Billions of
transactions in the form of programmatic advertising auctions happen on
a daily basis on ad-networks and exchanges, where advertisers compete
for the ad slots by bidding for that slot. The question: how much should I
bid? has lingered around and troubled many marketers from a long time.
Past strategies' formulation has been mostly based on targeting users by
analyzing their browsing behavior via cookies to predict the likelihood
that they will interact with the ad. But due to growing privacy concerns
where browsers are taking down cookies and recent regulations like General Data Protection Regulation (GDPR) in Europe, targeting users has
become difficult and these bidding methodologies fail to deliver. This
paper presents a novel approach to tackle the dual problem of optimal
bidding and finding an alternative to user-based targeting by focusing
on contextual-level targeting using features like site-domain, keywords,
postcode, browser, operating system, etc. The targeting is done at feature
combination level in the form Bid Decision Trees. The framework discussed in the paper dynamically learns and optimizes bid values for the
context features based on their performance over a specific time interval
using a heuristic Feedback Mechanism to optimize the online advertising
KPIs: Cost per Acquisition (CPA) and Conversion Rate (CVR). A comparison of the performance of this context-based tree-bidding framework
reveals a 59% lower CPA and 163% higher CVR as compared to other
targeting strategies within the overall campaign budget, which are clear
indicators of its lucrativeness in a world where user-based targeting is
losing popularity.

Keywords: Contextual targeting · Programmatic advertising · Real
time bidding · CPA · CVR · GDPR · Artificial Intelligence · Digital
marketing

1 Introduction

Though the origins of advertising [1] date back in time, digital advertising has
been booming for almost a decade now. In 2019, worldwide digital ad spend was

© Springer Nature Singapore Pte Ltd. 2020
M. Singh et al. (Eds.): ICACDS 2020, CCIS 1244, pp. 463–473, 2020.
https://doi.org/10.1007/978-981-15-6634-9_42

predicted to spring up by *17.6%* to $*333.25* billions, which is roughly half of the global advertising market [2]. In digital ad marketing, the traditional methods that were followed mainly involved targeting specific demographics and other behavioral tactics like direct cookie targeting as per the transactional history, which had been successful [3]. But the rising buzz and concerns on privacy issues due to cookie targeting led to the birth of GDPR (General Data Protection Regulation) in Europe. Now, user-specific targeting is strongly discouraged and hence finding an alternative is very crucial for advertising companies. Contextual targeting, which is a practice of displaying ads based on a website's contents ameliorated the damage control of GDPR restrictions and liabilities and hence is gaining huge popularity post GDPR.

However, it is equally important to show the ad at the right price to maximize Return on Investment (ROI) and this is where Real-time bidding (RTB) comes into the picture. RTB has changed the face of online advertising. It basically allows advertisers to bid on an ad slot on a website through an automated digital auction process [4] which happens within fraction of a second (*1/10*th of a second) [5]. Due to these constraints, it is essential to bid accurately and quickly to gain higher ROI and avoid the risk of losing the ad-auction respectively.

In this paper, we present a novel approach to tackle the dual problem of finding an alternative to cookie-based targeting and arriving at a dynamic optimum bid value using a Feedback Mechanism.

2   Ecosystem Description

2.1   Digital Advertising Landscape

Digital Advertising is known to have one of the most complex landscapes in the digital world with various entities. In simple words, it is more of a supply-demand entity in the digital world. Advertisers would want to show ads about their product to people on the ad spaces within a website. Hence, advertisers here act as demand and publishers who own ad spaces on websites act as supply partners. There are other crucial entities also, like Supply-Side Platform (SSP), Demand-Side Platform (DSP) and Ad exchanges that support this demand-supply process [6].

- SSPs help the publisher to connect their inventory to Ad Exchange. It also helps in making more revenue by selling premium inventory deals to advertisers/DSP's and figuring out other ways to reach the advertiser.
- DSPs help the advertiser in buying ad space and managing their ad campaigns.
- Ad Exchange is a marketplace where demand and supply get exchanged i.e., buying an ad slot on websites by the advertiser from the publisher through real-time bidding from multiple ad networks.

2.2   Real-Time Bidding

Real-time bidding [7] is an automated auction process where the multiple advertisers and DSPs compete to win an ad slot on a website to display their ad. Real-time bidding follows the second price auction model [8] i.e., each advertiser/DSP sets a bid amount they are willing to spend to win the ad slot and the advertiser with maximum bid wins the auction and buys the ad slot by paying next highest bid price i.e., second bid price. Recently, there has been a buzz around ad exchanges moving towards the first-price auction model.

Before delving into the proposed solution, let's define the terminologies used in the digital advertising industry. Each winning auction of an ad slot is considered as an **impression** [9]. A **click** will be recorded against an impression if the user clicks on the ad. Pixels are placed on the advertiser's page which gets triggered whenever the page is loaded. Conversions are tracked with the help of these pixels. A **conversion** is recorded against an impression when the user does a predefined activity like landing on the advertiser's page or making a transaction or downloading a brochure etc. after viewing the impression.

We, MiQ, represent and help advertisers in running their ad campaigns to reach their KPI within a specific budget. As per market standards, KPI of a display campaign is as follows:

- Cost per Acquisition (CPA): This can be defined as the cost spent by the advertisers for each conversion they've gained at a particular defined level.
- Conversion Rate (CVR): This is a ratio of number of conversions to number of impressions served at a particular defined level.
- Cost per Impression (CPM): This is defined as the cost spent by the advertisers to win 1000 impressions at a particular defined level.

The aim is to acquire a user at a lower cost alluded by the advertiser without compromising the quality and quantity of impressions delivered. Owing to GDPR [10], targeting a cookie/user directly is not recommended anymore and hence targeting contextually is a preferred way out. Hence, to reiterate, we need to show the ad at the **proper place, proper time and proper bid**. In this paper, we will focus on proper place and proper bid.

3   Data Preparation

The data used here is a typical portrayal of activity and bidder data in the digital advertising market. Bidder data comprises of the details on the available pandemic inventory whereas activity feed data comprises of the campaign level characteristics on the inventory MiQ managed to deliver. All the input variables except keywords are readily available in these two data sets. Keywords are therefore extracted from the URL present in bidder data. For example, keywords like "art" will be extracted from the URL https://www.ebay.com/b/Art/550/bn_ 1853728. Purchase history data is being used as well. These three data sources are merged and used further for feature combination extraction and ranking. Table 1 shows a snippet of sample data showing some of the features that can be used to construct the tree. There are other features like *country, hour, day, device* etc. which can also be included.

Table 1. Sample data

$f_c(id)$	Domain	Keyword	Browser	Postcode	Placement	Vendor	os	Creative size
1	ebay.com	shoe	chrome	99501	p1	v1	windows	720 × 90
2	msn.com	sports	safari	86003	p2	v2	iOS	300 × 600
3	yahoo.com	news	chrome	99501	p3	v3	windows	160 × 600
4	nytimes.com	washington	chrome	98435	p4	v4	windows	300 × 250
5	zillow.com	mortgage	lynx	85001	p5	v5	linux	300 × 250

We formulate a utility function that gives a score for each feature combination using total user count and converted user count. This score is treated as a metric for ranking the features from best to worst. However, before we go ahead with ranking, it is crucial to maintain the representation of a contextual feature combination only once irrespective of possible combinations from the context. Available inventory from the bidder data is used to judge each contextual feature combinations. Niche feature combination (the combination with a maximum number of features for precise targeting) is picked since it represents the context better and this process is called feature combination optimization. Once feature combinations are optimized, we proceed with the ranking of features using the utility function value of each feature combination denoted as f_c.

$$Utility\ Function\ (U) = h(c, t | f_c) \tag{1}$$

where $c \rightarrow$ converted user count, $t \rightarrow$ total user count

Table 2. Sample data with Utility Function values

$f_c(id)$	U
1	0.21
2	0.63
3	0.05
4	0.55
5	0.89

We now have a utility function value which acts as a factor of significance against each contextual feature combination and hence we have attained the list of proper contextual places to display the ad. The next step involves the calculation of a winnable bid value against each contextual feature. With upper and lower bounds fixed, initial bids are calculated in a linear fashion with respect to utility function values.

$$Bid\,Function\,(B) = g(U) = g(h(c,t|f_c)) = \varphi(c,t|f_c) \qquad (2)$$

We estimate the bid value of a ith feature combination f_c denoted as B_i using upper bound (B_{max}) and lower bound (B_{min}) as follows.

$$B_{i,f_c} = \frac{d\varphi}{dh} * h(c,t|f_c) + k \qquad (3)$$

Utility function acts as a proxy in defining the optimal bid value along with defining the importance of features. Once we have optimal bids against each contextual feature combination, these are fed into DSP in the form of programmable decision tree-like structure called bid decision trees where each feature combination and optimal bid value act as a leaf in the tree. Sample bid decision tree looks like as follows:

if every domain = "zillow.com", keyword = "mortgage", browser = "lynx", postcode = 85001, placement = p5, vendor = v5, operating system = "linux",creative size = "300 × 250": value: 5

elif every domain = "yahoo.com", keyword = "news", browser = "chrome", postcode = 99501, placement = p3, vendor = v3, operating system = "windows",creative size = "300 × 250": value: 4.3

.

.

.

elif every domain = "trulia.com", placement = p6, vendor = v6: value: 1.856
elif every domain = "ebay.com", postcode = 12345, vendor = v3: value: 1.342
else: value: nobid

4   Feedback Mechanism

Once these contextual bid decision trees are formed, they are deployed onto Demand Side Platforms (DSP) as one of the strategies in the form of campaigns for targeting. The Feedback module analyses the campaign's performance and improves it using proportional feedback by controlling bids and impression deliveries over feature combinations in the tree. There are a plethora of critical measurement metrics on which the performance of the campaigns in digital advertisement can be modeled and optimized. The choice of metric largely depends on the type of campaign. For example, if the campaign is a video campaign, we would want to improve metrics like video completion rate, reduce the cost per completed view, or if the campaign is to create brand awareness then we would want to serve more impressions to a large audience. Generally, the metrics that stand out to optimize bids are traditional KPIs: CPA, CVR [11]. The feedback module operates to minimize the CPA and also maximize the CVR of the campaigns.

We denote the ith ad strategy starting to serve as a campaign on t^{th} day by s_i^t in a set of strategies called Bag of strategies, B. Let there be n strategies in a Bag of a strategy, B. We form an ordered set S, $\{s_1^1, s_2^1, s_3^2, \ldots s_k^t, s_{k+1}^1, \ldots s_n^t\}$, of all the s_i^t such that the first k strategies correspond to contextual tree strategies and the rest $n - k$ strategies correspond to other targeting strategies. Furthermore, let $s_{i(cost)}^t$, $s_{i(conv)}^t$ and $s_{i(tot)}^t$ denote the cost, number of convert users and total number of impressions associated with strategy s_i^t respectively, which we keep track of on a daily basis. We sum aggregate the performance of contextual tree strategies and other targeting strategies separately over a period of T days. Note that T here is a parameter that can be tweaked as per the need. To track the performance of other targeting strategies (B_{nc}), we define custom CPA and CVR metrics for B_{nc} in the following way:

$$B_{nc,cpa} = \frac{\sum_{t=1}^{T} \sum_{i=k+1}^{n} s_{i(cost)}^t}{\sum_{t=1}^{T} \sum_{i=k+1}^{n} s_{i(conv)}^t} \quad where \ s_i^t \in S \tag{4}$$

$$B_{nc,cvr} = \frac{\sum_{t=1}^{T} \sum_{i=k+1}^{n} s_{i(conv)}^t}{\sum_{t=1}^{T} \sum_{i=k+1}^{n} s_{i(tot)}^t} \quad where \ s_i^t \in S \tag{5}$$

The performance metrics of each feature combination occurring in contextual tree strategies are also calculated in a similar fashion. Let a feature combination f_c occur in a set of m contextual tree strategies S_c, $\{s_1^1, s_2^1, s_3^2, \ldots s_m^t\}$, out of k contextual tree strategies in the set S and let $f_{c,i(cost)}^t$, $f_{c,i(conv)}^t$ and $f_{c,i(tot)}^t$ denote the cost, number of convert users and total number of impressions associated with f_c present in strategy s_i^t, then its custom performance metrics in terms of CPA and CVR for a window period of T days are defined in the following manner:

$$f_{c,cpa} = \frac{\sum_{t=1}^{T} \sum_{i=1}^{m} f_{c,i(cost)}^t}{\sum_{t=1}^{T} \sum_{i=1}^{m} f_{c,i(conv)}^t} \quad where \ s_i^t \in S_c, \ S_c \subset S \tag{6}$$

$$f_{c,cvr} = \frac{\sum_{t=1}^{T} \sum_{i=1}^{m} f_{c,i(conv)}^t}{\sum_{t=1}^{T} \sum_{i=1}^{m} f_{c,i(tot)}^t} \quad where \ s_i^t \in S_c, \ S_c \subset S \tag{7}$$

We devise two flag metrics based on the performance of feature combination f_c and performance of other targeting strategies B_{nc}. The flag metrics are indicators of whether the feature combination's performance was better or worse than the other targeting strategies' performance. The flag metrics are defined as follows:

$$flag_1 = \begin{cases} 1 & f_{c,cpa} \leq B_{nc,cpa} \\ 0 & otherwise \end{cases} \tag{8}$$

$$flag_2 = \begin{cases} 1 & f_{c,cvr} \geq B_{nc,cvr} \\ 0 & otherwise \end{cases} \tag{9}$$

Note that a feature combination f_c may perform poorly not only because it's a bad inventory to serve an ad, but also because we may not be spending enough (under-spending) amount of budget on it. Hence, we also keep track of the spending that we do on each feature combination. To capture the spending on feature combinations over the last window of T days, we perform percentile binning of spends on features combinations with a bin size of h bins. The first bin (*spend bin* 1) comprises of feature combinations with lowest spends and the last bin (*spend bin* h) comprises of feature combinations with highest spends. We opted for using percentile-bins in order to club similar spending features into one category. The bin size parameter can be tweaked according to different use-cases. Each feature combination f_c is also marked according to its life-cycle in the tree by another flag called *iteration*. The *iteration* flag denotes the window in which the feature combination has occurred. If a feature combination remains in the tree for 2 windows of T days each, then it's *iteration* flag will be 2. Note that *iteration* flag resets after a window period of D, which again is a parameter that can be tweaked.

The Feedback Module starts by doing a logical OR operation between the $flag_1$ and $flag_2$ binary-valued flags and takes subsequent decisions based on the outcome.

$$flag_1 + flag_2 = \begin{cases} 0 \; flag_1 = 0 \; and \; flag_2 = 0 \\ 1 \; otherwise \end{cases} \tag{10}$$

The following flow chart describes the feedback module (Fig. 1).

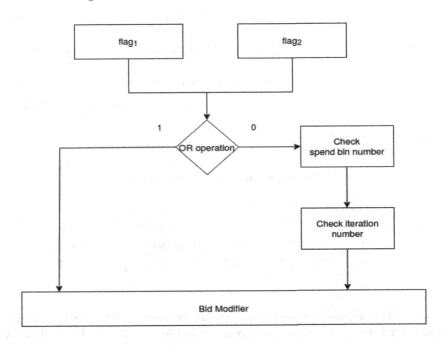

Fig. 1. Feedback module flow

One of the main components of the Feedback Module is the Bid Modifier. The Bid Modifier either increases, decreases or makes no change in the bid values associated with each feature combination. The Feedback Module works in such a way that when the logical OR operation for f_c is 1 then the Bid Modifier increases the bids on that f_c irrespective of its *spend bin*. Otherwise, the Feedback Module checks the *spend bin number* and the *iteration number* for f_c to make subsequent decisions. For example, if f_c is in *spend bin* 1 that means there is a possibility of under-spending on f_c due to which it has poor performance as compared to other targeting strategies. In this case, the Feedback Module increases the bid in *iteration* 1 and monitors the performance change over the next iteration i.e. after a window of T days. If f_c still does not perform up to the mark in *iteration* 2, then the Feedback Module decreases the bid and monitors the performance till *iteration* 3. In *iteration* 3, if the poor performance continues, then f_c is removed and new feature combination f_n replaces f_c. Similarly, other decisions are taken to make the bid change dynamic according to the feature combination's performance with respect to other targeting strategies' performance.

The percentage by which the old bid values (bid_{old}) from the previous iteration of f_c is increased or decreased is called the *bid update parameter* or b_u. The Bid Modifier incorporates a custom-defined business rule estimator, ψ, which is formed using the metrics defined above and the utility function value associated with each feature combination f_c, to give the value of b_u.

$$b_u = \psi(s^t_{i(cost)}, f_{c,cpa}, s^t_{i(conv)}, s^t_{i(tot)}, f_{c,cvr}, U, iteration, spendbin, flag_1, flag_2) \tag{11}$$

The range of b_u is $[-1,1]$. The new bid value (bid_{new}) is then calculated as:

$$b_{new} = bid_{old} + bid_{old} * b_u \tag{12}$$

Table 3 shows how the Feedback Module recommends new bid value (bid_{new}) for 3 feature combinations ids shown in Table 2 of a contextual tree campaign s_1 for a single window T days.

Table 3. Feedback module output for a campaign

$f_c(id)$	$s_{1(cost)}$	$f_{c,cpa}$	$s_{1(conv)}$	$s_{1(tot)}$	$f_{c,cvr}$	U	bid_{old}	b_u	b_{new}
1	313	10	21	98224	0.02	0.21	3	0.4	4.2
2	313	8	21	98224	0.03	0.63	4	0.5	6
3	313	20	21	98224	0.001	0.05	3	−0.2	2.4

Notice how Feedback Module suggests an increase in bid for $f_c(id)$ 2 with a low CPA value and suggests a decrease in bid for $f_c(id)$ 3 with a high CPA value, thus promoting a higher bid for better performing feature combinations.

5 Results

Various strategies for an advertiser, within a DSP, are examined to judge the performance by the end of the advertisement cycle. We have observed that the contextual tree strategy campaign has a very distinguishable performance compared to other strategies' campaigns. In terms of KPIs, the contextual tree strategy campaign has an *59%* lower CPA and *163%* higher CVR than the other strategies' campaigns. A CPA and CVR comparison of two types of strategies for an advertiser is shown below.

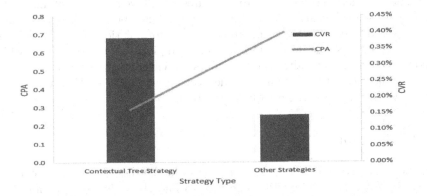

Fig. 2. Performance comparison

From Fig. 2, the performance of the contextual tree strategy in terms of CVR and CPA is visible in comparison to other strategies.

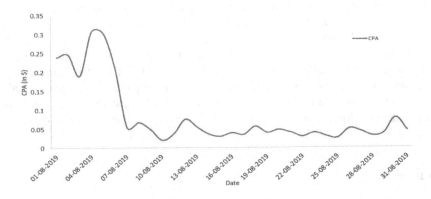

Fig. 3. Temporal CPA performance

Figure 3 clearly shows how the Feedback mechanism is dynamically trying to reduce the CPA by altering the bids on feature combinations in a contextual tree as the day passes in each *iteration*.

6   Limitations and Future Scope

Though the Feedback Module is dynamically able to promote the bids on feature combinations performing well and simultaneously penalize the bids on feature combinations performing poor, we have observed a slight decrease in impressions delivery in subsequent iterations. This decrease is attributed to the skewed performance of feature combinations i.e. only a fraction of feature combinations performs well, and majority of feature combinations don't perform well. This trade-off between bids and impressions delivery is pervasive in digital advertisement and is something that needs to be addressed. There is a scope of improving the Feedback Mechanism in order to attain a balance between bid values and impressions delivery. We are trying to modify the formulation of Feedback Module Flow to incorporate impressions delivery and giving more weight to it while making the decisions to alter the bid values on feature combinations. This will allow the algorithm to make informed decisions to win more auctions even when the bid values decrease in subsequent iterations.

7   Conclusion

The most important challenge for advertisers in the digital advertisement has always been whether or not they are able to find the right audience for their ad within a specified budget. Cookie-based targeting has been the center of attraction for a long time in the digital advertisement but as we move towards a privacy-first world where there are increased restrictions in cookie-based targeting, advertisers need to find other innovative ways to find these right set of audiences. The approach we have proposed shows how the advertisers can leverage the novel context-based targeting strategy to find the right audience and also generate higher returns on the spending of their advertisement budget. This new methodology was tested on real-world advertisement data and is designed to work across all the categories of campaigns. The efficiency of this methodology is conspicuous from the fact that there has been a significant uplift in the performance indicator metrics like CPA and CVR of the campaigns as compared to other campaigns which utilize other targeting strategies. We also discussed the limitation of this approach in terms of slight decrease in impressions delivery which paves the way for further improvements in the described framework.

References

1. Yogesh, S., Nallasivam, S.: Digital marketing and its analysis. Int. J. Innov. Res. Comput. Commun. Eng. **5**, 201957007 (2019)
2. Enberg J.: Global digital ad spending (2019). https://www.emarketer.com/content/global-digital-ad-spending-2019, Accessed 31 Jan 2020
3. Aziz, A., Telang, R.: What is a digital cookie worth? (2016). https://ssrn.com/abstract=2757325, https://doi.org/10.2139/ssrn.2757325

4. Donnellan, B., Helfert, M., Kenneally, J., VanderMeer, D., Rothenberger, M., Winter, R. (eds.): DESRIST 2015. LNCS, vol. 9073. Springer, Cham (2015). https://doi.org/10.1007/978-3-319-18714-3

5. Sayedi, A.: Real-time bidding in online display advertising. Market. Sci. **37**(4), 553–568 (2007)

6. Lee, K.-C., Orten, B., Dasdan, A., Li, W.: Estimating conversion rate in display advertising from past performance data. In: ACM SIGKDD Conference on Knowledge Discovery and Data Mining, pp. 768–776 (2012)

7. Wang, J., Zhang, W., Yuan, S.: Display advertising with real-time bidding (RTB) and behavioural targeting. Found. Trends Inf. Retr. **11**, 297–435 (2017)

8. Edelman, B., Ostrovsky, M., Schwarz, M.: Internet advertising and the generalized second-price auction: selling billions of dollars worth of keywords. Am. Econ. Rev. **97**(1), 242–259 (2007)

9. Braun, M., Moe, W.W.: Online display advertising: modeling the effects of multiple creatives and individual impression histories. Market. Sci. **32**, 753–767 (2013)

10. Sfetcu, N.: Legal aspects of big data - GDPR: SetThings (2019). https://www.setthings.com/en/legal-aspects-of-big-data-gdpr/, Accessed 18 March 2020

11. Kong, D., Shmakov, K., Yang, J.: Demystifying advertising campaign for CPA goal optimization. In: Companion Proceedings of the the Web Conference, pp. 83–84 (2018)

MOOC Performance Prediction by Deep Learning from Raw Clickstream Data

Gábor Kőrösi[✉] and Richard Farkas[✉]

Institute of Informatics, University of Szeged, Szeged, Hungary
{korosig,rfarkas}@inf.u-szeged.hu

Abstract. Student performance prediction is a challenging problem in online education. One of the key issues relating to the quality Massive Open Online Courses (MOOC) teaching is the issue of how to foretell student performance in the future during the initial phases of education. While the fame of MOOCs has been rapidly increasing, there is a growing interest in scalable automated support technologies for student learning. Researchers have implemented numerous different Machine Learning algorithms in order to find suitable solutions to this problem. The main concept was to manually design features through cumulating daily, weekly or monthly user log data and use standard Machine Learners, like SVM, LOGREG or MLP. Deep learning algorithms could give us new opportunities, as we can apply them directly on raw input data, and we could spare the most time-consuming process of feature engineering. Based on our extensive literature survey, recent deep learning publications on MOOC sequences are based on cumulated data, i.e. on fine-engineered features. The main contribution of this paper is using raw log-line-level data as our input without any feature engineering and Recurrent Neural Networks (RNN) to predict student performance at the end of the MOOC course. We used the Stanford Lagunita's dataset, consisting of log-level data of 130000 students and compared the RNN model based on raw data to standard classifiers using hand-crafted commulated features. The experimental results presented in this paper indicate the RNN's dominance given its dependably superior performance as compared with the standard method. As far as we know, this will be the first work to use deep learning to predict student performance from raw log-line level students' clickstream sequences in an online course.

Keywords: RNN · MOOC · Clickstream sequence analysis

1 Introduction

The approximate number of Massive Open Online Courses (MOOC) students who enrolled or took part in a single or more courses is around one hundred million students [31]. This means that MOOCs can be considered as the prime method of knowledge acquisition online, their main advantage being the fact that neither geographic location, not monetary issues play a role in opting for this type of education [7]. As the MOOCs became quite popular among students, they sparked a great deal of research interest in MOOC data analytics [8].

© Springer Nature Singapore Pte Ltd. 2020
M. Singh et al. (Eds.): ICACDS 2020, CCIS 1244, pp. 474–485, 2020.
https://doi.org/10.1007/978-981-15-6634-9_43

For all their benefits, the quality of MOOCs has been the target of criticism [9, 10, 11, 15]. Almost all research has pointed to their low completion rates (below 7–10% on average) as a property preventing more widespread adoption of these courses [10, 12]. Stakeholders would benefit from knowing whether or not a given student was expected to complete the course, especially in view of the low completion rates. To solve these problems there has been two new research fields established: 1. Educational data mining [17], 2. (online) learning analytics [16]. Within the fields of learning analytics and educational data mining, we are able to create automated MOOC "dropout detectors" or "forecast performance" [5, 15].

In terms of log data collection in the form of clickstream or social network measures, the MOOC systems offer a treasure-trove of data. The system gathers this data when students watch the various video lectures, try their hands at quizzes and interact with their peers in the available forums, discussing the learning materials during the course [11]. The MOOC log data can be leveraged for prediction tasks through Machine Learning approaches [20]. Eventually efficient student models were created which would serve as a forecasting tool for estimating how many students were likely to drop out, or preferably complete the course. This was made possible by extensive research into comprehending and hopefully increasing the registration and completion rate, ultimately contributing to a better all-round learning experience in MOOCs [19]. These issues involved the application of different supervised machine learning approaches so as to obtain an estimation for future learning results in MOOCs [7, 29]. The majority of conventional methods is primarily based on generalized linear models, incorporating logistic regression, linear SVMs and survival analysis. Every model takes into account various kinds of behavioral and predictive characteristics gleaned from a number of raw activity records, such as clickstream, grades, forum, and final grades [14].

Another solution is the Deep Learning (Deep Neural Networks) which could algorithmically find structure in log data and carry out prediction on various tasks in MOOCs [16, 18].

However, the promise of Deep Neural Networks is to learn the temporal context of input sequences in order to make better predictions, only a few prior works explore this opportunity. Our extensive literature survey indicates that all studies in Educational Data Mining following the Deep Learning approach used cumulated data (daily, weekly, etc.) of feature engineering [8, 10, 12, 13, 15, 20, 30]. We didn't find any research where the input of the Deep Learning model was the raw log-line-level activity data.

The main contribution of our study is to investigate this challenge, i.e. whether Deep Neural Networks can benefit by using directly raw log data, rather than hand-designed statistics as commulated features. In this paper, we use log-line-level data and Recurrent Neural Networks (RNNs) to predict student performance at the end of the MOOC course as both a multiclassification and a regression task. The Recurrent Neural Network models each element of an activity log sequence (line by line), from the beginning of student's activity until a certain point in time, and predicts the student's final performance at the very end of the MOOC.

To the best of our knowledge, this will be the first work to use RNNs to predict user performance from log-line level students' clickstream sequences in an online course.

The rest of the paper is organized as follows: In Sects. 1 and 2, we begin with the design of the problem, which involves the different hypothesis we make in the problem definition. Section 3 describes the main workflow, feature extraction, RNN-GRUs, and the prediction model especially. In addition, Sect. 3. details the key technical points of the research and steps to implement it. Section 4 details the key technical points of the research and steps to implement it. Afterward, we specify the details of the datasets and experimental setup we used in Sect. 4. Section 5 presents the experimental results of the prediction models on datasets provided by Stanford Lagunita's MOOCs. Finally, Sect. 6 presents our key findings and conclusions.

2 Background

Specific fine-tunings were made by stakeholders during the course of education so as to be able to deal with both low MOOC outcome results considerable dropout rates. These modifications included email reminders to students or offering constructive feedback to specific learners at risk of leaving the course [13].

While this may sound like a workable solution, indeed, beneficial to the students themselves, given the reality of learner numbers of over ten thousand at a time, it could hardly be implemented in practice. Out of those 10,000 students there are probably an estimated 9,000 learners who would need such online support lest they drop out [10]. To address this issue, researchers recommend an automated system capable of reliably predicting the students' future performance in real-time [9, 14].

The fact that the system is automatic, basing its work on learning analytics, makes it possible to both monitor and recognize those students who are in danger of leaving the course. At the same time, it will also be able to support early intervention design [10]. Hence, there is a considerable and ever-increasing amount research available about predictive modeling in MOOCs, especially focusing on models on the likeliness of a given student's dropout, stop out, or overall failure of completing a MOOC.

There have been earlier studies on student outcome forecast which were based on a wide range of characteristics obtained from clickstream data and the natural language used during postings in discussion forums, social networks, and assignment grades and activity [19]. These works relied on trace data from the introductory week or other specific times which would then serve as the basis for predicting students' outcome by way of the created prediction models [10]. Such prediction method enables the efficient detection of whether or not specific students are likely to drop out in the initial education phase, which, however, requires considerable time for feature extraction. The most popular feature representations are the measures of the distances among log events (time, points, etc.), aggregating the clickstream logs on a weekly basis and/or applying Natural Language Processing (NLP) on discussion forum content.

In terms of Machine Learning architecture, most prior approaches have used generalized linear models (including linear SVMs), survival analysis (e.g., Cox proportional hazard model), and Logistic Regression [13, 20]. From another perspective, scientists compare classification against regression approaches [21]. For instance, J. He et al. [22] dealt with the Support Vector Machine (SVM) and Least Mean Square (LMS) algorithms in order to identify what the learners' dropout rates were or how well they were performing in the MOOCs during the course period. Work was also conducted on clustering techniques, in which students were put into groups, clustering them on the basis of their student behavioral patterns [23].

The paper of Jo et al. (2018) [24] reviews the variety of solutions in existing studies on modeling student behaviors via clickstream logs. It categorizes the solutions into top-down and bottom-up. Top-down approaches predefine a set of student behaviors of interest, such as disengagement and sequential navigation, and corresponding click patterns. These approaches provide interpretability, but analyses are focused only on predefined behaviors and patterns. In contrast, bottom-up approaches aim to find meaningful click patterns from clickstream data and interpret behaviors they mean [24].

In order to discover the bottom-up approaches, several publications have been published in the last few years, and the most common solutions operate with Feedforward Neural Network or Self Organized Map (SOM), while other papers employ Recurrent Neural Networks (RNNs), like Long Short-Term Memory (LSTM) Networks. In this paper, we used Gated Recurrent Unit (GRU) which is introduced by Cho, et al. in 2014. GRU is a variation on the LSTM (Long Short-Term Memory) and it is a specific RNN architecture designed to model temporal sequences and their long-range dependencies more accurately than conventional RNNs [13].

Although, during the last few years several papers have been published which started to use Deep Learning to predict MOOC outcomes, the main concept has not change. Kim et al. (2018) [14] collected these studies, so far these have shown low accuracies. The fact that accuracy tends to be low can be related to the model's continued reliance on feature engineering to decrease input dimensions which seems to hamper the development of greater and improved Neural Network models [14]. A good example is R. Al-Shabandar's et al. (2017) [7], in which numerous characteristics were obtained from learners' historical data, including how many sessions they took, how often they watched the videos, how many courses they participated in, which was then all fed into a Feedforward Neural Network. Because of the information loss of feature extraction, current dropout or student performance prediction model's accuracy is limited, and Deep Learning methods could not give much better performance than classic Machine Learning methods.

We could think that the RNN network is not robust enough for raw log-line level data, but on the other hand, other fields – like anomaly detection – some experts applied Long Short-Term Memory (LSTM) Networks straight on unprocessed data. Zhang et al. [25] for instance, opted for using clustering techniques for the raw text from numerous log sources so as to create feature sequences fed to an LSTM for hardware and software failure predictions. Du et al. [26] implemented special parsing methods on the unprocessed text of system logs to create sequences for LSTM Denial of Service attack identification [27].

We address this challenge here by using RNNs directly on raw log-line level data, rather than hand-designed feature extracted statistics.

3 Methodology

In terms of algorithm, this work studied the power of Deep Learning in the context of education. The authors' prediction algorithm is based on raw clickstream data in order to provide a prediction of how well students will perform in an online course. We use the students' final assessment quiz responses to define the course performance measure. To do this, we measure the user performance prediction in two ways:

- as a regression problem (0–100%), and
- a multiclass problem (0–9 point). To get labels for the multiclass, we used uniform discretization of continuous data (0–10% = 0, 11–20% = 2, etc.).

In order to compare our research with other solution, we implemented the Baseline solutions, i.e. feature extractors on raw clickstream and used traditional classifiers and regressors on these feature set. We compare the Baseline solution to a GRU-based model on raw log-line level data (GRU).

3.1 Evaluation

To measure the accuracy of the prediction approaches, we build temporal performance prediction models on a weekly basis. The log data observed on the second week were directly added to the dataset on the first week, and similarly for other weeks as well. We are using the data collected until the current week to predict the student's outcome in the very end of the course as shown in Fig. 1. We extract the commulated features from the collected data for the Baseline solutions and train a GRU model on the raw data.

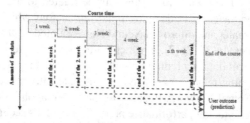

Fig. 1. Formulation of performance prediction problem

After the training step, we evaluate the accuracy of the prediction models on the hold-out set of students. The predictions on this hold out set is compared against the true value (either the actual score or the performance class) of the students' final course performance. Following Willmott and Matsuura (2005) [6], for measuring the accuracy of the solutions, we use accuracy scores (ACC) for classification and root mean squared error (RMSE) for regression. We split the students into four groups randomly and employ a 4-fold cross-validation at each week.

Certainly, one of the obstacles to be overcome when creating predictions in MOOCs is how to deal with sparse data [12]. In Stanford Lagunita's datasets, the majority of learners failed to provide answers to every quiz question in the relevant MOOC, which in turn, resulted in a sparse set of quiz replies for each individual student. To handle this problem, we only use those users who has filled second quiz after the first week, or filled third quiz after the second week, etc. To avoid under or over learning problem, we moved extreme outliers [28] from datasets.

3.2 Baseline Solutions

For the baseline solutions, we extracted commulated features from the raw data (see in Table 1). All of these features are normalized by min-max scaling with the maximum and minimum values of each feature in training dataset.

Table 1. Feature set of Stanford's course

Feature	Explanation (feature aggregated on a weekly basis)
Lecture view	Number of lecture videos viewed by a student
	Number of times a student visits a lecture site
	Number of plays, stop, pause, forward, backward
	Number of plays, stop, pause, forward, backward
	The distance of time between two log events
Quiz attempt	Number of quizzes attempted by a student

In this study, three popular Machine Learning algorithms were employed from the gradient boosting framework, and Ridge Regression to train and predict over the commulated feature set. We used the gradient boosting (XGboost) for classification and Ridge Regression, XGBregression for regression tasks. Our main goal is not to find the most accurate model, so we did not perform any hyper-parameter tuning.

3.3 Deep Learning Architecture

We propose a Deep Learning method based on raw data, to compare the prediction performance with baseline algorithms.

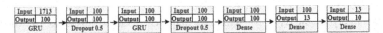

Fig. 2. Architecture overview of the proposed RNN model.

Our Recurrent Neural Network architecture consists of GRU cells as hidden units. In the model training, hidden states learned raw log-line by raw log-line to catch sequential information, which is able to absorb past information. Our GRU model also takes the dropout technique to prevent overfitting in training process. As the experiment result has shown in the result section, the usage of GRU and dropout technique helps us to achieve the training goal. Although there are various suitable algorithms in terms of this prediction, the authors opted for time series/sequence-based neural network predictor, mainly based on its frequent implementation in numerous research fields, thus also student knowledge tracing [12]. Sequence-based neural networks are Recurrent Neural Networks, with feedback connections enclosing several layers of the network. Gated

Recurrent Unit (GRU) networks is an example of Recurrent Neural Networks. GRU is a good choice at solving problems that require learning long-term temporal dependencies. Figure 2 depicts the architecture of our proposed Deep Learning model consisting of nine layers.

The input layer of our deep network uses a flat feature structure (one hot encoded 3-dimensional data), as we can see in Fig. 3. The rows in our 3D dataset contain user-generated log data. These rows are representing a user-generated 2D sequence of actions, which set up in chronological order. In this 2D sequence, a line is a 1714 length vector, which represents one of the 1713 possible actions which illustrated with a One Hot Encoded vector, along with a single feature which is the time elapsed since the last action. An action encodes either the type of action (e.g. "video stop") or the item the was accessed. For example, when the student opens "Lecture1 Part3" which is a webpage containing a lecture video we log this event. Next, when the student plays the video, we add a new action to the sequence, but we only store its action type "video play". While this condensation of data is necessary for keeping the input space at tractable size, we expect from the RNN, that it can learn the representation of items in its hidden states.

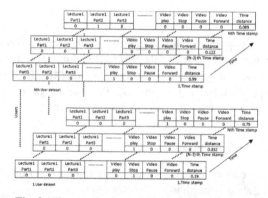

Fig. 3. Figure1, Formulation of 3-dimensional data

RNN is likely to quickly overfit a training dataset. To reduce the chance of overfitting, we used dropout layers which offers a very computationally cheap and remarkably effective regularization method to reduce overfitting and improve generalization error in our model.

Following the GRU and dropout layers we applied fully connected (dense) hidden layers with different number of neurons. We use the same network architecture for the regression and classification tasks besides the output layer, At the regression task, we use the result from the single output neuron without any transformation. At the multi-class classification task, the output layer consists of a vector that contains values for each class along with a SoftMax activation function.

4 Experimental Setup

In our experiments, we used the activity log dataset of the Stanford Lagunita's MOOC Computer Science 101 from the summer of 2014. The MOOC ran for six weeks, with video lectures, optional homework assignments, discussion forums, and quizzes. The original main dataset contains 39.6 million action items from around 142,395 students, where each action represents accessing a particular event in the course, like video view, assignment view, problem view. From the 142,395 students, 28,368 were active (went through almost every curriculum) and 12,015 completed enough assignments and scored high enough on the exams to be considered "certified" by the instructors of the course. The certified students accounted for 17.79 million of the original 39.6 million actions, with an average of 1,135 action items per certified student. In our research the set of 12,015 students were used.

To assess our model's ability to spin up rapidly and predict final outcome, we limit our scope to first 5 weeks, and we make a prediction after every week with the baseline and GRU models. At each week, we compared baseline and Deep Learning solutions for both a regression and classification tasks. In the regression task, the goal was to predict student performance over a range of 0–100%, while in the multiclass classification problem we targeted seven classes of student performance. In each experiment, we employed a 4-fold cross-validation over students. In the GRU, our learning rate is fixed to 0.001; for training we used the ADAMAX optimizer; we didn't use padding, instead we create variable sequence length for all user with fit generator function; our model used two GRUs layer with 100 hidden units; two Dropout layers (ranged 50%), two Dense layers with 100 hidden units each, and one Dense with 13 hidden unit.

The experiments were implemented in python by using Keras [1], Google's TensorFlow [2], Sckikit learn [3] and XGBoost [4] package.

5   Empirical Results

We ran our models at the end of each week, i.e. after a quiz, and based on that information, we predicted the final completion of the course. For example, *week1* represents all collected log data from the start of the course until the end of the first week, and week2 represents the collected data until the end of the second week. The results of the proposed GRU method and the baseline methods on Computer Science 101 dataset are shown in Table 2 respectively.

Table 2. Results of Gru and by baseline methods

	RMSE			MAE			Accuracy	
	XGBreg	*Ridge*	*GRU-reg*	*XGBreg*	*Ridge*	*GRU-reg*	*XGBoost*	*GRU-class*
Week1	16.010	18.451	10.001	11.706	12.854	7.743	0.361	0.496
Week2	16.217	19.173	9.831	11.784	13.564	7.451	0.378	0.486
Week3	15.730	20.779	9.378	11.251	13.101	7.117	0.39	0.545
Week4	15.346	29.207	8.746	10.202	15.378	6.096	0.405	0.559
Week5	15.265	25.355	8.653	10.585	15.280	6.568	0.405	0.551

The results of the two experiments in Table 2 shows that the GRU model is generally better than the XGBoost-regression and XGBoost and both RNN based models increase their prediction quality week by week. This shows that RNNs with raw datasets have more sensitivity to "catch" patterns than XGBoost or XGBoost-regression. This was not surprising, as standard machine learning methods without deep feature investigation cannot make evaluable results. In addition to basic transformations (sum, avg, normalization), we also used other methods to increase the performance of the XGBoost model, as discussed in Sect. 3, but apparently this solution was not sufficient. Remarkably, for all week in the proposed method outperforms the best baseline by 15% of ACC or 30 of RMSE. As expected, performance starts out poor (low ACC or high RMSE) but steadily improves as more data are fed into the model. Predictions continue to improve until approximately week 3, after which performance levels out. After 3 weeks of observations, our model achieves an accuracy of 54%, significantly better than the 39% baseline of predicting the class role.

Our empirical results demonstrate the feasibility of using Recurrent Neural Networks on raw log-line level dataset to predict MOOC students' performance. We do expect, however, that a more extensive search for the optimal choices of number of units and hidden layers (through e.g., hyper-parameter tuning, embedding layer) will improve our prediction quality further.

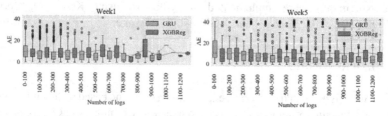

Fig. 4. Boxplots of average error (AE) achieved by GRU-reg and XGBReg in the function of student's log sequence length

To better understand our results, we performed other examination (see in Fig. 4), where the relation between the number of log-data and the absolute error (AE) of the model is plotted. In this process we calculated AE for every user falling in a particular bin of log-data sequence length and made a boxplot form this data to compare two models' outputs. In the first two weeks, the GRU network provides better results than the traditional approach on any log size. In weeks 3, 4, 5 traditional approaches outperform RNNs at short log sequences and GRU is only superior when it has got more data of longer sequences that provide extra information.

Other problem is length of the sequence, which is not same long for all users. For example, some users finish 6-week course very fast and short way (exp. Doesn't check additional information), while other users look into everything and spend more time checking the details. These results also show that there is a significant correlation between the length of the sequences and the prediction quality of the RNN.

RNN model do temporal processing and learn from sequences, e.g., perform sequence recognition or temporal association, which is impossible to solve with regression. On the other hand, XGBreg is more sensitive to overfit. To avoid XGBoost-regression disadvantages, we experimented with various regression models, but they provide almost the same or worse result (Ridge regressor was among the best one, see its results in Table 2).

This fact is also supported by our prediction density distribution graph of real and predicted course outcomes (Fig. 5). The diagram shows that XGBReg was not able to find any useful relationship among its features. On the other hand, GRUreg was capable to identify different class-like regions in its prediction space. The results are not exhaustive, but they do encourage us to explore further.

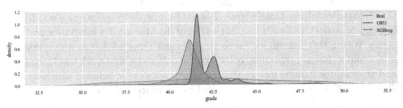

Fig. 5. The distribution of student's final outcomes - the test results of the proposed GRU method and the baseline (XGBRegression) methods on computer science 101 dataset at fourth week (n = 2159)

6   Conclusions

In this study, we propose a Recurrent Neural Network for solving outcome performance prediction problem in online learning platform. The main task of this paper is to build prediction model which could use raw datasets, and get same or better results than regular prediction models. The key advantage of our model is that, there is no manual feature engineering is needed, because it could be automatically extracted from the raw log-line level records. In this way, this approach could save a lot of time and human force, and ignore the possible inconsistency introduced by the hand-made process. Experimental results on Stanford Lagunita's dataset show that the expected model can achieve significantly better than the baseline models. The results for our model are sufficient to demonstrate the feasibility of using Recurrent Neural Networks. Additionally, architectural insights can be gleaned from the RNN applied, which may in the future lead to designing more effective models. The methodology in this study aimed to predict the outcome performance of student participations in MOOCs. In Table 2, we see that our GRU-Reg and GRU-Class algorithms are especially useful for predicting the performance of students who answer quizzes. Our neural networks give much better results than regular solutions. On the other hand, we compare a high-end solution with poor predicators, an important step for the feature work would be to implement a well-designed baseline model which could be enough powerful for the "fair play". In this paper our main goal, is not to find the most accurate model, but in the next possible interventions could be to improve these parameters from 55% to 80%. In the future, we still need to take some optimization techniques to adjust hyper parameters of the proposed RNN model, which is helpful to improve the accuracy. There are several ways that we plan to extend this ongoing work. First, we would like to use a hyper-parameter tuning to find the best number of units and layers in GRU network. We also plan to address the role imbalance by resampling and improve generalization, and layer normalization. Finally, we plan to use embedding layer which could give us an information what is happened inside of the neural network, this work also includes evaluating the system on different datasets which will provide by Stanford Lagunita.

Acknowledgment. We are grateful to the Leland Stanford Junior University, The Office of the Vice Provost for Online Learning, Stanford Lagunita for providing access to their MOOC log dataset. This work was supported by the National Research, Development and Innovation Office of Hungary through the Artificial Intelligence National Excellence Program (grant no.: 2018-1.2.1-NKP-2018-00008) and TUDFO/47138-1/2019-ITM of the Ministry for Innovation and Technology, Hungary. We are also grateful for Tamás Hegedűs and Tamara Sehovac who conducted data preprocessing and technical preparation of the experiments and whose work was funded from the "Integrated program for training new generation of scientists in the fields of computer science", no EFOP-3.6.3-VEKOP-16-2017-0002, supported by the European Union and co-funded by the European Social Fund.

References

1. Chollet, F., et al.: Keras (2015)
2. Martín, A., et al.: TensorFlow: large-scale machine learning on heterogeneous systems (2015)
3. Pedregosa, F., et al.: Scikit-learn: machine learning in Python. JMLR **12**, 2825–2830 (2011)
4. Chen, T., Guestrin, C.: XGBoost: a scalable tree boosting system. In: Proceedings of the 22nd ACM SIGKDD International Conference on KDD 2016, pp. 785–794 (2016)
5. Gardner, J., Brooks, C.: Student success prediction in MOOCs. User Model. User Adap. Inter. **28**(2), 127–203 (2018). https://doi.org/10.1007/s11257-018-9203-z
6. Cort, J.W., Kenji, M.: Advantages of the mean absolute error (MAE) over the root mean square error (RMSE) in assessing average model performance. Clim. Res. **30**, 79–82 (2005)
7. Al-Shabandar, R., et al.: Machine learning approaches to predict learning outcomes in massive open online courses. In: 2017 International Joint Conference on Neural Networks (IJCNN), pp. 713–720 (2017)
8. Fei, M., Yeung, D.Y.: Temporal models for predicting student dropout in massive open online courses. In: Proceedings of the 2015 IEEE International Conference on Data Mining Workshop, pp. 256–263 (2015)
9. Whitehill, J., Williams, J., Lopez, G., Coleman, C., Reich, J.: Beyond prediction: toward automatic intervention to reduce MOOC student stopout. In: Educational Data Mining (2015)
10. Xing, W., Du, D.: Dropout prediction in MOOCs: using deep learning for personalized intervention. J. Educ. Comput. Res. **57**(3), 547–570 (2018)
11. Kloft, M., Stiehler, F., Zheng, Z., Pinkwart, N.: Predicting MOOC dropout over weeks using machine learning methods. In: Proceedings of 2014 Conference Empirical Methods in Natural Language Process, pp. 60–65 (2014)
12. Yang, T.S., Brinton, C.G., Chiang, M.: Behavior-based grade prediction for MOOCs via time series neural networks. IEEE J. Sel. Top. Sign. Process. **11**(5), 716–728 (2017)

13. Liu, Z., Xiong, F., Zou, K., Wang, H.: Predicting learning status in MOOCs using LSTM. In: Proceedings of the ACM Turing Celebration Conference – China, ACM TURC 2019, article no. 74, pp. 74–81 (2019)
14. Kim, B., Vizitei, E., Ganapathi, V.: GritNet: student performance prediction with deep learning. In: Educational Data Mining (2018)
15. Whitehill, J., Mohan, K., Seaton, D., Rosen, Y., Tingley, D.: MOOC dropout prediction: how to measure accuracy? In: ACM Conference on Learning, pp. 161–164 (2017)
16. Fiaidhi, J.: The next step for learning analytics. IT Prof. **16**, 4–8 (2014)
17. Baker, R.S., Inventado, P.S.: Educational data mining and learning analytics. In: Larusson, J.A., White, B. (eds.) Learning Analytics, pp. 61–75. Springer, New York (2014). https://doi.org/10.1007/978-1-4614-3305-7_4
18. Tang, S., Peterson, J.C., Pardos, Z.A.: Predictive modelling of student behavior using granular large-scale action data. In: Lang, C., et al. (eds.) Handbook of Learning Analytics. Society for Learning Analytics, Alberta (2017)
19. Gardner, J., Brooks, C., Andres-Bray, M.L.J, Baker, R.: Replicating MOOC predictive models at scale. In: Proceedings of the Fifth Annual ACM Conference on Learning at Scale, L@S 2018, article no. 1, pp. 25–37 (2018)
20. Whitehill, J., Mohan, K., Seaton, D., Rosen, Y., Tingley, D.: Delving deeper into MOOC student dropout prediction (2017)
21. Gavai, G., Sricharan, K., Gunning, D., Hanley, J., Singhal, M., Rolleston, R.: Supervised and unsupervised methods to detect insider threat from enterprise social and online activity data. J. Wirel. Mob. Netw. Ubiquit. Comput. Dependable Appl. **6**(4), 47–63 (2015)
22. He, J., Bailey, J., Rubinstein, B.I.P.: Identifying at-risk students in massive open online courses. In: Proceedings of 29th AAAI Conference on Artificial Intelligence, pp. 1749–1755 (2015)
23. Kizilcec, R.F., Piech, C., Schneider, E.: Deconstructing disengagement: analyzing learner subpopulations in massive open online courses. In: LAK 2013 (2013)
24. Jo, Y., Maki, K., Tomar, G.: Time Series Analysis of Clickstream Logs from Online Courses. Language Technologies Institute, Carnegie (2018)
25. Zhang, K., Xu, J., Min, R.M., Jiang, G., Pelechrinis, K., Zhang, H.: Automated IT system failure prediction: a deep learning approach. In: 2016 IEEE International Conference on Big Data (Big Data), pp. 1291–1300. IEEE (2016)
26. Du, M., Li, F., Zheng, G., Srikumar, V.: DeepLog: anomaly detection and diagnosis from system logs through deep learning. In: Proceedings (2017)
27. Brown, A., Tuor, A., Hutchinson, B., Nichols, N.: Recurrent neural network attention mechanisms for interpretable system log anomaly detection. In: First Workshop on Machine Learning for Computer Systems. ACM HPDC (2018)
28. Tukey, J.W.: Exploratory Data Analysis. Addison-Wesley, Boston (1977)
29. Mourdi, Y., Sadgal, M., El Kabtane, H., Berrada Fathi, W.: A machine learning-based methodology to predict learners' dropout, success or failure in MOOCs. Int. J. Web Inf. Syst. **15**(5), 489–509 (2018)
30. Pigeau, A., Aubert, O., Prié, Y.: Success prediction in MOOCs a case study. In: Educational Data Mining 2019 (2019)
31. Friedl, C., Zur, A., Staubitz, T.: Moocs for business use: six hands-on recommendations. In: The 2019 OpenupEd trend report on MOOCs, pp. 10–14 (2019)

UDHR - Unified Decentralized Health Repository

Premanand P. Ghadekar, Anant Dhok, Anuj Khandelwal, Ayush Tejwani[✉],
Sonica Kulkarni, and Srivallabh Mangrulkar

Department of Information Technology, Vishwakarma Institute of Technology,
Pune 411 037, Maharashtra, India
{premanand.ghadekar,anant.dhok17,anuj.khandelwal17,
ayush.tejwani17,sonica.kulkarni17,
srivallabh.mangrulkar17}@vit.edu

Abstract. The healthcare domain is always upgrading itself in order to provide better care. The use of digital media for medical purposes has been on the rise. Naturally, these methods are being used to store and retrieve medical records. These records are widely known as electronic health records. The proposed system, Unified Decentralized Health Repository (UDHR) is inspired by Electronic Health Records (EHR). The project aims to integrate and digitize medical records. It also aims to provide wellness-oriented treatment to patients. Earlier the goal of the medical community was to provide the traditional illness-oriented treatment. The innovativeness in the system lies within the blockchain. It is modified to cater to the needs for security and data searching efficiency. Using natural language processing, text summarization is achieved for medical records. A summarized document is generated at the end. It is hashed to attain compression.

Keywords: Blockchain · Decentral · Digital · Electronic health records ·
Healthcare · Natural language processing · Secure · Seq2seq · SHA-256 · Text summarization

1 Introduction

Health is the greatest form of wealth a person possesses. Too often, it was observed that people and even hospitals find it difficult to maintain records over a long period. Paper-based documents were widely used ever since they were first introduced. They provided a hardcopy of the data. They required intense documentation. But they have been prone to calamities like arson, floods, and even degradation.

Additionally, these paper-based documents turn out to be bulky and consume space. Patients are reluctant to keep such massive records at home. There is always a chance that the patient might need such records again. Hence, they don't always understand the importance of preserving documents. When it comes to a hospital, they need a very sophisticated hierarchy to store documents. It takes a lot of time and effort to find one document. Apart from this, they also take up space in a hospital. Such difficulties gave rise to what is now known as electronic health records [1] or EHR. Such records store the health information of patients systematically in digital format. It includes various documents. For example, prescriptions, blood test results, diagnosis result documents, etc. These kinds of documents help in explaining the health status of a patient.

UDHR is a revolutionary initiative to overcome the strenuous operation of collecting, maintaining, and referring paper-based health records. The proposed system promotes digitization in the health domain for the overall development of society. Identified by a person's unique identification number, AADHAAR UID, UDHR provides security as well. Blockchain [2, 3] technology contributes to various features of the system, namely, security, search efficiency, and decentralization [17] of data. It also means that the system has an "append-only" feature and keeps valuable data safe from any potential attacker. The system has a simple GUI intended for all types of users and their capabilities to use the platform. It also has a well-defined and specialized access for doctors, pharmacists, pathologists, radiologists, etc. The proposed system provides the facility of viewing or adding records to a patient's profile. Apart from this, the system also gives users a summary of all the previously stored documents. This is as opposed to scrolling through countless numbers of records. It aids in saving time and act as a quick guide to a patient's medical history for doctors. Especially in cases of emergency, every minute counts. If a doctor can understand the patient's health in less time, it can prove useful. Text summarization [4] is achieved through the process of natural language processing.

The paper is organized as follows. Section 2 contains a literature survey; the proposed model is discussed in Sect. 3, Sect. 4 provides an algorithm, Sect. 5 comprises of the experimentation, and the conclusion is given in Sect. 6.

2 Literature Survey

In New Zealand [6], the electronic health records [1] work in such a way that every patient is assigned a unique identifier which uses health services for error-free identification. Similarly, the medical practitioner is assigned a unique index for secure access to such records. These records are associated with a Medical warning System that warns practitioners about risk factors while making clinical decisions with regards to a particular patient.

UDHR is taking inspiration from Electronic health records [1], plans to use AADHAAR Card number as a unique identifier for Indians so that they don't have to maintain various numbers and IDs. It provides ease of access to users. Doctors identified using their license numbers on similar lines. Along with this, the proposed system is capable of including the medical warning system as a part of itself and not a standalone system. Estonia, being the first country in the world who successfully implemented the e-governance system, the Estonian e-health portal has set a benchmark for completely taking healthcare online. However, their centralized database will not work in the Indian system due to its vast area & population. UDHR decentralizes [17] the database and enhances security. Also, in the Estonian health portal, healthcare professionals need to do extra work when patient's reports were concerned. They need to write different reports manually for themselves and the patient. UDHR overcomes this shortcoming by summarizing the medical reports by using NLP [7, 8] algorithms. Owing to the benefits of electronic health records [1], the government has also gotten involved. Under NITI Aayog, it has proposed a National Health Stack (NHS) [9]. This NHS [9] system proposes a central database that provides access to primary health care centers, doctors, hospitals. UDHR has some key differences with NHS [9], namely, UDHR has a decentralized [17] database to reduce searching time and optimizes storage. While NHS [9] focuses more on hospital occupancy and data for insurers, UDHR focuses on patient and doctor ease and satisfaction. Practo [10] is a revolutionary product for the healthcare sector. Practo [10] intends on providing a one-stop healthcare platform. It allows a patient to book or cancel appointments, search for doctors/clinics, order medicines online. UDHR aims to unify the electronic health records [1], whereas Practo [10] aims to provide a healthcare platform that will deliver everything online.

3 Proposed Model

The proposed system uses a person's AADHAAR as a unique identification number for user authentication and verification. The user able to log in using the UID (AADHAAR) and a password. There are special login credentials for the individuals concerned with the treatment and diagnosis of the patient. Like every doctor, pharmacist, laboratory specialist will be a patient, but vice versa will not be true every time. So, to reduce ambiguity and provide required functionalities to the persons who will be a part of the treatment process of the patient, the individuals allow access to the system using their own practice license numbers for medical personnel authentication and verification. They need to enter the patient's login credentials in order to fetch and access their profile along with the digital reports.

When a patient logs into his profile, he has viewing access to his records. All the reports and prescriptions are available to the user in read-only format exclusively, whereas the medical personnel has access to the patient's records, depending on their fields. A doctor has viewing and writing access to a patient's records while a pharmacist only has viewing access to a patient's prescription, which prescribed by the doctor. Similarly, a radiologist or pathologist can view recommended tests and add the records generated of those performed tests. All these records are added in the blockchain [2, 3]. In the blockchain [2, 3], a structured is defined in such a way that the general details are available to everyone, but the reports are accessible to only users with necessary rights. The newest block creates the first one to be retrieved. The system is decentralized [17] among the hospitals as they are the nodes of the system. Even if one hospital's server fails, it gets the data from the next hospital node.

Another part of the system is generating summarized records of a patient. This is the first document that pops up when a patient logs in. This document serves as a quick guide for the medical history of a specific patient. Every time a report is generated, the document is stored in the blockchain [2, 3]. The most recently uploaded report and the summarized report generated after that is stored in the blockchain [13] as an encrypted key [4]. AES algorithm [4] was used for file encryption. The encryption key is formed by hashing the password from the user using SHA 256 [11]. In order to decrypt that file for a doctor, the encrypted file and the hash [11] of the user's password are used to generate the decrypted file using AES [4] algorithm. The password of the user is stored in blockchain after using bcrypt [12] hashing algorithm. Once the text summarization [5] is complete, the output of that which exists in a text file, its hash value is calculated and again stored in the blockchain [2, 3] (Figs. 1 and 2).

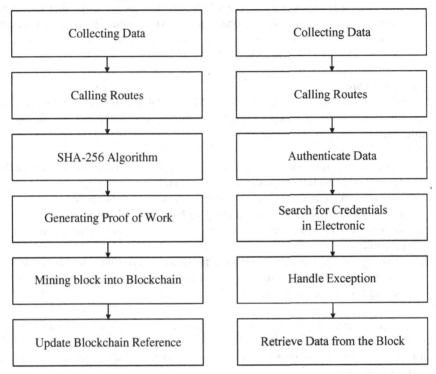

Fig. 1. Store in blockchain   **Fig. 2.** Retrieve from blockchain

4 Algorithm

The following algorithms used for implementation for report summarization and hashing data stored in the blockchain [13, 14] in order to create a hash number that is stored in the blockchain.

4.1 Sequence to Sequence Model

The sequence to sequence model [7, 8] tries to map input and output text of fixed length. But, the length of input and output may differ. Variants of Recurrent neural networks like Long short-term memory or Gated Recurrent Neural Network (GRU) are the method widely used. They can overcome the issue of vanishing gradient. This can be used for Speech Recognition, Machine Language Translation, Name entity/Subject extraction.

Seq2Seq Model
The seq2seq model [7, 8] is divided into two phases:

1. *Training Phase*
   a. Encoder - In the encoder, at each timestamp, a word is read or processed, and the contextual information is captured. This is passed to the next block as the previous context is needed for every word.
   b. Decoder - The decoder reads the whole target sequence to predict the same sequence, but with an offset of one timestep. The decoder is trained to predict the next word based on the previous word. Specific tokens are designated to mark the start and end of the sequence.

2. *Inference Phase*
   After training the encoder-decoder or Seq2seq model, the model is tested on input sequences for which the target sequence is not known. After this, the following inference architecture is developed:

   a. First, encode the entire input sequence. This is followed by the initializing of the decoder with the internal states of the encoder.
   b. Then pass the token, which is specified as an input to the decoder.
   c. Then, run the decoder for one timestep with the internal states.
   d. The output is the probability for the next word. Select the word with the maximum probability.
   e. This word is then passed as an input to the decoder and update the internal states.
   f. The steps 3–5 is repeated until the token to end or reach the maximum length of the target sequence is generated.

   It becomes difficult for the encoder to memorize the entire sequence into a fixed-sized vector and to compress all the contextual information from the sequence so it will be efficient for a short sequence. However, it will fail for a long sequence due to the large size of the vector.

Attention Mechanism

The concept of attention mechanism to overcome the long sequence problem is used. So, in this, importance to specific parts of the sequence is given, which are important rather than the entire sequence to predict that word. In the attention, for the decoder to give the output result, the intermediate form of the encoder state is utilized for context vector generation from all states. Attention is used to utilize all the contextual information from the input sequence so that the target sequence can be decoded. So, instead of going through the entire sequence, attention is given to specific words from the sequence and produce the result based on that.

The methodology of attention mechanism:

1. Computing score of each encoder state: Train the feed-forward network (encoder-decoder) using all encoder states and initialize the decoder initial state with the encoder final state. Train the network to generate a score for the states for which attention is used. Ignore the ones with a low score.

2. Computing attention weights: After generating scores, apply SoftMax on these scores to obtain the corresponding weights. SoftMax gives a probability with the summed up the value of all the weights in the range of 0–1.

3. Computing Context Vector: After computing attention weights, calculate the context vector, which is then used by the decoder in order to predict the next word in sequence. This contains information based on the weight of states which are considered.

4. Addition of context vector with previous output: Add the context vector with the start token since, for the first timestamp, there is no previous timestamp output. After this, the decoder generates output for a word in the sequence, and similarly, prediction every word in the sequence is generated. Once the decoder produces the end token, stop generating a word (Figs. 3, 4, 5 and 6).

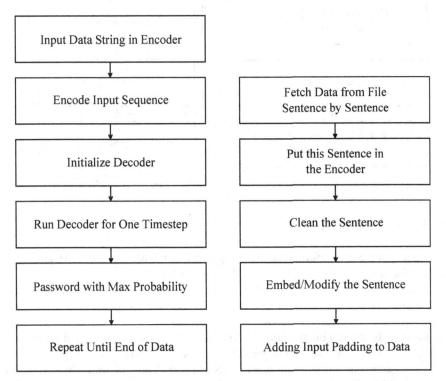

Fig. 3. Summarization of report using NLP   **Fig. 4.** Encoder in NLP input

4.2 SHA 256

In the proposed system, a secure hash algorithm 256 [11] (SHA256) is used to enhance security and feasibility. While registration, the data of the patient and the summarized reports [15] are stored in the form of the hash in the blockchain [2, 3]. After entering the data, by using SHA 256 [11], a 16 bit hash is generated, and it gets stored in the blockchain [2, 3].

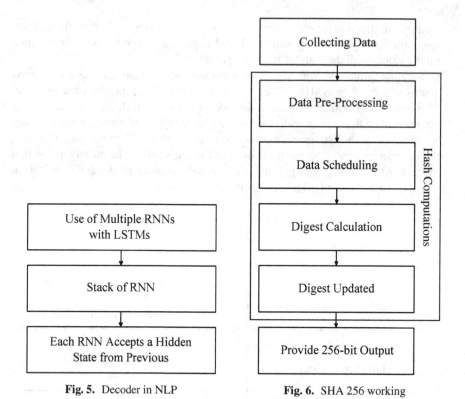

Fig. 5. Decoder in NLP   **Fig. 6.** SHA 256 working

5 Experimentation

The details of the user information are being stored in the blockchain [2, 3], and a new node is formed in the following manner (Figs. 7 and 8).

The system must be tested on different kinds of medical reports. As the user input here is of an image kind, there can be some blur in the images which may generate some noise after the processing of the OCR. The image above is the result of an OCR, and we must apply text summarization [15] on this. As seen above, the input text size in multiple paragraphs and the output generated is of only 10 lines.

The system was tested on different kinds of inputs, some of which had noise, but mostly, it didn't affect the output until the noise resembled medical terminology (Figs. 9 and 10).

Hospitals with a limited amount of resources have less chances of becoming a miner as in the proposed system, and mining requires high processing power. If a patient visits the doctor after a long time, the delay may occur in searching the patient's UID as its block may be at a certain distance in the blockchain. Keeping data security as the utmost priority, fetching the data may take a longer time as decryption of SHA 256 encrypted data can be time-consuming.

The language used for the user interface, as well as report summarization, is English. To make it more user-friendly and widely used, regional languages can be added

```
127.0.0.1:5001/get_chain

{
  "chain": [
    {
      "Information": [],
      "index": 1,
      "previous_hash": "0",
      "proof": 1,
      "timestamp": "2020-03-14 18:14:08.945020"
    },
    {
      "Information": [
        {
          "Aadhar Number": "2132-2132-3122",
          "Birth Place": "Pune",
          "Blood Group": "B+",
          "City": "Pune",
          "Date Of Birth": "1999-03-23",
          "Doctor": "Dr. Ayush Tejwani",
          "Gender": "Male",
          "Node Address": "192.168.0.105",
          "Password": "$2b$12$zWUMWUubv9EHIW7HRlK/aeSFdLTBWJK1w/.lbB8/J5v30bZEiNele",
          "Patient": "Srivallabh Mangrulkar",
          "Report": "89ca95e3d5e816ac30114b6f2c052965e25dfc2ef44a90a2dd868a831ade473c",
          "Summarized": "4ef5402b796987f4cf28c9dea56beb965a4938c1b6bcd4bfa048129663f65101",
          "Type": "User"
        }
      ],
      "index": 2,
      "previous_hash": "2cfc25280a5548dc14fac94d1dbfbe3279da6806c1af1ae0c0452c15a8b0da5a",
      "proof": 308,
      "timestamp": "2020-03-14 18:15:09.788487"
    }
  ],
  "length": 2
}
```

Fig. 7. Blockchain structure

```
127.0.0.1:5001/mine_block

{
  "index": 2,
  "message": "Congratulations, you just mined a block!",
  "previous_hash": "2cfc25280a5548dc14fac94d1dbfbe3279da6806c1af1ae0c0452c15a8b0da5a",
  "proof": 308,
  "timestamp": "2020-03-14 18:15:09.788487"
}
```

Fig. 8. Mining a block

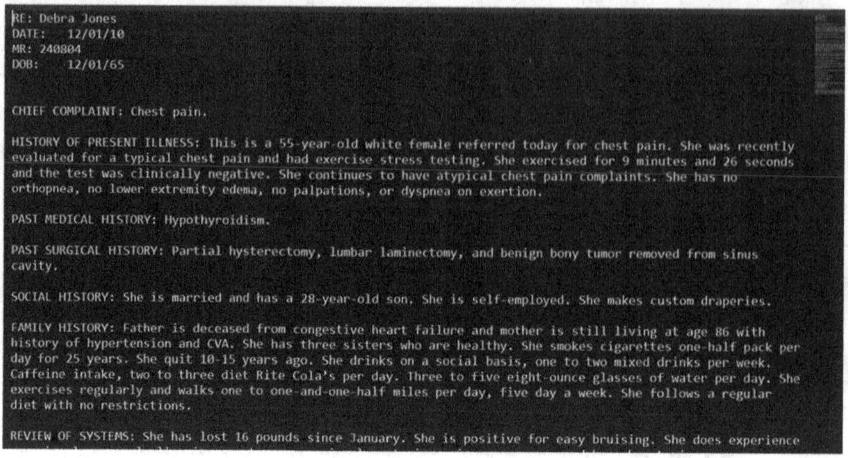
RE: Debra Jones
DATE: 12/01/10
MR: 240804
DOB: 12/01/65

CHIEF COMPLAINT: Chest pain.

HISTORY OF PRESENT ILLNESS: This is a 55-year-old white female referred today for chest pain. She was recently evaluated for a typical chest pain and had exercise stress testing. She exercised for 9 minutes and 26 seconds and the test was clinically negative. She continues to have atypical chest pain complaints. She has no orthopnea, no lower extremity edema, no palpations, or dyspnea on exertion.

PAST MEDICAL HISTORY: Hypothyroidism.

PAST SURGICAL HISTORY: Partial hysterectomy, lumbar laminectomy, and benign bony tumor removed from sinus cavity.

SOCIAL HISTORY: She is married and has a 28-year-old son. She is self-employed. She makes custom draperies.

FAMILY HISTORY: Father is deceased from congestive heart failure and mother is still living at age 86 with history of hypertension and CVA. She has three sisters who are healthy. She smokes cigarettes one-half pack per day for 25 years. She quit 10-15 years ago. She drinks on a social basis, one to two mixed drinks per week. Caffeine intake, two to three diet Rite Cola's per day. Three to five eight-ounce glasses of water per day. She exercises regularly and walks one to one-and-one-half miles per day, five day a week. She follows a regular diet with no restrictions.

REVIEW OF SYSTEMS: She has lost 16 pounds since January. She is positive for easy bruising. She does experience

Fig. 9. Input summarized text

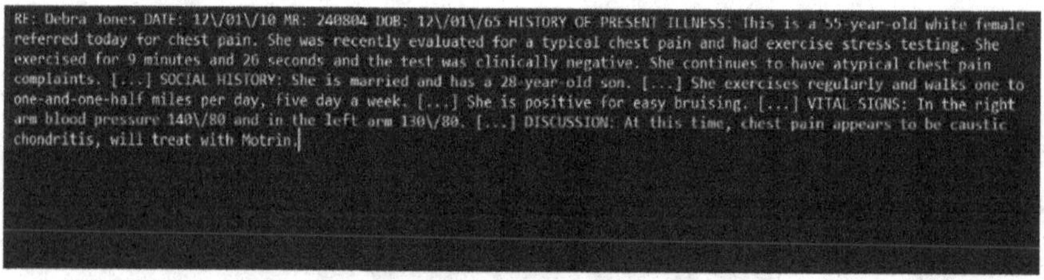
RE: Debra Jones DATE: 12\/01\/10 MR: 240804 DOB: 12\/01\/65 HISTORY OF PRESENT ILLNESS: This is a 55-year-old white female referred today for chest pain. She was recently evaluated for a typical chest pain and had exercise stress testing. She exercised for 9 minutes and 26 seconds and the test was clinically negative. She continues to have atypical chest pain complaints. [...] SOCIAL HISTORY: She is married and has a 28-year-old son. [...] She exercises regularly and walks one to one-and-one-half miles per day, five day a week. [...] She is positive for easy bruising. [...] VITAL SIGNS: In the right arm blood pressure 140\/80 and in the left arm 130\/80. [...] DISCUSSION: At this time, chest pain appears to be caustic chondritis, will treat with Motrin.

Fig. 10. Output summarized text

6 Conclusion

Paper-based records have been in action for a very long-time, causing inefficiency and overall environmental degradation. Electronic Health Records (EHR) [1, 16] makes the process of storing and maintaining medical records of patients a hassle-free venture. The system focuses on wellness-oriented medical care. Logging in and out of the system is very easy. Patients can view their medical history, prescriptions, test, etc. Medical professionals can complete their domain required tasks efficiently and effectively. Blockchain [2, 3] technology makes the system more secure and blocks attacks from happening. It also allows for the system to be decentralized, making the search time of medical records more efficient. The system generates well-summarized reports using the Sequence-2-Sequence model and hence saves time. Doctors are able to easily access these reports to get a brief overview of the patient's medical history. The use of the AADHAAR number makes the authentication process easy. It also enables every Indian citizen to get the most out of medical care. The doctors and other medical practitioners are held responsible for all medical actions taken on a patient making the whole process of giving and receiving medical care transparent. Medical records of patients are no longer prone to various natural calamities. Patients can access their records anytime and anywhere.

References

1. Shahnaz, A., Qamar, U., Khalid, A.: Using blockchain for electronic health records. IEEE Access **7**, 147782–147795 (2019)
2. Crosby, M., Nachiappan Pattanayak, P., Verma, S., Kalyanaraman, V.: Blockchain technology: beyond bitcoin. Corp. Couns. Bus. J. **2**, 71 (2015)
3. Olleros, F., Zhegu, M.: Research Handbook on Digital Transformations. Edward Elgar Pub, Northampton (2016). Chapter 11
4. Abdullah, A.: Advanced Encryption Standard (AES) Algorithm to Encrypt and Decrypt Data (2017)
5. Zhang H., Xu, J., Wang J.: Pretraining-Based Natural Language Generation for Text summarization, College of Computer, National University of Defense Technology, Changsha, China (2019)
6. Selwyn Jebaraj, A.: Implementing electronic health records in New Zealand: a critical appraisal of literature. Health Inf. – Int. J. (HIIJ) **5**(4) (2016)
7. Nallapati, R., Zhou, B., dos Santos, C.N., Gulcehre, C., Xiang, B.: Abstractive text summarization using sequence-to-sequence RNNs and beyond. In: Proceedings Of The 20th SIGNLL Conference On Computational Natural Language Learning (2016)
8. Sutskever, I., Vinyals, O., Le, Q.: Sequence to sequence learning with neural networks. In: Neural Information Processing Systems Conference (2014)
9. Aayog, N.: National Health Stack Strategy and Approach (2018)
10. Roshan, K.: Innovation in healthcare delivery a way to consumer delight: a case study of Practo.com. Int. J. Manag. Soc. Sci. **5**, 346–353 (2017)
11. Selvakumar, A., Ganadhas, C.: The evaluation report of SHA256 crypt analysis hash function. IEEE (2009)
12. Sriramya, P., Karthika, R.: Providing password security by salted password hashing using bcrypt algorithm. ARPN J. Eng. Appl. Sci. **10**, 5551–5556 (2015)
13. Mahmoud, Q., Lescisin, M., AlTaei, M.: Research challenges and opportunities in blockchain and cryptocurrencies. Int. Technol. Lett. **2**(2), e93 (2019)
14. Zheng, B., et al.: Scalable and privacy-preserving data sharing based on blockchain. J. Comput. Sci. Technol. **33**(3), 557–567 (2018)
15. Kumar, Y., Goh, O., Basiron, H., Choon, N., Suppiah, P.: A review on automatic text summarization approaches. J. Comput. Sci. Technol. **12**, 2–4 (2016). https://doi.org/10.3844/jcssp. 2016.178.190
16. Novikov, S., Kazakov, O., Kulagina, N., Azarenko, N.: Blockchain and smart contracts in a decentralized health Infrastructure. In: International Conference on Quality Management, Transport and Information Security, Information Technologies (2018)
17. Thimmaiah, S., Disha, S., Nayak, D., Gururaj, L., Diya, B.B.: Decentralized electronic medical records. Int. J. Res. Anal. Rev. **6**(1), 2–4 (2019)

Mining Massive Time Series Data: With Dimensionality Reduction Techniques

Justin Borg and Joseph G. Vella[✉]

Faculty of ICT, Department of Computer Information Systems, University of Malta,
Msida, Malta
{justin.borg.12,joseph.g.vella}@um.edu.mt

Abstract. A pre-processing step to reduce the volume of data but suffer an acceptable loss of data quality before applying data mining algorithms on time series data
is needed to decrease the input data size. Input size reduction is an important step
in optimizing time series processing, e.g. in data mining computations. During
the last two decades various time series dimensionality reduction techniques have
been proposed. However no study has been dedicated to gauge these time series
dimensionality reduction techniques in terms of their effectiveness of producing
a reduced representation of the input time series that when applied to various data
mining algorithms produces good quality results. In this paper empirical evidence
is given by comparing three reduction techniques on various data sets and applying their output to four different data mining algorithms. The results show that it
is sometimes feasible to use these techniques instead of using the original time
series data. The comparison is evaluated by running data mining methods over
the original and reduced sets of data. It is shown that one dimensionality reduction technique managed to generate results of over 83% average accuracy when
compared to its benchmark results.

Keywords: Time series · Data mining · Dimensionality reduction techniques

1 Introduction

Vast amounts of data are generated every second from various applications hailing from
such fields as meteorology, science, engineering, telecommunication networks, sensor
data, and location-based services. The majority of this data is in the form of a time
series as it is collected either chronologically or sequentially. Time series data is categorized as large in data size, characterized with high dimensionality, and the need of
continuous update [1]. Time series data has been given importance from researchers
during the last two decades with accentuated efforts in anomaly detection, motif discovery, query by content, prediction, classification, and clustering. However with the
increase of research efforts, reoccurring computational issues in dimensionality reduction, similarity measures and indexing on time series data proved to be consistently hard
[2].

Most time series data are of high dimensionality so it is very computationally expensive for one to compute on the original raw format in terms of both processing and

© Springer Nature Singapore Pte Ltd. 2020
M. Singh et al. (Eds.): ICACDS 2020, CCIS 1244, pp. 496–506, 2020.
https://doi.org/10.1007/978-981-15-6634-9_45

storage costs. As a result, a new branch of research in data processing emerged to deal with processing high dimensional time series data. Various time series dimensionality reduction techniques have been developed and implemented during the last twenty years, with their main aim being of reducing a time series of length N to length m where $m <<$ N ($<<$ denotes much smaller than) such that there is an acceptable loss of data quality in the reduced time series. Dimensionality reduction techniques are considered highly important as most related research has to address computing within reasonable resources and tolerable response time, and therefore research requiring time series analysis can benefit from it.

Data mining applications working on time series data sets are faced constantly with this problem of high dimensionality of time series data. Researchers seem to agree that a pre-processing step is needed before applying different data mining techniques on time series data and not to work directly on the original raw data as this requires even higher computational costs in terms of processing speed and storage. However, in literature, no 'standard' pre-processing step has been proposed but each researcher has proposed his method to be applied as a pre-processing step. Thus, various time series dimensionality reduction techniques have been proposed in literature but there exists no study which is dedicated on comparing these techniques together in terms of their ability of getting results of an acceptable accuracy from data mining algorithms, by comparing them with results one would gain if one would opt to use the original version of data.

The aim of this paper is that of implementing different dimensionality reduction techniques to serve as a pre-processing step before applying data mining techniques on a variety of publicly available time series data sets, which have already been used in literature for diverse time series data mining tasks. Effort is more dedicated on the aspect of numerosity reduction of the data dimensionality reduction problem i.e. the ability to reduce a time series of length N to length m. Each dimensionality reduction technique is evaluated depending on its ability to produce results of an acceptable accuracy when compared with the results obtained when applying data mining techniques on the original time series data. The indications from the benchmarks are extracted to produce insights that are really required by time series end users so as to decrease a data mining exercise turnaround project time and produce execution programs with tolerable time of execution.

The remaining part of this paper is organized as follows: in Sect. 2 the literature review describes what have been achieved so far in the areas of time series dimensionality reduction and time series data mining. Section 3 describes the methodology followed in order to be able to benchmark the different time series dimensionality techniques investigated. Results are shown and evaluated in Sect. 4. A discussion about the results generated is given in Sect. 5 and the conclusion is found in Sect. 6.

2   Literature Review

2.1   Time Series Processing

Dealing with the original data is very expensive in terms of computation and normally such data is very dirty in terms of structure and may contain missing values. Therefore, a pre-processing step is needed before the actual data mining exercise, which is a computationally expensive task itself, i.e. requiring an excessive amount of memory and processing power. An instance of this can be seen from a complex query required to join different time series data to calculate moving average deviances. Dimensionality reduction is the pre-processing step addressing this issue. Dimensionality reduction is considered in terms of compression, dimensionality reduction and numerosity reduction. Compression is the process of processing and storing a time series with fewer bits. Dimensionality reduction is the process of reducing the amount of attributes of a time series data item and keeping the nature of the original time series but not necessary the time series data entries' actual attributes. Numerosity reduction is the process of reducing the length of a given time series from N to m, in which m $<<$ N. This paper is going to focus more on the third term of dimensionality reduction; i.e. numerosity reduction.

2.2 Types of Dimensionality Reduction Techniques

A plethora of dimensionality reduction techniques have been developed and proposed in literature throughout the last twenty years. By making use of empirical evidence, some researchers have tried to identify which dimensionality reduction techniques produce a reduction of the highest similarity when compared to the original data. The taxonomy adopted by Ding in [4] categorizes the different types of dimensionality reduction techniques. The representation techniques are classified into two different groups, these being *data adaptive* and *non-data adaptive*.

Non-data Adaptive Reduction Techniques: The parameters of the transformation remain the same for every time series regardless of its nature. One such representation is Discrete Fourier Transform (DFT) which projects a time series as a superposition of sine and cosine functions basis in the real domain [5]. Other non-data adaptive techniques which makes use of *wavelets* are for instance the Discrete Wavelet Transform (DWT), which uses a scaled and shifted versions of a mother wavelet function [6]. An approach which was specifically proposed for time series data is the Piecewise Aggregate Approximation (PAA) which represents a time series as a vector of all the means of the fixed segments of a time series [7].

Data-Adaptive Reduction Techniques: The parameters of the transformation are set depending on the data available. Almost all non-data adaptive techniques can become data adaptive by adding an extra data-sensitive selection step. Examples of such an approach have been applied to DFT in [8], DWT in [9] and PAA in [10]. Adaptive Piecewise Constant Approximation (APCA) is a technique which is similar to the PAA technique, however in the APCA technique the segments can be of arbitrary lengths, which enables the technique to produce more segments in those parts of the time series where a lot of activity is present [11]. Techniques which transform a numerical time series into a discrete set of symbolic strings have also been proposed. The most widely used and cited technique in literature is the Symbolic Aggregate Approximation (SAX), in which PAA values are transformed to symbolic strings using a breakpoint lookup table [12]. A survey comparing different symbolic representations was done in [32].

2.3 Mining Time Series Data

One of the biggest issue with mining of time series data is the huge amount of instances i.e. points in a time series data set. This problem has encouraged researchers to propose and implement different dimensionality reduction techniques to serve as a pre-processing step in order to reduce the length of the time series whilst keeping the original nature and quality of the original data. Throughout the years, various data mining techniques have been applied by the data mining community to time series data. Various works in the literature have been dedicated to time series clustering in different domains such as in energy [13], finance [14] and medicine [15]. In [30] the authors compared clustering results after using the DFT and DWT techniques to dimensionally reduce the original dataset. Another data mining technique which has been applied on time series data is classification. Given a set of time series, each defined with a label, the classification task consists in training a classifier and this classifier is used to test and classify new time series data [3]. The authors of [31] proposed a new additive representation technique which generates more accurate time series classification results. A data mining technique which has not been given a lot of attention from researchers is discord discovery. The discord of a time series is the subsequence which is the most different from all other non-overlapping subsequences [16]. The problem of discord discovery was derived from the anomaly detection and similarity search problems found in the data mining community. Discord discovery have been applied in various domains such as to discover abnormal heartbeats [17], in electricity consumption data [18], and to discover unusual shapes [19].

3 Methodology

Three different dimensionality reduction techniques were chosen to be investigated on four data mining algorithms. The chosen techniques were the Piecewise Aggregate Approximation (PAA), the Symbolic Aggregate Approximation (2 different alphabet sizes) and the Discrete Wavelet Transform (DWT). By choosing these three techniques it was ensured that a wavelet, a piecewise and a symbolic based technique were investigated in this study. Moreover, it was also assured that both data adaptive and non-data adaptive techniques were chosen. Due to space limitations, the exact method of how one can compute each technique is not given in this paper but the reader is referred to the original paper where each technique was first proposed.

Data mining results produced after applying the computed representations by making use of each of these three techniques were compared to the data mining results produced from applying the original time series data to the data mining algorithms, i.e. benchmark results. The PAM clustering algorithm, an agglomerative clustering algorithm, the kNN algorithm and a discord discovery algorithm were the data mining algorithms chosen for the techniques to be applied to. Each dimensionality reduction technique produced reductions of 35%, 50%, 65%, 80% and 95% of the original length of the inputted time series data.

3.1 Piecewise Aggregate Approximation (PAA)

The PAA technique proposed by Keogh in [7] (submitted independently as Segmented Means in [20]), enables a time series to be represented as a series of segments. Each segment contains the computed average value of all the values that fall within that segment. The only input parameter required to compute a PAA representation is w which is equal to the number of segments of the resulting PAA representation. This parameter was determined by the percentage of reduction that needed to be computed.

3.2 Symbolic Aggregate Approximation (SAX)

SAX is a symbolic representation in which a PAA representation is further reduced and represented by a series of strings obtained from a Gaussian distribution table [12]. Another input parameter that is required by the SAX technique is a, which is equal to the alphabet size. It has been shown in literature that an alphabet size in the range of 5 to 8, yields the best results and offers a reasonable balance between space requirements and the tightness of lower bounds [12]. As a result of this, two SAX representations were chosen to be investigated in this study, i.e. a SAX representation with $a = 5$ and another representation with $a = 8$. It is also required that before transforming an input time series to a PAA representation, the input series would be normalized to have a mean of zero and a standard deviation of one.

Furthermore, the custom MINDIST distance function [12] was used by the data mining algorithms so as to be able to calculate the distance between different SAX representations.

3.3 Discrete Wavelet Transform (DWT)

The Discrete wavelet transform (DWT) transforms an input time series using a set of basis functions called wavelets. Wavelets are a set of mathematical functions used to decompose data into different components [21]. The Haar wavelet was used to compute the DWT representations as one of the properties of the Haar wavelet is the ability to allow a good approximation of the original data by using a subset of coefficients.

It has been proven that to achieve a more accurate representation, keeping the first k coefficients and approximate the rest with zero yields better results [22], so this option was applied in this study. In order to compute a DWT representation, an input time series was first transformed using the Haar wavelet and then the resulting series was sorted in descending order. Finally, the first k coefficients were chosen depending on the percentage of reduction that needed to be computed.

3.4 PAM and Agglomerative Clustering Algorithms

The PAM and an agglomerative clustering algorithms were used as the clustering techniques in this study. Both algorithms require the value of k to be inputted by the user which represents the number of clusters to consider by the algorithm. To reduce implementation and data bias different values of k were passed to both algorithms and these were: 2, 4, 6 and 8 clusters.

The Rand Index [23], the Jaccard Index and the Fowlkes-Mallows Measure [24] external evaluation measures were applied to compare both sets of results (original and reduced). The class labels generated by each algorithm for the original set of data were considered to be the true class labels and these were compared to the class labels generated from using the reduced sets of data for each dataset used, for each computed reduction by each dimensionality reduction technique and for each number of clusters considered as input by both algorithms. The datasets chosen to be used by both clustering algorithms were a Transactions dataset [25], a Household Power Consumption dataset [26] and a Gas Sensors dataset [27].

3.5 KNN Classification Algorithm

The kNN algorithm was the classification algorithm chosen for representations to be applied to. An input parameter of the kNN algorithm is k which is equal to the number of neighbors considered by the algorithm. Once again different values of k were passed to the algorithm to reduce implementation and data bias. The values of k considered as input were: 3, 5 and 7 neighbors.

The class labels predicted for the testing set of data by the classifier were compared to the true class labels of the testing set in order to calculate the accuracy of the classifier for both the original and reduced data. Then, the accuracy of the classifier of each reduced set of data was compared to the accuracy of the classifier produced by the original set of data for each dimensionality reduction technique, for each number of neighbors considered as input by the kNN algorithm and for each dataset. The CinC_ECG_Torso, HandOutlines and StarLightCurves from the UCR time series classification archive [28] were the datasets chosen.

3.6 Discord Discovery Algorithm

The last data mining algorithm investigated in this study was the discord discovery algorithm. In discord discovery sequences which are the least similar to all the other sequences are discovered. A brute force discord discovery algorithm was implemented based on the work done by [29] to extract the top k most significant discords. The top k most significant discords were defined as the top k time series which are the least similar to all other time series in a time series database. The value of k was taken as the real value 5.

In order to reach the aim of this study, the discords discovered from the original data were compared to the discord discovered from the reduced data for each reduction computed by each dimensionality reduction technique and for each dataset used. The InlineSkate and MALLAT datasets taken from the UCR time series classification archive [28] and a generated random-walk dataset were used for discords discovery.

4 Results

The results reported here are generated after following the methodologies indicated earlier for time series data reduction. The main aim is to compute and compare results produced by applying reductions computed by the investigated dimensionality reduction techniques to different data mining algorithms over different datasets. The results obtained for the different runs undertaken have been compared to the benchmark results, i.e. running the data discovery over the original time series. The results obtained over the reduced datasets enable identification of better data reduction techniques across data mining algorithms and conversely, for each data mining algorithm investigated here, which is the better dimensionality reduction technique to transform with. Another aim, albeit secondary, is reporting the different input parameters required by each data mining algorithm effect on discovery accuracy. Table 1 below shows the average accuracy of results obtained by each dimensionality reduction technique for each data mining algorithm. These are further explained in the following sections.

Table 1. The average accuracy of results obtained by each dimensionality reduction technique when compared to the benchmark results

	PAA	SAX-5 ($a = 5$)	SAX-8 ($a = 8$)	DWT
PAM	72.52%	46.57%	47.67%	36.09%
Agglomerative	78.14%	54.43%	62.63%	40.78%
kNN	100.09%	82.62%	87.25%	84.04%
Discord discovery	61.33%	13.33%	13.33%	6.67%

4.1 Clustering Algorithms

The average accuracy of results when grouping by each dimensionality reduction technique was computed by calculating the average values of all evaluation measures computed for all number of clusters considered as input by the PAM and agglomerative algorithms, for all reductions computed and for each dataset used.

For both the PAM and agglomerative clustering algorithms, the PAA technique produced the highest accurate results with an average accuracy of 73% for the PAM algorithm and 78.14% for the agglomerative algorithm.

When compared to the benchmark results, both SAX representations managed to produce results with an average accuracy between 46.57% and 62.63% for both clustering algorithms. However, the SAX representation with the larger alphabet size ($a = 8$) managed to produce higher accurate results than the other SAX representation.

The DWT technique produced results with the lowest accuracy with just an average accuracy of 38% for both algorithms when compared to their benchmark.

4.2 kNN Algorithm

The average accuracy of results when grouping by each dimensionality reduction technique was computed by calculating the average values of all accuracy measures computed for all number of neighbors considered as input by the kNN algorithm, for all reductions computed and for each dataset used.

All time series dimensionality reduction techniques produced results of high accuracy when applied to the kNN classification algorithm. All techniques managed to produce results with an average accuracy of more than 82%. The PAA technique managed to outperform the benchmark results by a 0.09% in accuracy, which is a very surprising result to be achieved by a dimensionality reduction technique. The computed results accuracy of the PAA representation when comparing the predicted class labels with the true class labels was higher than the accuracy when comparing the predicted class labels with the true class labels for the original set of data.

The SAX representation with a larger alphabet size ($a = 8$) outperformed the other SAX representation by an average accuracy of 5%. The DWT technique produced results with an average accuracy of 84%.

4.3 Discord Discovery Algorithm

The average accuracy of results when grouping by each dimensionality reduction technique was computed by calculating the average number of similar discords discovered from the reduced data when compared to the discords discovered from the original data, for each dataset used.

Once again, the PAA technique outperformed all other dimensionality reductions techniques with a results' average accuracy of 61% when compared to the benchmark results. Both the SAX and DWT dimensionality reduction techniques are not recommended to be applied to the discord discovery problem as both techniques produced results with low accuracies, with an average accuracy of 13% for both SAX representations and just 7% for the DWT technique.

5 Discussion

The change in the results' accuracy for both the clustering and classification algorithms to the discord discovery algorithms is explained from the fact that the former algorithms are not affected that much with a change in the data values of the input time series when computing the reductions. Since both the original nature and quality of the data are still mostly preserved in the reduced series, and both the clustering and classification algorithms generate results on the whole nature of the series rather than only part of the whole series, results are not affected that much, thus, results' accuracy remains high. On the other hand, the discord discovery algorithm is more data dependent and even the slightest change in the nature and quality of the original series made during the computation of the reduced series, affect negatively the algorithm, thus, a lower results' accuracy.

After analyzing and evaluating the results generated, the PAA technique produced the highest accurate results when compared to the benchmark results out of all dimensionality reduction techniques investigated in this study when applied to all data mining algorithms implemented. The fact that for one to compute the PAA representation, one just need to calculate the averages of the data points that fall within each segment of the resulting PAA representation, the majority of the nature and quality of the original data is preserved in the reduced set. Since data values which fall within the same segment are normally on the same wavelength, this results in a small variance between the average value computed for the respective segment and the original data values that fall within that segment. As a result of this, the majority of the original data's nature and quality is preserved in the computed reduced set of data.

The SAX dimensionality reduction technique is built on the PAA technique and thus, requires an extra step to transform the PAA values to symbolic strings by making use of a Gaussian distribution table. By having to compute this extra step apart from computing the PAA representation of an input time series, the nature and quality of the original data is further lost between the first and the required second step. This issue can be considered as the main reason why both SAX representations failed to match the accuracy of the results produced by the PAA technique. An important result that was generated in this study was that when using a larger alphabet size, one would yield better data mining results than using a smaller alphabet size. When using a larger alphabet size, data points taken from the Gaussian distribution table are better distributed. As a result, there would be less variance between the PAA values and the values taken from the table, thus, more accurate results are produced when using a larger alphabet size. As a result, one would be preserving more of the nature and quality of the original time series data in the computed reduced representation.

The DWT dimensionality reduction technique produced the least accurate results out of all the dimensionality reduction techniques investigated in this study when compared to the benchmark results. Although the DWT manages to preserve the nature and quality of the original data with just a subset of coefficients, it is still behind both the SAX and PAA representations in terms of producing a good accurate reduced representation. By keeping only the largest k coefficients to compute the reduction, a major part of the original data's nature and quality is evidently lost. Thus, the DWT was not able to produce results of high accuracy as both the PAA and SAX techniques. Another major drawback of the DWT technique is that the reduction can only be computed for time series whose lengths are equal to an integral power of two. This is not practical at all in the real world as most of the time series data produced are of an arbitrary length and not equal to a length which is an integral power of two. As a result, more time and computational resources are wasted on the extra step needed to pad zeros at the end of an input time series when its length is not an integral power of 2.

6 Conclusions

This study can be considered as the first of its kind as different time series dimensionality reduction techniques were compared in terms of their effectiveness of producing high quality results when applied to various data mining algorithms by comparing the produced results to the benchmark results, i.e. data mining results produced by using the original time series data. Three dimensionality reduction techniques, the PAA, SAX (2 representations with different alphabet sizes) and DWT, were implemented and applied to a PAM clustering algorithm, an agglomerative clustering algorithm, a kNN classification algorithm and a discord discovery algorithm. Moreover, publicly available time series datasets which have been previously used in literature for time series analysis were used as the main source of time series data.

After analyzing and evaluating all results generated in this study, it was proven that the PAA technique performed the best out of all investigated techniques for all data mining algorithms in terms of the achieved results' accuracy. The least performer was the DWT technique, while the SAX representation with the larger alphabet size ($a = 8$) produced results of higher accuracy than the SAX representation with the smaller alphabet size ($a = 5$).

This study can be used and serve as a guide to time series end users who are looking for empirical evidence which indicates what is feasible to make use of time series dimensionality reduction techniques to produce an accurate reduced representation to apply it to various data mining applications. This study has shown that some time series dimensionality reduction techniques are able to produce high accurate representations, which in turn can generate high accurate data mining results when compared to the benchmark results.

As future work, the accuracy of results generated from applying dimensionality reduction techniques on multi-dimensional time series can be investigated.

References

1. Fu, T.: A review on time series data mining. Eng. Appl. Artif. Intell. **24**(1), 164–181 (2011)
2. Keogh, E., Kasetty, S.: On the need for time series data mining benchmarks: a survey and empirical demonstration. Data Min. Knowl. Discov. **7**(4), 349–371 (2003)
3. Esling, P., Agon, C.: Time-series data mining. ACM Comput. Surv. **45**(1), 1–34 (2012)
4. Ding, H., Trajcevski, G., Scheuermann, P., Wang, X., Keogh, E.: Querying and mining of time series data: experimental comparison of representations and distance measures. Proc. VLDB Endowment **1**(2), 1542–1552 (2008)
5. Agrawal, Rakesh, Faloutsos, Christos, Swami, Arun: Efficient similarity search in sequence databases. In: Lomet, David B. (ed.) FODO 1993. LNCS, vol. 730, pp. 69–84. Springer, Heidelberg (1993). https://doi.org/10.1007/3-540-57301-1_5
6. Chan, K., Fu, A.W.-c.: Efficient time series matching by wavelets. In: Proceedings of the 15th International Conference on Data Engineering, ICDE 1999, Washington DC (1999)
7. Keogh, E., Chakrabarti, K., Pazzani, M., Mehrotra, S.: Dimensionality reduction for fast similarity search in large time series databases. Knowl. Inf. Syst. **3**(3), 263–286 (2001)
8. Vlachos, M., Gunopulos, D.: Indexing time-series under conditions of noise. In: Data Mining in Time Series Databases, pp. 67–100. World Scientific Press (2004)
9. Struzik, Z., Siebes, A.: Measuring time series similarity through large singular features revealed with wavelet transformation. In: Proceedings of the 10th International Workshop on Database and Expert System Applications (1999)
10. Megalooikonomou, V., Li, G., Wang, Q.: A dimensionality reduction technique for efficient similarity analysis of time series databases. In: Proceedings of the Thirteenth ACM International Conference on Information and Knowledge Management, CIKM 2004, Washington DC (2004)
11. Chakrabarti, K., Keogh, E., Mehrotra, S., Pazzani, M.: Locally Adaptive Dimensionality reduction for indexing large time series databases. ACM Trans. Database Syst. (TODS), pp. 188–228 (2002)
12. Lin, J., Keogh, E., Wei, L., Lonardi, S.: Experiencing SAX: a novel symbolic representation of time series. Data Min. Knowl. Discov. **15**, 107–144 (2007). https://doi.org/10.1007/s10 618-007-0064-z

13. Bode, G., Schreiber, T., Baranski, M., Müller, D.: A time series clustering approach for Building Automation and Control Systems. Appl. Energy **238**, 1337–1345 (2019)

14. Caiado, J., Crato, N., Poncela, P.: A fragmented-periodogram approach for clustering big data time series. Adv. Data Anal. Classif. **14**, 117–146 (2020)

15. Wismuller, A., et al.: Cluster analysis of biomedical image time-series. Int. J. Comput. Vis. **46**(2), 103–128 (2002)

16. Luo, W., Gallagher, M., Wiles, J.: Parameter-free search of time-series discord. J. Comput. Sci. Technol. **28**(2), 300–310 (2013)

17. Chuah, M.C., Fu, F.: ECG anomaly detection via time series analysis. In: Thulasiraman, Parimala, He, Xubin, Xu, Tony Li, Denko, Mieso K., Thulasiram, Ruppa K., Yang, Laurence T. (eds.) ISPA 2007. LNCS, vol. 4743, pp. 123–135. Springer, Heidelberg (2007). https://doi.org/10.1007/978-3-540-74767-3_14

18. Keogh, E., Lin, J., Fu, A.W., Van Herle, H.: Finding unusual medical time-series subsequences: algorithms and applications. IEEE Trans. Inf Technol. Biomed. **10**, 429–439 (2006)

19. Wei, L., Keogh, E., Xi, X.: SAXually explicit images: finding unusual shapes. In: Sixth International Conference on Data Mining, 2006, ICDM 2006, Hong Kong (2007)

20. Yi, B., Faloutsos, C.: Fast time sequence indexing for arbitrary Lp norms. In: Proceedings of the 26th International Conference on Very Large Databases, San Francisco, VLDB 2000 (2000)

21. Chaovalit, P., Gangopadhyay, A., Karabatis, G., Chen, Z.: Discrete wavelet transform-based time series analysis and mining. ACM Comput. Surv. (CSUR) **43**(12), 1–37 (2011)

22. Gunopulos, D.: Tutorial Slides: Dimensionality Reduction Techniques (2001)

23. Rand, W.: Objective criteria for the evaluation of clustering methods. J. Am. Stat. Assoc. **66**, 846–850 (1971)

24. Fowlkes, E., Mallows, C.: A method for comparing two hierarchical clusterings. J. Am. Stat. Assoc. **78**, 553–569 (1983)

25. Alcala-Fdez, J., et al.: KEEL data-mining software tool: data set repository, integration of algorithms and experimental analysis framework. J. Multiple-Valued Logic Soft Comput. **17**, 255–287 (2011)

26. Lichman, M.: UCI Machine Learning Repository. University of California, School of Information and Computer Science, Irvine (2010). http://archive.ics.uci.edu/ml

27. Fonollosa, J., Sheik, S., Huerta, R., Marco, S.: Reservoir Computing compensates slow response of chemosensor arrays exposed to fast varying gas concentrations in continuous monitoring. Sens. Actuators B: Chem. **215**, 618–629 (2015)

28. Chen, Y., et al.: The UCR Time Series Classification Archive, July 2015. http://www.cs.ucr.edu/~eamonn/time_series_data/

29. Keogh, E., Lin, J., Fu, A.: HOT SAX: efficiently finding the most unusual time series subsequence. In: Proceedings of the Fifth IEEE International Conference on Data Mining, ICDM 2005, Washington (2005)

30. Bahadori, S., Charkari, N.M.: Increasing efficiency of time series clustering by dimension reduction techniques (2018)

31. Sirisambhand, K., Ratanamahatana, C.H.: A dimensionality reduction technique for time series classification using additive representation. In: Third International Congress on Information and Communication Technology. Advances in Intelligent Systems and Computing, Singapore (2019)

32. Wang, Lin, Lu, Faming, Cui, Minghao, Bao, Yunxia: Survey of methods for time series symbolic aggregate approximation. In: Cheng, Xiaohui, Jing, Weipeng, Song, Xianhua, Lu, Zeguang (eds.) ICPCSEE 2019. CCIS, vol. 1058, pp. 645–657. Springer, Singapore (2019). https://doi.org/10.1007/978-981-15-0118-0_50

Comparative Analysis of Data Mining Techniques to Predict Heart Disease for Diabetic Patients

Abhishek Kumar[1][(✉)], Pardeep Kumar[2], Ashutosh Srivastava[3], V. D. Ambeth Kumar[4], K. Vengatesan[5], and Achintya Singhal[1]

[1] Department of Computer Science, Banaras Hindu University, Varanasi, India
abhishek.maacindia@gmail.com, achintya.singhal@gmail.com
[2] CSE Department, G H Raisoni College of Engineering, Nagpur, India
pardeepsep2@gmail.com
[3] Department of Electrical Engineering, Indian Institute of Technology, IIT BHU, Varanasi, India
ashutosh.rs.eee@iitbhu.ac.in
[4] Department of Computer Science Engineering, Panimalar Engineering College, Anna University Chennai, Chennai, India
ambeth_20in@yahoo.co.in
[5] Department of Computer Engineering, Sanjivani College of Engineering, Savitribai Phule University, Kopargaon, MH, India
vengicse2005@gmail.com

Abstract. The healthcare sectors have many difficulties and challenges in finding diseases. Healthcare organizations are collecting bulk amount of patient data. The Data mining methods are utilized to decide covered data that is valuable to healthcare specialists with effective analytic decision making. Data mining strategies are utilized in the field of the healthcare industry for different purposes. The objective of this paper is to assess and analyze using three unique data mining arrangement methods, for example, Naïve Bayes (NB), Support Vector Machine (SVM) and Decision Tree to decide the potential approaches to predict the possibility of heart disease for diabetic patients dependent on their predictive accuracy.

Keywords: Data mining · Naïve Bayes · Support Vector Machine · Decision Tree

1 Introduction

This Data mining is widely spread in many areas. It assumes a significant job in the medical field. The number of patient records stored in the hospital administration database is increasing day by day. The algorithms for data mining are used to extract particular data from medical records. Traditionally, heart disease was a problem in developed countries, but today it also causes headaches for developing countries, what's more, this is one of the main sources of death. The heart is the most critical strong organ in people, which

M. Singh et al. (Eds.): ICACDS 2020, CCIS 1244, pp. 507–518, 2020.
https://doi.org/10.1007/978-981-15-6634-9_46

circulates blood through the veins of the circulatory system. Human life depends on the proper function of the heart. The malfunctioning of the heart affects other parts of the human body I.e. the brain, kidneys, etc. On the off chance that the blood flow in the body is inefficient, it influences both heart and cerebrum. Generally, blood arrest in the heart is called an attack and a blood clot in the brain is called a stroke.

The symptoms of heart disease are largely dependent on the nervousness a person feels. Some symptoms are not normally distinguished by normal people. The most common symptoms are chest pain, breathlessness, and shivers. Common chest pains in various types of heart disease are called angina or angina pectoral and occur when part of the heart is not getting enough amount of oxygen Angina can be activated by distressing occasions or physical effort and typically keeps going under 10 min. Cardiovascular failures can in like manner happen on account of different sorts of coronary illness. The indications of a heart attack are like those of angina, besides that, they can occur to rest and will all in all be progressively extraordinary. The symptoms of a heart attack can once in a while look like indigestion. It can cause heartburn and stomach upset, as well as a feeling of heaviness in the chest. Different symptoms of heart attack to incorporate pain in the body example of the neck, back, chest to the arms, dizziness, abdomen or jaw, nausea, profuse sweating and vomiting.

Heart failure is also the result of heart disease and breathing difficulty can happen when the heart turns out too low blood circulation. Some heart disease has no symptoms, particularly in the elderly and diabetics. The expression "congenital heart disease" covers an assortment of conditions. Notwithstanding, general symptoms sweating, severe fatigue, quick heartbeat, and fast breathing, chest pain. However, these symptoms cannot occur before the age of 13 years. In such cases, the diagnosis becomes a complex task requiring much experience and skill. If the risk of a heart attack or the possibility of heart disease is detected early, it can help patients take precautionary measures and take regulatory action. More recently, the healthcare industry has generated bulk amounts of patient data, and its diagnostic reports on the disease are being prepared specifically for the purpose of global prediction of heart attacks.

This paper aims to solve the challenges of improving the predictive model for predicting heart disease in diabetic's patients and provide the timely response to predicting disease based on the most accurate data mining technique (Fig. 1).

2   Literature Review

Haleh Ayatollahi et al. [1] presents the investigation was directed by utilizing data mining methods. The sample data was the patient health records with coronary vein disease who were hospitalized in three hospitals partnered to AJA University of Medical Sciences, Tehran. The dataset and the anticipating factors utilized in this investigation was the equivalent for the two data mining methods. Absolutely, 25 factors influencing CAD were chosen and related data were removed. Subsequent to normalizing and cleaning the data, they were examine into SPSS and Excel.

Monther Tarawneh et al. [2] proposed another heart disease expectation framework that join all systems into one single calculation, it called hybridization. The outcome confirm that exact analyze can be taken by utilizing a consolidated model from all methods.

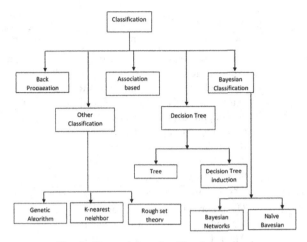

Fig. 1. Data mining classification method

Priyan Malarvizhi Kumar et al. [3] Proposed integrated cloud computing and IoT-based innovation to build adaptability and accessibility. The design utilizes Apache HBase to store the large volume of the sensor data in the cloud. The people's wellbeing data is gathered with the support of RFID and 5G mobile systems. And more, Apache Mahout is utilized in the proposed health observing framework for structure the logistic regression-based preidentification model for heart diseases. At last, the exhibition of the estimation model is similarly investigated with the assistance of different execution measurements. The processed outcomes, for example, throughput, affectability, f-measure, and accuracy, are utilized for exhibiting the proficiency and execution of the proposed IoT-based health observing framework.

Apruv Patel et al. [4] presented a hybrid or particular of data mining calculations can be utilized to explore a few papers utilized in heart disease forecast to detect calculations with high accuracy for future research.

Reddy Prasad et al. [5] proposed the logistic regression calculations are utilized and the health care information which differentiates the patients whether they are having heart diseases or not as indicated by the data in the record. Additionally, it will attempt to utilize this data model which predicts the patient whether they are having heart disease or are not.

Approach	Year	Objective	Pros	Cons
Ching-seh et al. [6]	2019	Purpose of reporting about taking benefit of the different data mining methods and develop prediction models for heart disease patient survivability	Naïve Bayes and Logistic Regression have high accuracy when running on high dimensional algorithms and dataset	Decision-Tree gives lower accuracy than Random Forest classifier

(continued)

(*continued*)

Approach	Year	Objective	Pros	Cons
Jarrel C. Y. Seah et al. [7]	2018	Congestive heart failure features on chest radiographs learned by NNs can be recognized using Generative Visual Rationales, over-fitted prototypes, enabling detection of bias	A trained deep learning model will produce GVRs that do so frequently than an intentionally Over-fitted framework	Limited resolution prevented the evaluation of fine-image details
T. Nagama ni et al. [8]	2019	The proposed Mapreduce Algorithm's implementation in parallel and distributed frameworks was evaluated by using Cleveland dataset and compared with that of the predictable ANN method	The parallel Meta-heuristic method with a prepared neural network approach based on Mapreduce algorithm reduces the training time significantly	Hbase is used for storing resultant data. It has some latency due to batch processing
Proposed	2019	To identify the most significant classification framework which can help the doctors in predicting the risk of heart disease using diabetic attributes	To detect the patients at danger, with the aim of increasing the quality of care and to reduce cost of care	Decision tree Classifiers can help only in early detection of the vulnerability of a diabetic patient to heart disease

3 Research Methodology

See Fig. 2.

Comparing Support Vector Machine (SVM), Naive Bayes and Decision Tree analysis

Based on the data from the research presented, the experiments were performed by combining the three Naive Bayes classification techniques, decision tree, SVM, and using the Rapid miner tool.

The 4 models that use the selected data mining structure were created by classifying the dataset based on specific attribute-value pairs. The outcome of the experiments on the different models is concise in Table 1.

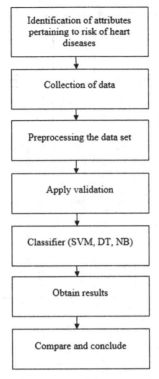

Fig. 2. Flow diagram of the proposed method

Table 1. Prediction Models of Heart Disease (PMHD) with categorization

Model No	Categorization
PMHD 1	Age 35 to 75 and Sex = M
PMHD 2	Age 35 to 75 and Sex = F
PMHD 3	Age > 75 and Sex = M
PMHD 4	Age > 75 and Sex = F

Model 1 prediction of heart disease risk for gender = M attribute and age between 35 > 75 years

The analysis report performed by combining Support Vector Machine, Decision Tree (DT) and Naive Bayes (NB) algorithms for Model 1 is concise in Table 2.

Acc - means Overall accuracy, terms of precision (PL, PM, PH) and recall (RL, RM, RH) of the three classes, (Low, Medium & High).

Certain rules derivative from the decision tree produced from model 1 are shown in what follows.

If VLDL > 27.50

Table 2. Prediction of heart disease Male age > 35 and age < 75, Model 1

Algorithm	TM	FM	TH	FH	TL	FL	PM%	PH%	PL%	RM%	RH%	RL%	Acc%
Decision Tree	69.0	23.0	44.0	8.0	352.0	23.0	75.0	84.6	93.8	74.5	78.5	95.1	89.5
Naive Bayes	55.0	32.0	34.0	15.0	349.0	34.0	63.2	69.3	91.1	59.1	60.7	94.3	84.4
Support Vector Machine	24.0	17.0	15.0	20.0	344.0	99.0	58.5	42.8	77.6	25.8	26.7	92.9	73.8

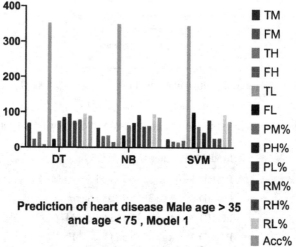

Prediction of heart disease Male age > 35 and age < 75 , Model 1

&
LDL > 153.50
BP = 110/80 then high {high = 1, low = 1, medium = 0}
If VLDL > 27.50 & LDL > 153.50
BP = 120/80 then high {high = 10, low = 0, medium = 0}
If VLDL > 27.500 &
LDL > 153.500
BP = 130/80
&Fasting > 251 then high {high = 2, low = 0, medium = 0}

If LDL > 153.50
&VLDL > 27.50
BP = 130/80
Fasting ≤ 251 then medium {high = 0, low = 0, medium = 3}

Model 2 prediction of heart disease risk for gender = F attribute and age between 35 and 75 years

The consequences of the examination of this done by combining Support Vector Machine (SVM), Decision Tree (DT), Naïve Bayes (NB) and calculations over the model 2 are condensed in Table 3.

Table 3. Prediction of heart disease Female age > 35 and age < 75, **Model 2**

Algorithm	TM	FM	TH	FH	TL	FL	PM%	PH%	PL%	RM%	RH%	RL%	Acc%
Decision Tree	53.0	21.0	48.0	11.0	221.0	18.0	71.62	81.35	92.46	74.64	76.18	92.85	86.55
Naïve Bayes	30.0	28.0	45.0	21.0	217.0	31.0	51.70	68.17	87.50	42.24	71.42	91.17	78.48
Support Vector Machine	58.0	127.0	30.0	1.0	131.0	25.0	31.35	96.76	83.98	81.68	47.61	55.03	58.85

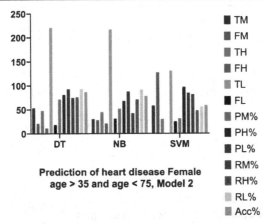

Prediction of heart disease Female
age > 35 and age < 75, Model 2

Some of the rules derived from the DT and generated from model 4 are shown in what follows.
If
VLDL > 33.50 and LDL > 155
& LDL > 127.50
and A1C > 8.60 then high {high = 37, low = 0, medium = 0}
If VLDL > 33.50
& LDL > 155 and LDL > 127.50
and A1C ≤ 8.60

& Age > 45.50 then high {high = 4, low = 0, medium = 0}
If
VLDL > 33.50
& LDL > 155 & LDL > 127.50
& A1C ≤ 8.60 & Age ≤ 45.50 then
medium {high = 0, low = 0, medium = 2}

Model 3 prediction of heart disease risk for gender = M attribute and age > 75
The results of the investigation done by consolidating Support Vector Machine (SVM), Decision Tree (DT), Naïve Bayes (NB) calculations over model 3 are condensed in Table 4.

Table 4. Prediction of heart disease Male age > 75, **Model 3**

Algorithm	TM	FM	TH	FH	TL	FL	PM%	PH%	PL%	RM%	RH%	RL%	Acc%
Decision Tree	0.0	2.0	1.0	2.0	24.0	1.0	0.0	32.3	97.0	0.0	33.3	100.0	82.3
Naïve Bayes	0.0	1.0	0.0	1.0	24.0	4.0	0.0	0.0	84.7	0.0	0.0	100.0	80.0
Support Vector Machine	1.0	2.0	2.0	16.0	6.0	3.0	33.3	11.1	66.6	33.3	66.6	25.0	30.0

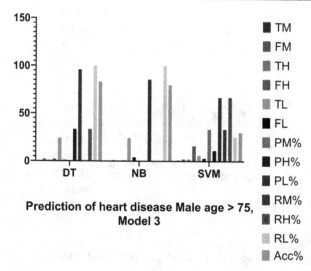

Prediction of heart disease Male age > 75, Model 3

Some of the rules resulting from the decision tree generated from model 3 are revealed in what follows:
If
LDL > 1 20.50
&

Fasting > 154 then
medium {high = 0, low = 0, medium = 2}
If
LDL > 120.50 and
Fasting ≤ 154 then
high {high = 3, low = 0, medium = 1} If LDL ≤ 120.50then
low {high = 0, low = 24, medium = 0}

Model 4 prediction of heart disease risk for gender = F attribute and age > 75
The result of the investigation done by combining SVM and Naïve Bayes, Decision Tree calculations over model 4 are outlined in Table 5.

Table 5. Prediction of heart disease Model 4

Algorithm	TM	FM	TH	FH	TL	FL	PM%	PH%	PL%	RM%	RH%	RL%	Acc%
Decision Tree	7.0	0.0	0.0	0.0	7.0	0.0	100	0.0	100	100	0.0	100	100
Naïve Bayes	4.0	2.0	0.0	0.0	5.0	3.0	66.6	0.0	62.5	57.1	0.0	71.4	70.0
Support Vector Machine	2.0	0.0	0.0	0.0	7.0	5.0	100	0.0	58.3	28.5	0.0	100	65.0

Prediction of heart disease Model 4

Some of the rules resulting from the decision tree generated from model 6 are revealed in what follows.

If
LDL > 126.50 then
medium {high = 0,low = 0, medium = 7}
If
LDL ≤ 126.500 then low {high = 0, low = 7, medium = 0}

Result of Comparing SVM and Decision Tree, Naive Bayes

To validate the final results obtained in the research presented, were done by consolidating the three methods and the exhibition of Bayes theorem, SVM and Decision tree have appeared in Tables 6, 7 and 8 separately.

Table 6. Performance of Bayes Theorem with an accuracy of 81.58%

	True low	True medium	True high	Class precision
pred. high	11	25	86	70.49%
pred. medium	39	98	26	60.12%
pred. low	631	62	21	88.38%
class recall	92.66%	52.97%	64.66%	

Table 7. Performance of Support vector machine with an accuracy of 61.26%

	True low	True medium	True high	Class precision
pred. high	0	5	59	92.19%
pred. medium	283	155	59	31.19%
pred. low	398	25	15	90.87%
class recall	58.44%	83.78%	44.36%	

Table 8. Accuracy of various classification techniques

Technique	Accuracy in percentage
Naïve Bayes (NV)	81.5
Support Vector Machine (SVM)	61.2
Decision tree (DT)	90.7

The decision tree utilizing different split techniques, for example, Gain ratio, Information addition, and Gini list has been attempted as appeared in Table 8 which gives various degrees of precision.

Accuracy of various classification techniques (High, Medium, Low)

The results of all three models, decision tree seems to be best as it has the most astounding level of right expectations (90.79%) for patients with heart diseases, trailed by pursued by naïve Bayes and SVM (Fig. 3).

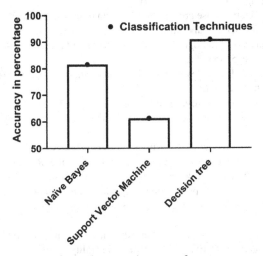

Fig. 3. Performance in terms of accuracy

4 Conclusion

Medical data mining plays a dynamic role in the finding of diseases and in making diagnostic decisions to save lives. The key purpose of this work is to use three different data mining algorithms, namely Support Vector Machine (SVM), Naive Bayes (NB), and Decision Tree (DT) to predict the risk of heart disease based on their precision. Therefore, a comparison of the results of the many grouping techniques was performed and higher accuracy of the decision tree was found. The results are displayed with average accuracy and repeatability. The outcome found that the accuracy of this proposed method is 90.79%, followed by a naive Bayesian process 81.58% and a carrier vector machine 61.26% to predict the heart disease in people with diabetes using diagnostic features.

References

1. Ayatollahi, H., et al.: Predicting coronary artery disease: a comparison between two data mining algorithms. BMC Publ. Health **19**, 1–9 (2019)
2. Tarawneh, Monther, Embarak, Ossama: Hybrid approach for heart disease prediction using data mining techniques. In: Barolli, Leonard, Xhafa, Fatos, Khan, Zahoor Ali, Odhabi, Hamad (eds.) EIDWT 2019. LNDECT, vol. 29, pp. 447–454. Springer, Cham (2019). https://doi.org/10.1007/978-3-030-12839-5_41

3. Kumar, P.M., et al.: A novel three-tier Internet of Things architecture with machine learning algorithm for early detection of heart diseases. Elsevier (2018)
4. Patel, A., et al.: A Literature review on heart disease prediction based on data mining algorithms. Int. J. Res. Trends Innov. **2**, 3003–3008 (2018)
5. Prasad, R., et al.: Heart disease prediction using logistic regression algorithm using machine learning. Int. J. Eng. Adv. Technol. (2019)
6. Wu, C.M., et al.: Heart disease prediction using data mining techniques. In: 2nd International Conference on Data Science and Information Technology (2019)
7. Seah, J.C.Y., et al.: Chest radiographs in congestive heart failure: visualizing neural network learning. RSNA **290**, 514–522 (2018)
8. Nagamani, T., et al.: Heart disease prediction using data mining with mapreduce algorithm. Int. J. Innov. Technol. Exp. Eng. (2019)
9. Benjamin, E.J., et al.: Heart Disease and Stroke Statistics (2018)
10. Rajpurkar, P., et al.: Radiologist-level pneumonia detection on chest x-rays with deep learning (2018)
11. Singh, V.K., Singhal, A., Rai, K.N., Kumar, A., Dwivedi, A.N.D.: Randomized key-based GMO-BCS image encryption for securing medical image. Int. J. Recent Technol. Eng. (2019). https://doi.org/10.35940/ijrte.C4453.098319
12. Kaur, A., et al.: Heart disease prediction using data mining techniques: a survey (2018)
13. Sahaya Arthy, A., et al.: A survey on heart disease prediction using data mining techniques (2018)
14. Kesavan, S., Kumar, E.S., Kumar, A., Vengatesan, K.: An investigation on adaptive HTTP media streaming Quality-of-Experience (QoE) and agility using cloud media services. Int. J. Comput. Appl. (2019). https://doi.org/10.1080/1206212X.2019.1575034
15. Nikhil Kumar, M., et al.: Heart Diseases using Data mining and machine learning algorithms and tools (2018)
16. Kishore, A., et al.: Heart attack prediction using deep learning (2018)
17. Ambeth Kumar, V.D., et al.: Exploration of an innovative geometric parameter based on performance enhancement for foot print recognition. J. Intell. Fuzzy Syst. 1–16 (2019). https://doi.org/10.3233/jifs-190982
18. Karras, T., et al.: Progressive growing of GANs for improved quality, stability, and variation. In: International Conference on Learning Representations (ICLR) (2018)

Author Index

Printed in the United States
By Bookmasters